W. Herbst, K. Hunger

Industrial
Organic Pigments

© VCH Verlagsgesellschaft mbH, D-6940 Weinheim (Federal Republic of Germany), 1993

Distribution:

VCH, P. O. Box 10 1161, D-6940 Weinheim (Federal Republic of Germany)

Switzerland: VCH, P. O. Box, CH-4020 Basel (Switzerland)

United Kingdom and Ireland: VCH (UK) Ltd., 8 Wellington Court,
    Cambridge CB1 1HZ (England)

USA and Canada: VCH, 220 East 23rd Street, New York, NY 10010–4606 (USA)

ISBN 3-527-28161-4 (VCH, Weinheim)   ISBN 0-89573-981-X (VCH, New York)

Willy Herbst, Klaus Hunger

Industrial Organic Pigments

Production, Properties, Applications

VCH   Weinheim · New York · Basel · Cambridge

05802192
CHEMISTRY

Dr. Willy Herbst
Frankfurter Straße 10
D-6238 Hofheim

Dr. Klaus Hunger
Hoechst AG
Postfach 80 03 20
D-6230 Frankfurt/Main 80

Originally published under the German title
Industrielle Organische Pigmente by VCH Verlagsgesellschaft mbH,
Weinheim, 1987

Published jointly by
VCH Verlagsgesellschaft mbH, Weinheim (Federal Republic of Germany)
VCH Publishers, Inc., New York, NY (USA)

Editorial Directors: Philomena Ryan-Bugler, Louise Elsam, Monika Pikart-Müller
Production Manager: Claudia Grössl

Library of Congress Card No. applied for

British Library Cataloguing-in-Publication Data:
A catalogue record for this book is available from the British Library

Die Deutsche Bibliothek – CIP-Einheitsaufnahme
Herbst, Willy:
Industrial Organic Pigments : Production, Properties,
Applications / Willy Herbst ; Klaus Hunger. –
Weinheim ; New York ; Basel ; Cambridge : VCH, 1993
Einheitssacht.: Industrielle Organische Pigmente <engl.>
ISBN 3-527-28161-4 (Weinheim ...)
ISBN 0-89573-981-X (New York)
NE: Hunger, Klaus:

Data conversion and printing: Zechnersche Buchdruckerei, D-6720 Speyer
Bookbinding: Großbuchbinderei J. Schäffer, D-6718 Grünstadt
Printed in the Federal Republic of Germany

Preface

Organic pigments – the increasingly most important group of organic colorants worldwide – have never yet been treated comprehensively with respect to their industrial significance and their application properties. In this book we have tried to give an account of the chemistry, the properties, and applications of all commercially produced organic pigments.

This book is intended for all those who are interested in organic pigments, especially chemists, engineers, application technicians, colorists, and laboratory assistants throughout the pigments industry and in universities and technical colleges. We have specifically avoided an in–depth discussion of the underlying scientific and theoretical framework, but there are references to the pertinent literature.

The initial part is devoted to chemical and physical characterization of pigments and discusses important terminology connected with pigment application. This is followed by three chapters describing the chemistry and synthesis, the properties and application of individual pigments. In these chapters pigments are classified according to their chemical structure and listed by their Colour Index Name instead of their trade name. The Colour Index, published by the Society of Dyers and Colourists, lists all those pigments and dyes which have been registered by the pigment and dye manufacturers. The products are listed by their Colour Index (C.I.) Generic Name, followed by a Constitution Number, provided the chemical structure has been published. An example is C.I. Pigment Yellow 1, 11680. The last chapter discusses questions of ecology and toxicology. The literature references listed at the end of each dual-numbered subchapter have been limited to a selection covering the most important topics. The appendix shows general structural equations for the syntheses of individual groups of pigments and lists all pigments mentioned in this book, including the respective CAS (Chemical Abstracts Service) registry numbers.

The technical and fastness properties of different pigments have been assessed by unified, usually standardized test methods. Lightfastness measurements, however, had to be carried out by comparison to the Blue Scale – despite serious objections which are explained in the text. This was the only technique which made it possible to list comparative values for all pigments described in this book.

After careful deliberation we have reluctantly refrained from listing data on pigment economics. Reliable data on organic pigments have only been published in a few countries. Moreover, many of the other data turned out to be either contradictory or so incomplete that it was impossible to elicit reliable information from them.

We are pleased to present here the English version of our book, which is an update of the German edition of 1987, supplemented by all appropriate and newly published data. Also included are those commercial organic pigments which have recently been introduced to the market.

We would like to thank Mrs. Barbara Hoeksema for her work in translating this book.

We would like to express our gratitude to the management of the Fine Chemicals and Colors Division of Hoechst AG for their support and for making the scientific and technical resources available to us. We would also like to thank the numerous colleagues, both at other companies – especially at BASF AG and at Ciba/Geigy AG – and in-house colleagues, who through their stimulation, critique, and suggestions supported us considerably. We would like to express our particular gratitude to Dr. F. Gläser, who wrote Chapter 1.6.1.

Our appreciation is also extended to our families and friends, without whose consideration and patience it would not have been possible to write this book.

It is a pleasure to express our gratitude to the VCH publishing company who helped us greatly through their stimulation and their compliance with many of our wishes.

Frankfurt–Höchst   W. Herbst
December 1992   K. Hunger

Contents

3   **Polycyclic Pigments**

1 General

1.1 Definition: Pigments—Dyes

Colorants are classified as either pigments or dyes. Pigments are inorganic or organic, colored, white or black materials which are practically insoluble in the medium in which they are incorporated. Dyes, unlike pigments, do dissolve during their application and in the process lose their crystal or particulate structure. It is thus by physical characteristics rather than by chemical composition that pigments are differentiated from dyes [1]. In fact, both are frequently similar as far as the basic chemical composition goes, and one structural skeleton may function either as a dye or as a pigment.

In many cases the general chemical structure of dyes and pigments is the same. The necessary insolubility for pigments can be achieved by avoiding solubilizing groups in the molecule or by forming insoluble organic structures. Carboxylic and especially sulfonic acid functional groups lend themselves to the formation of insoluble metal salts (lakes); formation of metal complex compounds without solubilizing groups and finally suitable substitution may decrease the solubility of the parent structure (e.g., carbonamide groups).

Pigments of many classes may be practically insoluble in one particular medium, yet dissolve to some extent in another. Partial solubility of the pigment is a function of application medium and processing conditions, especially of the processing temperature. Important application properties of pigments and/or pigmented systems, such as tinctorial strength, migration, recrystallization, heat stability, lightfastness and weatherability, are often determined by the portion of pigment that dissolves to a minor degree in the vehicle in which it is applied.

Monoazo yellow pigments of the Hansa Yellow type (e.g., P.Y.1, P.Y.3; Sec. 2.3.4) may serve as an example. Their solubility in air-dried alkyd resin systems is so negligible that they are considered insoluble; which explains their frequent use in such media. Since their solubility increases with increasing temperature, they migrate considerably in vehicles such as various oven-dried varnish systems or in plastics. This results in bleeding or blooming (Sec. 1.6.3). Strong migratory tendencies unfortunately preclude their use in such high temperature applications. Even slight temperature changes in the course of pigment incorporation into its application medium may often determine the commercial fate of a pigment. Moreover, the inherent tinctorial properties of a product in a particular vehicle system are sometimes compromised by difficulties such as recrystallization, which arises through a certain solubility of the pigment in its medium.

Under certain circumstances, it may even be advantageous to have a pigment dissolved to some degree in its binder system in order to improve certain application properties such as tinctorial strength and rheological behavior. Such conditions arise when special amine-treated diarylide yellow pigments are incorporated in toluene-based publication gravure inks (Sec. 1.8.1.2). In toluene, up to 5% of the amine-treated pigment may be either dissolved or dispersed to a nearly molecular level. This improves the tinctorial strength and decreases the viscosity, which in turn enhances the rheology of the pigmented ink. The performance of a colorant in its role as a commercial pigment is therefore defined by its interaction with the application medium under the conditions that govern its application.

1.1.1   Organic–Inorganic Pigments

In some application areas, inorganic pigments are also used to an appreciable extent, frequently in combination with organic pigments. A comparison of the respective application properties of inorganic versus organic pigments shows some fundamentally important differences between the two families.

Most inorganic pigments are extremely weatherfast (Sec. 1.6.6) and many exhibit excellent hiding power (Sec. 1.6.1.3). Their rheology is usually an advantage (Sec. 1.6.8), being superior to that of most organic pigments under comparable conditions. In white reductions, however, many inorganic pigments have much less strength than organic pigments. With the exception of Molybdate Red, Chrome Yellow, and cadmium-based pigments, inorganic pigments provide dull shades. Since there are only relatively few inorganic types, the spectral range that is accessible by inorganic pigments alone is very limited. Many hues cannot be produced in this manner.

Inorganic pigments not only exhibit coloristic limitations but also frequently present application problems. Ultramarine Blue, for instance, is not fast to acid; while Prussian Blue must not be exposed to alkalis. Such limitations preclude the application of especially Prussian Blue in paints that are to be applied to a basic substrate (e.g., exterior house paints). In the red range of the spectrum, iron oxide red pigments produce weak hues of comparatively little brilliance. Molybdate Reds and Chrome Yellows lend themselves to a host of applications but are nevertheless sensitive to acids and light. There are stabilized versions of such pigments which claim improved lightfastness and acid resistance. These products also claim to be chemically fast to hydrogen sulfide, which affects the brightness of a coating through sulfide formation. However, if the particle surfaces of such types are damaged in the course of the dispersion process, the above-mentioned deficiencies are apparent at the damaged site.

Poor tinctorial strength and lack of brilliance restricts the use of inorganic pigments in printing inks. There are areas of application, however, where it is hardly, if at all, possible to replace the inorganic species by an organic pigment. The ceramics industry, for example, requires extreme heat stability, which precludes the use of organic compounds. Thus, the organic and inorganic classes of pigments are generally considered complementary rather than competitive.

Reference for Section 1.1

[1] DIN 55 943: Farbmittel, Begriffe. ISO 4618-1-1984 (TC 35): Paints and varnishes—Vocabulary – Part 1: General terms.

1.2   Historical

The history of pigment application dates back to prehistoric cave paintings, which give evidence of the use of ocher, hematite, brown iron ore, and other mineral-based pigments more than 30,000 years ago. Cinnabar, azurite, malachite, and lapis lazuli have been traced back to the third millenium BC in China and Egypt. With Prussian Blue in 1704 the first inorganic pigment was synthesized. It was not until a century later that Thenard produced his Cobalt Blue. Ever increasing expertise and technology led to the production of Chrome Yellow, Cadmium Yellow, several synthetic iron oxides covering parts of the ranges of yellow, red, and black hues; Chrome Oxide Green, and Ultramarine.

Important twentieth-century developments include the addition of Molybdate Red to the series of inorganic synthetic coloring matters in 1936; Titan Yellow followed in 1960.

The beginning of organic pigment application dates to antiquity. It is certain that the art of using plant and animal "pigments" to extend the spectral range of available inorganic colorants by a selection of more brillant shades had been practised thousands of years ago. However, for solubility reasons, most of these organic colors would now be classified as dyes rather than pigments. Even in antiquity, they were used not only for dyeing textiles but also, due to their ability to adsorb on mineral based substrate such as chalk and china clay, were used for solvent resistant coatings for decorative purposes. These materials later came to be known as lakes or toners. For thousands of years, derivatives of the flavone and anthraquinone series have been the major source of natural colors for such applications.

The beginning of the era of scientific chemistry was marked by the synthesis of large numbers of dyes for textile related purposes. Some of these were also applied to inorganic substrates by adsorption, for use as pigment toners. The commercially available soluble sodium salts of acid dyes were rendered insoluble, an essential property of pigments, by reacting them with the water soluble salts of calcium, barium or lead to form lakes. Basic dyes (commercially available as chlorides or as other water-soluble salts), on the other hand, were treated with tannin or antimony potassium tartrate to yield insoluble colorants, that is, pigments. Some of the early commercially important lakes, such as Lake Red C (Pigment Red 53:1) and Lithol Rubine (Pigment Red 57:1), released on to the market in 1902 and 1903, respectively, are still commercially important products (Sec. 2.7).

Entering the market in the late nineteenth century, the first water insoluble pigments that did not contain acidic or basic groups were the red β-naphthol pigments (Para Red P.R.1, 1885). Falling into the same chemical class are Toluidine Red (P.R.3, 1905) and Dinitroaniline Orange (P.O.5, 1907), two members of this class of pigments which still enjoy commercial importance today. In 1909, Hansa Yellow (P.Y.1) was introduced to the market as the first monoazo yellow pigment. The first red Naphthol AS pigments followed in 1912, and the first commercial pioneers of the diarylide yellow pigment range, some of which had been patented as early as 1911, appeared in 1935. Phthalocyanine blue pigments also appeared in 1935, followed by phthalocyanine green pigments a couple of years later [1]. The rapid advances in pigment chemistry led to such important classes of pigments as disazo condensation pigments in 1954, quinacridones in 1955, azo pigments of the benzimidazolone series in 1960, the isoindolinone pigments in 1964 [2], and the diketopyrrolo pyrrole pigments in 1986.

References for Section 1.2

[1] G. Geissler, DEFAZET Dtsch. Farben Z. 31 (1977) 152–156.
[2] H. MacDonald Smith, Am. Ink Maker 55 (1977) 6.

1.3   Classification of Organic Pigments

Publications have over the course of the years proposed several classification systems for organic pigments. Basically, it seems appropriate to adopt a classification system by grouping pigments either by chemical constitution or by coloristic properties. Strict separation of the two classification systems is not very practical, because the categories tend to overlap; however, for the purposes of this book it is useful to list pigments according to chemical constitution.

A rough distinction can be made between azo and nonazo pigments; the latter are also known as polycyclic pigments. The commercially important group of azo pigments can be further classified according to structural characteristics, such as by the number of azo groups or by the type of diazo or coupling component. Polycyclic pigments, on the other hand, may be identified by the number and the type of rings that constitute the aromatic structure.

1.3.1   Azo Pigments

Azo pigments, subdivided into the monoazo and disazo pigments, have the azo group (−N=N−) in common. The synthesis of azo pigments is economically attractive, because the standard sequence of diazonium salt formation and subsequent reaction with a wide choice of coupling components allows access to a wide range of products.

1.3.1.1   Monoazo Yellow and Orange Pigments

Monoazo yellow pigments that are obtained by coupling a diazonium salt with acetoacetic arylides as coupling components cover the spectral range between greenish and medium yellow; while coupling with 1-arylpyrazolones-5 affords reddish yellow to orange shades.

All members of this pigment family share good lightfastness, combined with poor solvent and migration resistance. These properties define and limit their application. Monoazo yellow pigments are used extensively in air-dried alkyd resin and in emulsion paints, and certain inks used in flexo and screen printing. Other applications are in letterpress and offset inks, as well as in office articles.

1.3.1.2   Disazo Pigments

There is a dual classification system based on differences in the starting materials. The first and most important group includes compounds whose synthesis involves the coupling of di- and tetra-substituted diaminodiphenyls as diazonium salts with acetoacetic arylides (diarylide yellows) or pyrazolones (disazo pyrazolones) as coupling components. The second group, bisacetoacetic arylide pigments, are obtained by diazotization of aromatic amines, followed by coupling onto bisacetoacetic arylides.

The color potential of disazo pigments covers the color range from very greenish yellow to reddish yellow and orange and red. Most show poorer lightfastness and weatherfastness; but better solvent and migration fastness than monoazo yellow and orange pigments. Their main applications are in printing inks and plastics, and to a lesser extent in coatings.

1.3.1.3   β-Naphthol Pigments

β-Naphthol pigments provide colors in the range from orange to medium red. The typical coupling reaction with β-naphthol as a coupling component yields such well-known pigments as Toluidine Red and Dinitroaniline Orange. Their commercial application in paints requires good lightfastness. Solvent resistance, migration fastness and lightfastness are comparable to the monoazo yellow pigments.

1.3.1.4 Naphthol AS Pigments (Naphthol Reds)

These pigments are obtained by coupling substituted aryl diazonium salts with arylides of 2-hydroxy-3-naphthoic acid (2-hydroxy-3-naphthoic acid anilide = Naphtol AS). They provide a broad range of colors from yellowish and medium red to bordeaux, carmine, brown, and violet; their solvent fastness and migration resistance are only marginal. Naphthol AS pigments are used mainly in printing inks and paints.

1.3.1.5 Azo Pigment Lakes

These pigments are formed by precipitating a monazo compound which contains sulfo and/or carboxy groups. The coupling component in the reaction may vary: β-naphthol lakes are derived from 2-naphthol, BONA pigment lakes use 2-hydroxy-3-naphthoic acid (**B**eta-**O**xy-**N**aphthoic **A**cid); and Naphthol AS pigment lakes contain anilides of 2-hydroxy-3-naphthoic acid as a coupling component. Lakes may also be prepared from naphthalenesulfonic acids. Lake Red C is one of the commercially significant β-naphthol lakes. Limited lightfastness, which ranks far behind the non-laked β-naphthol counterparts, along with a tendency to migrate largely restricts their use mainly to the printing inks field.

Most BONA lake pigments provide an extra site for salt formation. Apart from the usual substituents, the diazo components of almost all BONA lake pigments contain a sulfonic acid function. Two acid substituents are thus available to form insoluble salts, which is the form in which these pigments are commercially available. Metal cations such as calcium, strontium, barium, magnesium, or manganese combine with the organic anion to produce shades between medium red and bluish red. Their use in printing inks exceeds their increasing use in plastics and paints.

The organic acid group of Naphthol AS pigment lakes is part of the diazo component; a second site for salt formation can be provided by the coupling component. The plastics industry is the main user of such lakes.

Naphthalenesulfonic acid lake pigments are based on naphthalenesulfonic acid as a coupling component; introduction of an additional SO$_3$H function as part of the diazo component is possible.

1.3.1.6 Benzimidazolone Pigments

Benzimidazolone pigments feature the benzimidazolone structure, introduced as part of the coupling component. The pigments that are obtained by coupling onto 5-acetoacetylaminobenzimidazolone cover the spectrum from greenish yellow to orange; while 5-(2'-hydroxy-3'-naphthoylamino)-benzimidazolone as a coupling component affords products that range from medium red to carmine, maroon, bordeaux, and brown shades. Pigment performance, including lightfastness and weatherability, is generally excellent. Pigments that satisfy the specifications of the automobile industry are used to an appreciable extent in automotive finishes. Benzimidazolone pigments are also used extensively to color plastics and high grade printing inks.

1.3.1.7 Disazo Condensation Pigments

These pigments can formally be viewed as resulting from the condensation of two carboxylic monoazo components with one aromatic diamine. The resulting high molecular weight pigments show good solvent and migration resistance and generally provide good heat stability and lightfastness. Their main markets are in the plastics field and in spin dyeing. The spectral range of disazo condensation pigments extends from greenish yellow to orange and bluish red or brown.

1.3.1.8 Metal Complex Pigments

Only a few azo metal complexes are available as pigments. Most of these are very lightfast and weatherfast. The chelating metal is usually nickel, and less commonly, cobalt or iron(II).

The azo group ($-N=N-$) may be replaced by the analogous ($-CH=N-$) moiety to form an azomethine complex pigment, usually with copper as a chelating metal. The number of commercially available products in this group is also restricted. They typically afford yellow shades. Those species that provide the required lightfastness and weather resistance are used in automotive finishes and other industrial coatings.

1.3.1.9 Isoindolinone and Isoindoline Pigments

Although of comparatively good light- and weatherfastness, solvent and migration resistance, only a few members of the isoindolinone and isoindoline families are commercially available as pigments. Chemically classified as heterocyclic azomethines, these pigments produce greenish to reddish yellow hues. Isoindolinone pigments are preferably supplied for the pigmentation of plastics and high grade coatings.

1.3.2 Polycyclic Pigments

Pigments with condensed aromatic or heterocyclic ring systems are known as poly-cyclic pigments. The several pigment classes that fall into this category do not re-flect their actual commercial importance; only few are produced in large volume. Their chief characteristics are good light- and weatherfastness and good solvent and migration resistance; but, apart from the phthalocyanine pigments, they are also more costly than azo pigments.

1.3.2.1 Phthalocyanine Pigments

Phthalocyanine pigments are derived from the phthalocyanine structure, a tetraaza tetrabenzoporphine. Although this basic molecule can chelate with a large variety of metals under various coordination conditions, today only the copper(II) com-plexes are of practical importance as pigments. Excellent general chemical and phy-sical properties, combined with good economy, make them the largest fraction of organic pigments in the market today. Copper phthalocyanine blue exists in several crystalline modifications. Commercial varieties include the reddish blue alpha form, as stabilized and nonstabilized pigments; the greenish blue beta modifica-tion; and, as yet less important, the intense reddish blue epsilon modification. Bluish to yellowish shades of green pigments may be produced by introduction of chlorine or bromine atoms into the phthalocyanine molecule.

1.3.2.2 Quinacridone Pigments

The quinacridone structure is a linear system of five anellated rings. These pig-ments perform largely like phthalocyanine pigments. Outstanding light- and wea-therfastness, resistance to solvents and migration resistance justify the somewhat higher market price in applications for high grade industrial coatings, such as auto-motive finishes, for plastics, and special printing inks. Unsubstituted trans-quina-cridone pigments are commercially available in a reddish violet beta and a red gam-ma crystal modification. One of the more important substituted pigments is the 2,9-dimethyl derivative, which affords a clean bluish red shade in combination with excellent fastness properties. Solid solutions of unsubstituted and differently sub-stituted quinacridones and blends with quinacridone quinones resulting in reddish to yellowish orange pigments are commercially available; while 3,10-dichloroquin-acridone as yet enjoys only limited success as a pigment.

1.3.2.3   Perylene and Perinone Pigments

Perylene pigments include the dianhydride and diimide of perylene tetracarboxylic acid along with derivatives of the diimide; while perinone pigments are derived from naphthalene tetracarboxylic acid.

Commercially available types provide good to excellent lightfastness and weatherability; some of them, however, darken upon weathering. A number of them have excellent heat stability, which renders them suitable for spin dyeing. They are also used to color polyolefins which are processed at high temperatures. The list of applications includes high grade industrial coatings, such as automotive finishes; and, to a lesser degree, special printing inks for purposes such as metal decoration and poster printing.

1.3.2.4   Thioindigo Pigments

4,4',7,7'-Tetrachlorothioindigo with a reddish violet shade reigns supreme as a pigment among the derivatives of this indigo. It can be used for bordeaux shades in automotive refinishes. Thioindigo pigments are generally used in industrial coatings and plastics for their good lightfastness and weatherfastness in deeper shades.

Anthraquinone Pigments

Apart from some nonclassified pigments such as Indanthrone Blue (P.Bl.60), the anthraquinone pigments, which are structurally or synthetically derived from the anthraquinone molecule, can be divided into the following four groups of polycyclic pigments.

1.3.2.5   Anthrapyrimidine Pigments

The commercially leading member of this class, Anthrapyrimidine Yellow in very light white reductions affords a greenish to medium yellow with excellent weatherfastness. It lends itself primarily to application in industrial coatings such as automotive metallic finishes or to modify the shades of automotive finishes.

1.3.2.6   Flavanthrone Pigments

Flavanthrone Yellow, the only commercially used flavanthrone, is a moderately brilliant reddish yellow. Excellent lightfastness and weatherfastness, combined with good solvent and migration resistance, make this pigment an attractive supplement to Anthrapyrimidine Yellow, mainly in the automotive finish industry.

1.3.2.7   Pyranthrone Pigments

Commercial attention focuses on the derivatives of the pyranthrone molecule at a varying level of halogenation. Most are orange; but others exhibit a dull medium to bluish red shade. Due to their good weatherfastness pyranthrone pigments are used for high grade industrial finishes.

1.3.2.8   Anthanthrone Pigments

Dibromoanthanthrone is the only commercial pigment within this group. Qualities such as outstanding light- and weatherfastness justify the relatively high cost for application in high grade industrial coatings such as automotive finishes. The transparent pigment provides shades of scarlet for metallic finishes.

1.3.2.9   Dioxazine Pigments

The dioxazine molecule is derived from triphenodioxazine, a linear system of five anellated rings. Apart from Pigment Violet 37, the commercially most representative one is Pigment Violet 23, an extremely lightfast and weatherfast compound with good to excellent solvent and migration resistance. Applications include the pigmentation of coatings, plastics, printing inks, and for spin dyeing. Apart from producing violet shades, the pigment also lends itself to the shading of phthalocyanine blue pigments in colorations, particularly in coatings. It is also used to tone the light yellowish shade of titanium dioxide in whites and in shading carbon blacks that have a brownish cast.

1.3.2.10   Triarylcarbonium Pigments

There are two groups of triarylcarbonium pigments: inner salts of triphenylmethane sulfonic acids, and complex salts with heteropolyacids containing phosphorus, tungsten, molybdenum, silicon, or iron.

The first group is characterized by poor lightfastness and limited solvent resistance. Alkali Blue is the only member of this group with considerable commercial value. To tone black printing inks, Alkali Blue is used in combination with the very high-absorbing carbon black pigment which increases its lightfastness considerably.

The second group includes the complex salts of basic pigments that are common in the dyes industry, such as Malachite Green, Methylene Violet, Crystal Violet or Victoria Blue with certain heteropolyacids. Despite the disadvantages of comparatively poor solvent resistance and limited lightfastness, these pigments are used for their excellent color brilliance and clarity of hue; properties which exceed those of any of the other known organic or inorganic pigments. These are features which make those types whose lightfastness satisfies the commercial requirements suitable candidates for the printing inks industry and especially for packaging inks.

1.3.2.11   Quinophthalone Pigments

Quinophthalone pigments have a polycyclic structure derived from quinaldine and phthalic anhydride.

A few members of this class have gained commercial recognition for their very good temperature resistance. The main markets for their mostly greenish yellow shades are in the plastics and coatings industries.

1.3.2.12   Diketopyrrolo Pyrrole Pigments

The basic skeleton of this newly developed group of pigments consists of two anellated five-membered rings each of which contains a carbonamide moiety in the ring.

This recently introduced class of pigments has as yet only one commercially used representative. In full shade, the pigment affords medium shades of red, while the white reductions are somewhat bluer. At present, the pigment is used primarily in automotive finishes and to a lesser extent in plastics.

1.4   Chemical Characterization of Pigments

In this chapter, the correlation between chemical constitution and pigment performance is outlined in terms of empirical rules. These correlations essentially apply independently from the application medium for all industrial uses of pigments.

While the properties of (soluble) dyes are determined almost exclusively by their chemical constitution, application characteristics of pigments — which are by definition insoluble in the medium in which they are applied (see Sec. 1.1) — are largely controlled by their crystalline constitution, i.e., by their physical characteristics. This is discussed in the next chapter.

The application properties of a pigment are basically governed by its chemical constitution, which in turn has a bearing on the physical parameters of crystal geometry. This seemingly straightforward correlation is complicated by the fact that a variety of crystal structures (modifications, see Sec. 1.5.3) may evolve from one and the same chemical constitution. Apart from knowledge about the chemical constitution of a compound, only extensive insight into its solid-state physics thus allows certain predictions as to the application properties of the pigment.

The object of this chapter is to discuss the influence of the chemical constitution on the hue, tinctorial strength, lightfastness, weatherfastness, solvent resistance, and migration resistance of a pigment. The systematic synthesis of a pigment with certain defined target properties is only possible to a very limited extent. Studies on how the relative position of a pigment backbone and its substituents within the crystal lattice determine the pigment properties are few and far between. Unfortunately, it is not possible to generalize these results.

1.4.1   Hue

The appearance of color in a molecule is associated with electronic excitation [1,2,3,4,5] caused by absorption of incident electromagnetic radiation in the ultraviolet and visible regions of the spectrum. Electrons are elevated from the ground state energy level to an excited state by absorbing selected frequencies of incident visible light, thereby giving the molecule the shade of the resulting complementary color. The fact that each electronic excitation is accompanied by a battery of rotational and vibrational transitions is responsible for the appearance of more or less broad absorption bands. An absorption band is said to undergo a bathochromic shift if a comparison of spectra shows that it has moved to longer wavelengths; a hypsochromic shift involves movement to shorter wavelengths.

The hue is primarily defined by the pattern of chromophores, a system of conjugated double bonds (π-electronic system) which is responsible for the absorption of visible light.

Azo pigments, as all azo compounds, can be viewed as systems in which electron donors and electron acceptors are connected via a chromophore—the azo group ($-N=N-$)—to form a system of conjugated double bonds.

Substituents with lone electron pairs, such as alkoxy, hydroxy, alkyl, and arylamino groups, are known as electron donors. The CH_3 group, despite the absence of such free electron pairs, is also considered an electron donor. Functional groups with conjugated π-electron systems, such as NO_2, $COOH$, $COOR$, SO_2, or SO_2Ar act as electron acceptors.

In discussing substituents to azo pigments, both electron donors and electron acceptors are effective particularly as parts of the coupling component; that is, they are located in the conjugated part of the system. There they usually cause a bathochromic shift of the absorption band with the longest wavelength. Evaluation of the empirical bathochromic effects of differently substituted derivatives against a standard conjugated system can provide a sequence of increasingly effective substituents. In azobenzene systems, for instance, electron donors are more bathochromically active than electron acceptors. Since this effect, however, also depends on the electron distribution between donor, acceptor, and azo function, a (usually less pronounced) hypsochromic shift may be observed, especially if electron acceptors are involved. Bathochromic or hypsochromic behavior is also determined by the electron distribution at the substituted site within the conjugated system. If a bonding electron pair is promoted to a nonbonding excited state, a $\pi \rightarrow \pi^*$ transition ensues which determines the frequency of the dominant absorption with the longest wavelength.

The intensity of an $n \rightarrow \pi^*$ transition (transition of one p-electron of a lone electron pair from a nonbonding (n) to an antibonding (π^*) orbital) is about two magnitudes less than the dominant $\pi \rightarrow \pi^*$ transitions and thus has no effect on the color.

Today, the Witt substitution rules, originally derived from empirical data on conjugated systems such as azo derivatives, can be approached quantum mechanically. Azo pigments basically obey the rules for azo dyes, although deviations through interactions within the crystal lattice must be taken into account.

The basic hue of an azo pigment is primarily defined by the structure of the coupling component, since pigment manufacturers focus almost exclusively on substituted anilines as diazo components. Shades of yellow, for instance, are preferably produced by using acetoacetic arylides CH_3COCH_2CONH-Ar or heterocyclic coupling components based on the structure

in a cyclic conformation. Products include barbituric acid (1), 2,4-dihydroxy quinoline (2), and 2-hydroxy-4-methyl quinoline (4-methyl carbostyril) (3). Such compounds absorb mainly in the short wave (blue) region of visible light. The reddish

yellow to orange shades produced by monoazo pigments obtained from 1-arylpyrazolone-5 derivatives as coupling components have to be considered exceptional. A more bathochromic red shift is provided by monoazo pigments featuring the enlarged conjugated system of 2-hydroxynaphthalene (β-naphthol); and particularly its 3-carboxylic acid and 3-carboxylic arylide derivatives.

Diazo components may also be of some significance in defining absorption frequencies. An amine with a considerably enlarged conjugated system, made, for instance, by dimerization, can contribute to a considerable bathochromic shift. Examples include the following:

Diazo component		Coupling component	Shade
Aniline derivatives	→	Pyrazolone-5 derivatives	yellow to orange
3,3'-Dichlorobenzidine	→	Pyrazolone-5 derivatives	yellowish red
Aniline derivatives	→	Naphthol AS derivative	red
3,3'-Dichlorobenzidine	→	Naphthol AS derivative	violet to blue

Azo pigments range in shade from greenish yellow to orange, red, blue, violet, and brown. The chemical constitution of the pigment, especially the substitution pattern of the coupling component, determines the basic color of a pigment; different shades within this color are influenced by physical characteristics, such as crystal geometry, particle size and shape, particle size distribution, and polymorphic crystal modification [6].

Substitution patterns, especially that of the diazotized aromatic amine, determine the color of a pigment to some extent; but empirical data do not lead to unambiguous conclusions as to the exact influence of a particular substituent on the shade. The problem is intricate, since the substitution pattern also has a bearing on the size and orientation of a pigment molecule and therefore on its crystal structure, including all the interactions associated with it.

An exchange of substituents on the arylide moiety of the coupling component (arylides of acetoacetic acid and 2-hydroxy-3-naphthoic acid) fails to afford a consistent influence on the shade. In this case, intramolecular and intermolecular interactions within the crystal lattice gain more significance, since the conjugation of π-electrons is not expected to extend far beyond the carbonamide bridge.

The same general considerations apply to nonazo pigments. Attempts have been made to theoretically interpret absorption spectra of polycylic compounds by quantum mechanics. These studies compared the absorption of systems such as indigo [1] with known spectra of polymethine dyes. However, the investigation of solid state spectra is aggravated by interactions within the crystal lattice and has rarely been studied.

The coloristic effect of substituting a polycyclic pigment system is expected to parallel that afforded by employing substituted diazo components in the manufacture of azo pigments. Considerable chlorine substitution may even change the basic color of a pigment. In contrast to the blue copper phthalocyanine pigments, copper polychlorophthalocyanine pigments appear in green shades (Sec. 3.1); chlorination of tetrachloroisoindolinone pigments shifts the absorption bands from the yellow region towards orange or red (Sec. 2.11).

The hue of a red azo pigment lake carrying sulfonic acid functions is determined to a considerable extent by the metal ion. In the series Na→Ba→Sr→Ca→Mn the shift of hue from yellowish to bluish red increases in the order in which they are listed. The complex correlation between chemical constitution and color in pigment molecules poses a quantum mechanical challenge. This is complicated by interactions within the crystal lattice and by the contribution of intermolecular and intramolecular hydrogen bonds. Such parameters have an insufficient theoretical understanding.

Questions about such interactions could only be resolved by knowledge of the exact geometry of the atoms of a pigment molecule in its unit cell and the relative position of each individual molecule within the crystal lattice. This is elucidated through three dimensional X-ray diffraction analysis of single crystals [7].

The complexity of the problem is aggravated by the fact that the absorption bands of a pigment in solution differ from those produced by the corresponding solid. It is impossible to predict with any certainty whether the hue associated with a certain set of chromophores within a polycyclic aromatic system will shift towards deeper colors if the conjugated system is extended.

Insertion of a heteroatom into a polycyclic system, on the other hand, restricting the options for electron delocalization, is frequently accompanied by a hypsochromic shift. Thus, for example:

Pyranthrone

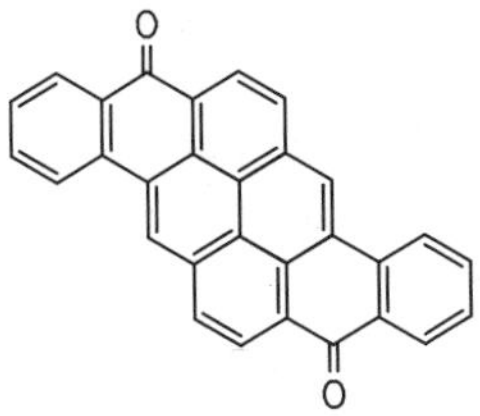

orange

Flavanthrone

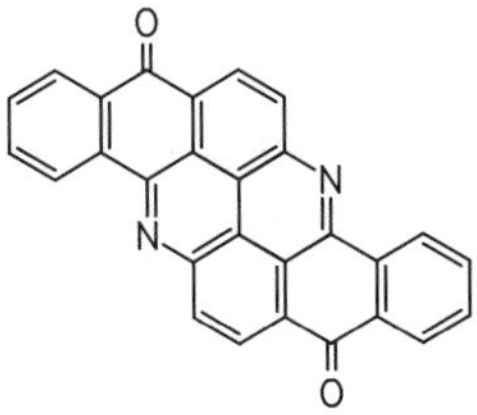

yellow

Particularly large color changes are associated with the transition from solution to solid state in heterocycles such as the cross-conjugated indigo system or the quinacridone skeleton. Quinacridone pigments, dissolved in concentrated sulfuric acid or in DMF, exhibit a pale yellow shade; the intense red color appears only in the solid state. This is a particularly distinctive example to demonstrate the correlation between crystal lattice interactions and hue [8].

Such simplified considerations can only serve as approximations to a more complex problem; even the shade of a polycyclic pigment basically depends on crystal characteristics, such as particle size, particle size distribution, and crystal modification.

Hydrogen bonds are important structural elements of organic pigments: intramolecularly, they enforce planarity in a molecule; intermolecularly, they may even play a role in determining the basic color of a pigment.

Quinacridones, for example, depend on hydrogen bonds to define their crystal structure during crystal formation.

1.4.1.1   Crystal Modification and Crystal Structure

Some pigments are polymorphic; in other words, there are different ways in which molecules with one and the same chemical composition may be arranged within the crystal lattice, at the unit cell level.

Considering the fact that the X-ray diffraction pattern of a crystal depends on its lattice structure, pigment powders can be analyzed with a Debye–Scherrer diffraction camera to establish a correlation between X-ray diffraction and crystal modification. It is synthetically not possible to produce a defined crystal modification of a new pigment. Attempts to modify the preparative procedure or to apply different aftertreatment may result in a pigment of two or more crystalline forms, different not only in lattice structure, but also in color and performance.

Crystal structure and molecular geometry of a pigment can be elucidated by three dimensional X-ray diffraction analysis of a single crystal. In the azo family, some of the monoazo yellow pigments, some β-naphthol and Naphthol AS pigments, and two benzimidazolone pigments have been studied. In the polycyclic series, besides quinacridone and dioxazine pigments, publications about copper phthalocyanine pigments predominate [9]. Details are presented under the respective classes of pigments.

There is some hope that a comprehensive data bank on the three dimensional crystal structure of one class of pigments might afford an empirical correlation between color and geometry in a pigment. However, the difficulty of growing single crystals is mainly responsible for the fact that few studies have been performed to elucidate the crystal structure of organic pigments, despite improved equipment and state-of-the-art computer programs. This is particularly true for high grade, solvent resistant pigments, which are practically insoluble in solvents in which single crystals can be grown.

Comparative three dimensional X-ray diffraction studies have been carried out on a red and a brown representative of the Naphthol AS pigment series, differing only by the presence or absence of one methoxy group in the anilide function of the coupling component (4).

R = H: red
R = OCH₃: brown

4

The almost completely planar structure of the red pigment is distorted in the brown pigment by an additional methoxy group, which forces portions of the molecule out of the molecular plane. As a result, not only is the conjugation pattern disturbed, but the entire crystal geometry is changed. Fig. 1 shows the structure of the two pigments **4** (R = H and OCH$_3$, respectively), as determined by three dimensional X-ray diffraction studies. The molecules are shown individually and also in their crystal lattice, seen vertically to the molecular plane. The hydrogen bonds are exclusively intramolecular and do not contribute to intermolecular interaction, which relies on weaker forces [10].

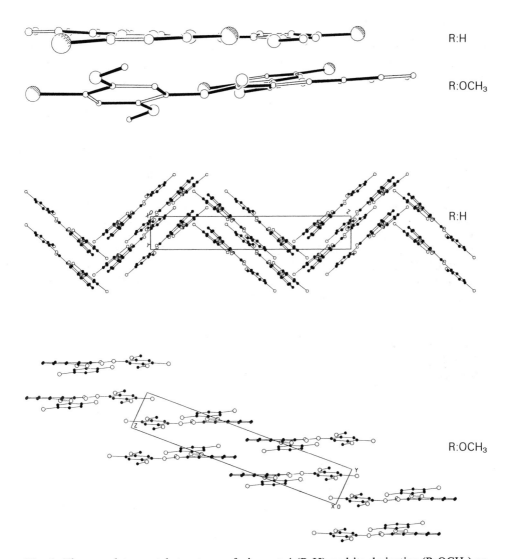

Fig. 1: The complete crystal structures of pigment **4** (R:H) and its derivative (R:OCH$_3$) as derived from three dimensional X-ray diffraction analysis. The molecules are shown both as single units (above) and within the crystal lattice (below), seen perpendicular to the molecular plane.

Both molecules favor the o-quinone-hydrazone form over the o-hydroxyazo form. Corresponding studies on a selection of monoazo pigments and several β-naphthol and Naphthol AS pigments are described in the respective chapters.

1.4.2   Tinctorial Strength

Since the color strength of a pigment is defined by its tendency to absorb light, the maximum molar extinction coefficient ε_{max} can be used to estimate the relative strength of a pigment, which is, however, also dependent on the physical parameters of the pigment crystal (Sec. 1.5). A better approach is to measure the absorption connected with a single electron transition, especially the one with the longest wavelength, and to integrate the peak between the lower limit v_1 and the upper limit v_2 ($f = \int_{v_1}^{v_2} \varepsilon_v d_v$) to calculate the total area f under the peak. This is known as the oscillator strength. For practical purposes, $\varepsilon_{max} \cdot \Delta v_{1/2}$, with the half width $\Delta v_{1/2}$ provides a good approximation. To be precise, the term oscillator strength refers to a free molecule or, as a first approximation, to a molecule in a solvent which does not noticeably interact with the chromophore system. Thus, for all practical purposes, it seems much more useful to approach the problem by focusing on the correlation between extinction and molecular parameters.

The primary chemical aspect of a mesomeric pigment system concerns the correlation between tinctorial strength and extent of electron delocalization. A higher degree of conjugation in a molecule is associated with a bathochromic shift; color strength improves with the intensity of absorption. A resonating system proliferates by:

– enhancing the aromaticity of a polycyclic pigment, transition from monoazo to disazo compounds (doubling of the azo moiety in azo pigments),
– promoting planarity in a molecule,
– incorporating substituents — preferably π-electron carriers — into the conjugated region of a pigment molecule. The π-orbitals of a nitro function, for instance, are being oriented perpendicularly to the plane of the aromatic moiety of the pigment molecule so that the π-orbitals of chromophore and substituent are arranged parallel to each other.

Molecular dimerization, as in the transition from Hansa Yellow to disazo yellow pigments, increases the color strength considerably. A comparison of a Naphthol AS/disazo condensation pigment pair gives similar results:

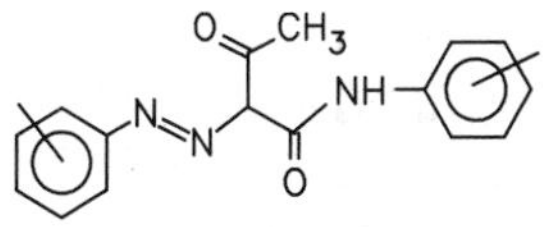

Monoazo yellow pigment

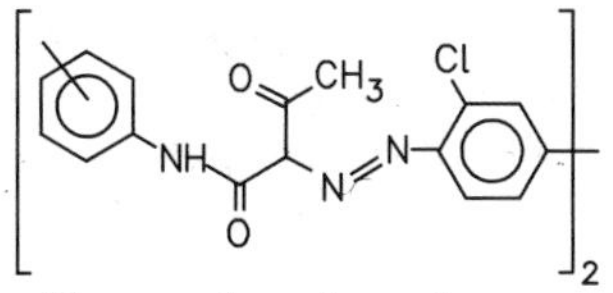

Disazo yellow pigment

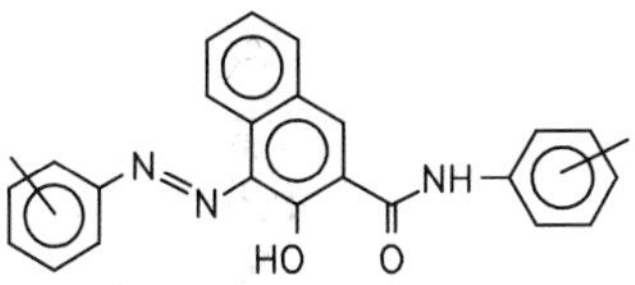

Naphthol AS pigment

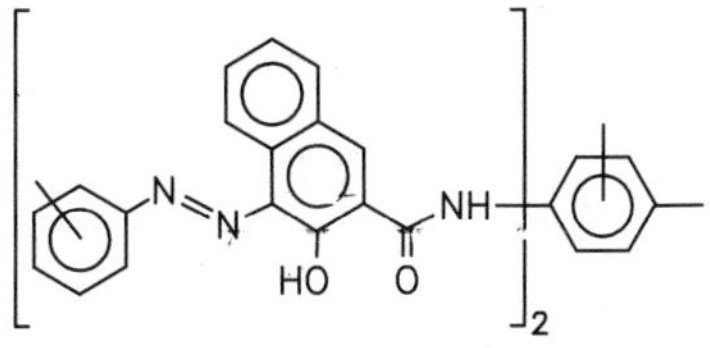

Disazo coridensation pigment

G. Eulitz [11] studied the correlation between chemical constitution and tinctorial strength directly on dissolved pigment samples. He was able to exclude physical crystal parameters, such as particle size distribution and degree of dispersion, by achieving molecular dispersion. His analysis demonstrated the following:

The transmission spectra of constitutionally similar pigment pairs, such as Pigment Yellow 1 / Pigment Yellow 3, or the pair Pigment Yellow 12 / Pigment Yellow 13, were evaluated according to Lambert–Beer.

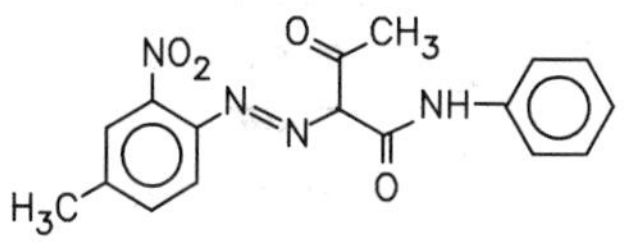

Pigment Yellow 1

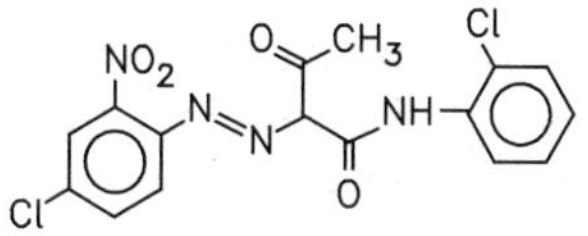

Pigment Yellow 3

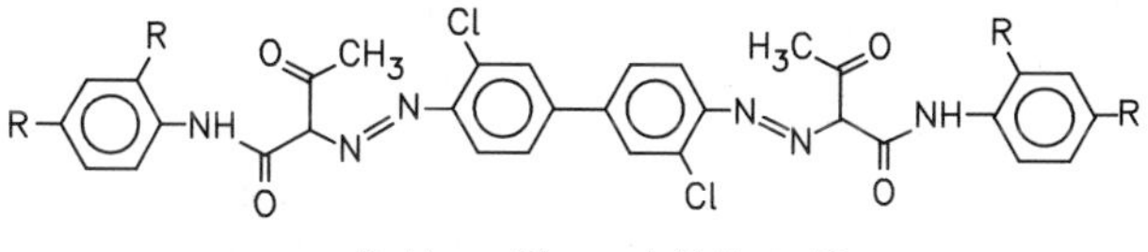

R: H   Pigment Yellow 12
R: CH$_3$ Pigment Yellow 13

The resulting extinction coefficient remained constant within the accuracy of the measurements. In the transition from monoazo to disazo pigment, however, the value of the maximum molar extinction coefficient is more than doubled, because the effect of the two azo linkages is enhanced by additional interaction via the diphenyl moiety. This, however, does not improve the conjugation; the shade does not shift remarkably:

Pigment solution	Maximum molar extinction coefficient (in 10^4 L/mol·cm)
Pigment Yellow 1	1.97
Pigment Yellow 3	2.08
Pigment Yellow 12	6.57
Pigment Yellow 13	6.70

Considering the fact that pigments exist as crystals dispersed in the application media, these studies are only of theoretical value. However, it has been shown that the maximum specific extinction of Pigment Red 112 is independent of whether it is measured in solution or in crystalline form. The study compared transmission spectra of the dissolved pigment with the solid-state spectra of crystals in a transparent film.

A certain correlation between the maximum extinction in solution and in the solid state has also been established for other classes of pigments.

Three dimensional X-ray diffraction data point to a largely planar structure in all of the molecules studied; this ensures optimum conjugation (overlapping of the π-electrons). Intramolecular hydrogen bonds probably contribute considerably towards supporting the planar conformation.

π-electron interactions between the aromatic ring and adjacent conjugated free electron pairs, either nonbonding electron pairs or double bonds of substituents, improve electron delocalization, which in turn increases absorption intensity, that is, color strength.

The reverse is true for large substituents within a molecule which do not contribute to the resonating system. These may have an adverse effect on the tinctorial strength.

1.4.3   Lightfastness and Weatherfastness

Little is known about the relation between chemical constitution and lightfastness/weatherfastness of pigments, although it is clear that fastness is primarily defined by the chemistry of the molecule. Various pigments, for example, undergo similar degradation upon exposure to light, irrespective of whether they are crystalline or in solution. The surrounding medium — the vehicle in which a pigment is applied — also affects the stability of a compound to light: the tendency of optically excited pigment molecules to react with their surroundings varies with the vehicle in which they are applied.

High grade pigments cannot be studied both in solution and in the solid state, because they are insoluble in almost all solvents.

In azo and some polycyclic pigments, a straightforward correlation may be established between the lightfastness and weatherfastness on the one hand and the substitution pattern on the other hand. Resistance to light and weather is generally improved by electron acceptors that are introduced through the diazo component (e.g, halogen atoms, nitro groups, carbo-alkoxy groups), even more so in combination with electron donors (methoxy or methyl groups) in the phenyl moiety of the coupling component. Obviously, the substitution pattern of the aromatic system, especially that of the diazo component, has a major impact on the molecule: properties tend to improve in the sequence meta → para → ortho substitution. In the same order, the formation of intramolecular or intermolecular hydrogen bonds is probably facilitated.

Lightfastness and weatherfastness of quinacridone pigments (Sec. 3.2) deteriorate in the order 2,9→3,10→4,11 substitution. It is assumed that decreasing the distance between substituents and NH function disturbs the formation of hydrogen bonds [12]; a tendency which culminates in the very poor light and weather resistance of 5,12-N,N'-dimethyl quinacridone.

Introducing additional carbonamide groups frequently improves the lightfastness of azo pigments; a trend which should not be confused with the generally observed tendency of carbonamide groups to enhance the migration fastness of their basic structure (Sec. 1.6.3). Monoazo yellow pigments, for instance, although more prone to migrate, survive exposure to light better than the almost insoluble diarylide yellow pigments. Disazo condensation pigments, on the other hand, are much more lightfast and show less of a tendency to migrate than monoazo pigments.

Metal complex formation considerably improves both lightfastness and weatherfastness in o,o'-dihydroxyazo and o,o'-dihydroxy azomethine pigments, usually compromised by distinctly dull color.

Likewise, the lightfastness of an azo pigment lake is controlled to a certain extent by the metal cation: manganese salts are usually most lightfast.

1.4.4   Solvent and Migration Fastness

The influence of the chemical constitution as related to solvent and migration resistance opens up a number of options to improve pigment performance:

a) Increasing the Molecular Weight

A comparison between related monoazo yellow pigments and diarylide yellow pigments clearly demonstrates the impact of increased molecular weight on solvent and migration resistance. Pigment Yellow 12, for instance, migrates much less than a comparable monoazo yellow pigment such as Pigment Yellow 1. The fact that Naphthol AS pigments migrate much more readily than comparable disazo condensation products adds to the evidence for the advantage of "doubling" the molecular structure.

b) Avoiding Solubilizing Substituents

Substituents such as long-chain alkyl, alkoxy, or alkylamino groups and also sulfonic acid functions tend to increase solubility.

Insolubilizing substituents are carbonamide groups and possibly nitro groups or chlorine for azo pigments, hetero atoms, especially nitrogen and to a lesser extent also chlorine or bromine in polycyclic pigments. Polar substituents generally decrease the solubility of a pigment in common nonpolar solvents. Naphthol AS pigments, for instance, tolerate solvents much better and tend to migrate less easily than β-naphthol pigments (demonstrated by the pair Pigment Red 13 / Pigment Red 3). Introducing a second carbonamide function adds to the advantage, as in Pigment Red 170; a principle which culminates in the production of almost completely migration resistant compounds such as Pigment Red 187. Solvent and migration fastness thus improve in the following order:

Pigment Red 3	Pigment Red 13
Number of CONH–groups 0	1

Pigment Red 170	Pigment Red 187
Number of CONH–groups 2	3

Consequently, there is remarkable advantage in introducing carbonamide groups into various positions on heterocyclic five- and six-membered rings anellated to a benzene nucleus in order to improve the migration fastness of the basic structure (see structures 5–9). Of the various heterocyclic systems, the benzimidazolone moiety (5, X:NH), especially as part of the coupling component in an azo pigment, has proven to be most effective, giving rise to an entire product line of benzimidazolone pigments (Sec. 2.8).

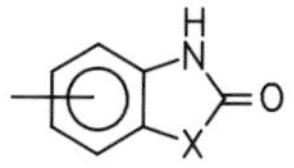

5

Benzazolone
(X : NH, O, S)

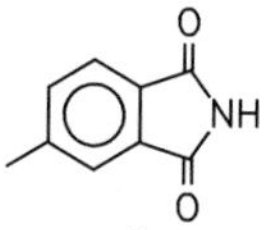

6

Phthalimide

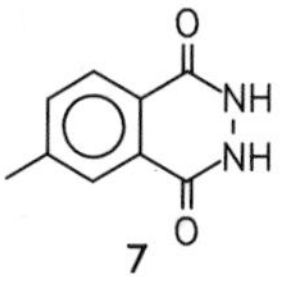

7

Tetrahydrophthalazinedione

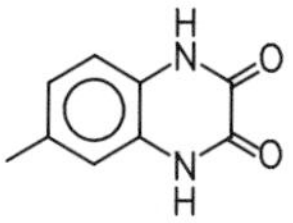

8

Tetrahydroquinoxalinedione

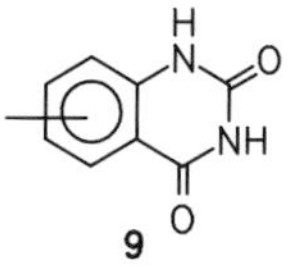

9

Tetrahydroquinazolinedione

The supportive function of hydrogen bonds towards decreasing the solubility of a pigment is particularly striking in benzimidazolone pigments: the good solvent and migration resistance of these pigments compared to their monoazo yellow and Naphthol AS counterparts ([13,14], see also Sec. 2.9) relies to a large extent on the comparatively small benzimidazolone group and its capacity to form intermolecular hydrogen bonds.

c) Forming Insoluble Polar Salts by Laking

Azo dyes containing sulfo and/or carboxy groups form pigment lakes if they are precipitated with the salts of alkaline-earth metals such as calcium, barium, strontium, or with manganese. The polar (salt) character of such pigment lakes makes for good solvent and migration fastness. Consequently, an increasing number of sulfo groups for salt formation is associated with improved solvent fastness, as demonstrated by Pigment Red 53:1 and Pigment Red 48:4:

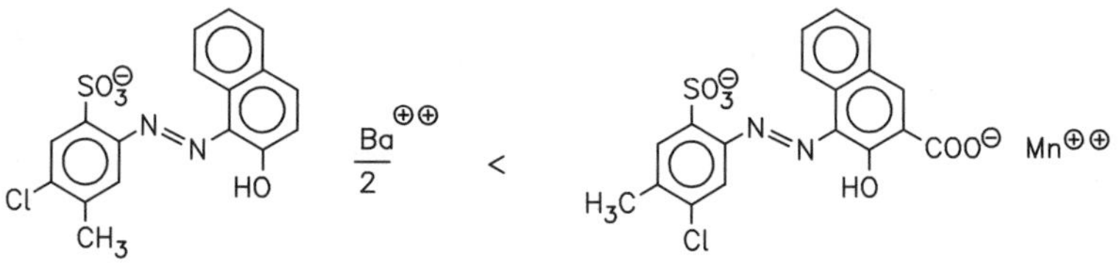

Pigment Red 53 : 1   Pigment Red 48 : 4

d) Forming Metal Complexes

Complexation of o,o'-dihydroxyazo compounds or their corresponding azomethine derivatives affords extremely solvent and migration resistant pigments (Sec. 2.10).

The correlation between constitution (substitution pattern of the diazo component) and solubility of an azo compound has been studied on a selection of Naphthol AS pigments [15]. Substituents being identical, their position on the aromatic ring of the diazo component ultimately controls pigment solubility; that is, the difference in enthalpy between crystalline and dissolved pigment for a given substituent is largest for meta substitution and smallest for para substitution.

References for Section 1.4

[1] M. Klessinger, Chem. unserer Zeit 12 (1978) 1–11; further references ibid.
   S. Dähne, Z. Chem. 10 (1970) 133, 168; Science 199 (1978) Nr. 4334, 1163–1167.
[2] P. Rys and H. Zollinger, Farbstoffchemie. 3. Suppl., Verlag Chemie, Weinheim, 1982;
   H. Zollinger, Color Chemistry, 2. Ed., VCH, Weinheim, 1991.
[3] J. Fabian and H. Hartmann, Light Absorption of Organic Colorants, Springer-Verlag, 1980.
[4] J. Griffiths, Rev. Prog. Color. Vol. 11 (1981) 37–57.
[5] J. Griffiths: Colour and Constitution of Organic Molecules, Academic Press, London, 1976.
[6] A. Pugin, Chimia, Suppl. (1968) 54–68.
[7] E.F. Paulus in: Ullmann's Encyclopedia of Industrial Chemistry, 5th ed., Vol. 20 (1992) p. 412.
[8] G. Lincke, Farbe + Lack 86 (1980) 966–972.
[9] J.R. Fryer et al., J. Chem. Technol. Biotechnol. 31 (1981) 371–387;
   D. Horn and B. Honigmann, Congr. FATIPEC XII, Garmisch, 1974, 181–189.
[10] D. Kobelt, E.F. Paulus and W. Kunstmann, Z. f. Kristallogr. 139 (1974) 15–32.
[11] G. Eulitz, Hoechst AG, private communication.
[12] G. Lincke, Farbe + Lack 76 (1970) 764–774.
[13] E.F. Paulus and K. Hunger, Farbe + Lack 86 (1980) 116–120.
[14] K. Hunger, E.F. Paulus and D. Weber, Farbe + Lack 88 (1982) 453–458.
[15] F. Gläser, XIII. Congr. FATIPEC, Juan les Pins, 1976, Kongreßbuch 239–243.

1.5   Physical Characterization of Pigments

The relevance of the chemical constitution of a pigment in controlling its application properties (Sec. 1.4) is embedded in the context of physical parameters, such as geometry of the unit cell, crystal lattice pattern (modification), and crystal shape; along with particle characteristics such as specific surface area, crystallinity, particle size distribution, and surface structure. Pigment properties as defined by certain target applications can be approached or optimized by adjusting the physical parameters. Particle size modifications, for instance, frequently facilitate the use of a pigment in an area in which its applicability would normally be compromised by processing problems or in which the pigmented medium might exhibit poor application properties.

A significant comparison of the application properties of pigments with different chemical constitutions can only be meaningful if it focuses on compounds whose physical characteristics are somewhat similar (particularly particle size).

There is an increasing tendency among manufacturers to provide their pigments with physical specifications. This facilitates comparisons between pigments and provides the user with some indication about application properties that may be relevant for his specific needs. Considering the importance of such information, it seems useful to discuss a few physical characteristics and methods for their determination, including the inherent limitations of such methods. The discussion focuses on organic pigments and the implications of the resulting data to the pigment chemist. There are a number of publications and text books that can familiarize the reader with the underlying theory and the application of such methods [1].

All pigments have some degree of particle size distribution. The ultimate particle size of most commercial pigments is less than 1 μm, often even smaller than 0.3 to 0.5 μm (anisometric particles are measured lengthwise). Depending on the application for which a pigment is targeted, the smallest particles may be as much as one to more than two orders of magnitude smaller.

For surface energy reasons, the tendency of these small particles to agglomerate and form crystallites increases with decreasing particle size. This is particularly true for the final phase of pigment manufacture, the drying and milling processes.

Organic pigment powders therefore comprise a mixture of such crystallites and single crystals. Incorporation into the application medium, be it a plastic, a printing ink, or a paint, relies on dispersion, which involves an effort to break down the agglomerates as far as possible. The quality of the pigment dispersion determines the majority of the application properties of a pigment, especially regarding the tinctorial and rheological potential of a pigment-vehicle system (Sec. 1.6.5).

The particle size distribution of an organic pigment powder is usually different from that found in the pigment-vehicle system; and since both have practical importance, methods were developed for their determination.

Pigments are morphologically described by a standardized terminology [2]; they can occur as single or primary particles, aggregates, or agglomerates.

The final crystals that ultimately constitute the crude pigment product are known as **primary particles**. These are true single crystals with the typical lattice disorders or combinations of several lattice structures that appear as units under X ray.

Primary particles may assume a variety of shapes, such as cubes, platelets, needles, or bars, as well as a number of irregular shapes. Fig. 2 shows some examples. Combinations between different shapes often make the unequivocal assignment to a particular class impossible.

Aggregates are primary particles that are grown together at their surfaces. The total surface area of an aggregate is smaller than the sum of the surfaces of the individual particles. Aggregates are not broken down by dispersion processes.

Agglomerates are groups of single crystals and/or aggregates, joined at their corners and edges but not grown together (Fig. 3), which can be separated by a dispersion process. The surfaces of the individual crystals are readily available to adsorption. By definition, the total surface area of an agglomerate may not differ

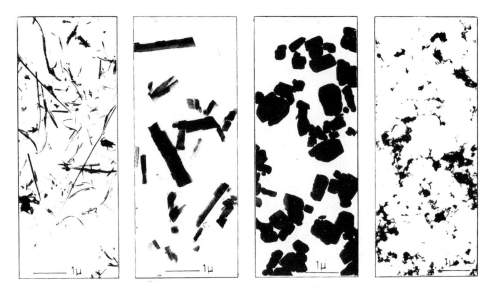

Fig. 2: Examples of crystal forms of organic pigment particles.

considerably from the sum of the surfaces of the individual particles. The dispersibility of a pigment is largely determined by the nature and density of the agglomerates, which in turn depend on particle shape and density. Platelets or rodlike particles, for instance, generally make for more voluminous and more easily dispersible agglomerates than isometric species.

Already dispersed pigment particles may for various reasons reassemble and form loosely combined units with various shapes. The most important among these are **flocculates** (Fig. 4), assemblies of wetted crystallites and/or aggregates or smaller agglomerates. They usually form in a low viscosity medium which fills the interior cavities of the pigment flocculates. Flocculates are therefore mechanically more labile than agglomerates and can usually be broken up by weak shear such as stirring.

1.5.1   Specific Surface Area

The specific surface area of a pigment is the surface in m^2 per 1 g of pigment. Typical values for organic pigments range between about 10 and 130 m^2/g.

The surface area of a pigment, however, is not a definitive value: it is controlled by factors such as the method of determination and the experimental parameters. Depending on the procedure, the experimental results may either reflect the exterior geometry of the pigment particles or also account for additional interior surface area, which is hard to define. Parameters such as the molecular size of the ad-

Fig. 3: Primary particles, aggregates, and agglomerates according to DIN 53 206, part I [2].

sorbed substance and pigment geometry, especially its porosity, determine how much of the total surface area is accessible to measurement.

Adsorption techniques are based on the assumption that the amount of gas, liquid, or solute that is adsorbed by a surface is proportional to the surface area. The resulting linear correlation makes it possible to calculate the total surface area of a pigment from the molecular volume of the adsorbed material and the amount that is necessary to completely cover the surface with a monolayer. The amount of adsorbate may be determined

- directly from the pigment sample,
- from the change of pressure in the adsorbed gas,
- from the concentration of the adsorbate dissolved in a carrier gas or in a liquid, or
- from the heat of reaction produced or required by the adsorption process.

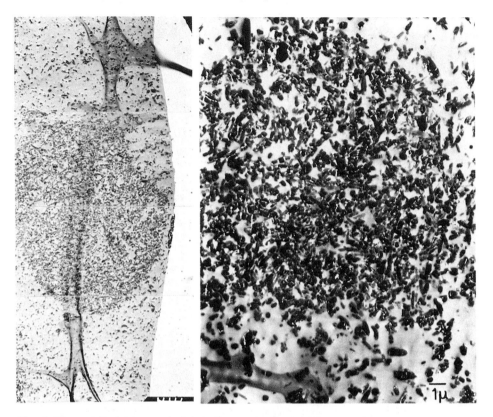

Fig. 4: Pigment Red 3 in an air-dried alkyd resin film. Electron micrograph of an ultrami-crotome-cut thin layer of a pigment flocculate. Right: Magnified image.

To determine the surface area of a pigment, it is more common to enforce an adsorption process by working at low temperatures and high pressure and concentration than to rely on desorption, which can be brought about by reversing the conditions.

The BET method, an adsorption method working with nitrogen, argon, or krypton gas, is the most widely used technique to measure specific surface areas of organic pigments. The isotherms are evaluated according to the Brunauer, Emmett and Teller equation [3], which is decribed satisfactorily in text books [4–7]. The market of routine determinations is dominated by commercial instruments which focus on a single point on the adsorption line close to the upper limit of the BET equation instead of plotting the entire adsorption isotherm.

Likewise, the Haul and Dümbgen method of measuring specific surface areas through nitrogen adsorption has developed into an industrial standard [8].

The interior surface area of a pigment powder may be assessed through a series of determinations, using a range of gaseous, liquid, or dissolved substances of various molecular sizes with different spatial demands, and measuring the amount of adsorbed material gravimetrically as well as volumetrically [9].

A number of phenols have proven to be valuable adsorbates for surface area determinations in organic pigments [10]. Phthalocyanines in particular have been evaluated successfully by phenol adsorption [11].

All of the gas adsorption methods rely on the availability of the entire interior surface of the pigment, including the inside of the agglomerates, to the gas molecules. In organic pigments, however, this is not always true, so that the results often seem considerably smaller than the actual values. Two varieties of Pigment Red 168 whose surfaces had not undergone treatment prior to the experiment have been used to demonstrate this effect [12]. Superior hiding power in sample 2 seemed to be connected with a larger specific surface area. Initial measurements afforded values of 20.8 m^2/g and 36 m^2/g for samples 1 and 2, respectively, which remained essentially constant even after the samples were rinsed thoroughly with organic solvents. Interestingly, the BET results contradict the electron microscopic photographs of the two samples (Fig. 5), which indicate that the particle size in sample 1 is relatively small, whereas sample 2 appears to be comparatively coarse.

Sample 1: 20.8 m^2/g   Sample 2: 35.9 m^2/g

Fig. 5: Electron micrographs of Pigment Red 168. The pigment was dispersed in ethanol by means of ultrasound and vaporized onto the electron microscope slide.

The experimental results are in keeping with the tendency of the agglomerates of sample 1, which are unavailable to wetting by polyolefins, to resist dispersion in nonwetting application media even in the presence of vigorous shear. Sample 2, however, disperses easily under the same conditions, which considerably improves its tinctorial strength. The agglomerates in sample 1 can be broken down by predispersing the sample in a molten low molecular weight polyethylene wax and simultaneously applying shearing force to the material before incorporating the pigment into the polyolefin. It is not surprising that, after the same treatment has been applied to both, the tinctorial strength and transparency of sample 1 exceeds that of sample 2. A similar effect occurs in the process of slurrying both samples in dioctyl phthalate and subsequently working them into plasticized PVC.

Clearly, there is little sense in experimentally determining specific surface areas of highly agglomerated organic pigment powders with comparatively small particle sizes; it is not advisable to rely on such values alone. Unfortunately, since both manufacturers and the literature fail to provide specifications about degrees of agglomeration in commercial products, the user faces the problem of dealing with values for specific surface areas that give little or no indication of the application properties of a particular product.

The same effect arises from a phenomenon known as "overmilling" organic pigments: inadvertently reducing both specific surface area and tinctorial strength of a pigment powder by milling it too long or too intensely, so that particles are fractured only to reassemble and produce increasing amounts of agglomerates.

Agglomeration, however, is not the only factor to falsify the experimental results of surface area determinations; there can be a number of substances adhering to the surface of the pigment particles that affect gas adsorption. Azo pigments, for instance, are manufactured using auxiliary agents, such as emulsifiers, coupling agents, etc. (Sec. 2.2), which are partially adsorbed on the fresh surface of a newly synthesized pigment crystal. Measurements may therefore reflect specific surface areas that appear smaller than the expected value.

Frequently, large amounts of additives such as rosin are used in manufacturing highly transparent types of azo pigments for application in process color printing inks. Pigment particles that emerge from such a process are comparatively small, because such resins tend to inhibit crystal growth [13]. Although they may occur separately, these substances are largely adsorbed on the surface of the pigment particles, where they more or less interfere with the surface area determinations.

The extent to which measurement results can be distorted by additives is particularly severe in the case of disazo yellow pigments, whose preparation involves fatty amines (Sec. 1.8.1.2, 2.4.1.1). Measurements are commonly carried out under conditions that give rise to the cleavage of volatile amines or ammonia, resulting in seemingly even negative specific surface area values.

It is not known to what extent surface area determinations are distorted by traces of water that cling to the particles despite extensive drying.

There are also techniques that utilize flow phenomena to measure parameters such as surface area or average particle size. Information about the latter is obtained by analyzing the amount of gas (air) or wetting liquid that is pulled or pressed through a defined pressing of pigment powder. This approach is based on the principle that the flow rate is controlled by the interior cavities of the pressing, which in turn depend on the particle size: the larger the particles, the higher the number of sizable cavities. The experimental results obey the Karman-Kozeny equation [14], which may be used to calculate the specific surface area. Surface-to-volume ratios of less than 1.2 m^2/cm^3 may be determined by permeability techniques, among which the Blaine method and instrumentation has developed into an industrial standard in Germany [15]. Flow methods tend to be unsatisfactory, however, in connection with organic pigments. Extremely large specific surface areas and exceptionally wide particle size distributions compared to inorganic pigments and extenders provide ample opportunity for surface coating materials and other surface processing additives to distort the measurements.

It seems reasonable to conclude that experimentally determined specific surface areas can only qualitatively relate to the physical characteristics of organic pigments. Instead, their value emerges in combination with other physical or physico-chemical parameters or in the context of application properties such as oil absorption [16] or wettability (Sec. 1.6.5).

1.5.2   Particle Size Distribution

The particles of a synthetic pigment, far from being uniform, cover a more or less wide range of sizes. Normally discontinuously produced pigment batches are usually combined so as to yield mixtures that meet the technical standards of certain target applications. This explains why it is possible for different batches of the same pigment to exhibit somewhat divergent particle size distributions.

There are various methods for the determination of the size distribution of organic pigment particles, the most common are sedimentation techniques in ultracentrifuges and specialized disk centrifuges as well as electron microscopy. These methods require considerable experimental skill, since the results depend largely on sample preparation and especially on the quality of the dispersion.

The agglomerates that constitute organic pigment powders (Sec. 1.5) are more or less broken apart during the dispersion process, e.g., incorporation into the application medium, which leaves a pigmented system comprised of primary particles, aggregates, and smaller agglomerates. The dispersion process is a very complex phenomenon and its outcome depends largely on a variety of factors (Sec. 1.6.5). It is therefore not surprising that the degree of pigment dispersion that is achieved under certain conditions is by no means constant. Experimentally determined particle size distribution information can only relate to the application properties of a pigmented system if the particle size distribution of the pigment powder exhibits roughly the same values as in the pigment-vehicle system. This, however, unfortunately poses somewhat of a challenge.

Size analyses are commonly carried out by mixing the pigment powder with an organic solvent or with water and adding appropriate surfactants to enhance the dispersibility of the powder. Aqueous dispersions frequently undergo size separation in ultracentrifuges, while organic solvents are more appropriate for electron microscopic techniques.

1.5.2.1   Determination of Particle Size by Ultrasedimentation

Discrete particle size separation or fractionation is carried out in the strong gravitational field of rapidly rotating ultracentrifuges. Theoretical background and experimental technique are described in depth in the literature [17, 18, 19]. While there is some difficulty in preparing a completely homogeneous pigment suspension, the two-layer version of the Marshal method suffers particularly from separation problems, including a phenomenon known as the streaming effect*.

The different ball mills that are currently used to prepare aqueous pigment suspension for sedimentation analyses share the disadvantage that mutual abrasion between particle, ball, and container surface can hardly be avoided. Suitable corrections must also be applied to account for incomplete and inconsistent dispersion of the very fine agglomerates of organic pigment particles through ultrasonic dispersion techniques. Ultrasound dispersion is therefore often associated with a certain trend towards larger portions of coarse particles at the expense of the finer grains. As a result, the pigment powder that is dispersed by the ultrasound technique is not dispersed to the same degree as in its application medium. The values for comparatively coarse pigments which disperse easily in aqueous media agree much better. No such problems exist for specialized pigment preparations in which the pigment is predispersed so as to distribute easily in the medium of application without extensive shear, such as emulsion paints. With these, even post-preparative flocculation can safely be neglected [21]. The degree of pigment dispersion in the analyzed suspension is thus the same as in the preparation and therefore in the medium of application.

Particle size distribution analyses in organic solvents may show dissolution and recrystallization effects when pigments with poor solvent fastness are measured.

There are several methods to monitor the sedimentation process, but optical techniques have prevailed since the 1970s. Joyce Loebl [22] paved the way for this analytic novelty by designing transparent disk centrifuges. Today it is possible to study organic pigments by photosedimentometry with conventional ultracentrifuges. The entire particle size distribution is therefore determined during one sedimentation process only, while older methods required a separate gravimetric analysis for each point on the distribution curve.

The pigment concentration in a photosedimentometry sample is very low. The pigment suspension is placed into a small cell which is inserted into a rotor. By directing a beam of light through the windows of the cell, a transmission-time curve is recorded during the run of the centrifuge which provides particle size distribution information through the quantitative correlation between specific extinction coefficient and particle size. White light is most commonly used, although it may be replaced by a monochromatic source or a laser beam [23]. Photometric methods are not susceptible to reagglomeration, because the experiments are carried out at pigment concentrations as small as $10^{-5}\%$ by weight.

Recent ultrasedimentation techniques for pigment analysis have abandoned the two-layer principle. Measurements can now be taken directly on the aqueous dispersion itself, which alleviates a number of problems, such as the streaming effect. However, both the intricate instrumentation and the complex mathematical evalua-

---

* As centrifugation proceeds, filaments of suspension may break through the interface between the suspension and the separating liquid and "stream" down. The problem is alleviated or diminished by more sophisticated techniques, such as the three-layer method [20] or the gradient technique.

tion require considerable skills as well as familiarity with sedimentation theory and practice.

1.5.2.2   Determination by Electron Microscopy

The difficulties that accompany the dispersion of a pigment for ultrasedimentation analysis (Sec. 1.5.2.1) parallel those associated with electron microscopy with its high demands on sample preparation.

Even a routine operation, such as applying the pigment suspension to a microscope slide by ultrasound vaporization, may be a source of complications. Pigment particles, previously completely dispersed in suspension and even in a vaporized droplet, may reagglomerate in the suspension whose volume is reduced as the carrier (organic solvent or water) evaporates.

There are a number of organic solvents whose value as dispersants, although they may be more suitable than water or alcohol, is compromised by side effects. Such agents may dissolve materials such as collodion, which is used as a thin film on support grids, or they may distort the particle size distribution by promoting recrystallization during dispersion and application. Such solvents are therefore not useful for the electron microscopic analysis of organic pigments.

It seems more important to focus on problems that may arise in the course of a quantitative evaluation of electron photomicrographs of an organic pigment. Automatic image analyzers cannot extract information from images which indicate more or less agglomerated, nonisometric, or even platelet-shaped particles.

Particles arranged on top of each other can be detected individually by semiautomatic analyzers; but the difficulty remains with studies of agglomerated units, and results tend to be inaccurate.

It is obviously not easy to measure the variable dimensions of platelets, rodshaped, or acicular particles that differ in length, width, and height. The shape outlined in the image reflects the orientation of a particle relative to the analyzing beam; a graphical grid [24] allows a two dimensional examination of such a profile. The third dimension, however, which illustrates particle size and shape, remains elusive. There are several approaches to a more complete image interpretation by introducing additional factors to represent the particle shape, however accompanied by a fresh set of problems. A recent technique [25] affords comprehensive two dimensional or three dimensional descriptions of anisometric particles through scattering ellipses and/or ellipsoides.

Scanning electron microscopy is a common and very useful device to determine particle size distributions in organic pigments; its advantage lies in the recognition of individual particles lying on top of each other as well as horizontally agglomerated units. It is therefore possible for a pigment sample to display different particle size distributions (Fig. 7), depending on whether the graphical grid is applied to a photograph obtained by TEM or by SEM techniques (Fig. 6).

Pigment particles to be studied by electron microscopy are sometimes incorporated into ultrathin membranes. At a first glance, a quantitative image analysis of such ultrathin layers seems convenient, since it appears to reflect the distribution of

Fig. 6: Images of samples of the same organic pigment, taken with a transmission electron microscope (left) and a scanning electron microscope (right) at equal magnification.

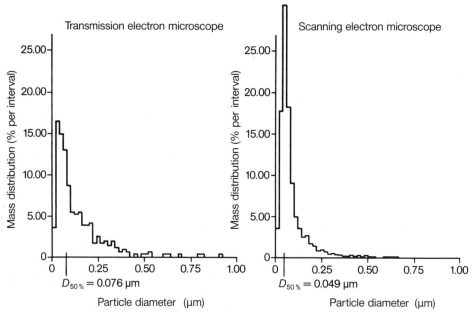

Fig. 7: Particle size distributions of the same pigment sample as derived from images taken with a transmission electron microscope (left) and a scanning electron microscope (right).

pigment particles in the medium in which they are applied. This argument is as compelling as it is deceptive, because during the preparation of the ultrathin layers both isometric and platelet-shaped or acicular crystallites are severed at random and ultimately reproduced in various spatial orientations, as shown in Fig. 8. Depending on the method of application, platelets or other pigment particles may be oriented along a certain axis when they are dispersed in their medium. In this case,

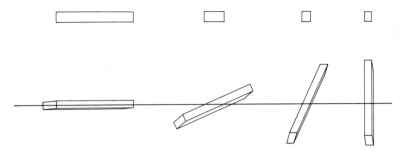

Fig. 8: Cross-sections through rod-like pigment particles in relation to the cutting angle.

the values obtained by electron microscopy are largely determined by the geometric arrangement of the acicular pigment particles within the sample membrane; i.e., the results depend on whether their orientation is parallel or perpendicular to the plane of the ultrathin layer (Figs. 9–11).

A phenomenon known as the rub-out effect (Sec. 1.6.5) frequently evolves from a certain set of conditions. Most often, it occurs during the application of a paint film. Particle size separation occurs when pigment particles accumulate along the substrate or on the surface of the coating, reversing the dispersion process. A suitable experimental technique must account for this effect by considering the entire layer.

There are various problems that might be solved if the particle size distribution within the medium of application can be elucidated. This is especially true where the undiluted paint system is used. In 1968, Geymeyer and Grasenick of the Technische Hochschule in Graz broke new ground in this area by proposing an experimental procedure, a novelty at the time. They designed specialized equipment to rapidly cool the liquid pigment-vehicle system to very low temperatures by inserting it into liquid nitrogen, where the material is dissected into ultrathin layers. The resulting sections are freeze-dried to remove all traces of solvent and may then undergo etching with nascent oxygen. Finally, vaporization techniques are used to cover the dried sections with thin layers of graphite and gold for observation with a scanning electron microscope. During the initial phase of the cooling process, however, interaction between pigment particles or between pigment and vehicle particles tends to result in a certain degree of aggregation. The structure of these units is a function of the pigment and the vehicle composition (see Figs. 12–14). Methods to reliably predict the exact distribution of a pigment in its carrier thus remain to be developed.

Particular difficulties similar to those associated with specific surface area determinations (Sec. 1.5.1) also accompany particle size distribution studies on surface coated organic pigments. The identity of the additive, be it an amine, a hard resin, or some other material, is less of a concern than the question of its concentration. Unfortunately, there is no information on the concentration limits above which an additive may distort the measurements; but one can expect this value to be defined largely by the specific surface area and the average particle size of each individual pigment. For pretreated surfaces, sizing the pigment particles by electron

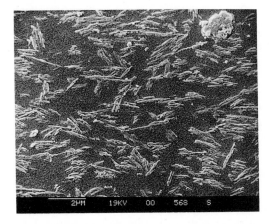

Fig. 9: Electron micrograph of an ultramicrotome-cut thin section of a layer of paint, seen parallel to the orientation of acicular pigment particles and perpendicular to the surface of the layer.

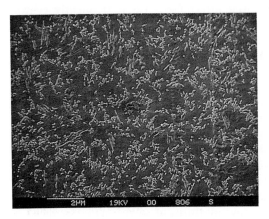

Fig. 10: Electron micrograph of an ultramicrotome-cut thin section of a layer of paint, seen perpendicular to the orientation of acicular particles and to the surface of the layer.

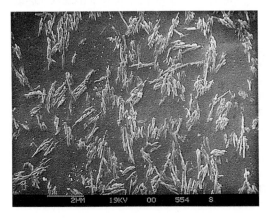

Fig. 11: Electron micrograph image of an ultramicrotome-cut thin section of a layer of paint, seen parallel to the orientation of acicular pigment particles and to the surface of the layer.

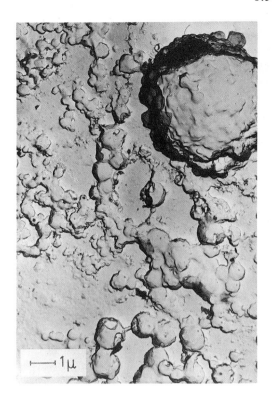

Fig. 12: Pigment Yellow 3 in an air-
dried alkyd resin paint.

microscopy thus appears more reliable than to determine the particle size distribution by ultrasedimentation. The additive typically functions as an adhesive that attaches pigment particles to each other, releasing them by dissolution processes only through dispersion which more or less distributes the pigment throughout the application medium. It is therefore impossible even for powerful dispersing agents to break down such units in an aqueous suspension for ultracentrifugation purposes. Electron microscopy, on the contrary, allows the use of resin dissolving agents to facilitate dispersion. The only disadvantage is that artefacts may develop during evaporation.

Other instrumentation to determine particle size distributions, such as the coulter counter, tends to be unreliable and problematic for organic pigments. This is also true for methods such as analyzing laser light passing through a pigment dispersion. The intensity of the resulting light pulses reflects the frequency of Brownian Motion within the sample, which in turn is a function of the particle size. This technique works for particle sizes between 40 and 3000 nm.

1.5.2.3   Data Representation

Particle size distribution may be characterized in terms of a spread of values in a table, a graphical representation in the form of a histogram (block diagram), or as

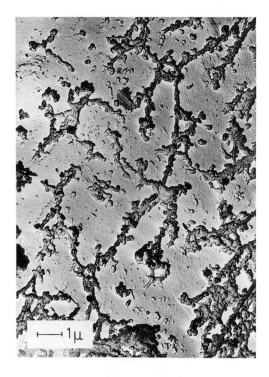

Fig. 13: Pigment Red 53:1 in a publication gravure printing ink based on calcium resinate.

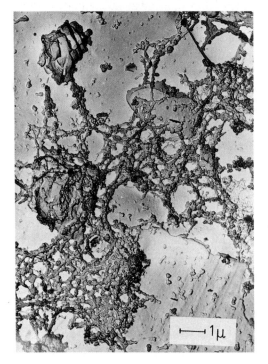

Fig. 14: Pigment Yellow 12 in a publication gravure printing ink based on metal resinate.

a continuous curve [26] (Fig. 15). Nonlinear representations allow plotting the data such that particle size distributions are well arranged and facilitate interpolation and extrapolation. The most common among these are defined by standards, such as the linear and the logarithmic representation for distribution sums [27], the exponential system [28], the logarithmic normal distribution [29], and the RRSB grid, which derives its name from the authors Rosin, Rammler, Sperling, and Bennett [30]. This list includes also:

– **Density Distribution.** The percentage of particles with the same given parameter is plotted against this parameter. A property that lends itself excellently to this purpose is the equivalent spherical diameter (D), which is the diameter of a fictitious spherical particle. The parameter determined by sedimentation techniques is usually the equivalent spherical diameter or the Stokes diameter (D_{ae} and D_{ST}, respectively), which is the diameter of a sphere of the equal density that would have the same sedimentation velocity in the medium as the non-spherical particle being measured. Another commonly used property is the light scattering equivalent diameter D_L, which is the diameter of a sphere that has the same light scattering ability as the particle in question. All these parameters, however, should be treated with some caution, because the concept of the equivalent spherical diameter is based on an ideal spherical particle which remains fictitious. According to a study [31], however, laminar flow effects that prevail during sedimentation render the velocity of rod-shaped particles in a medium independent of their orientation.
– **Sum Distribution.** A cumulative presentation of equivalent diameters, which converts the density distribution curve to a plot that represents percentages of particles which are smaller than a given equivalent diameter D.

It is possible to graphically present both the density distribution and the sum distribution according to either one of the following mathematical representations (Fig. 16):

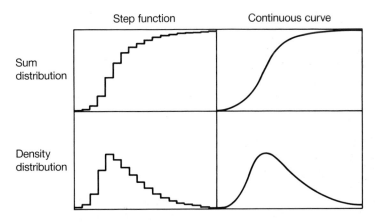

Fig. 15: Representation forms of particle size distributions.

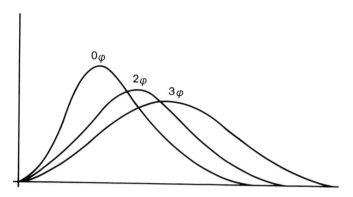

Fig. 16: Numerical distribution, surface distribution, and mass distribution.

- **Numerical Distribution** n(D), the percentage of particles with a given equivalent diameter,
- **Surface Area Distribution** s(D), the percentage of particles with a certain surface area plotted against the equivalent diameter,
- **Volume or Mass Distribution** v(D), the percentage of particles with a certain volume or mass, respectively, plotted against the equivalent diameter.

The choice of the method depends both on the purpose for which the representation of the particle size distribution is intended and on the experimental technique by which it is determined.

The numerical distribution is important for organic pigments; this is particularly true for electron microscopy, which affords results in terms of numbers of particles (Sec. 1.5.2.2).

Surface area distributions are especially adapted to reflect the dispersion characteristics or solvent resistance of pigment particles, since properties such as surface energy and dissolution rate are proportional to the surface area of a particle. Lightfastness and tinctorial features of pigment particles, such as tinctorial strength, hiding power, or transparency, are best compared by using mass distribution or volume distribution representations. This is true because the extinction of an organic pigment powder with a particularly fine particle size is largely proportional to the mass of the particles. The different mathematical representations of particle sizes are interconvertible.

There is a standardized terminology to designate the mean values for the different particle size distribution representations [2]. Among these are (Fig. 17):

- the arithmetic mean (D_a). This is the mean diameter of either the numerical (D_{an}), the volume (D_{av}), or the surface area distribution (D_{as}).
- the most frequent diameter D_{mf}. This is the diameter at the maximum of the frequency distribution.
- the median or central value $D_c = D_{50\%}$. This is a cumulative value which represents the diameter that lies exactly between the upper half and the lower half of all particle sizes and thus corresponds to a sum distribution of $0.5 = 50\%$.

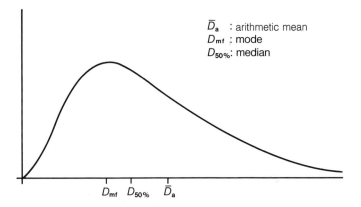

Fig. 17: Different mean values for the particle diameter *D*.

In spite of the fact that all of these values are quoted freely in the literature, it is only the median that seems to be of any physical significance, considering the typically unsymmetric particle size distributions of organic pigments.

Standards have also been developed for the statistical evaluations of frequency or numerical distributions, means and standard deviations, confidence intervals, regression, and correlation [26].

1.5.3 Polymorphism

Forces of mutual attraction between ions or molecules of many organic compounds which prevail as solid compounds frequently transform the individual particles into differently structured crystals. A large number of organic pigments of very different chemical classes are polymorphous; i.e., they occur in more than one modification or crystal structure. Despite identical chemical composition, more than one crystal modification may thus exist for a given organic pigment. The most versatile pigment in this respect appears to be copper phthalocyanine blue, which occurs in no less than five forms (α, β, γ, δ, ε). Quinacridone pigments and various azo pigments, on the other hand, display at least three different modifications.

The thermodynamic stability is a feature unique to each of the individual crystal structures of a chemical compound. Transforming an unstable modification into a stable one may require a certain amount of activation energy, but the process always offers an ultimate energetic advantage. Energy levels may vary considerably between pigments and even between crystal modifications.

The physical properties of a chemically uniform pigment are controlled to a large extent by its crystal lattice structure. This is true for particle size and shape, light absorption, specific weight, and melting point. Application properties such as color, tinctorial strength, cleanness, rheology, hiding power/transparency, heat and solvent resistance, lightfastness, and weatherfastness can thus differ very much within a given chemical constitution of an organic pigment.

Molecular and crystal lattice geometry vary with the crystal modification of a pigment. Important details are discussed in connection with the corresponding classes of pigments.

The crystallinity of organic pigment powders makes X-ray diffraction analysis the single most important technique to determine crystal modifications. The reflexions that are recorded at various angles from the direction of the incident beam are a function of the unit cell dimensions and are expected to reflect the symmetry and the geometry of the crystal lattice. The intensity of the reflected beam, on the other hand, is largely controlled by the content of the unit cell; in other words, since it is indicative of the structural amplitudes and parameters and the electron density distribution, it provides the basis for true structural determination [32].

Any crystal modification is practically fingerprinted by its X-ray diffraction spectrum. Another factor determined by the same instrumentation is isomorphism in chemically different pigments, which is associated with almost equal diffraction angles and X-ray intensities in both experiments. It is also important to ensure that the intensity of the incident beam is approximately the same for both measurements.

Isomorphous pigments; compounds which, for instance, chemically differ only in a single substituent but whose spatial requirements are similar, are frequently more uniform in their physical and application characteristics than is true for polymorphous pigments.

The advantage of a three dimensional X-ray diffraction study, which reflects a pigment's entire molecular and crystal structure by determining its atomic and molecular distances, bond angles, and crystal geometry, is diminished by the fact that the measurements have to be carried out on single crystals. Growing single crystals is a very difficult technique, especially with typically insoluble pigments. Complete X-ray data for a set of crystal modifications of a certain chemical compound are only obtained if each of the thermodynamically different crystal structures of the compound can be represented by a single crystal. The technique of growing such a series of crystals in different solvent systems has been perfected with the three modifications of Pigment Red 1 [33], whose three dimensional structures have thus been elucidated.

X-ray diffraction is not the only spectroscopic technique to determine geometries of different modifications. The range of options includes IR and UV studies as well as NMR and mass spectroscopy [34].

A change of modification can also be deduced from different application properties of a pigment with a given chemical structure but different synthesis.

Manufacturing processes afford azo pigments with, for instance, certain defined physical forms. Thermal treatment usually increases the particle size, reduces the tinctorial strength, and may improve the hiding power of a pigment. Reversal of this process often indicates a change of modification: the particle size is reduced and tinctorial strength and possibly also transparency are enhanced. The signs that point to structural variation may be as drastic as a significant change of color during manufacture. Observations such as these commonly accompany the processing of polycyclic pigments.

There are various approaches to the synthesis of a defined target modification. The following parameters are the ones that largely determine the outcome:

- diazotization and coupling methods, direct or indirect (for azo pigments),
- thermal conditions during synthesis and aftertreatment,
- pH during synthesis and aftertreatment,
- addition of a surface active agent or resin,
- choice and timing of surface active agent,
- presence of solvents during synthesis or aftertreatment,
- choice of solvent,
- method of precipitation or reprecipitation,
- drying process and temperature,
- milling method,
- mixed coupling (azo pigments),
- chemical modification.

The extent to which each of these factors controls the formation of a particular target crystal structure depends on the individual case.

It is even possible for one batch of pigment to comprise two different modifications, provided the two are energetically similar.

Table 1 lists a number of distinct differences in the properties of polymorphous pigments:

Table 1: Properties of several polymorphous pigments.

	Hue	Stability towards solvents	Migration fastness	Lightfastness and weatherfastness	Heat stability	Hiding power	Tinctorial strength
P.R.12	×			×			
P.R.170	×			×		×	
P.R.187	×	×	×	×	×		×
P.V.19	×	×	×	×			
P.B.15	×	×					×

1.5.4 Crystallinity

Little is known as yet about the crystallinity of organic pigments and its determination. Most pigments, such as the members of the azo series, emerge from the manufacturing process as fine powders of poor crystallinity. Drying of the crude press-cake immediately after synthesis without further finishing usually results in considerable agglomeration of the fine particles, which may later resist even the most determined dispersion efforts.

Considering the crystal imperfections that are typically found in all crystals, the crystal quality of organic pigments is a major concern. The external surface of any crystal exhibits a number of defects, which expose portions of the crystal surface to the surrounding molecules. Impurities and voids permeate the entire interior structure of the crystal. Stress, brought about by factors such as applied shear, may change the cell constants (distances between atoms, crystalline angles). It is also possible for the three dimensional order to be incomplete or limited to one or two dimensions only (dislocations, inclusions).

Untreated pigment samples are frequently closer to the amorphous state than to the idealized crystal. Lattice imperfections and voids "heal out" during the finishing that commonly follows pigment synthesis. The crystallinity of an azo pigment is thus improved by thermal treatment; while the crystallinity of nonazo compounds may also be improved by reprecipitation and grinding. It is generally true that the application properties of a pigment which consists of high grade crystals are superior over those of a crude product whose crystal structure is disturbed. The argument is of an energetic nature: the lower lattice energy of a regular crystal lattice enhances its fastness. More perfect crystals, for instance, are less prone to agglomerate and therefore disperse more easily.

X-ray diffraction of pigment powder lends itself to the determination of pigment crystallinity. It is thus possible not only to determine the chemical configuration of a crystalline compound, but also the lattice system of the crystal through the diffraction pattern, in other words, the crystal quality: size of crystallites, structural defects (Fig. 18).

The 500 nm size is a limit value; crystallites below this size tend to broaden the diffraction peaks in a spectrum, while size distributions above this value produce particularly sharp signals whose half width is a function only of the wavelength of the X-ray beam and the equipment. Signal broadening is at its maximum in materials known as X-ray amorphous substances, featuring particle size distributions below 8 nm. These afford flattened, washed-out spectra of little analytical value.

Since the properties of organic pigments are closely related to the quality of their crystals, X-ray diffraction provides the instrumentation that is necessary to monitor those properties during synthesis and, even more important, during finishing.

Another phenomenon to be detected by X-ray crystallography is the formation of mixed crystals, as observed in the mixed coupling of azo pigments or the solid solutions of quinacridone pigments. A change in the angles of the reflected X-rays of a mixed crystal indicates a transition from one crystal phase to another. If, however, one component assumes the lattice pattern of the other, then only the reflexes of the host lattice will be detected. Characteristic alterations in the spacings (less dense lattices: shift towards smaller reflection angles) or intensities of certain lines may completely eliminate certain signals or create new ones, indicating the subtle conversion to a new crystalline phase.

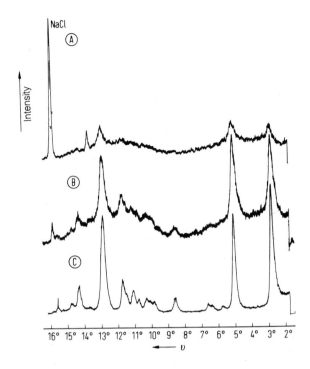

Fig. 18: Different crystal qualities of organic pigments made from the same compound. X-ray diffraction spectra, taken with Cu-K$_\alpha$ radiation. (A) 'overmilled' pigment, (B) moderately finished pigment, (C) well finished, i.e., recrystallized pigment. (From: Ullmanns Enzyklopädie der technischen Chemie, 4th edition, vol. 5 (1980), 256.)

References for Section 1.5

[1] e.g. T. Allen: Particle Size Measurement, Chapman and Hall, London, 1968.

[2] DIN 53 206 Teil 1: Teilchengrößenanalyse, Grundbegriffe.

[3] S. Brunauer, P.H. Emmet and E. Teller, J. Am. Chem. Soc. 60 (1938); S. Brunauer: The Adsorption of Gases and Vapours, Oxford Univ. Press, London, 1943.

[4] J.H. de Boer: Structure and Properties of Porous Materials, Butterworth, London (Editor Everett and Stone) 1958.

[5] D.H. Everett and R.H. Ottewill (Ed.): Surface Area Determination, Butterworth, London 1970.

[6] C. Wedler: Adsorption, Verlag Chemie, Weinheim, 1970.

[7] DIN 66 131: Bestimmung der spezifischen Oberfläche von Feststoffen durch Gasadsorption nach Brunauer, Emmet and Teller (BET).

[8] DIN 66 132: Bestimmung der spezifischen Oberfläche von Feststoffen durch Stickstoffadsorption.

[9] E. Robens, Lab. Pract., 18 (1969) 292; S.J. Gregg and K.S.W. Sing, Academic Press, New York, 1967, 330.

[10] C.H. Giles and S.N. Nakhwa, J. Appl. Chem. 12 (1962) 266.

[11] M. Herrmann and B. Honigmann, Farbe + Lack 75 (1969) 337–342.

[12] K. Merkle and W. Herbst, Farbe + Lack 82 (1976) 7–14.

[13] K.H. List and J. Weissert, Chem.Ing.Techn. 36 (1964) 1051–1053.

[14] P.C. Carman, J. Soc. Chem. Ind. (London) 57 (1938) 225 and 58 (1939) 1. see also DIN 66 126 Page 1.

[15] DIN 66 127: Bestimmung der spezifischen Oberfläche pulverförmiger Stoffe mit Durchströmungsverfahren; Verfahren und Gerät nach Blaine.

[16] DIN 53 199: Bestimmung der Ölzahl (Spatelverfahren). ISO 787 T5: General methods of testing for pigments and extenders – Part 5: Determination of oil absorption value.

[17] J.K. Beddow, Particulate Science and Technology, Chemical Publishing Co. Inc., New York, 1980.

[18] R.R. Irani and F.C. Callis: Particle Size: Measurement, Interpretation and Application, Wiley, New York, 1963.

[19] R.D. Cadle: Particle Size, Reinhold, New York, 1965.

[20] B. Scarlett, M. Ripon and P.J. Lloyd, S.A.C. Particle Size Analysis Conference, Paper 19, Loughborough, Sept. 1966.

[21] M.A. Maikowski, Prog. Colloid Polym. Sci. 59 (1976) 70–81.

[22] J. Beresford, J. Oil Colour Chem. Assoc. 50 (1967) 594–614.

[23] P. Hauser and B. Honigmann, Farbe + Lack 83 (1977) 886–890.

[24] Graphisches Tablett, MOP, halbautomatisches Bildanalyse-System. Manufactured by Fa. Kontron, D-8057 Eching/München.

[25] H. Völz and G. Weber, XVII. Congr. FATIPEC, Lugano 1984, Congressbook, Vol. IV, 77–90 und Farbe + Lack 90 (1984) 642–646.

[26] DIN 55 302: Häufigkeitsverteilung, Mittelwert und Streuung.

[27] DIN 66 141: Darstellung von Korn-(Teilchen-)größenverteilungen; Grundlagen. ISO 9276-1-1988: Representation of particle size analysis – Part 1: Graphical representation.

[28] DIN 66 143: Potenznetz.

[29] DIN 66 144: Logarithmisches Normalverteilungsnetz.

[30] DIN 66 145: RRSB-Netz.

[31] Personal communication by G. Eulitz, Hoechst Aktiengesellschaft.

[32] E.F. Paulus in: Ullmanns Enzyklopädie der technischen Chemie 4. ed., Vol. 5 (1980) 235–268.

[33] C.T. Grainger and J.F. McConnell, Acta Crystallogr. Sect. B 25 (1969) 1962–1970. A. Whitaker, Z.f.Kristallogr. 152 (1980) 227–238 ibid. 156 (1981) 125–136.

[34] K. Kobayashi and K. Hirose, Bull. Chem. Soc. Japan 45 (1972) 1700–1704; A. Kettrup, M. Grote and J. Hartmann, Monatsh. Chem. 107 (1976) 1391–1411; R.A. Parnet, J. Soc. Dyers Col. 92 (1976) 368–370, ibid. 371–377; R. Haessner, H. Mustroph und R. Borsdorf, Dyes and Pigments 6 (1985) 277–291.

1.6   Important Commercial Properties and Terminology

The application properties that define the commercial performance of a pigment in its vehicle system include all those characteristics that are determined or at least largely controlled by the organic pigment itself. Heading the list are features known as fastness properties, which reflect the ability of a pigment-vehicle system to withstand a variety of factors, such as light, weather, heat, or solvents. It is also possible for an organic pigment to adversely affect the processing of its vehicle system or even its performance in use, giving rise to effects such as plate-out, chalking, flocculation, etc. There is a certain amount of emphasis on the dispersibility and the rheology of a pigment in its medium, both of which are prime considerations to the user. Consumer demand includes certain standards regarding depth of shade, tinctorial strength, hue, and hiding power/ transparency, which will be discussed in addition to the aspects of application.

1.6.1   Tinctorial Properties (by F. Gläser)

The term 'coloristics of a pigment' describes those properties of a pigment that are connected with its appearance after application, i.e., the shade, tinctorial strength, and hiding power or transparency, features that define the basic commercial value of a pigment.

It is not useful to determine the coloristics of a pigment, not even its color, directly upon delivery by the manufacturer. Before being analyzed, a pigment has to undergo processing similar to that which is required by its intended use. The results of one test simulating the conditions of a specific application do not necessarily apply to other applications.

For a long time, visual judgment has been the basic method underlying all coloristic evaluation, depending solely on the trained eye of a professional colorist. Recent theories about the optical characteristics of pigmented systems have innovated the field of color evaluation techniques so as to reach modern standards of perfection.

In the following sections we will introduce some of the basic properties connected with optical features of pigmented systems, followed by a list of the basic properties used in colorimetry. Since discussing more than the most fundamental questions would be beyond the scope of this book, we have to refer to the literature for more details and specific problems [1].

1.6.1.1   Basic Properties

Color

The color of a sample is an individual perception produced by a color stimulus. A color stimulus is the light that is received by the eye, physically described as electromagnetic radiation with a wavelength between about 400 and 700 nm. The color perception that is produced by a stimulus depends on the energy distribution over different wavelengths within the visible spectrum. To emphasize the highly personal nature of color vision, it might be useful to point out that the perception that a sample may create depends on the observer's individual ability to perceive color, adjustment of the eye to a changing illumination, and the reaction to other colors that are perceived simultaneously with the sample.

Color vision, however, is not just a matter of illumination. The color stimulus produced by an object which does not emit radiation by itself is a dual function of illumination and observed object. The spectral power distribution within the visible range of the spectrum constitutes given illumination, while the influence of the object itself can be described by the fraction of light that it reflects at each wavelength from its source to the eye of the observer. It is possible to control the extent of reflection and its dependency on the wavelength by either choosing another pigment or by changing the pigment concentration.

The perception of a given color can be definitively and completely characterized by three different variables. Visual evaluation by a colorist is commonly based on the attributes hue, depth, and cleanness, terms derived from traditional visual assessment. The hue characterizes the color of a sample as red, blue, yellow, or green; the depth of shade indicates whether the color is deep or pale; and the cleanness refers to the brilliance or dullness of shade.

Other terms, such as brightness or colorfulness, are also used occasionally. The brightness of a color is loosely defined by its equivalent shade of gray. The colorfulness of a sample, on the other hand, describes the sensation varying from gray to a very pronounced hue, for example a pure full red. Since verbal specification is frequently inadequate, defined color samples and descriptions referring to such samples continue to play an important role even today.

The most important colorimetric identification system is based on three standardized values X,Y,Z, which are related to the experimentally determined remission values by the following equations:

$$X = \sum_{n=1}^{L} J_n R_n \bar{x}_n \Delta\lambda$$

$$Y = \sum_{n=1}^{L} J_n R_n \bar{y}_n \Delta\lambda$$

$$Z = \sum_{n=1}^{L} J_n R_n \bar{z}_n \Delta\lambda$$

X, Y, and Z can be calculated by dividing the visible spectrum into L intervals of the size $\Delta\lambda$ ($L = 16$ or more). The index n indicates the individual wavelength intervals; J_n characterizes the spectral power distribution of the illuminant, and R_n stands for the spectral reflectance factor in the interval n. $\bar{x}_n$, $\bar{y}_n$, and $\bar{z}_n$ describe the reaction of the human eye to the incident radiation within the wavelength interval n. J_n, $\bar{x}_n$, $\bar{y}_n$, and $\bar{z}_n$ are defined by international standards for a number of standard illuminants and, respectively, two standard observers. Two stimuli are said to produce the same color sensation if each of the tristimulus values X, Y, and Z of one of the stimuli is equal to the corresponding tristimulus value of the other stimulus. The system thus allows for the possibility that with a given illuminant, two samples may match merely on the basis of identical tristimulus values. Two such samples are said to form a metameric pair. A change of illuminant, however, may destroy the match.

It is comparatively difficult to visualize a color as characterized by the normative values X, Y, and Z only. For this and other reasons, systems for the characterization of color have been developed. Heading the list is the CIELAB system, which is now the leading system used to characterize color and especially to assess color differences [2]. The pertinent color coordinates are derived from the X, Y, and Z values. These values, despite their disadvantages, continue to dominate the field of colorimetric evaluation, largely because they are closely associated with the reflectance of a sample.

Colorimetric evaluation is also used to measure differences between colors, a very common problem. Currently the most important method is to quantify the difference between two colors by referring to the distance between two points within the CIELAB space, representing the colors of two samples (Fig. 19). It is possible to improve the description of a given color difference by splitting the total dis-

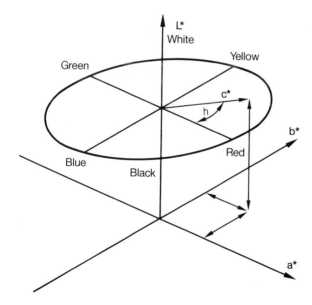

Fig. 19: Characterization of colors in the CIELAB system.

tance into segments, each of which relates to a specific attribute of color. This is commonly done within the frame of the CIELAB system by projecting the resulting segments of the total color difference either on the axes of the rectangular system L*a*b* or on the axes of the equivalent polar system L*C*h* (see Fig. 19) [3].

Depth of Shade

While the concepts of hue and cleanness are easily accessible to verbal description, the depth of a color is a more intricate phenomenon. Like all aspects of color, it is illustrated and evaluated best by comparative studies relative to samples with specific properties. As a consequence, a substantial body of standard samples with visually the same depth of shade were tabulated around the year 1935. These developed into a reference for a consistent judgment of the fastness properties of various colorants. Rabe and Koch [4] and Schmelzer [5] established that the extent to which the color of a sample deviates from white correlates closely with its depth of shade. On the continuous scale of possible color depths, certain defined levels that have proven to be commercially important are known as the Standard Depths. These are used to prepare samples for fastness evaluations and to provide a uniform chart for color illustration purposes.

Gall and Riedel [6] developed a colorimetric technique to measure the depth of shade in a sample in terms of the extent to which it deviates from one of the standard levels. H. Schmelzer [5] modified this system by scaling depth of shade according to a continuous range of values between 0 and 10. The availability of this consistent method of assigning a definitive depth of shade to each sample according to a continuous range of values has much improved the applicability of the depth of shade as a colorimetric parameter.

1.6.1.2   Optical Properties of Pigmented Layers, Absorption, Scattering

The interaction of incident light with a pigmented layer, including its surface and its background, can be described by a more or less intricate combination of elementary processes, including mirrorlike reflection, scattering, and absorption (Fig. 20). The portion of light which is reflected (and thus not absorbed by the material) and the direction in which it propagates depend on the relative impact of each of these processes. The portion which is reflected at the surface depends on the ratio of refractive indices at the interface as well as on the direction of the incident radiation and the geometry of the surface structure. The reflection of light at a surface is closely related to the gloss of a sample.

The light that enters a layer may deviate from its original direction of propagation within the layer by being scattered. The portion that is absorbed by the material dissipates its energy as heat. Absorption therefore weakens the intensity of light, while scattering changes the direction of propagation, letting light pass out again through the surface of the layer. The portion of incident radiation which is

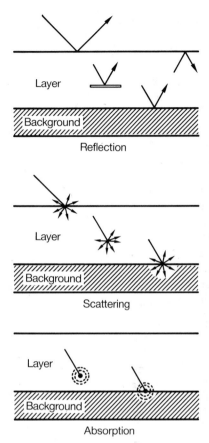

Fig. 20: Schematic representation of light absorption, scattering, and reflection in pigmented layers.

reflected depends both on the efficiency of scattering and on the extent to which it is absorbed. Another factor is the internal reflection of radiation from the surface back into the pigmented layer.

If a pigmented layer is applied to a substrate, the optical properties of the system are more or less affected by the background. Scattering and absorption within the pigmented layer influence the portion of radiation which reaches the background. The background determines the portion and direction of incident radiation which is reflected at the interface between pigmented layer and substrate. This radiation is again scattered and absorbed as it passes back through the pigmented layer. The amount of light that again penetrates the illuminated surface, i.e., that which is reflected after reaching the substrate, is a function of scattering and absorption within the pigmented layer.

If only a small portion of the total amount of reflected radiation penetrates the pigmented layer, reaches the background, and passes out again through the illuminated surface, the contribution of substrate reflectance can be neglected. Such layers are referred to as hiding, i.e., differences in the reflectance of the background are no longer discernable. Extensive scattering within a layer provides good hiding power. The hiding power of a layer also increases with its power of absorption, since the route of the incident light to the substrate and back to the illuminated surface exceeds the pathway of radiation that is scattered closer to the surface of the object. The longer route involves more extensive absorption and the light propagating along this path therefore has less influence on the total amount of reflected light.

Absorption and scattering obey simple laws in a very thin layer which provides nearly constant flux in a perpendicular direction. Since the resulting absorption is proportional to the thickness of the layer, the resulting proportionality constant can be used as a measure of absorption. Likewise, the portion of scattered light is proportional to the thickness of the layer. In this case, however, the proportionality constant depends on the angle of deflection. In other words, the extent of scattering cannot be characterized by a single constant only but is a function of the angle between incident and remitted radiation.

Isotropic scattering as a model of the behavior of a pigment is a special case of great practical importance. Isotropically scattered light is scattered with equal intensity into all directions. The scattering function which is normally dependent on the scattering angle is therefore reduced to a single constant, which greatly simplifies the mathematical treatment of this model. The extent of absorption is defined by another single constant.

Scattering and absorption are generally dependent on the wavelength. Effective scattering and absorption coefficients within a pigmented layer can be calculated from the properties of the components using the following equations [7]:

$$K(\lambda) = k_1(\lambda) \cdot c_1 + k_2(\lambda) c_2 + \ldots + k_n(\lambda) c_n + K_B$$
$$S(\lambda) = s_1(\lambda) \cdot c_1 + s_2(\lambda) c_2 + \ldots + s_n(\lambda) c_n + S_B$$

λ: wave length

$K(\lambda)$, $S(\lambda)$: total absorption coefficient, total scattering coefficient

$k_1(\lambda)$, $k_2(\lambda)$, ..., $k_n(\lambda)$: specific absorption coefficients (per concentration unit) of pigments no. 1 to n

$s_1(\lambda)$, $s_2(\lambda)$, ..., $s_n(\lambda)$: specific scattering coefficients (per concentration unit) of pigments no. 1 to n

c_1, c_2, ..., c_n: concentrations of pigments no. 1 to n

K_B: absorption coefficient of the substrate

S_B: scattering coefficient of the substrate

The reflection of a system can thus be predicted from the absorption and scattering coefficients of the layer, its thickness, and the optical parameters of the vehicle. At a given wavelength λ, the reflectance factor $R(\lambda)$ indicates the portion of incident light that is remitted by a sample. For a completely opaque layer with isotropic scattering and diffuse illumination, the Kubelka–Munk theory quantifies the simple correlation between the reflectance $R(\lambda)$ and the scattering and absorption coefficients.

$$\frac{(1 - R(\lambda))^2}{2R(\lambda)} = \frac{K(\lambda)}{S(\lambda)}$$

At a given wavelength λ, the reflectance $R(\lambda)$ of a pigmented layer equals the portion of incident radiation which is reflected by the layer. This equation, in combination with the above relations for the absorption coefficient $K(\lambda)$ and the scattering coefficient $S(\lambda)$, makes it possible to calculate the reflectance of the layer $R(\lambda)$ from known concentrations, specific absorption coefficients, and specific scattering coefficients. With some expertise, the expression may also be used to determine specific absorption and scattering from the reflection coefficients and concentrations; or, alternatively, to find the appropriate ratio in concentrations between two systems containing the same pigment.

The same type of calculations may be carried out for a transparent layer which is applied to a scattering substrate (e.g., a printing ink on white paper) by using the equations derived by K. Hoffmann [8].

1.6.1.3   Assesssment of Coloristic Properties

Characterization of Color and Color Differences

Nowadays color specification relies primarily on color samples, because verbal descriptions inherently lack precision. The choice is to either define a reference sample for a specific purpose or to refer to a sample in a color-order system. Manufacturers in Germany frequently refer to a color-order system known as the RAL color scale to define certain shades for industrial equipment, machinery, etc. Paint manufacturers and the automotive industry issue sets of color samples which list the range of available colors. Other sets of samples show the shades which are expected to be fashionable in the future. These sets normally include a limited number of shades which are more or less randomly distributed throughout the entire color space. They are chosen according to the intended use. There are also systematic bodies of samples covering the entire color space, such as the Munsell system, the DIN color chart [9], the Natural Color System, and the Eurocolor system.

Visual evaluation of color differences relies on more or less standardized verbal descriptions. Verbal descriptions can be far more precise in characterizing color differences than absolute assessments of individual colors. Terms that are agreed on by all concerned parties make it possible for different observers to arrive at a consistent evaluation.

Fastness properties are commonly evaluated through comparison with the Gray Scale, which shows the magnitude of the differences between pairs of standardized gray samples.

Industrial analyses, especially those concerning fastness properties, rely on the Gray Scale [10] as a basis for the visual evaluation of color differences. The extent to which the colors of two test objects deviate from each other is indicated by matching these differences with pairs of very similar shades of gray.

In colorimetry, color is usually characterized through a system of tristimulus color values X, Y, Z, or the CIELAB values L^*, a^*, b^*. It is possible to either determine the tristimulus values or to refer to the following terms:

$$Y, \ x = \frac{X}{X+Y+Z}, \quad y = \frac{Y}{X+Y+Z}.$$

Comparatively systematic graphical representations showing the relations between various colors can be drawn up by plotting the chromaticity coordinates x and y, provided that the information provided by different y values is of no importance for the problem in question. Figure 21 shows such an x, y chromaticity diagram, in which the respectively brightest shades of color as related to (x, y) pairs are illustrated.

Determining distances within the CIELAB space (Fig. 21) is currently the most widely used method to evaluate differences between colors [3]. This space also makes it possible to distinguish between differences in hue, colorfulness, and brightness, and thus to better visualize the numeric results.

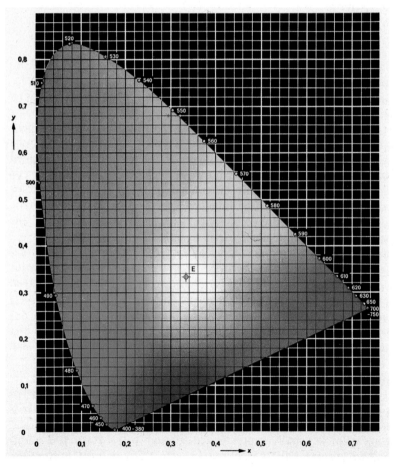

Fig. 21: *XY*-Chromaticity diagram according to DIN 5033.

It is important to note that equally perceived differences between pairs of samples do not necessarily represent equal intervals in the color space, especially if the shades of two samples of a pair are very different. Colorimetric determinations of color differences provide a basis on which tolerance values can be defined; but meaningful tolerance limits cannot be derived from strictly colorimetric principles alone.

Tinctorial Strength

Like other optical characteristics, the tinctorial strength of a pigment, i.e., its ability to add color to a substrate, is controlled by the conditions under which it is applied. Tinctorial strength may be defined either as an absolute value or relative to another pigment. Attempts to assess absolute values by visual methods at best afford semiquantitative results; while the relative strength of one pigment compared to another is more easily assessed by a human observer. It is determined quantitatively by comparing a series of samples featuring varying amounts of a pigment incorporated in a substrate with a corresponding series of pigment/substrate systems containing the reference pigment. Test and reference samples are then compared and matched as far as possible. The ratio between the known pigment concentration of the respective test sample of a matching pair and the pigment concentration of the corresponding reference sample is used as a measure of the relative tinctorial strengths of the two pigments.

If no exactly matching sample can be found, the concentration which might lead to a match is approximated through visual interpolation from the two nearest reference samples. If it is impossible to arrive at an exact match for any concentration, then the concentration is found which experience shows to be equivalent. The corresponding tinctorial strength is supplemented by a verbal description of the remaining differences in hue and cleanness.

Colorimetrically, the absolute tinctorial strength of a pigment in a vehicle system is best described by referring to the depth of shade that is achieved under specific conditions of sample preparation. It is possible to either measure the amount of pigment that is necessary to produce a certain depth of shade, or to refer to the depth of shade that can be produced with a certain amount of pigment. The second approach, which is based on a defined amount of pigment, appears particularly convenient for the user who is interested in estimating the maximum depth of shade that can be achieved under different working conditions. Such values are illustratively plotted by using the continuous depth of shade scale designed by H. Schmelzer [5].

Other approaches to determine absolute tinctorial strength in a sample are based on the assumption that the tinctorial strength of a pigment is defined by its light absorbing properties. The absorption coefficient for a known concentration is usually measured at the wavelength of maximum absorption; an alternative method of determining the tinctorial strength involves summarizing the absorption coefficients over the entire range of the visible spectrum. In the latter case, absorption coefficients are sometimes weighted using different weights for each wavelength.

Absorption methods may occasionally present problems, particularly in trying to compare pigments which are very different in shade. Depending on the approach, large parts of the spectrum might simply be ignored (if only the absorption at the absorption maximum is used), or they are represented somewhat indiscriminately (if the sum of absorption coefficients is used). However, those parts of the spectrum that are not correctly included in the calculations obviously contribute to the perception of color by the human eye. Weighting the absorption coefficients according to their wavelength does not entirely alleviate the problem, because the correct choice of weights is not always obvious.

The relative tinctorial strength is defined in relation to a standard rather than being an absolute property of a pigment. For colorimetric assessment, the absolute tinctorial strength of a sample pigment and that of a reference pigment are determined by the same method and the ratio between two such absolute values is then calculated. Despite the variety of options, however, both the absorption at maximum wavelength and the absorption sum over the visible range of the spectrum remain the most common criteria for the determination of the relative tinctorial strength.

It is also possible to find the amount of pigment which is necessary to produce a match of a reference sample containing a known amount of pigment. The relative tinctorial strength is then defined as the ratio between the two amounts. In this case, two samples are already considered to match if they exhibit the same depth of shade; an exact match is not always found. After all differences in tinctorial strength are accounted for, any remaining deviation is due to the inherent difference in shade between the two pigments and cannot be eliminated by simply adapting the amounts. It is useful to describe the remaining color difference by referring to the CIELAB system.

A quantitative comparison of numerical values from several different sources requires some caution, because the tinctorial strength of a given pigment is largely defined by the working conditions which govern its application, as well as by the methods of determination and evaluation.

Hiding Power

In discussing the performance of a pigment within a vehicle system, it should not be forgotten that there are certain aspects that relate to the substrate onto which it is applied. The hiding power of a pigment, for instance, refers to the ability of a layer of pigmented medium to conceal any differences in substrate coloration so that they become invisible. It is defined as the area over which a certain amount of pigmented paint can be spread without losing its opacity. It is also possible to refer to the minimum thickness of layer which is necessary to conceal a substrate.

A layer which effectively conceals its substrate must be scattering. The necessary amount of scattering depends on the thickness of the layer, the absorption of light within the layer, and the magnitude of the color differences of the substrate. The less absorbing the paint or the larger the color differences of the substrate, the higher the scattering power needed to adequately hide the substrate. The hiding power depends also on the wavelength of the incident radiation. The hiding power of a chromatic pigment exhibits opacity in relation to certain substrate colorations [11].

The hiding power of a pigmented layer is usually determined by applying it to a substrate with differently colored patches. Black stripes on a white background or a black/white or grey/brown checkerboard surface with a standardized reflectance (e.g., $R = 80\%$ for white and $R = 5\%$ for black) are commonly used. To quantify the hiding power of a layer in relation to a standardized substrate, the maximum area may be specified over which the test paint can be spread without losing its opacity. Alternatively, it is also possible to find the minimum thickness of layer which just effectively hides to the eye any difference in color between the differently colored patches of substrate surface. According to definition, this is the point at which the color difference equals one in the CIELAB system.

According to ASTM D 2805-70, the hiding power of a layer is characterized by the ratio of the Y values (as defined in the CIE color system) over a black and a white patch of substrate. If this ratio is larger than 0.98, the layer is referred to as hiding. This criterion occasionally fails if applied to brightly colored pigments.

Transparency

The list of system-dependent parameters includes the transparency of a pigmented layer. It is usually determined by applying the pigmented system to a black background whose darkness is retained or reduced according to the transparency of the layer. Scattering increases the opacity of a layer. Quantitative evaluation relies on the difference in color between the pigment-vehicle system spread over a black surface and a black surface that is covered with pure vehicle only. The amount of pigment that is necessary to create a difference in coloration of $\Delta E = 1$ using the CIELAB system for color differences provides an indication of pigment transparency.

1.6.2 Fastness to Solvents and Special Application Fastness

1.6.2.1 Organic Solvents

Although according to definition, the ideal pigment is practically insoluble in its medium of application, organic pigments may in reality deviate more or less from this postulate of insolubility, depending on the medium in which they are applied and the conditions under which they are processed. Since a pigment that is to a certain extent soluble in its carrier is expected to perform poorly and may even recrystallize, bleed, or bloom, it is important to prevent pigment dissolution. Fac-

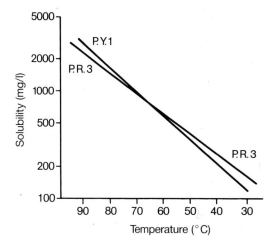

Fig. 22: Relation of the solubility of Pigment Yellow 1 and Pigment Red 3 in dibutyl phthalate to the temperature.

tors that control the solubility of a pigment in its vehicle include the choice of solvent, chemical structure, particle size (Sec. 1.7.7), and temperature. To demonstrate the correlation between solubility and temperature in organic pigment/solvent systems, Fig. 22 traces the solubility of P.Y.1 and P.R.3 in dibutyl phthalate at temperatures between 25 and 90°C [12].

There are certain accepted tests used to determine the extent to which a given organic pigment tolerates solvents. Experimental procedures commonly involve enclosing a certain amount of pigment powder into a piece of fluted filter paper, which is inserted into a test tube containing the organic solvent. The sample is then kept in the organic solvent for 24 hours at room temperature, after which the results are evaluated. The extent to which the solvent is colored indicates the fastness of the pigment to this particular solvent. Although laboratory studies cannot reliably imitate a commercial environment, the analytical results do allow general conclusions and assist in selecting the suitable pigment for a particular application. In this stability test, the tendency of a pigment to lose color in a certain medium, for instance, is usually associated with reduced fastness to overspraying.

The way in which a pigment behaves toward a specific solvent has a considerable impact on how it can be employed by the user. If a pigment which is partially soluble in a solvent is processed in a system containing that solvent, then recrystallization may occur. This in turn alters the coloristic, rheological, and fastness properties of the product.

A number of potential users, such as the printing industry, are interested in the solvent fastness not only of a pigment but of an entire pigment-vehicle system. Standardized tests are available. A proof print of a certain size is placed inside a test tube and allowed to remain in the target solvent at 20°C for 5 minutes. The change in solvent color is determined and the print dried and compared with an untreated specimen. Standard solvents [13] are ethanol or the following mixture:

30 vol% ethyl acetate
10 vol% ethyl glycol

10 vol% acetone
30 vol% ethyl alcohol
20 vol% toluene

However, the results do not necessarily reflect the fastness to overlacquering in actual application, because overprint varnishes may contain other solvents as well. Besides, a series of additional factors such as the effect of plasticizers, uneven flow of the pigmented lacquer, etc., play a role in pigment performance. The fastness to overlacquering is therefore most realistically determined in actual application.

Pigment performance also includes fastness to transparent lacquer coatings ("silver lacquer"), that is, fastness to transparent enamels which are applied to metal deco prints to give them rub and scratch fastness.

1.6.2.2   Water, Soap, Alkali, and Acids

The test methods discussed in this context are broad in scope and relate to a variety of applications. In some such tests, a pigment powder is extracted with water and its contents analyzed; other procedures reveal the stability of a pigmented system with regard to water, acids, or bases. A number of these tests have developed into industrial standards.

A variety of techniques are used to determine the water-soluble content of a pigment.

The cold extraction method [14] involves moistening defined amounts of pigment powder (between 2 and 20 g) with small amounts of water, alcohol, or a suitable wetting agent. 200 ml of freshly distilled or completely deionized water are then added and the sample allowed to remain in this solvent for 1 h at ambient temperature. After thorough shaking, the liquid is removed by filtration.

The hot extraction method is a variation of this procedure. The aqueous pigment suspension, prepared as above, is refluxed for a certain time, usually 5 minutes, cooled rapidly, and filtered. A known amount of extract is then dried by evaporation and the weight of the residue determined, or the extracted pigment weighed and the dissolved portion determined by calculating the difference.

The acid number of a pigment is determined by titrating the aqueous extract of 100 g of pigment with a 0.1 N alkali (acid) solution [15].

Whenever organic pigments are analyzed, it is not uncommon for the results of the two extraction methods to deviate considerably. Depending on the pH of the medium in which the synthesis is carried out, such as the coupling slurry for azo pigments, occlusion or adsorption effects may trap portions of the soluble substances inside or on the agglomerates. It is impossible to remove such impurities by pigment filtration and intensive washing, and even the effect of hot extraction procedures tends to be slow and unsatisfactory. Considerable amounts of such soluble species may remain within the pigment even after hours of refluxing and repeated filtration with freshly distilled water. A pH or conductivity analysis of the extracts will readily reveal this unsatisfactory condition.

It is also possible to determine the pH of the aqueous pigment extract from either the cold or the hot process. A frequently used modification of this method involves measuring the pH of the pigment slurry with a glass electrode directly after reflux [16], without prior filtration. There are certain disadvantages to this technique if it is used on pigments whose surface is covered with resin or has been treated with other additives such as amines: resins and similar materials tend to occlude portions of the surrounding medium more than do untreated pigment surfaces.

There are commonly accepted experimental standards for the determination of conductivity and specific resistance in aqueous pigment extracts [17]. The electrical conductivity γ is calculated from the electrical conductance; its inverse is the specific resistance $\xi = 1/\gamma$, derived from the electrical resistance. Additional experimental methods have been developed for the determination of soluble sulfates, chlorides, and nitrates [18].

From the user's point of view, the ability of a pigmented system to withstand exposure to water, alkali, or acids is frequently of more importance than an evaluation of pigment extracts.

The printing industry uses standardized proof prints for all printing methods to examine the fastness of a pigment to water or alkali. The sample prints are placed between or on top of filter paper soaked with water or with a sodium hydroxide solution. Letterpress, offset, or screen prints require a 2.5% aqueous sodium hydroxide solution; a 1% solution satisfies the standard for flexo and gravure printing [19]. The resulting soaked print with its filter paper is sandwiched between two glass plates, which are then placed underneath a 1 kg weight. Water fastness is evaluated after 24 hours; alkali fastness tests require only 10 minutes. Prints or printing inks which neither migrate into the filter paper nor undergo a coloristic change are said to be water or alkali resistant.

There are no similarly defined standards to indicate pigment performance in connection with coatings; but a variety of methods have gained acceptance throughout the automotive finishes industry. A generally accepted technique to determine fastness to alkaline agents involves applying a 5% sodium hydroxide solution to the coating and allowing the covered specimen to remain in a circulating air chamber at 70°C for 1 hour. The purpose is to evaluate the properties and fastness of a coating under conditions that reflect those in a commercial car wash. Acid tests commonly use 1 N sulfuric acid with a certain amount of iron(II)sulfate at 70°C ± 1°C for 1 hour. The effect on the color of the test coating should not exceed 2 CIELAB units, and its gloss should not be reduced by more than 10%.

Although it is uncommon for organic pigments to be affected by acid or alkali, color changes are occasionally observed with laked pigments or pigments containing free acid or amino functions. These may form salts if they are exposed to acids or bases.

The requirements of the coatings industry depend on the actual context in which a pigment is to be used. Exterior house paints, for instance, are expected not only to exhibit good weatherfastness but also to be fast to lime and concrete. Fastness to lime is frequently tested by coloring freshly slaked lime with the pigment dispersion or pigment preparation, applying the mixture to asbestos concrete sheets, and after 24 hours comparing the color of the test sample with that of a freshly prepared specimen. Only very few organic pigments can be mixed directly with cement, but several meet the requirements for paints to be applied to concrete surfaces. The outcome of such a test is crucial, because the color of a pigment that is not sufficiently fast to its carrier material will deteriorate or change rapidly when it is exposed to weather.

Likewise, the evaluation methods for polymer binders are designed specifically for certain target materials. One such procedure is known as the "tropics test" for polyurethane coatings. The ability of a colored test sample to resist hydrolysis is determined by exposing it for 7 days to a climate of 100% humidity at 70°C and then extracting it with ethyl acetate; or, alternatively, by submitting the hydrolyzed specimen to a crock test. Coatings may become more prone to hydrolyze due to the influence of certain pigments; if a pigmented system is used, the binder may be responsible for hydrolysis.

1.6.2.3   Pigment Performance in Special Applications

There is no precise definition for the frequently used term "application fastness" of a pigment. It usually refers to the behavior exhibited by a finished product used in accordance with the specifications. The term may thus refer to a print, a coated object, or a plastic product, and the list of features ranges from properties such as lightfastness and weatherfastness to migration fastness and fastness to solvents. In this context, there is a certain amount of emphasis on features which play a role in connection with packaging materials and packed articles.

Test methods for prints are broad in scope to suit a variety of practical purposes. Users are frequently interested in the fastness of a product to soap and detergents [19]. To determine the fastness of a print to a freshly prepared 1% aqueous solution of a special test soap, the printed sample is placed on top of several layers of filter paper soaked with the test solution. The resulting test specimen is then sandwiched between two glass plates, mounted underneath a 1 kg weight, and held at 20°C for 3 hours in a chamber saturated with water vapor. The printed sample is then removed, rinsed with distilled water, and dried. The resulting color change of the test specimen is evaluated by comparing the color with an untreated print. Any coloring of the filter paper is noted as well.

Likewise, pigment performance in connection with detergents frequently decides the commercial range of a product. It is generally considered useful to test the target detergent directly as opposed to the more common approach of simultaneously evaluating alkali and soap sensitivity. The latter approach has a major impact on the range of available pigments because the demand for perfect fastness to soaps under the conditions of actual processing and use eliminates so many types. Because of this, a large number of shades are not accessible for this use.

The requirements for prints used in the packaging industry are determined by the packaged material, which explains the wide variety of standards that govern pigment application in this area. The list includes fastness to materials such as cheese, grease, oil, paraffin, wax, and certain spices [19]. Tests are carried out by placing the test print on top of the medium in question, printed side down. A weight is then placed on the sample. It is important to remember that any fastness determined by this method is only valid for the material which was actually used in the test. Spices, for instance, may act very differently, depending on their age, storage temperature, and milled form.

Likewise, the behavior of a print in connection with sausage, ham, bacon, or fish, or its sensitivity to cleaning agents, disinfectants, bath soaps, essential oils, or fertilizers may be determined accordingly.

Colored household items made of plastics and other polymers [20] are tested similarly. The effect of food may be imitated by using coconut oil or peanut oil ("coconut oil test"). Strips of filter paper soaked with these materials typically remain in contact with the test sample for 5 hours at 50°C.

The packaging industry is particularly interested in the stability of a pigment to sterilization and to heat sealing, as well as its effect on the heat sealability of a printed material. In a sterilization test, a pigment may be exposed to hot air, steam, or water. Wet tests are carried out either at ambient pressure or in an autoclave. Color-change comparisons with reference specimens and, for wet methods, the extent to which the water is colored are indications of the fastness of a print to sterilization. A pigment is said to be heat-sealable if under certain defined conditions appreciable cohesion is attained between printed film and substrate. The test is thus standardized with respect to temperature, time, and pressure, and the test conditions are expected neither to weaken the sealing site nor to induce a color change in the print. Sealability tests comprise both print/print adhesion and print/substrate interaction, including pigment application to coated substrates. Printed layers are heat-sealed at defined temperatures with grooved heated jaws; both time (a few seconds) and pressure are standardized. Factors that may possibly preclude pigment application in this packaging sector include color change, physical damage to the printed film, or adhesion to the heat-sealing equipment.

Different demands are placed on a pigment which is targeted for covered or laminated printing. In this technique, a reverse impression is printed onto a plastic film and laminated by means of one or two component adhesives, laminating resins and wax onto another substrate such as paper, aluminum foil, or polymer foil. These substances may adversely affect pigment performance. For high gloss applications, a normal impression is covered with a clear plastic film.

1.6.2.4 Textile Fastness Properties

The term "textile fastness properties" is used in this context primarily to designate the ability of spin dyed or printed textiles to retain their color value throughout processing, application, and use. The national body active in the development of standardized tests in Germany is the Deutsche Echtheitskommission DEK (DEK = German Commission for Fastness). The latter is a member of the European Fastness Convention (ECE) and is represented by the Deutscher Normenausschuss DNA (DNA = German Standards Committee), which follows the recommendations of the International Standards Organization (ISO).

To determine the colorfastness of a dye or a pigment, it is useful to maintain a constant depth of shade (Sec. 1.6.1.1), usually 1/1 SD (SD = Standard Depth of

Shade). The tendency of a sample of spin dyed or printed textile material to change color may either be determined on an individual swatch of fabric; or, if there is also some interest in migration, the test sample may be sewn onto a colorless control material. The results are reported with respect to the specified test, describing the extent of color change and staining of the control sample and expressed by reference to the Gray Scale [10, 21].

A pigment targeted for the spin dyeing or textile printing market must be fast to water [22] and seawater [23], irrespective of its method of application. Laundering tests are designed to evaluate the washfastness of textiles. Test cycles with standard detergents can last 30 minutes to 4 hours at temperatures between 40 and 95 °C [24]. Laundering methods also include techniques to determine the colorfastness of a sample to peroxide containing washing agents such as sodium perborate [25], fastness to bleaching with hypochlorite [26], or to chlorinated water [27]. The fastness to perspiration is simulated through alkaline or acidic test solutions which contain histidine [28]. There are other tests concerning fastness to wet or dry rubbing [29], ironing [30], dry cleaning [31], solvents [32], acid [33] or alkali [34], heating with sodium carbonate [35], including the fastness of a pigmented textile sample to bleaching with peroxide, hypochlorite, or chlorite [36]. It is possible to simulate the conditions at each stage of the manufacturing process through a series of determinations on test solutions containing varying amounts of the active agent in combination with suitably adapted experimental parameters. Out of the large number of other standardized tests, the hot pressing test should be mentioned. It consists of a series of experiments, carried out for 30 seconds at 150, 180, and 210 °C under pressure. This test was designed specifically for synthetic fibers [37]. Occasionally, there is some interest in the vat fastness of a pigment that is used to spin dye fabrics such as viscose-rayon or cellulosics. This is usually determined in an alkaline dithionite solution containing ammonia at about 60 °C for 30 minutes. Samples are classified according to change of color and the extent to which the color is transferred onto other pieces of cloth.

1.6.3 Migration

The term migration refers to the occurrence of bleeding and blooming. Dissolved portions of pigment may migrate from their medium of application to the surface or into a similar material that their system is in contact with. A number of seemingly related phenomena that may arise during the processing or application of a pigmented system which shows similar results, such as plate-out or chalking, are caused by different effects and are mentioned elsewhere (Sec. 1.6.4.1, 1.6.4.2).

The theoretical and application aspects of the mechanisms leading to migration have been studied particularly in plasticized PVC [38,39]. From the results and mechanistic concepts, which are not limited to the materials and binders that were actually used in these studies, the following picture emerges: bleeding and/or blooming is only observed in a supersaturated solution of a pigment in its vehicle system. There is a number of conditions that promote blooming and bleeding:

- Processing of the pigmented system at higher temperatures, i.e., creating a large difference between processing and application temperature.
- Partial or complete dissolution of the pigment in its medium (a polymer, a plasticizer, or a mixture of these) at the processing temperature.
- The actually dissolved portion of pigment exceeds its solubility at the storage or application temperature. This condition occurs very frequently.
- Supersaturation after cooling, as mentioned above.
- Free mobility of the dissolved pigment particles in their medium at the temperature of application. This is a condition that frequently occurs if pigments are dispersed in plasticized PVC, polyolefins, or other polymers which are applied at a temperature – usually room temperature – above their glass transition point. Polystyrene and polymethacrylate, on the other hand, whose glass transition temperatures are above 100 °C, do not migrate at ambient temperature. There are exceptions, however; typically stable pigments may start to migrate at the high concentrations and under the harsh experimental conditions under which tests are typically carried out (see e.g., [40]). Migration is more likely to occur at high storage temperatures (120 or 80 °C). At ambient temperature, the rigid polymer matrix prevents dissolved pigment particles from moving, which makes it possible to even use soluble dyes without facing the problem of migration. The plasticizing effect of "pigment dissolution" on polymers such as polyester may decrease the glass transition temperature and the crystallinity.
- Satisfactory crystal formation; i.e., sufficiently high probability for new seed crystals to form. This implies that the already dissolved particles should not crystallize preferably or exclusively at the surface of the undissolved pigment particles.

There is some doubt about the frequently quoted migration or diffusion of solid pigment particles in plasticized PVC. The Kumis and Roteman equation [41], which describes the correlation between the free volume inside a polymer matrix and the diameter of diffusing substances, seems to preclude the possibility of migration of solid pigment particles. It is likely that the transportation of these solid particles is entirely unrelated to migration mechanisms based on diffusion phenomena. The effect probably arises in connection with a general streaming of plasticizer to the surface of the pigmented material, in which case it should be classified as a plate-out phenomenon [39].

Finally, the degree of migration is also controlled by the chemical constitution of the pigment (Sec. 1.4.4) and by its particle size distribution (Sec. 1.7.7).

1.6.3.1   Blooming

The term refers to the migration of dissolved pigment particles from the inside of a pigment/medium system to its surface, where they are deposited as a layer of pigment crystals. Even if it is rubbed away, this film will form again. The process is completed only after years, when most of the organic pigment has crystallized either on the surface of the pigmented material or inside it. Blooming is observed in a variety of pigmented media whose application or processing involves high temperatures, such as baking enamels used in the coatings industry, metal deco printing for the printing industry, plasticized PVC, polyethylene, or rubber mixtures containing high-boiling naphthenic oils as plasticizers for the plastics market.

In a given medium and at constant temperature, the tendency of an organic pigment to bloom is a function of its concentration. Many systems have a concentration limit below or above which blooming does not occur. The range of concentrations at which blooming is observed widens with increasing processing temperature, which subsequently enhances the solubility of a pigment. This is indicated in Table 2, which shows the correlation between blooming, concentration, and processing temperature for Pigment Red 170, a Naphthol AS pigment, in PVC containing 30% dioctyl phthalate (DOP) as a plasticizer. Specifications about the temperature dependence of pigment solubility are given in Table 3.

Blooming is one of the features that differentiate pigments from dyes. Fluorescent dyes can only be applied in polymers such as plasticized PVC or polyolefins up to a certain concentration limit. Dyes dissolve in these polymers even at ambient temperature, preventing crystallization and thus precluding blooming. The solubility limit at room temperature can only be exceeded if dye concentrations are very high. In that case, blooming may occur.

Table 2: Blooming (+) of Pigment Red 170 in plasticized PVC in relation to the pigment concentration and the processing temperature (by F. Gläser).

Processing temperature (in °C)	Pigment concentration (in wt%)					
	0.005	0.01	0.025	0.05	0.1	0.5
140	+	+	−	−	−	−
160	+	+	+	−	−	−
180	+	+	+	+	−	−
200	+	+	+	+	−	−

Table 3: Solubility of Pigment Red 170 in plasticized PVC (by F. Gläser).

Temperature (in °C)	Solubility (in wt%)
20	$8 \cdot 10^{-7}$
50	$1.2 \cdot 10^{-5}$
100	$4.0 \cdot 10^{-4}$
120	$1.3 \cdot 10^{-3}$
140	$3.7 \cdot 10^{-3}$
160	$9.7 \cdot 10^{-3}$
180	$2.3 \cdot 10^{-2}$
200	$5.1 \cdot 10^{-2}$

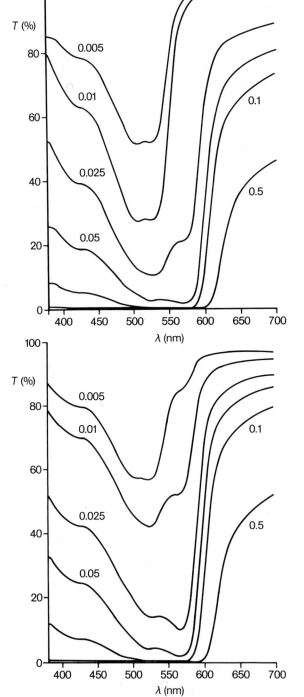

Fig. 23: Transmission spectrum of P.R.170 in plasticized PVC; processing temperature: 180 °C.

Fig. 24: Transmission spectrum of P.R.170 in plasticized PVC; processing temperature: 140 °C.

Pigment concentration is not the only system-dependent migration determinant; the entire composition of the colored medium plays a role. The tendency of plasticized PVC to promote blooming, for instance, depends on a series of parameters, such as the polymer itself, its synthesis including possible use of emulsifiers, the K value, choice of plasticizers and stabilizers, and their concentration. Specialized agents such as polymer plasticizers or intermediate layers designed to block particle movement can reduce or prevent migration.

Figs. 23 and 24 show the mechanisms underlying the dissolution and crystallization processes during manufacture and storage. As above, the example chosen to illustrate this principle is Pigment Red 170 in plasticized PVC [38].

Transmission spectra were taken immediately after allowing the pigmented samples to cool down to room temperature. Interpretation of the characteristic 570 nm band shows that at 180 °C, all of the pigment crystals are dissolved, provided their concentration is small (0.005 and 0.01%, respectively). At 140 °C, however, not all crystals are dissolved. Fig. 25 demonstrates that crystallization, taking place inside a sample rather than on its surface, is not completed even after years of storage [38].

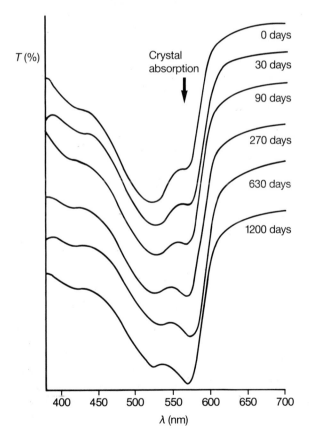

Fig. 25: Transmission spectra of Pigment Red 170 in plasticized PVC in relation to the storage time (pigment concentration: 0.01%, processing temperature 140 °C). The curves are shifted relative to each other in the direction of transmission.

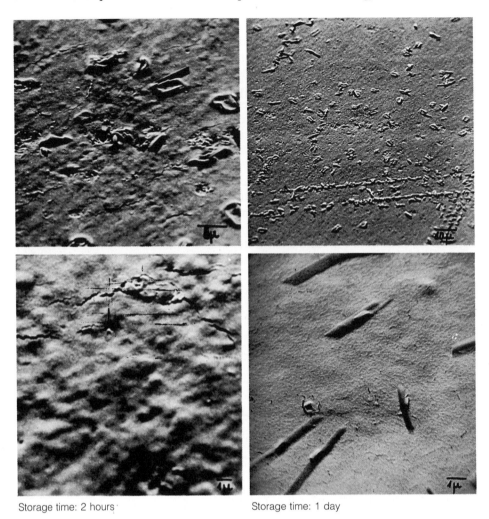

Storage time: 2 hours   Storage time: 1 day

Fig. 26a: Scanning electron micrographs of a plasticized PVC surface with bloomed Pigment Yellow 1 after different storage times. Pigment concentration: 0.2%. Below: Magnified images.

Scanning electron microscopy makes it possible to trace the time curve of blooming on the surface of a plasticized PVC sample [42]. Pigment Yellow 1 develops detectable surface crystals within a period of only a few hours, and the area is densely covered within a day (Fig. 26). Even a small space may be sufficient for a pigment to develop a variety of apparently different crystalline forms (Fig. 27), although only one crystal modification appears by X-ray diffraction analysis. The tendency to bloom renders a series of pigments unsuitable for use in such systems.

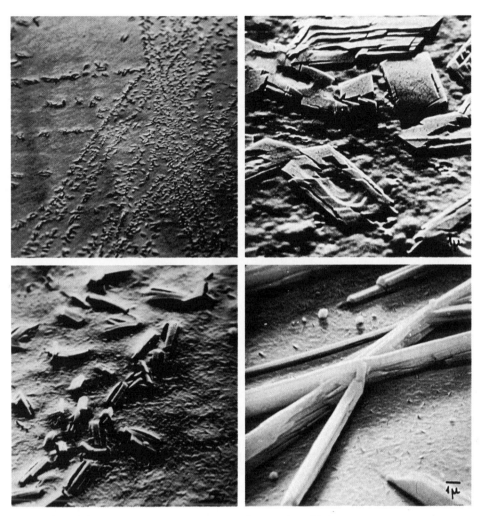

Storage time: 2 days   Storage time: 84 days

Fig. 26b: Scanning electron micrographs of a plasticized PVC surface with bloomed Pigment Yellow 1 after different storage times. Pigment concentration: 0.2%. Below: Magnified images.

Basically, blooming and bleeding are thought to be a function of pigment constitution. Monoazo pigments of comparatively low molecular weight, such as P.Y.1, migrate easily. It is unfortunately not possible to extrapolate the experimental results concerning one pigment to predict the behavior of another, even if they are chemically related. This is especially true for concentration limits.

Pigments which bloom also bleed easily. The reverse is not always true: there are a number of pigments which are somewhat prone to bleeding but which do not bloom.

Fig. 27: Scanning electron micrograph of a plasticized PVC surface with bloomed Pigment Red 3 after 84 days. Pigment concentration 0.5%. Processing temperature: 180°C.

1.6.3.2   Bleeding/Overspraying Fastness

A pigment is said to bleed if some portion of it migrates from the application medium in which it is partly dissolved to a similar colorless or differently colored medium with which it is in contact. This phenomenon is of particular concern to the plastics and coatings industry.

It is therefore not surprising that the phenomenon of bleeding has given rise to a variety of industrial standards, including generally accepted terminology [43,44], standardized tests to determine the tendency of a material to migrate and the rate of plasticizer migration [45], evaluations of bleeding into colored paper, cardboard, and carton [46], plastics in general [47], and plasticized PVC in particular [40]. The samples used in each of the standardized tests must be prepared according to exact specifications. This is particularly important considering the fact that both the entire thermal treatment and the surface structure of the polymer may influence the tendency of a dissolved pigment to migrate.

The extent of bleeding in a plastic is evaluated with a test known as the sandwich method. The test sample is placed between a white, smooth surface of plasticized PVC on the top and a standardized sheet of filter paper underneath. The resulting sandwich is then placed between layers of foam material, which in turn are covered by glass plates. The resulting sandwich is allowed to remain in an exposure chamber at 50°C for 72 hours [47].

Plasticized PVC is tested by covering both sides of the pigmented test sheet with white plasticized PVC films, which are placed between glass or aluminum plates. With a certain weight on it, this sandwich is exposed for 24 hours at a temperature of 80 °C (or for 15 hours at 100 °C) [40]. The thickness of the test sample does not affect the test results.

The degree of bleeding is estimated either by colorimetrically determining the staining of the contact sheet and/or the filter paper [48], or visually by reference to the Gray Scale [21]. It is important to evaluate the results immediately after testing, because pigment diffusion into the contact sheet may make it impossible to maintain the color during storage.

Tests to determine the bleed fastness of a paint system—also called overpainting fastness—are not subject to industrial standards. A commonly used technique involves overpainting a white paint film of defined thickness to a full shade coloration of the test pigment dispersed in the same binder. Baking is carried out under defined conditions; heat-set paints, for instance, remain in a drying oven at 140 °C for 30 minutes. Pigment migration into the white material is determined visually or colorimetrically.

The commercial demand for pigments which tend to bleed is limited. A major consideration is the processing temperature in a particular medium. There are a number of baking paints, for instance, which do not bleed into a white coating if they are cured at 120 °C; while a certain extent of bleeding is observed at temperatures between 140 and 160 °C.

Likewise, in the paint and printing ink field, the properties of an entire system are equally important, including the composition of binders and the solvents. Among others, important variables are the rate of solvent release and the crosslinking of the binding agent, as well as the extent and duration of heat exposure. Pigments designed for use in such systems should be tested together with their vehicles and under conditions that relate to actual processing and use.

As mentioned above, increased pigment concentrations are associated with a tendency to bleed, especially in materials such as plastics, which makes it necessary to work below a certain limit value. Some media, such as plasticized PVC, can accommodate modest amounts of a bleed-sensitive pigment. Increasing amounts of plasticizer in plasticized PVC enhance the tendency of a pigment to bleed. Another concern is the choice of plasticizer. There are phosphoric acid triesters, for instance, which solubilize pigments somewhat more than compounds such as the frequently used dioctyl phthalate, which in turn exceeds the action of polymer plasticizers. The latter may even be used to reduce or prevent otherwise migrating pigments from bleeding. The use of a pigment with a certain tendency to bleed may present an economic advantage, if it is applied under conditions which preclude bleeding. There is also a number of applications that are not susceptible to bleeding; the pigmented polymer for instance, may not come into contact with another suitable plastic material; or there may be no necessity to cover a pigmented paint film with a second (also pigmented) layer.

1.6.4   Disturbances During the Processing of Pigmented Systems

There are many ways in which an organic pigment may interact with its medium; and both user and manufacturer frequently find that a pigment adversely affects the performance of an entire system. Polymer products are among the most sensitive in this respect. In pigment processing and application, where surface phenomena are important, the effect of pigment-vehicle interaction may be extremely aggravating, and it is advisable to avoid concentrations and conditions that are prone to give rise to problems. Pigmented systems are particularly susceptible to the temperature, and knowledge of the mechanisms underlying pigment-vehicle effects might prevent an unwelcome surprise.

1.6.4.1   Plate-Out

The term refers to the deposition of pigment on the surface of the processing equipment or on the system itself. In contrast to similar phenomena caused by blooming, plate-out does not reappear once the pigment film is rubbed off.

Plate-out is commonly observed on the calendering and rolling equipment that is used to process pigmented PVC. It is caused by an incompatibility between components such as lubricants, stabilizers (especially those of a barium/cadmium basis), plasticizers, or other materials by the PVC mixture. These materials, in migrating to the surface of the system, carry pigment particles along with them. Plate-out is thus a direct result of the incompatibility of certain additives with the PVC medium. As a consequence, rejected particles are deposited on the surface of the calender rolls or other processing equipment, covering it with a colored film.

Conditions to be avoided are high temperatures and high pigment concentration, both of which increase plate-out effects.

The phenomenon appears to be particularly sensitive to the wettability of the pigment particles by the polymer. Types with higher specific surface areas are more subject to plate-out than varieties with a larger particle size [49,50]. Likewise, plate-out phenomena are also observed in powder coatings, which frequently contain pigment particles or other solid particles that are not suitably wetted. Such coatings emerge from the curing process covered by a layer of pigment, which may also include crystals of the curing agent. Poor rheology must be expected to present problems with regard to the curing process by making it more difficult for the binder to wet the surfaces of the particles. Within seconds, the molecules begin to cross-link and the material is thus cured at a temperature at which it is expected only to melt and sinter. This phenomenon not only causes a rapid viscosity increase but also fails to allow sufficient time for the vehicle to completely wet all particles. Attempts to remedy this problem have resulted in the development of specialized hardeners which react only at elevated temperatures, considerably reducing the tendency of a material to exhibit plate-out.

1.6.4.2   Overpigmentation/Chalking

Surface degradation of a pigment-vehicle system results in an effect known as chalking, a familiar phenomenon in connection with the artificial or natural weathering of paint films which contain titanium dioxide. The chemical reactions and micromorphological processes underlying this effect have been studied extensively [51,52]. In the presence of moisture, the photoactive titanium dioxide pigment, acting at the interface between pigment and binder, adds to the natural decomposition of the binder through daylight UV radiation. Increasing weathering time thus separates the pigment from its medium and degrades the surface of the pigment-vehicle system to such an extent that previously dispersed pigment particles come to the surface. Chalking is facilitated by using insufficiently processed pigment types or by working pigments into carrier materials that do not have good weatherfastness. The degree of chalking on a given surface can be determined by the Kämpf method [53] or the adhesives technique [54], both of which also lend themselves to the analysis of degraded paint films.

Similar phenomena are observed in enamels which contain certain organic pigments, but it should be noted that the chemical and micromorphological processes that are responsible for this form of molecular degradation are not known.

Despite the fact that a number of lightfastness and weatherability studies have reported that coatings made from chalking pigments absorb appreciably more heat than do comparative nonchalking systems, the exact role of the temperature in decomposing the vehicle is not completely clear. For paints containing organic pigments, the degree of chalking may be determined in analogy to TiO_2-containing coatings.

In the plastics industry, exceeding the limit of pigment incorporation into its vehicle is likely to give rise to chalking phenomena in the broader sense of the word, particularly with organic pigments which exhibit a high specific surface area. In this case, the typically long-chain polymers or PVC plasticizers are unable to suitably coat and fix the pigment particles, a problem which has an added degree of complexity in the presence of additives or other pigments, such as TiO_2 or extenders.

The effect of chalking is rapid and marked even after a short weathering time. There appears to be no clear distinction between chalking in the broader sense of the word and plate-out; the two phenomena are very much related. As in plate-out and unlike in blooming, the particles that adhere to the polymer surface through chalking can be removed comparatively easily without reappearing.

1.6.4.3   Distortion/Nucleation in Polymers

Many organic pigments, like a number of other compounds, are able to induce nucleation in partially crystalline polymers, a phenomenon which is commonly observed in certain polyolefins. The surfaces of foreign particles such as pigments that are contained within a polymer material provide crystallization nuclei which initiate crystal growth. It is not only the crystallization rate, however, which is influenced by nucleating agents, but also the morphology and thus the mechanical properties of their host compound. Nucleating agents can even have a devastating effect on extrusion molded plastics articles. This may be exemplified by the comparatively drastic consequences that it has for strong, thick, usually large, nonrotation-symmetrical extrusion parts, such as bottle crates. A distorted crate that will not fit onto any other is worthless. Rotation-symmetrical parts, on the other hand, may suffer from increased internal stress. The consequence of abnormal shrinkage can range from stress crack corrosion to long-term damage through weathering. The performance of a material in terms of tensile strength, breaking strain, and impact resistance is more or less affected by nucleation [55,56], which is as much a function of the time-temperature curve as it is of pigment concentration and of the choice of polymer. The shape of the pigment particles is another major factor; acicular particles are usually oriented along the direction of the polymer melt and give rise to uncommon nucleation effects. It appears that the surface structure of the pigment particles forces the adsorbed polymer molecules to grow into spherulites, as demonstrated in Fig. 28.

Fig. 28: Image of a crystallization nucleus triggered by a pigment particle in polyethylene in polarized light, taken with an optical microscope.

1.6.5   Dispersion

1.6.5.1   General Considerations

Organic pigment powders are normally sold as agglomerates (Sec. 1.5), ready to be incorporated into the medium of application. This process, known as pigment dispersion, refers to the distribution of a pigment throughout the application material, accompanied by a reduction of the agglomerate size to afford primary particles and aggregates, or at least smaller agglomerates. A pigment is referred to as being more or less easily dispersible, depending on the mechanical forces that have to be applied to achieve optimum dispersion. Reducing the size of agglomerates is necessary in order to develop the best possible application properties in a given pigment or pigment-vehicle system, whose performance is largely controlled by the particle size distribution. The effect of dispersion on the behavior of a pigment-vehicle system is comparable to the effect of simply reducing the average pigment particle size: The pigmented system will

- · exhibit increased tinctorial strength, particularly in white reductions,
- undergo a change of shade,
- be more transparent/less opaque,
- provide enhanced gloss,
- increase in viscosity,
- and finally, feature a reduced critical pigment volume concentration, which is important in mill bases.

A certain amount of shearing forces have to be applied in order to overcome the surface forces that maintain the adhesion between agglomerated pigment crystals. In practice, the shearing forces that are necessary to reduce the particles in a given pigment sample to smaller or even optimal particle size, i.e., the dispersibility of a pigment powder, depends on a number of factors:

- Pigment-related aspects, which involve the chemical constitution, crystalline modification, particle size distribution, particle shape, surface structure, preparation, and processing of the pigment powder, especially in terms of drying and milling.
- The chemistry and physics of the vehicle and its components, including factors such as polarity, molecular weight or molecular weight distribution, and viscosity. The dispersibility of a system which contains several different components such as resins depends on the solubility of each individual component in the medium and their compatibility with each other.
- The chemical or physical characteristics of solid or liquid additives, such as extenders, white pigments, dispersing agents, plasticizers, or pigment/extender combinations.
- Interactions at the pigment-vehicle interface.
- Design of the dispersing agent and its mode of operation in combination with the dispersion conditions.

- Pretreatment of the pigment prior to the dispersion process, including time, temperature, wetting, and other parameters connected with incorporation of the pigment in the vehicle.
- Formulation of the mill base, including the pigment volume concentration (PVC).

The dispersion process may be viewed as simultaneously pursuing four different objectives:

1. Agglomerates are broken down through mechanical shearing forces: Particle size reduction: **Desagglomeration**.
2. The surface of the pigment particles is wetted by the binder and by other components of the medium: **Wetting**.
3. The resulting wetted pigment particles are distributed throughout the entire medium: **Distribution**.
4. The pigmented system that is comprised of dispersed particles is stabilized to preclude reagglomeration and/or flocculation: **Stabilization**.

These four facets of the dispersion process are discussed below.

1.6.5.2   Desagglomeration of Pigment Particles

According to Rumpf [57], there are four ways in which agglomerates can be broken down:

- between two solid surfaces (dry milling in roll mills or roll crushers),
- by impingement, i.e., by imparting kinetic energy to a material and crushing it through the impact on a solid surface (dry milling in impact mills),
- by mechanical energy that is transferred onto the material through the surrounding medium (wet grinding),
- by thermal crushing.

Pigments that are incorporated in a medium are practically always crushed by mechanical means, i.e., the agglomerates are broken down through shearing forces. The pigment particles are thus broken down through their medium and not through direct impact between each other.

Sometimes the dispersion process may inadvertently result in destruction of the aggregates and the primary particles, which happens particularly with pigment powders consisting of coarse or acicular particles – not to forget the effect of intensive shear [58].

1.6.5.3   Wetting of Pigment Particle Surfaces

There are two aspects to the wetting of pigment particles:

- the spreading of a liquid or a binder system over the surface of the pigment particles, and

- the soaking of the pigment powder, especially of the agglomerates, with liquid components.

The wetting operation is primarily a function of [59 ff.]:

- the energetics of the interaction between the pigment surface and the medium,
- kinetic parameters (such as diffusion and wetting rate),
- the steric sizes and geometry of the components, including the porosity of the agglomerates and the molecular size of the wetting agents,
- the rheology of the wetting medium [59 ff.].

The interactions between the surface of the pigment particles and their surrounding medium are of prime consideration in the wetting process and are optimized in the development and production of easily dispersible pigments.

The energetics of the wetting of particles have been discussed extensively [60, 61, 62]. The specific thermodynamics of interfaces have been studied for a number of organic pigments and other materials [63, 64, 65]. This approach provides information as to whether a liquid will spread over the surface of a solid on its own, whether the operation requires energy, or whether a surface is not wettable at all. It is possible to find the surface tension of a pigment y_1 by measuring and comparing contact angles α, which reflect the relative attraction of the liquid for the solid compared to the attraction of the liquid for itself. The experimentally determined surface tension of the liquid y_2 is the basis from which to calculate the interfacial tension y_{12} and the wetting tension β_{12} by using the following equations:

$$\text{Contact angle:} \quad \cos\alpha = \frac{y_1 - y_{12}}{y_2}$$

$$\text{Wetting tension:} \quad \beta_{12} = y_{12} - y_1 = -\varkappa_2 \cdot \cos\alpha$$

Thus, the greater the value of β_{12}, the easier it is to wet the surface. The kinetic aspect, i.e., the actual wetting rate or wetting time, is quantified by the modified Washburn equation, a simplified version of the Hagen–Poiseuille equation for cylindrical capillaries in the case of incomplete wetting [66, 67]. In using this equation, wetting is always assumed to be thermodynamically allowed [64]:

$$t = \frac{V \cdot l \cdot 2\eta}{r^3 \pi y_2 \cdot \cos\alpha}$$

in which:
t = penetration time
α = contact angle
y_2 = surface tension of the liquid
$y_2 \cdot \cos\alpha$ = wetting tension
l = capillary length
η = viscosity
r = radius of the pores within the agglomerates
V = transported volume of liquid

According to the above equation, the time t that is required to wet a powder is proportional to the viscosity of the surrounding medium, the depth of the pores, and the transported volume. The wetting time, on the other hand, is inversely proportional to the wetting tension — the driving force that promotes wetting — and to the cube of the radius of the pores in the agglomerates.

The viscosity of the binder is thus the only variable wetting time determinant during pigment processing: both r and l are pigment-specific structural parameters and are therefore constant, which is also true for the wetting tension. The viscosity, on the other hand, can be changed systematically by modifying the composition of the mill base or by increasing the dispersion temperature.

The efficiency of any dispersion operation is essentially a dual function of the breakdown of the agglomerates and the wetting of the pigment surface. One or the other of these two barriers will usually dominate, depending on the medium and the dispersion equipment. Dispersing a pigment in a polymer such as PVC or a polyolefin, for instance, is largely a question of breakdown of the agglomerates through mechanical action.

To demonstrate this tendency, samples of the gamma form of Pigment Violet 19 (Sec. 3.2.4) were dispersed on a roll mill into PVC containing varying amounts of plasticizer, as shown in Fig. 29 [68]. The tinctorial strength of the pigment-

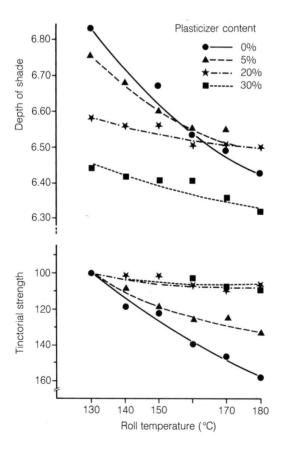

Fig. 29: Dispersion of Pigment Violet 19, γ-modification, in PVC on a roll mill in relation to the plasticizer content (DOP) and the dispersion temperature. PVC: K value 70.

vehicle system is plotted against the dispersion temperature. The tinctorial strength is an indication of the quality of pigment dispersion in the end product. The resulting graphs illustrate the correlation between pigment dispersion, temperature, and plasticizer content. Increasing the amount of plasticizer within a polymer appears to reduce the tinctorial strength, reflecting poorer dispersion. The second diagram, showing the tinctorial strength that is attained through a certain plasticizer content at the lowest possible temperature (130 °C = 100), demonstrates the decreasing influence of heating as more plasticizer is added. It is obvious that the quinacridone pigment being used in this study does not disperse easily in its medium. The result is somewhat surprising, since this particular plasticizer is known to wet organic pigments easily. One might therefore expect an increased dioctyl phthalate concentration to increase rather than decrease the dispersibility of the pigment. It can thus be concluded that in this case the amount of shear being imposed on the system appears to be the primary determinant. If the plasticizer content in a polymer is increased, its rheology will change, which is associated with a stronger plastification of the polymer and consequently reduced shearing forces. This is in aggreement with a number of rheological studies on a series of these PVC-plasticizer mixtures, in which additional shear was applied through temperature changes (Fig. 30).

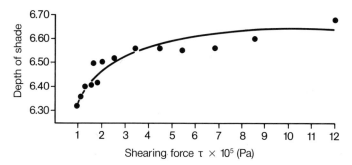

Fig. 30: Dispersion of Pigment Violet 19, γ-modification, in PVC at varying plasticizer content. Relation between the shearing forces in the plasticized PVC during pigment dispersion and the degree of pigment dispersion (depth of shade) after 8 minutes of dispersion.

Wetting out a pigment for several days by simply storing the manually prepared pigmented PVC paste (DOP content: 39%) makes for almost optimum dispersion, which requires very little shear (see Fig. 90, p. 171). The wetting of the surface of the pigment particles by the plasticizer molecules thus determines the outcome of the dispersion process.

For all practical purposes, shear continues to be of prime consideration in pigment dispersion, even if the wetting tension of the medium is high. It is therefore not surprising that the degree of agglomeration in a dispersion product varies according to the design of the equipment with which it is produced. Most of the dispersion equipment that is used in industry today has received appreciable theoretical and experimental consideration regarding both machinery and operation and has been evaluated on a variety of pigment-vehicle systems [69, 70, 71].

Increasing the temperature frequently leads to better wetting, while the reverse is true for the mechanical breakdown of agglomerates. Similar observations have been made regarding the effect of wetting and dispersion agents [72].

To illustrate this trend, a sample of the β-modification of Pigment Violet 19 was dispersed in an offset vehicle on a three-roll mill. Two series of determinations, one at constant temperature and one at constant shearing force, afforded data on the dispersibility of the pigment powder in relation to the shearing force (applied in the nip between primary and secondary rolls) and to the dispersion temperature. The resulting printing inks were more or less thixotropic; their viscosities were measured at 23 °C [69] after the thixotropic effect had leveled off. (Sec. 1.6.8.1). Figs. 31 and 32 plot viscosity values derived from flow curves that were taken at a con-

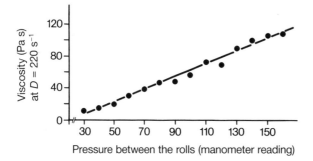

Fig. 31: Effect of the pressure between the rolls during isothermic dispersion (23 °C) of 20% Pigment Violet 19, β-modification, in a sheet offset vehicle (6 passes) on the viscosity of the printing ink.

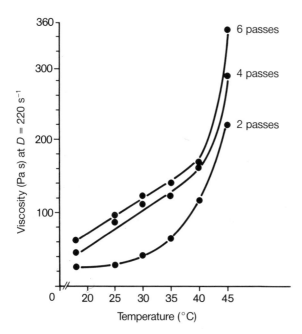

Fig. 32: Effect of the roll temperature and the number of passes during the dispersion of 20% P.V.19, β-modification, in a sheet offset vehicle at constant pressure between the rolls on the viscosity of the printing ink.

stant shear rate of $D = 220$ s^{-1} versus the pressure between the rolls and the temperature, respectively. Evidently, the effect of the temperature increase during dispersion is more significant than the shearing stress or shearing force created by the rolls. The plots demonstrate that the degree of dispersion afforded by even a slight temperature increase exceeds the effect of several additional passes on three-roll mills, even at maximum shearing force. However, the shape of the graph also demonstrates that the degree of dispersion is still far from being optimum: the tinctorial strength of prints made from these inks continues to increase considerably under comparable conditions.

Average temperatures (40–70 °C) in combination with high-shear dispersion equipment, of which three-roll mills are a good example, appear to afford the best results in terms of pigment dispersion in wetting systems, such as offset varnish. Fig. 33 shows a curve in which the degree of dispersion reaches a distinct maximum at one particular temperature.

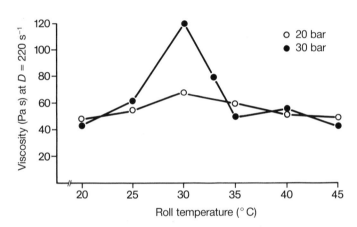

Fig. 33: Effect of the dispersion temperature (roll temperature) and the pressure between the rolls during dispersing of Pigment Red 53:1 in a sheet offset varnish (pigment concentration: 20%) in a three-roll mill on the degree of pigment dispersion. This correlation is illustrated by showing the viscosity of the printing inks at a shearing rate $D = 220$ s^{-1}.

1.6.5.4   Distribution of the Dispersed Pigment in its Medium

Achieving a statistically uniform distribution of pigment particles throughout all volume units of the entire application medium is a question of the equipment and its mode of operation and also of the conditions under which the dispersion process is carried out.

1.6.5.5   Stabilization

One of the primary concerns in the present discussion is the necessity to stabilize the system of wetted pigment particles in their vehicle system so that the dispersion process does not reverse itself. Stabilization, i.e., protection against reagglomeration and/or flocculation, is a somewhat complex matter (Sec. 1.7.5) [73] and depends on a variety of parameters, including the composition of the medium, especially the choice and amount of solvents, the chemical and physical features of the different components, the viscosity of the vehicle system, and the effect of adsorptive layers covering the surface of the pigment particles. The complexities of attraction and repulsion are discussed at great length in the literature; three of these concepts should be mentioned. All of them regard adsorption as a key process in stabilization, but differentiate between ionic and steric forces:

- The DLVO theory, a quantitative theory of colloid fastness based on electrostatic forces, was developed simultaneously by Deryaguin and Landau [74] and Verwey and Overbeek [75]. These authors view the adsorptive layer as a charge carrier, caused by adsorption of ions, which establishes the same charge on all particles. The resulting Coulombic repulsion between these equally charged particles thus stabilizes the dispersion. This theory lends itself somewhat less to nonaqueous systems.
- A mechanism claiming steric interaction between the molecules adsorbed at the solid–liquid interface [76], was first proposed by Crowl [77], and
- The entropic repulsion theory by Clayfield and Lumb [78], who suggest that the bulk of the protective macromolecular layer prevents other particles from approaching close enough for the attractive forces to cause agglomeration [79].

1.6.5.6   Dispersion and the Critical Pigment Volume Concentration

The term Pigment Volume Concentration (PVC) refers to the percentage of the volume of a pigment that constitutes the nonvolatile portion in a formulation. The Critical Pigment Volume Concentration (CPVC), on the other hand, indicates the proportion of the volume of a pigment to the volume of the vehicle at which there is just sufficient binder to wet all pigment particles and to fill the voids between them, leaving no excess vehicle [80, 81]. The CPVC has received considerable attention in connection with operations such as optimizing the formulation of mill pastes, a process in which cost-benefit considerations are a primary concern [82]. The CPVC decreases with increasing degree of dispersion and hence increase of the surface area which is available for wetting by the binder.

This principle has been illustrated in a study designed to show that the CPVC is affected by changes in the dispersion conditions (Fig. 34) [83]. A sample of Pigment Yellow 13 was dispersed in an offset vehicle. A series of printing inks, each with a different pigment concentration, were produced on a three-roll mill in four passes at temperatures of 15, 30, and 45 °C, respectively. Flow curves were taken after 24 hours at 23 °C; the respective viscosities for $D = 220 \text{ s}^{-1}$ had to be determined indirectly by interpolation. The CPVC was then determined by plotting viscosities against pigment volume concentration and finding the point at which the viscosity values level off [83]. According to definition, this is the point at which all binder molecules are adsorbed at the pigment surface and are thus unavailable to wet additional pigment surfaces: the system is now immobile.

A number of parameters change abruptly at the CPVC. Experimental data on the influence of binders, additives, and suitable pigment preparation are shown in Fig. 35, which is a plot of the viscosity versus PVC curves for three samples of one pigment, incorporated in three different offset varnishes [84].

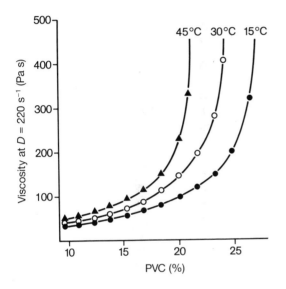

Fig. 34: Effect of the dispersion temperature and the degree of dispersion of P.Y.13 on the viscosity and CPVC of sheet offset inks. Viscosity measurement at 23 °C.

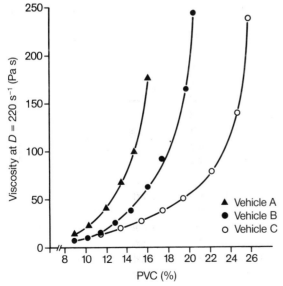

Fig. 35: Influence of the binder on the CPVC and of the binder and the CPVC on the viscosity of P.Y.13 in sheet offset vehicles of different composition at constant dispersion conditions in a three-roll mill.

Dispersed pigment formulations must be adequately prepared to afford products with optimum commercial properties. This concerns both the application of concentrates and the conversion of concentrates to end use products, which should preferably exhibit good flow properties in combination with high tinctorial strength. Knowledge of the above correlations and mechanisms aid the manufacturer in achieving an optimum in product performance.

1.6.5.7   Test Methods

Manufacturers of organic pigments frequently use additives in an attempt to improve parameters such as the wetting time and wetting volume and thus to enhance the dispersibility of their products. The decision about each individual approach is usually made on the basis of empirical data concerning the application properties of the products. This is understandable in view of the fact that the dispersibility of a pigment is known to depend on its medium and on the conditions under which it is applied, a problem that is so complex that it is solved best by carrying out a series of pilot scale experiments in different vehicles.

Where pigment dispersibility is to be quantitatively related to other parameters, it is often useful to indirectly determine the degree of dispersion and the particle size. The technique of measuring the quality of dispersion that is achieved at certain marked points in the procedure — which may be defined in terms of time, passes, or cycles — is a feature common to all of these methods. Alternatively, it is also possible to relate the dispersibility of a material to a series of different processing temperatures, which is particularly interesting to the plastics industry. The list of possible criteria includes tinctorial strength, gloss, transparency, viscosity, and other application properties of the pigment-vehicle system.

A different principle is pursued by methods which rely on direct particle size determination by optical microscopes or other equipment. However, it is not always possible for direct methods to recognize all of the particles in organic pigments, because they can be much smaller than 1 µm. Beyond a certain degree of dispersion, the only information that can be extracted by direct methods relates to the still existing coarser portion of the pigment agglomerates which is not normally recognized when application properties are determined. Additional information may be obtained by similarly measuring the particle size in a paint or an ink by examining the respective pigmented application media with grindometers [85, 86]. To control the production and to test new formulations, grindometers are useful as long as the particle size is larger than 1–2 µm. No useful information can be given for pigments with a degree of dispersion smaller than 1 µm, but this range provides the primary determinants of tinctorial strength, gloss, and other aspects of pigment application properties, which restricts the use of grindometers. They are, however, used to advantage in predicting possible processing problems caused by the pigment and are employed, for instance, by producers of offset printing inks to analyze layers around 1 µm (Sec. 1.8.1.1).

The manufacturer who wishes to characterize the dispersibility of a given organic pigment batch frequently refers to curves plotting the tinctorial strength versus dispersion time. Industrial standards [87] have been developed for a number of dispersion processes and instruments, their primary operational parameters, and data evaluation. Other standard tests provide information regarding individual applications: specific tests have been designed for pigment dispersion in low-viscosity alkyd resin systems which are dried by oxidation, in heat-set alkyd/melamine resin systems, and also in highly viscous binders dried by oxidation [88]. The mill base is sampled at certain points during the dispersion process, which are either referred to simply by time or defined through equipment-related units such as the repeated passage of material through the mill. The resulting samples are combined with a white pigment or a white pigment paste and the white reductions evaluated photometrically. The tinctorial strength presumably reflects the degree of dispersion. Although this is a seemingly straightforward method, the results may be deceptive if particles flocculate as white pigment is added or if the dispersion process is reversed, distorting the measurements. Such problems are readily revealed by the rub-out test, which determines the difference in tinctorial strength between a rubbed and a nonrubbed area. A deviation of more than 10% suggests that the determination be carried out at the rubbed area.

The tinctorial strength of a pigment is not normally proportional to the dispersion time, but generally increases hyperbolically [89]. Since all pigment batches are expected to contain a more or less large portion of agglomerates, held together by weak forces, it is not surprising that the initial dispersion rate is comparatively high. The agglomerates are thus the first units to be broken down, leaving a material composed of less easily dispersible parts. This explains why the dispersion rate slows down and the dispersion-time curve asymtotically approaches an upper limit (Fig. 36).

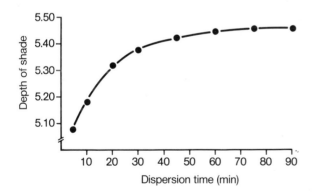

Fig. 36: Dispersion of Pigment Violet 19, β-modification, in an alkyd-melamine resin baking enamel with a paint shaker. Reduction 1:2 with TiO_2. Tinctorial strength–time curve.

This limit, which might be referred to as the "ultimate tinctorial strength", reflects the maximum degree of dispersion which can be achieved in a particular vehicle system under a certain set of conditions. However, experimental results may deviate more or less from the theoretical concepts; and an ideal dispersion is not normally realized: not all agglomerates are broken down entirely. This, however, is of no consequence, because even the experimentally determined ultimate tinctorial strength is by no means considered a standard for industrial application: technical operations are not always allowed to go to completion, and the dispersion process is often discontinued, mainly for economical reasons.

Although a number of concepts can be illustrated conveniently by plotting the tinctorial strength versus time, the resulting curves do not necessarily reflect a quantitative relationship. Not only is it impossible to accurately predict the ultimate tinctorial strength, but the curve also fails to reflect the shearing forces that are necessary to afford a certain degree of dispersion. Moreover, as dispersion time increases, the shape of the curve is often increasingly controlled by side effects.

A graph more suitable for quantitative evaluation is easily drawn up by transforming the above correlation into a linear one. The inverse of the tinctorial strength is simply plotted versus the inverse of the degree of dispersion, referred to in terms of time or some other unit [90, 91, 92]. The resulting doubly reciprocal presentation can be extrapolated to the intercept with the ordinate. This is the point at which the dispersion time is infinite and its inverse is 0, so that the height of the intercept equals the inverse of the ultimate tinctorial strength. A low gradient is associated with easily dispersible pigments, while a steep slope reflects increased shear requirements.

A characteristic number may be derived from the difference between two points reflecting the increase in tinctorial strength, e.g., between two dispersion periods.

This characteristic, known as the dispersion hardness DH, indicates the effort that is necessary to disperse a pigment [93]:

$$DH = 100 \frac{F_2}{F_1} - 1$$

in which F_1 is the K/S value measured at an early point on the curve, preferably chosen such that the degree of dispersion just allows the pigment-vehicle system to be completely homogeneous.

F_2 represents the K/S value at the second degree of dispersion, at which at least 90% of the ultimate tinctorial strength F_∞ is reached.

The tinctorial strength in white reductions is thus quantitatively defined by the Kubelka–Munk relation between the spectral absorption coefficient K and the spectral scattering coefficient S: in which β_∞ refers to the reflection of a completely opaque layer. The ratio K/S is proportional to the tinctorial strength.

$$\frac{K}{S} = \frac{(1-\beta_\infty)^2}{2\beta_\infty}$$

The dispersibility of a pigment in a particular vehicle system is also reflected by the viscosity of the pigmented medium and by the gloss that it produces in application. Viscosity and gloss are frequently considered more suitable parameters to indicate the state of dispersion of a pigment in an application medium than the tinctorial strength, which reaches its optimum at a point at which other parameters, such as gloss, can be improved.

Standardized tests have been designed to determine the development of gloss in a pigment-vehicle system with time [94]. The reflectance values [95] of a coating or a print are plotted against the dispersion energy, measured in units of dispersion time or some other parameter. Since the resulting curve is as unsuitable for quantitative evaluation as the one derived from tinctorial strength determinations, plotting the inverse values is the approach that first comes to mind, and it is similarly useful in this case. The resulting linear correlation can be interpolated graphically to afford information regarding the dispersion energy that is necessary to produce a certain reflectance value, or to indicate the reflectance value that is realized at a particular point in the dispersion process.

Other effects, such as pigment recrystallization, also frequently influence the different partial processes of dispersion. Recrystallization may be defined in this context as a growth of larger particles at the expense of the smaller ones. Recrystallization is commonly associated with coloristic changes; it particularly improves the opacity and accordingly reduces the transparency of a material. Fig. 37 outlines a tinctorial strength–time curve that is affected by recrystallization. In this particular experiment, Pigment Orange 43, a naphthalenetetracarboxylic acid pigment (Sec. 3.4.2.4), was dispersed in an oscillating shaking machine (paint shaker) at 20, 40, and 60 °C, respectively, in a toluene-based publication gravure printing ink. Printed samples (cell depth = 18 μm) made from these inks were evaluated colorimetrically. As was expected, longer dispersion times evidently increase the particle size through recrystallization, decrease the tinctorial strength, and cause a shift to more yellowish shades.

Moreover, the crystal modification of a polymorphous pigment may change during the dispersion process. This was exemplified by incorporating a sample of the epsilon modification of copper phthalocyanine blue into a thermosetting coating system. The mill base contained pigment, alkyd resin, xylene (56%), and aliphatic hydrocarbons (10%); a paint shaker was used to disperse the pigment in the medium. Thermostats maintained a nearly constant temperature so that the difference between the readings for the mill base and the thermostated bath immediately after the dispersion process did not exceed 2 °C. The increase in tinctorial strength with time was determined colorimetrically after the coatings had been reduced 1:50 with TiO_2, applied, and dried (Fig. 38).

Likewise, Fig. 39 illustrates the corresponding change in shade with increasing dispersion time.

The complex mechanism underlying this dispersion process of the epsilon modification of phthalocyanine blue is conceptually resolved into three steps:

- the dispersion itself, which consists of breaking down the agglomerates and covering the pigment surface with the binder components: a fast step in this example. The tinctorial strength thus increases rapidly. As the dispersion process proceeds, other mechanisms gain importance, such as
- a change in the crystal modification from the very reddish blue epsilon form to the greenish blue beta modification. The hue shifts accordingly, which is possibly also connected with a change in tinctorial strength
- recrystallization of the pigment particles. This is also associated with a decrease in tinctorial strength and a shift from a very greenish blue to a reddish blue shade.

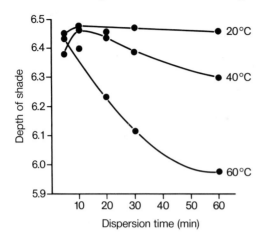

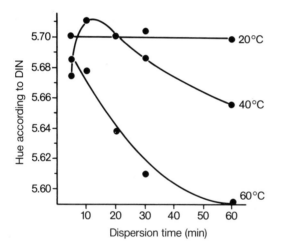

Fig. 37: Dispersion of 5% Pigment Orange 43 in a toluene-based gravure printing vehicle with a paint shaker. Effect of the dispersion time and the dispersion temperature on the depth of shade (top) and on the hue of the prints (below).

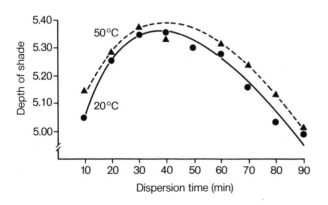

Fig. 38: Dispersion of Pigment Blue 15:6 in an alkyd-melamine resin baking enamel. Effect of the dispersion time and the dispersion temperature on the coloristic properties (depth of shade). White reduction 1:50 TiO_2.

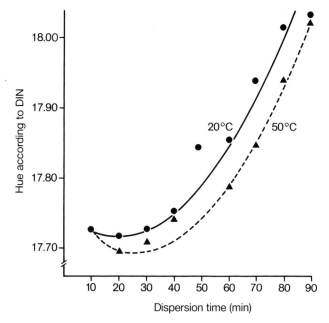

Fig. 39: Dispersion of Pigment Blue 15:6 in an alkyd-melamine resin baking enamel. Effect of the dispersion time and the dispersion temperature on the coloristic properties (hue). White reduction 1:50 TiO_2.

1.6.5.8 Flush Pastes

Flushing is sometimes considered an alternative to the dispersion process, because it is the direct transfer of pigments in an aqueous phase, as they emerge from the synthesis, to a nonaqueous phase without previously drying and milling the colorant. The nonaqueous phase commonly consists of binders, such as alkyd resins, mineral oils, celluloseacetobutyrate, or other suitable water-insoluble vehicles. The idea is to displace the adsorbed water from the surface of the pigment particles and to replace it by the vehicle components.

There are certain advantages to flushing, especially since it avoids drying the pigment presscake and milling the dry pigment; steps that are known to promote reagglomeration. The technique has been extensively investigated [96].

Alkali Blues (Sec. 3.8.1.3), which are highly polar, are particularly prone to reagglomerate and are difficult to disperse in dried form. Therefore printing ink manufacturers prefer to use them as flush pastes at a pigment concentration around 40%. Flush pastes of other organic pigments, such as diarylide yellow pigments for letterpress and offset printing, are less common in Europe. They are used for these purposes only in the USA and to a lesser extent in Japan, as well as by European branches of US companies. Since flushing generally produces dispersions of greater gloss and transparency than formulations produced with dry methods, products targeted for the letterpress and offset market were used also in Europe in flushed form before it became possible to manufacture highly transparent and glossy organic pigment powders for such printing inks.

Presscakes used for the flushing process typically contain between 15 and 20 wt% pigment. It is now also possible to purchase highly concentrated presscakes, containing 50 wt% and more. Despite the fact that they are recommended for use in aqueous media throughout the printing inks and coatings industry [97], such products only enjoy minor commercial recognition. It is difficult to coloristicallly standardize presscakes and to safeguard them against bacteria and fungi.

1.6.5.9 Pigment Preparations

The pigment industry today provides the user with a wide variety of pigment preparations to suit all purposes. In these pigment preparations, the pigment is in an already dispersed form. Purchasing a pigment in such a form considerably facilitates its application. Details regarding the composition, synthesis, and application of such preparations are described under the respective applications (Sec. 1.8.1, 1.8.2, 1.8.3).

1.6.6   Lightfastness and Weatherfastness

1.6.6.1 Definition and General Information

The lightfastness of a material is defined by the inherent ability of a given pigment-vehicle system to retain its initial color value upon exposure to daylight. It is thus a system-related parameter which cannot be determined in connection with pure pigment only.

Although a few inorganic pigments are stable enough to withstand light completely and almost indefinitely, most inorganic and all organic pigments change their shade more or less readily upon illumination. The sensitivity of a given pigment-vehicle system is a function of pigment-related factors such as chemical constitution (Sec. 1.4.3), concentration, physical parameters (particle size distribution, Sec. 1.7.4 and crystal modification Sec. 1.5.3) on the one hand, and, last but not least, the chemical nature of the surrounding vehicle on the other hand.

Light is not the only power that may destroy or alter a pigment within its vehicle system. Even traces of water, gases, or industrial effluents in the atmosphere may adversely affect the application properties of a material. The combined effects of weather and light will generally destroy a pigment faster than light alone. Weatherfastness has thus been defined in connection with the ability of a pigment-vehicle system to withstand the chemical and physical factors inherent in weather, which also includes light. The exposure may be alternating or simultaneous. It is frequently recommendable and more common to refer comprehensively to the weatherfastness of a pigmented system instead of focusing on the effect of light alone.

Weatherfastness is defined as the simultaneous or alternating effect of irradiation and atmospheric impacts. Weather is difficult to express quantitatively, since it is a complex phenomenon and the response of a pigment-vehicle system to outdoors exposure depends largely on environmental conditions, such as the intensity of the sun, temperature, humidity, precipitation, oxygen concentration, and the gaseous composition of the air. To complicate matters, all of these parameters are in turn functions of the time of day, the season, and the location at which the determinations are carried out, including latitude, longitude, altitude, proximity of industrial plants, etc. It is evidently difficult to compare two or more sets of data, unless they are taken simultaneously in the same location, and there are no universally accepted weatherfastness standards. Both the lightfastness and the weatherfastness of a pigment-vehicle system are thus considered relative indications of pigment behavior in a very strictly defined environment compared to a standard.

1.6.6.2   Evaluation Techniques and Equipment

A number of lightfastness and weatherability test methods have in several countries been developed into industrial standards.

The Blue Scale, also known as the Wool Scale, which was originally developed for the evaluation of exposed textile prints, serves as the standard of comparison mainly for graphic prints [98, 99]. It consists of eight blue-colored woolen samples that differ in their fastness to light. The standards are prepared with chemically defined dyes whose chemical constitution is listed in the Colour Index. The swatches are graded according to numbers between 1 and 8, 1 indicating extreme sensitivity to light, while highly resistant samples qualify for number 8. If lightfastness is to be evaluated in terms of a color-change comparison, a defined standard and a sample are exposed together until the standard shows a significant change in color, irrespective of whether it is the shade, the brightness, and/or the intensity of the color that is affected. A change equal to step 3 on the Gray Scale for Evaluating Color Change is considered significant [10]. Several swatches may be tested simultaneously by exposing all samples to the same light source until one of the standards on the Blue Scale, e.g., the number 3, shows a noticeable change in color. After partly covering the swatches (about 1/4), exposure is continued until the next higher standard fades appreciably. The lightfastness test may be continued over the range of the Blue Scale up to step 6; the time is doubled in going from one standard to the one next in line.

The comparison with the Blue Scale may lead to different results, depending on how the standard and/or the pigment reacts to light. Blue dyes, for instance, are not at all uniform in their performance towards UV radiation [100]. This effect has been studied by using UV filters. The results point to the fact that although samples do not absorb differently over the visible range of the spectrum, steps 3 to 5 on the Blue Scale are considerably more sensitive to wavelengths under 335 nm and 360 nm than steps 1 and 2. Fading is therefore largely a function of the power distribution of the illuminant; an aging light source or aging filters may inadvertently influence the UV radiation that may distort the results of an otherwise standardized test. Accelerated exposure devices thus have to be carefully controlled (see below), and location and even date and time should be recorded meticulously. Extremely lightfast materials, which correspond to steps 7 to 8 on the Blue Scale, present some unusual problems with regard to color assessment. Differences between these samples are frequently subtle, difficult to determine accurately, and rarely reproducible. The problem is aggravated by the fact that the textile carriers may not always be sufficiently stable.

Attempts have been made to replace the Blue or Wool Scale by physical devices that measure the radiation of the light source. To determine the radiation of a given light source, the irradiance (in W/m^2) is multiplied by the exposure time (in s). The equipment necessary to carry out such measurements is commercially available. Measurement in the UV region would be desirable. It has recently become possible to purchase an instrument which can measure spectrum-related irradiance in the important UV range.

Very light and weather resistant systems require very long outdoor exposure times. In order to accelerate the testing of pigmented systems, test methods have been developed which simulate outdoor conditions. The evaluation of such a test is not only conveniently accelerated, but the resulting data are easy to reproduce and independent of location, climate, date, and time.

Thus, the compulsory requirement for automotive coatings of a two-year Florida exposure can be reduced to a period of less than 6 months. This is advantageous for the developmental process.

Today, xenon arc lamps are the most commonly used light sources for such tests. Xenon tubes are equipped with combinations of filters, which make it possible to approximate the radiation of natural sunlight. The filters can be combined as desired to alter the spectral distribution of the radiation, which may influence the outcome of degradation processes such as chalking. Fig. 40 shows the energy distribution curve provided by a filtered xenon arc light compared to natural daylight. The filtered radiation emitted at high intensities by a xenon tube provides a good approximation to the energy distribution of normal daylight. Certain filter combinations screen out portions of UV radiation at specific wavelengths and thus approach daylight even more closely.

An industrial standard method has been developed to test the lightfastness of polymers in accelerated test equipment [102]. The apparatus consists of a quartz-xenon tube with a special optical filter between the light source and the specimen to produce light that resembles window glass-filtered daylight [103]. Samples are mounted at a specific distance from the arc and are supported on a frame which revolves around the arc 1 to 5 times per minute for uniform exposure. A blower unit in the base provides a flow of air which makes it possible to maintain a black panel temperature of 45 °C, measured by a black panel thermometer which is positioned at level with the samples. A black panel unit consists of a bimetallic thermometer mounted on a steel frame. Both faces of the frame plate and also the stem of the

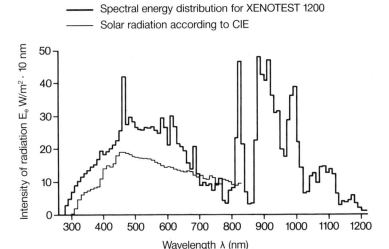

Fig. 40: Energy distribution of solar radiation (according to CIE, No. 20) and filtered xenon arc light (Xenotest 1200) [101].

thermometer are coated with a heat-resistant glossy black enamel. The relative humidity level in the exposure cabinet is closely controlled.

Lightfastness is evaluated colorimetrically by comparing exposed colored samples or portions of colored samples with the original (unexposed) sample [104]. The color difference between the two test specimens is expressed in CIELAB units [2,3]. Instead of referring to the Blue Scale, results are described in terms of the radiation energy that is necessary to achieve a certain color difference. Although ample data are available on the lightfastness of a large number of organic pigments in various media, the information usually relates to the Blue Scale. For the sake of uniformity and comparison, it therefore appears more appropriate to refer to the Blue Scale as a classification and evaluation system to describe specific pigments in parts 2 and 3 of this book.

Accelerated exposure equipment may also be used to test for weatherfastness in plastic materials [105]. The natural destructive agents inherent in weather are approximated by filtering the radiation emitted by the xenon arc lamp and by spraying the sample with water under standardized conditions [105]. Test programs are designed to relate to actual outdoor exposure to rain and humidity. In a standard program, a 3 minute wet cycle typically alternates with a 17 minute dry period. Weatherfastness tests are carried out and evaluated like lightfastness tests: the black panel temperature and other parameters are the same in both procedures.

Weatherfastness tests on coatings have shown that prolonging the dry and wet periods, i.e., extending the usual 17 minute dry cycle to a full 107 minutes and the wet period to 13 minutes, affords results which correlate much better with the Florida outdoor exposure tests. The humidity that penetrates a layer is known to interact with the various components in the coating, adversely affecting the mechanical properties of the material. The above mentioned cycle apparently corresponds much better with the climatic conditions in areas like Florida. If recirculated water is used in accelerated exposure equipment, suspended particles may adhere to the surface of the coating, producing a thin film. This obviously has a detrimental effect on the test results.

It might be interesting to note how far the results of accelerated test methods agree with full-length outdoor exposure and the effect of natural daylight. The correlation between outdoor or daylight exposure, including factors such as location, elevation, proximity of industrial plants, maritime climate, etc. on the one hand and accelerated methods on the other hand have been studied extensively on a wide variety of pigment-vehicle systems.

The results show that although the xenon arc radiation is adjusted by filters, there is still some problem in comparing coloristically similar pigments with different chemical structures. Such pigments frequently respond differently to the source of radiation (daylight or xenon arc lamp), irrespective of the surrounding vehicle system. The following studies demonstrate this very clearly. In letterpress proof prints and in long-oil alkyd vehicles, Pigment Yellow 17 is more lightfast than Pigment Yellow 13 if both are exposed to daylight. A xenon lamp, however, corrected with filters, reverses the results, and it is now Pigment Yellow 13 that is more stable. Pigment Red 57:1 and Pigment Red 184 react in the same manner. This behavior reflects a difference in the spectral sensitivity of pigments, the effects of which are observed especially in the UV range and in the difference between natural daylight and artificial light from the xenon tubes (see Fig. 40).

Test specimens can only be compared in terms of lightfastness or weatherfastness if all samples show the same depth of shade. Up until a few years ago, such determinations were carried out at equal pigment concentrations.

1.6.6.3   Factors Determining the Lightfastness

The lightfastness of a pigment-vehicle system is a function of
- the vehicle,
- the substrate,
- the pigment volume concentration,
- the thickness of the layer,
- the additives.

Vehicle

Comparative studies show the considerable impact attributed to the medium. Certain types of fat dyes, for instance, have lightfastness equal to step 8 on the ISO Scale (Blue Scale) if they are incorporated in polystyrene or similar polymers, while in other media their lightfastness drops to step 1 or 2.

Similar observations have been made on offset printing inks containing 15% Pigment Yellow 13 (Fig. 41).

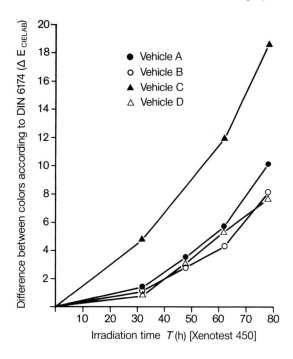

Fig. 41: Effect of the binder on the lightfastness. Pigment content in the printing ink: 15% Pigment Yellow 3.

Substrate

To test for lightfastness with respect to the substrate, the same letterpress ink containing 20% Pigment Red 53:1 was printed on different papers (Fig. 42). The first sample was printed on a standardized (DIN 16519) coated art paper without any optical brightener; the second substrate was a slightly fluorescent glazed magazine gravure paper, and the third was regular wood pulp paper for newsprint purposes, also slightly fluorescent.

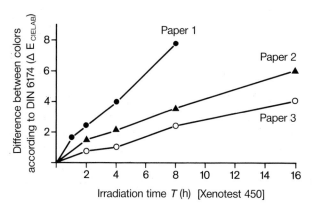

Fig. 42: Effect of the substrate (paper) on the lightfastness of letterpress proof prints. Pigmentation in the printing ink: 20% Pigment Red 53:1.

Pigment Volume Concentration

Lightfastness and weatherfastness of a pigmented system improve with the depth of shade; i.e., the system becomes more stable as the pigment concentration increases, up to the critical pigment volume concentration. Since the pigment particles in the uppermost layer of the pigmented system are the first to be affected by incident light, increasing the pigment volume concentration in a layer can be expected to provide better fastness. This is clearly demonstrated by comparative studies on 1/3 and 1/25 SD letterpress prints containing Pigment Red 53:1 (Fig. 43). 1/25 SD prints, which contain much less pigment per cm^2 than 1/3 SD prints and therefore provide a lower pigment surface concentration, fade much faster than 1/3 SD prints. The pigment volume concentration in a 1/3 SD print is much higher than that of a 1/25 SD print. The color change upon irradiation, expressed according to DIN 6174 in CIELAB units, is shown in Fig. 43.

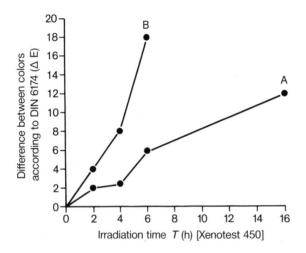

Fig. 43: Dependance of the lightfastness of letterpress proof prints with the standard depth of shade. Pigment: Pigment Red 53:1. Curve A: 1/3 SD print. Curve B: 1/25 SD print.

Effect of the Thickness of the Layer

It is interesting to note that in spite of equal composition and equal pigment surface concentration, two samples may respond differently to light, depending on how thickly they are applied. The effect of the film thickness has been experimentally demonstrated on a series of offset printing samples made from a 15% Pigment Yellow 12 formulation. Appropriately diluted with vehicle, this paste affords a set of inks, formulated at 12, 9, 6, and 3% pigment concentration, respectively. All inks are printed at the thickness required to match the original 15% pigment paste in terms of pigment surface concentration. Measurements are thus carried out on a series of prints that differ only in film thickness. Fig. 44 shows that despite equal pigment surface concentration, the samples respond very differently to light. The effect of film thickness on the lightfastness of a print has not yet been

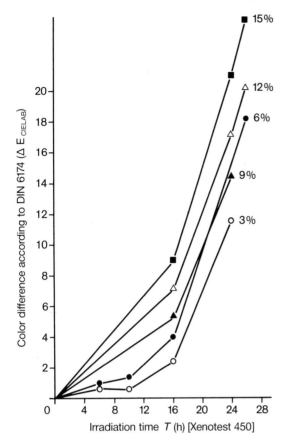

Fig. 44: Influence of the pigment volume concentration on the light-fastness at constant pigment surface concentration of the resulting letterpress proof prints (different thickness of layers). Pigment concentration in the printing ink: 3, 6, 9, 12, 15% Pigment Yellow 12.

explained mechanistically. The assumption is that the vehicle, which absorbs radiation in the longwave UV range and in the shortwave visible region, protects the pigment particles from incident flux. This protection improves with the thickness of the layer. Reducing the pigment volume concentration may be visualized as protecting the average pigment particle from incoming radiation by a thicker absorptive vehicle layer.

Effect of Additives

Since lightfastness and weatherfastness are properties of entire pigment-vehicle systems, additional components must be expected to have a more or less pronounced effect on the sensitivity of a material. In referring to test results, it is therefore necessary to mention the exact composition of the sample. In some systems, problems arise with peroxides, which may rapidly destroy organic pigments. UV absorbers, on the other hand, may improve the lightfastness of a system.

The chemical process underlying the degradation and/or destruction of organic pigments through light or atmospheric conditions is difficult to elucidate, since they comprise a large range of different chemical compounds, each of which must be expected to react on an individual basis. Atmospheric contaminants such as peroxides, which appear as products of radiation, provide a certain amount of common ground in that they frequently initiate the degradation process. Since each pigment reacts according to its own specific mechanism, it might be useful to study the photochemical degradation mechanism of pigments, although to our knowledge organic pigments have not yet been examined individually.

Obviously, there are not only external factors but also the components of a pigmented system that largely define weatherfastness and lightfastness. The chemical and micromorphological degradation mechanisms in systems containing titanium dioxide, for instance, have been investigated extensively [106]. The results, especially those concerning micromorphological degradation of layers, may also be applied to organic pigment systems.

The electron photomicrographs in Figs. 45–47 show the degradation of a number of pigment-vehicle systems in response to irradiation and weathering. These examples are borderline cases, in which either the pigment degrades considerably faster than its medium, or the vehicle is the first to decompose.

Rapid pigment degradation relative to the vehicle is shown in Figs. 45 and 46. These present ultrathin sections produced with an ultramicrotome. The sections were cut out of a gravure printed layer made from Pigment Yellow 12 and a vehicle based on a colophony-modified phenolic resin before and after irradiation. The decomposition of the pigment crystals in the irradiated sections is already far advanced: a large number of voids have replaced the pigment crystals. The degradation products apparently migrate into the vehicle. The chlorine content of these

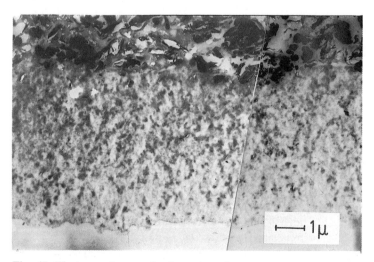

Fig. 45: Electron micrograph of an ultramicrotome-cut thin section of a nonirradiated gravure printed layer. Pigment concentration in the printing ink: 4% Pigment Yellow 12.

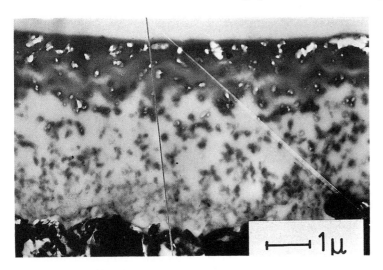

Fig. 46: Electron micrograph of an ultramicrotome-cut thin section of an irradiated gravure printed layer. Pigment concentration in the printing ink: 4% Pigment Yellow 12.

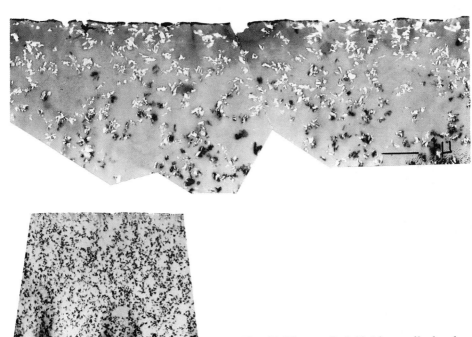

Fig. 47: Pigment Red 48:4 in an alkyd-melamine resin baking enamel. Electron micrographs of ultramicrotome-cut thin sections of an irradiated coating (above: detail).

compounds makes them easy to recognize by the concentric black rings that surround each pigment particle, as shown in electron photomicrographs. A closer look shows that pigment destruction, visible through these black spots, is most advanced on the exposed side of the ink film. The sections that are closest to the surface are almost entirely black, indicating advanced pigment destruction. Moreover, the vehicle has started to decompose.

Fig. 47 shows the surface of a corresponding P.R. 48:4 coating (Sec. 2.7.2.4) made of an alkyd-melamine system. The samples were examined after a 5000 hour exposure in an accelerated exposure unit (Xenotest 1200).

If, on the other hand, the vehicle is less stable to light and weather than the pigment, then lightfastness and weatherfastness of the entire system are largely a function of the vehicle sensitivity. Fig. 48 shows electron photomicrographs of ultrathin sections of coatings containing pigments embedded in an alkyd/melamine system, taken before and after exposure; the sections were cut with an ultramicrotome. Pigment crystals of unsubstituted γ-quinacridone are beginning to protrude from the surface of the coating. These are the particles that, being more lightfast, have survived the decomposition of the surrounding vehicle; they can be removed by polishing or rubbing.

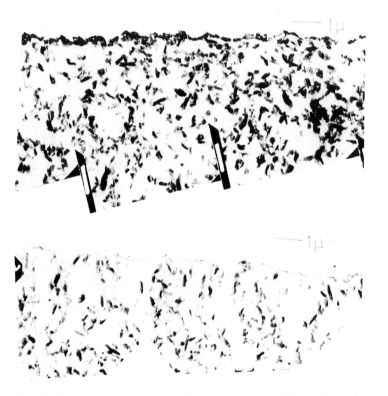

Fig. 48: Electron micrographs of ultramicrotome-cut thin sections of a nonirradiated and an irradiated layer (alkyd-melamine resin baking enamel) containing Pigment Violet 19, γ-modification. Above: irradiated, below: nonirradiated.

1.6.7   Thermal Stability

The fact that not all pigments are thermally stable considerably limits the choice of products being useful in systems which are processed at high temperatures. Thermal stability is a particular concern in connection with the spin dyeing of polypropylene, polyester, and polyamides (Sec. 1.8.3.8), and the mass coloration of low-density polyethylene and polypropylene (Sec. 1.8.3.2). The maximum temperature to which an organic pigment is subjected during such a process is between 260 and 320 °C, and the pigments are usually exposed for a few seconds to minutes only. Only few organic pigments resist these heating conditions. As a result, there are some shades that are inaccessible to organic pigments with such thermal requirements.

Insufficient thermal stability may affect the technical properties of a pigment, particularly its coloristics and fastness properties.

Thermal stability is always a system-dependent property. It is a function not only of the chemical composition of the medium but also of the processing conditions, degree of dispersion, and pigment concentration. For most practical purposes, for instance, the color change due to pigment decomposition in a highly pigmented system is hardly noticeable and can thus be tolerated.

Color change in a pigment-vehicle system may originate from the following phenomena [107]:

1. Thermal pigment decomposition
Degradation occurs if a pigment-vehicle system is processed above the decomposition temperature of the pigment. Pigments incorporated in a solid or melt sometimes decompose spontaneously by undergoing a reaction that is initiated at a distinct temperature. In these cases, differential thermal analysis may be applied to detect the thermal transitions associated with pigment decomposition (Fig. 49). Such abrupt changes, however, are rare. More commonly, pigment degradation is a subtle transition. Polycyclic pigments in particular undergo a continuous, slow exothermic process over a wide temperature range (Fig. 50). The thermographs, however, do not differentiate between pigment degradation and slow physical alterations of the pigment particles, such as phase transitions (change of crystal modification) or changes in the particle size.

2. Chemical interaction between pigment and application medium
The heat sensitivity of a system is rarely affected by reactions between pigment and vehicle, but two examples should be mentioned. As a powder coating is cured, the organic pigment may start to interact with other components in the system [108]. Pigment Red 168, Dibromoanthanthrone (Sec. 3.6.4.2), incorporated in an epoxy resin powder, for instance, will react with a basic hardener such as dicyandiamide. The reflection curves in Fig. 51 indicate that the system undergoes considerable coloristic changes. However, the system is fast to acidic hardeners, such as trimellitic acid. The coloristic changes are accompanied by drastically reduced lightfastness. A certain dullness of shade, sometimes observed at higher curing tempera-

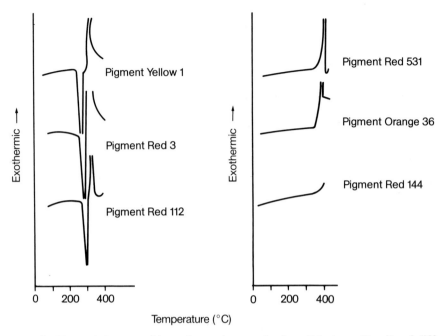

Fig. 49: Thermal decomposition of azo pigments in the solid phase. Results of differential thermoanalysis performed on pigment powders in a nitrogen atmosphere.

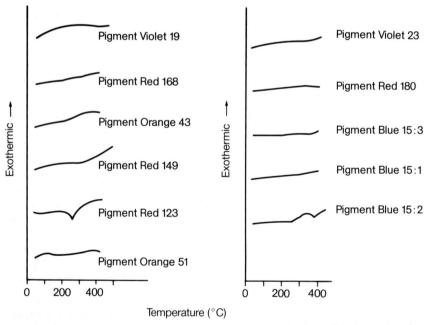

Fig. 50: Thermal decomposition of polycyclic pigments in the solid phase. Results of differential thermoanalysis performed on pigment powders in a nitrogen atmosphere.

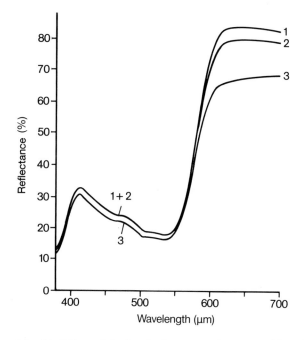

Fig. 51: Effect of the hardening system in an epoxide resin powder coating on the coloristic properties of the paint. Pigment: Dibromoanthanthrone (Pigment Red 168). Cross-linking conditions: 15 minutes at 180 °C, respectively.
Curve 1: Trimellitic anhydride, Curve 2: Imidazoline hardener, Curve 3: Accelerated dicyandiamide hardener.

tures, vanishes rapidly upon exposure to light; and the lightfastness of the resulting product equals that of the pigment component. Even under mild conditions, the bromine atoms in the anthanthrone molecule will be replaced by amino or cyano groups, resulting in grey to greyish blue pigments of poor lightfastness. Light exposure and other tests indicate that the pigment molecules that are closest to the surface of the material may interact with the hardener. It is very likely that under curing conditions, the pigment molecules undergo a substitution reaction, exchanging bromine atoms by cyano groups, which results in dull compounds with reduced lightfastness. The substitution products are thought to be responsible for the observed changes in the coloristics and the fastness properties of a pigment-vehicle system.

Special consideration should be paid to metal complexes such as azomethine pigments (Sec. 2.10). At high temperatures, the yellow copper complex with the chemical constitution **10**, incorporated in PVC, will exchange its chelated copper atoms with the metal atoms present in the application medium. Stabilizers containing barium/cadmium or lead produce yellow shades, while dibutyl tin thioglycolate or other tin compounds produce a brilliant medium red. Color change is slow at low temperatures, but at 160 °C the effect is rapid [107].

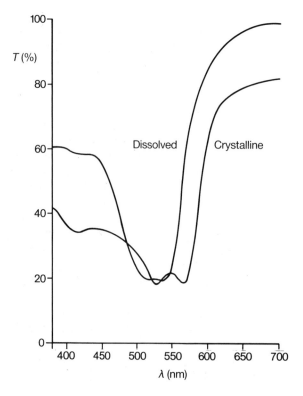

10

3. Dissolution in the application medium

A certain degree of pigment solubility in its medium is frequently responsible not only for poor heat fastness, but also for processing problems, such as bleeding (Sec. 1.6.3.2) or blooming (Sec. 1.6.3.1). A pigment that partially dissolves in its medium must be expected to change color, as shown in Fig. 52 [38]. Transmission curves for Pigment Red 170 (Sec. 2.6.4), dissolved in dimethylformamide or suspended in water, indicate the coloristic behavior of the system. Coloristic changes make it easy to monitor dissolution effects. Thermal stability frequently parallels the fastness of a pigment to migration, as shown in Table 4 [107]. The investigation was carried out on a number of azo pigments, incorporated in an alkyd/melamine resin paint reduced 1:20 with TiO_2. Thermal stability was established by baking the samples at 120 and 200 °C, respectively. The resulting coloristic changes were compared to known data on the migratory tendencies of each pigment in the test system

Fig. 52: Transmission spectrum of Pigment Red 170 dissolved in dimethylformamide, or aqueous suspension, respectively.

Table 4: Relation between the heat stability of azo pigments in an alkyd-melamine resin baked system, their fastness to overpainting in this paint system, and their bleed resistance in plasticized PVC.

Colour Index Generic Name	Temperature effect in CIELAB-units*	Fastness to overpainting on a fastness scale from 1 to 5	Bleed resistance on a fastness scale from 1 to 5
P.R.112	28.1	1	1
P.R.3	12.7	2	1
P.Y.1	10.9	2	1
P.Y.74	4.5	3	1
P.Y.12	4.4	4–5	1–2
P.R.170	3.1	4	2–3
P.Y.83	1.9	5	4–5
P.O.36	1.3	5	5
P.R.175	0.5	5	5
P.R.144	0.2	5	5

* Color difference between a coating baked at 200 °C and one baked at 120 °C. (Reduced 1:20 with TiO_2).

(overpainting fastness) and its fastness to bleeding in plasticized PVC, determined according to DIN 53775. Color changes are referred to in terms of CIELAB units. The results clearly indicate that solubility in these pigments parallels thermal stability.

4. Changes in the physical characteristics and particle size of a pigment

Elevated temperatures tend to modify the crystal structure of a pigment. Characteristic changes in the X-ray diffraction pattern include line sharpening, i.e., decreased half-width; while the relative intensities of the characteristic diffraction lines typically increase, indicating enhanced crystallinity and a higher portion of large-sized crystals. Analysis of the scattering pattern shows that there are now larger coherent areas which appear as units under X-ray (Sec. 1.5.4).

These are the changes that are observed in a pigment powder as it is processed or heated in the application medium. Modifications in the sizes of the crystallites of a polycyclic pigment which is treated in its application medium under varying conditions [107] may be derived from X-ray crystallographic data. Table 5 reflects a clear correlation between crystallinity and thermal stability: increased crystallinity leads to improved heat fastness.

Polymorphous pigments, on the other hand, may undergo phase transitions as the temperature rises (Sec. 1.5.3), indicating the fact that different crystal modifications respond differently to heat.

A number of thermal stability tests are available, some of which have developed into national (DIN) or international industrial standards (ISO). Pigments in ther-

Table 5: Changes in the crystallite size of some polycyclic pigments in coatings or plastics coloration as a result of heat exposure (calculated from the X-ray diffraction spectra).

Pigment	Temperature exposure	Size of crystallites
P.R.123	baked 5% in full shade in alkyd-melamin resin paint: 30 min, 120 °C 30 min, 240 °C	225×10^{-10} m 340×10^{-10} m
P.R.122	baked 5% in full shade in polyester-modified silicone resin system: 60 min, 200 °C 60 min, 300 °C	245×10^{-10} m 295×10^{-10} m
P.R.149	colored in 1/3 standard depth of shade in low pressure polyethylene: 5 min, 200 °C 5 min, 300 °C	270×10^{-10} m 380×10^{-10} m

moplastic systems, for instance, are studied under heat extrusion conditions [109]. The colorant to be tested, possibly together with titanium dioxide, is dispersed in the thermoplastic, using a mixer and a granulating extruder (Sec. 1.8.3). The pigmented test pellets are then fed into a screw extruder which ejects a standardized test specimen with defined dimensions [110]. Starting at the lowest possible temperature level, the extrusion temperature is increased by intervals of 10 or 20 °C between samples.

To assess the color change of the test films, each sample is compared colorimetrically with the standard, which is processed at the lowest temperature. The thermal stability of the colorant in the test medium is defined by the interpolated temperature value at which the color difference between sample and standard equals $\Delta E^*_{ab} = 3$. Determinations are carried out at various Standard Depths of Shade, common values are 1/3 and 1/25.

Pigments incorporated in plasticized or unplasticized PVC are assessed by different methods [111, 112]. The preparation of test dispersions is commonly performed on a standardized two-roll mill at temperatures between 190 and 195 °C or, respectively, between 180 and 185 °C. Samples are taken at intervals of 10, 20, and 30 minutes, molded into films, and evaluated for color change.

1.6.8   Flow Properties of Pigmented Systems

1.6.8.1   Rheological Properties

Newton's law states that for a liquid under shear, the shear stress τ is proportional to the shear rate. In this sense, most of the unpigmented vehicles used in the paint and printing ink industries are considered ideal or **Newtonian liquids**. The ratio of the shear stress τ to the shear rate D is thus a constant η, dependent only on temperature and pressure. This is not true for specialized gel varnishes and thixotropic systems, which are designed to have special rheological properties.

Ideal and non-Newtonian behavior is demonstrated by the flow curves in Fig. 53, in which the shear rate is plotted versus the shear stress. The first example describes a Newtonian fluid, represented by a linear correlation whose slope equals the viscosity of the material. When a pigment is dispersed in a vehicle, the behavior of the resulting liquid deviates more or less from Newtonian flow. The previously linear shear stress-shear rate function is transformed into a curve whose slope decreases with increasing shear rate and shear stress. This phenomenon is known as **structural viscosity** or pseudoplastic behavior. Shearing is thought to break up the pattern of pigment-vehicle interaction, deforming or rearranging the units and lining up the previously disordered vehicle molecules, thus reducing their effective cross-section and affording reduced viscosities. This reversible process is not time-dependent, i.e, there is a definitive correlation between shear stress and viscosity, no matter how long the stress is applied. A number of such systems have been studied empirically and their behavior has been described by flow curves. For practical purposes, the Casson flow model is a useful approximation to describe the structural viscosity of a pigment-vehicle system. It is expressed by the following equation:

$$(\sqrt{\tau} - \sqrt{\tau_o})^2 = \eta_\infty \cdot D$$

A Casson fluid is rheologically identified by two parameters: yield value and plastic viscosity. The plastic viscosity relates to the asymmetry of the flow particles and the yield value is connected with the forces of attraction between particles. The plastic or ultimate viscosity η_∞ in pascal seconds (Pa·s) may be determined from the slope of a shear stress versus shear rate plot, as the shear rate tends to infinity. The yield value τ_o (in N/cm^2), on the other hand, is determined by extrapolation to $D=0$.

The ultimate viscosity η_∞ of a pigmented medium is practically independent of the particle size, but increases appreciably with rising pigment volume concentration and higher vehicle viscosity [113].

Fig. 54 quantifies the relation between the yield value τ_∞ and a number of system-inherent parameters. The curves show that the yield value τ_∞ decreases with the cube of the particle size and increases drastically with the pigment volume concentration. Since the yield value is an indication of pigment-vehicle interaction, it is also proportional to the degree of pigment dispersion [113]. Interparticle attractions have received considerable theoretical and experimental treatment, and a

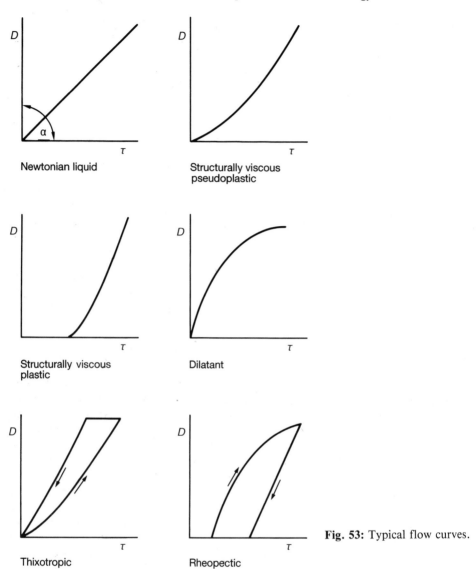

Newtonian liquid

Structurally viscous
pseudoplastic

Structurally viscous
plastic

Dilatant

Thixotropic

Rheopectic

Fig. 53: Typical flow curves.

large number of original publications and reviews are available (such as [114–116]).

Thixotropy is one of the reversible time-dependent effects that constitute non-ideal behavior (Fig. 54).

An increasing pigment concentration is associated with more or less pronounced **thixotropy** (Figs. 53 and 54). Thixotropic systems have a gel structure. Shear, however, breaks up the previously stable pigment-vehicle structures. The viscosity of a thixotropic fluid therefore decreases with time when shear is applied,

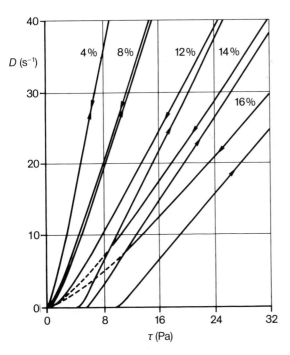

Fig. 54: Influence of the pigment concentration on the rheological behavior. P.Y.73 in air-drying alkyd resin paint.

even at constant shear rate, and finally approaches a minimum. The graphs in Fig. 54, read from left to right, exemplify the occurrence of thixotropy with increasing pigment concentration in a paint system. When the shearing action stops, the original gel structure reappears after a certain period of time. Thixotropic decay also depends on the shear rate. At a given shear rate D (in s^{-1}), an equilibrium is reached which does not normally coincide with the sol value. As mentioned above, thixotropy may thus be described as a time-dependent and reversible change in the consistency of a material, observed during and after the effect of shear. A number of processes are affected by the rate at which the matrix within a system is built up and the rate at which it is broken down with shear. Examples include the time which is needed for a dispersion to flow out of a ball mill or other dispersion unit and separate from the milling medium at different points in time after completion of the dispersion process. Thixotropy plays an important role at various steps throughout the printing process, such as color separation, and provides the basis for thixotropic paints and inks. Fig. 55 illustrates the time that is necessary for the thixotropic structures of two systems to rebuild on standing (Sec. 1.6.5.8).

Dilatancy occurs less frequently in pigmented systems. In flush pastes or pigment concentrates, which are formulated at high pigment levels, shearing may produce an increase in viscosity η. As the pigment concentration reaches the vicinity of the critical pigment volume concentration, and even more so if this point is exceeded, increasing the shear stress τ or the shear rate D will thicken the fluid (Fig. 53). This may create a problem with aqueous pigment preparations, which are sometimes automatically metered into the application system under considerable

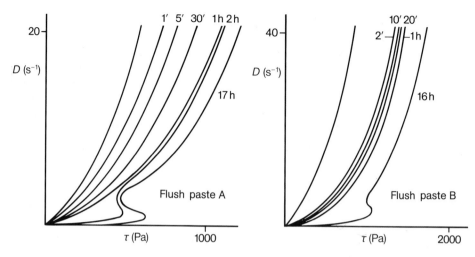

Fig. 55: Rate of thixotropy buildup. Paste A shows a continuous buildup over time. In paste B, buildup of the structure seems to be already completed after a few minutes; however considerable change is still observed 16 hours later.

shear. These pastes should therefore be tested for dilatancy at high shear rates before they are processed.

Rheopexy, a reversible time-dependent effect like thixotropy, is a rare phenomenon in pigmented systems. Rheopectic fluids increase in viscosity η with time when sheared at a constant shear rate D or a constant shear stress τ until they approach a viscosity maximum (Fig. 53).

1.6.8.2   Viscoelastic Properties

Most pigmented systems are considered viscoelastic. At low shear rates and slow deformation, these systems are largely viscous. As the rate of deformation or shear rate increases, however, the viscous response cannot keep up, and the elasticity of the material increases. There is a certain amount of emphasis on viscoelastic behavior in connection with pigment dispersion as well as ink transportation and transformation processes in high-speed printing machines (see below). Under periodic strain, a viscoelastic material will behave as an elastic solid if the time scale of the experiment approaches the time required for the system to respond, i.e., the relaxation time. Elastic response can be visualized as a failure of the material to flow quickly enough to keep up with extremely short and fast stress/strain periods.

The viscoelastic behavior of a printing ink or other material is largely a function of the polarity of pigment and medium [117]. However, it is difficult to quantify the complex rheology of a viscoelastic system, because structural viscosity, thixotropy, and dilatancy may all occur at the same time.

1.6.8.3   Influence on the Flow Properties

Rheology control is necessary in connection with the manufacture, handling, and application of pigmented systems; but the requirements of today's industry are so complex that only a limited number of aspects can be discussed within the scope of this book. Flow is defined by a variety of parameters. Pigment-determined factors include concentration (Fig. 54), specific surface area (Sec. 1.7.8), particle shape, and surface structure. The system also plays a major role. Flow is a function of the choice of components and composition of the entire system, interactions between the various components and between components and pigment surface, dispersion conditions, and the conditions that govern various stages during the application. Among these aspects, the emphasis is on the dispersion conditions of the pigment (Sec. 1.6.5, 1.7.5). The operation and effectivity of the dispersion equipment [71] as well as operational parameters have a considerable impact on the various facets of dispersion, especially concerning the breakup of the pigment agglomerates and wetting by the application medium. The degree of dispersion determines the surface area which is accessible to wetting by the various components in the system. Interactions at the interfaces between pigment and medium or between pigment and solvent also appreciably influence the response of a material to mechanical force.

1.6.8.4   Correlation between Flow Behavior and Rheological Parameters

Although the rheology largely determines the application properties of a pigment in its vehicle system, the exact correlation between rheological and performance parameters remains elusive. Both quality control and research face problems because of the difficulty of predicting the behavior of a pigmented system or systematically developing materials with certain defined flow properties. There are no standardized techniques to measure the rheological performance of a material under the conditions of practical use. To characterize a fluid, specialized instruments are used to simulate specific steps in the processing and application of a pigment-vehicle system. The printing inks industry, for instance, typically resorts to tackmeters [118]. The tack [119] or, its reverse, the shortness, are primary applicational factors in determining the printability and therefore the commercial fate of an offset printing ink. Defined in 1957 [120], the equation

$$\text{Shortness} = \frac{\tau_o}{\eta_\infty}$$

relates to the stepwise change of one variable only. A test series in which either the pigment concentration or the composition of the medium is changed systematically correlates closely with the results of a test that is carried out under actual application conditions (Table 6) [121]. However, the equation is medium-specific, and errors may occur if it is used to extrapolate data outside the range of experimental

Table 6: Relation between rheological parameters and properties related to the printing of offset inks.

Ink no.	$\tau_0 \cdot 10^5$ (Pa)	η_∞ (Pa·s)	Shortness (according to Zettlemoyer)	Printing test
1	9.2	13-5	680	too long
2	7.2	10.0	720	very long
3	6.2	7.4	835	long
4	4.8	5.3	905	long
5	4.5	3.3	1360	short
6	4.0	2.8	1430	too short

determination. Tackmeters simulate the conditions within the printing presses by rotating a test offset blanket cylinder versus an inked plate cylinder at different speeds. The generated force is measured by the deflection of the cylinder as it is pulled by the ink. The test results usually correlate well with the demands of practical offset printing.

The complicated rheology of a pigmented system under processing conditions is reflected in the flow behavior of publication gravure printing inks in high-speed printing machines, especially in the gap between doctor blade and printing cylinder [122]. Studies carried out on modern high pressure capillary viscometers [123] indicate that the flow curves may show unexpected deviations at high shear rates. This is especially true if the test parameters ($D = 10^6$ s^{-1} and flow times t of up to 10^{-4} s) closely approach actual printing conditions ($D = 10^7$ s^{-1} and $t = 10^{-6}$ s). Moreover, flow behavior as a function of the shear rate D may differ between inks, as shown by a study comparing the flow curves of two publication gravure printing inks at $D = 10$ s^{-1} and at $D = 10^4$ s^{-1}. The ink which displays higher viscosity at a low shear rate loses viscosity as the shear increases to 10^4 s^{-1}, and the relation is reversed again at $D = 10^5$ s^{-1}. The viscosity profiles thus cross each other and intersect at medium shear rates, at which both inks exhibit identical flow behavior. Such unpredictable shear effects make it necessary to collect a complete set of data points to draw up system-specific flow curves for publication gravure inks. There is little point in using a low shear rate viscometer to determine the behavior of a gravure printing ink which is to be processed at a high shear rate, since the flow curve cannot be extrapolated to higher values. Traditional viscometers or flow cups are inadequate even for qualitative purposes.

Publication gravure printing inks apparently show elastic behavior; at $D = 10^5$ s^{-1}, the modulus of shearing G is 10^3 Pa. This indicates that at high shear rates D, between 10^5 and 10^6 s^{-1}, gravure printing inks develop normal stress effects and thus show a hydrodynamic lubricating effect. These deductions are relevant to an understanding of the mechanical effects between doctor blade and cylinder. At $D = 10$ s^{-1} and an experimental time $t = \infty$, highly thixotropic gravure printing inks have an apparent viscosity η of 0.1 Pa·s; while at $D = 10^5$ s^{-1} and $t = 10^{-3}$ s, the viscosity assumes values between 0.05 and 0.06 Pa·s.

1.6.8.5   Rheological Measurements

A large number of different commercial viscometers, which provide a variety of geometries and a range of viscosity and shear rates, can be used to characterize a fluid. Specialized techniques have been developed to suit specific purposes [124, 125, 126].

Standards have been developed regulating the use of some of these commercially available viscometers. Such standards, for instance, cover the use of rotational viscometers with coaxial cylinders [127] if they are employed to measure flow curves and viscosities at shear rates between 5 s^{-1} and 1000 s^{-1}. Systematic errors distorting viscosity determinations on Newtonian liquids using such rotational viscometers and corrections that have to be applied to the resulting measurements are discussed in the literature [128]. Other standards cover the testing of paints, which have to be measured at very high shear rates in the range between 5000 and 20000 s^{-1} to reflect the conditions of actual paint application. These are tested according to industrial standards using rotational viscometers or standard cone-and-plate instruments [129]. Tests have been developed for letterpress and offset printing inks and similar systems which employ specialized falling rod viscometers [130]. These devices do not afford definitive flow curves for thixotropic systems but instead trace a line somewhere between the gel and the sol curve. There is extensive literature on techniques describing how to determine the rheology of special systems.

The determination of flow properties taken during the pigmentation and processing of thermoplastics are described in the literature on polymers [131, 132].

Since the rheology of many systems depends largely on the temperature, accurate and reproducible measurements require very careful temperature control. A 1 °C temperature drop, for instance, increases the apparent viscosity η of an offset printing ink by approximately 15%. To demonstrate the correlation between thixotropy and temperature, Figures 56 and 57 show the flow curves at different temperatures for two offset printing inks [133]. Both materials clearly lose thixotropy—indicated by the area under the thixotropic loop—as the temperature increases. This effect is much more pronounced in the first case (Fig. 56), while the second ink exhibits a very slow decrease thixotropic behavior (Fig. 57).

Thixotropy is frequently calculated by determining the area between the flow curves for increasing and decreasing shear rates. Standard procedures to establish this loop involve subjecting the tempered printing ink in a cone-plate viscometer to maximum tensile shear rates. After allowing the sample to rebuild its structure on standing at zero shear rate for five minutes, the up curve is determined by increasing the shear rate from 0 to maximum (300 s^{-1}) within 5 s and shearing the sample at this rate for 30 s. The down curve results from reducing the shear rate back to 0 within another 5 s. It is important to remember that the exact shape of the flow curve for a given material depends on the history of the sample and the experimen-

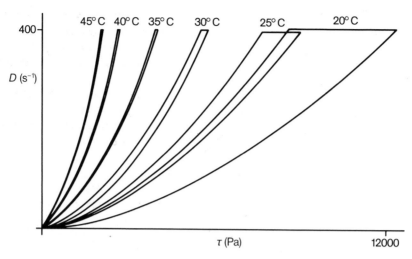

Fig. 56: Variation of thixotropy with temperature. Example of a printing ink whose thixotropy decreases rapidly with the temperature.

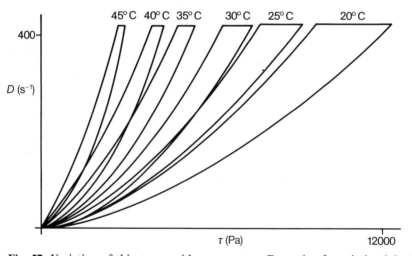

Fig. 57: Variation of thixotropy with temperature. Example of a printing ink whose thixotropy decreases slowly with the temperature.

tal procedure, e.g., the area under the curve characterizes the specific thixotropic properties in each case.

The flow behavior of a system may change significantly during the first few hours after the shearing action stops. This is particularly true for systems with a high pigment concentration. The plot in Fig. 58, for instance, represents the viscosity-shear-time profile of a toluene-based gravure printing ink. The pigment concen-

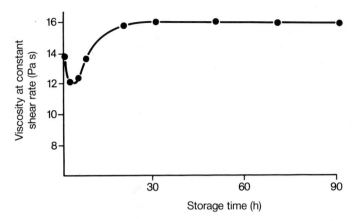

Fig. 58: Variation of the viscosity of publication gavure printing inks containing 15% Pigment Yellow 12 with the storage time after dispersion.

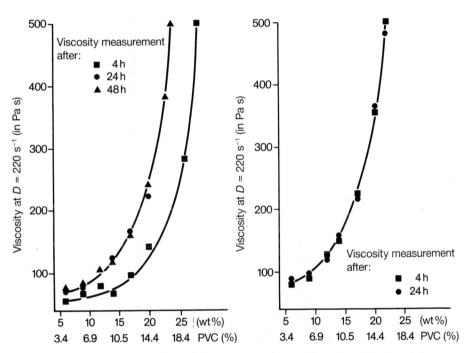

Fig. 59: Influence of the dispersion temperature (left: 15 °C, right: 50 °C) and the storage time on the viscosity of sheet offset printing inks containing Pigment Yellow 13. Measurement temperature 23 °C.

tration is 15% (Pigment Yellow 12). The viscosity drops at first and then increases as soon as the shearing action stops. This behavior is attributed to superimposed reagglomeration effects (Sec. 1.7.5). These effects include reduction of the wetted surface area of the pigment particles and delayed wetting effects, which increase the viscosity of the ink. Easily dispersible pigments with good wettability exhibit no such time-dependent flow behavior.

No time-dependent change of the rheological values is observed if the printing ink is produced at higher dispersion temperatures. Studies have been carried out on offset printing inks containing Pigment Yellow 13 which were dispersed at temperatures of 15 and 50 °C, respectively. Measurements were taken 4 and 24 hours after dispersion (Fig. 59). The effect of delayed wetting is negligible in the ink produced at 50 °C, because the surface of the pigment particles is wetted thoroughly at this high temperature.

References for Section 1.6

[1] D.B. Judd and G. Wyczecki: Color in Business, Science and Industry, J. Wiley & Sons, New York, 1975;
M. Richter: Einführung in die Farbmetrik, Walter de Gruyter, Berlin, 1976;
R.S. Hunter: The Measurement of Appearance, J. Wiley & Sons, New York, 1975.

[2] DIN 5033: Farbmessung. ISO 7724-1-1984: CIE Standards colorimetric observers; Paints and varnishes – Colorimetry. Part 1: Principles; ISO 7724-2-1984, Part 2: Colour measurement.

[3] DIN 6174: Farbmetrische Bestimmung von Farbabständen bei Körperfarben nach der CIELAB-Formel. ISO 7724-3-1984: Paints and varnishes – Colorimetry, Part 3: Calculation of colour differences. Ed. 1.

[4] P. Raabe and O. Koch, Melliand Textilber. 38 (1957) 173.

[5] G. Geißler, H. Schmelzer and W. Müller-Kaul, Farbe + Lack 84 (1978) 139.

[6] DIN 53 235: Prüfungen an standardfarbtiefen Proben.

[7] D.R. Duncan, Proc. Phys. Soc. London 52 (1940) 380.

[8] K. Hoffmann, Farbe + Lack 76 (1970) 665.

[9] DIN 6164: DIN-Farbenkarte.

[10] DIN 54 001: Herstellung und Handhabung des Graumaßstabes zur Bewertung der Änderung der Farbe. ISO 105-A02-1987 (TC 38).

[11] DIN 55 987: Bestimmung eines Deckvermögenswertes pigmentierter Medien; Farbmetrisches Verfahren. ISO 6504-5-1987: (TC 35) Determination of hiding power – Part 5: Colour difference method for dark coloured paints.

[12] F. Gläser, XIII. Congres FATIPEC, Juan les Pins, 1976, Congress book 239–243.

[13] DIN 16 524 T1: Widerstandsfähigkeit gegen verschiedene physikalische und chemische Einflüsse.

[14] DIN 53 197: Bestimmung des Gehaltes an wasserlöslichen Anteilen. ISO 787-8-1979: General methods of test for pigments and extenders – Part 8: Determination of matter soluble in water – cold extraction method.

[15] DIN 53202: Bestimmung der Aciditätszahl oder Alkalitätszahl. ISO 787-4-1981: General methods of test for pigments and extenders – Part 4: Determination of acidity or alkalinity of the aqueous extract.

[16] DIN 53 200: Bestimmung des pH-Wertes von wäßrigen Pigment-Suspensionen.

[17] DIN 53 208: Bestimmung der elektrischen Leitfähigkeit und des spezifischen Widerstandes von wäßrigen Pigment-Extrakten. ISO 787-14-1973: General methods of testing for pigments – Part 14: Determination of resistivity of aqueous extract.

[18] DIN 53 207 T1, T2: Bestimmung des Gehaltes an wasserlöslichen Sulfaten, Chloriden und Nitraten. ISO 787-13-1973: General methods of testing for pigments – Part 13: Determination of water-soluble sulphates, chlorides and nitrates.

[19] DIN 16 524: Prüfung von Drucken und Druckfarben des graphischen Gewerbes.

[20] Empfehlung des Bundesgesundheitsamtes (BGA) IX B II vom 1.7.72. (Recommendation of the German Federal Health Office)

[21] DIN 54 002: Herstellung und Handhabung des Graumaßstabes zur Bewertung des Anblutens. ISO 105-A03-1987: Textiles – Tests for colour fastness – Part A03: Grey scale for assessing staining.

[22] DIN 54 005, DIN 54 006: Bestimmung der Wasserechtheit von Färbungen und Drucken (leichte bzw. schwere Beanspruchung). ISO 105-E01-1989: Textiles – Tests for colour fastness – Part E01: Colour fastness to water.

[23] DIN 54 007: Bestimmung der Meerwasserechtheit von Färbungen und Drucken. ISO 105-E02-1989: Textiles for colour fastness – Part E02: Colour fastness to sea water.

[24] DIN 54 010 – 54 014: Bestimmung der Waschechtheit von Färbungen und Drucken. ISO 105-C03-1989: Textiles for colour fastness: Part C03: Colour fastness to washing: Test 3.
ISO 105-C04-1989: Part C04: Colour fastness to washing: Test 4.
ISO 105-C05-1989: Part C05: Colour fastness to washing: Test 5.
ISO 105-C02-1989: Part C02: Colour fastness to washing: Test 2.
ISO 105-C01-1989: Part C01: Colour fastness to washing: Test 1.

[25] DIN 54 015: Bestimmung der Peroxid-Waschechtheit von Färbungen und Drucken.

[26] DIN 54 016: Bestimmung der Hypochlorit-Waschechtheit von Färbungen und Drucken.

[27] DIN 54 019: Bestimmung der Farbechtheit von Färbungen und Drucken gegenüber gechlortem Wasser. ISO 105-E03-1987: Textiles for colour fastness – Part E03: Colour fastness to chlorinated water (swimming-bath water).

[28] DIN 54 020: Bestimmung der Schweißechtheit von Färbungen und Drucken. ISO 105-E04-1989: Textiles for colour fastness – Part E04: Colour fastness to perspiration.

[29] DIN 54 021: Bestimmung der Reibechtheiten von Färbungen und Drucken. ISO 105-X12-1987: Textiles – Tests for colour fastness – Part X12: Colour fastness to rubbing.

[30] DIN 54 022: Bestimmung der Bügelechtheit von Färbungen und Drucken. ISO 105-X11-1987: Textiles – Tests for colour fastness – Part X11: Colour fastness to hot pressing.

[31] DIN 54 024: Bestimmung der Trockenreinigungsechtheit von Färbungen und Drucken. ISO 105-D01-1987: Textiles – Tests for colour fastness – Part D01: Colour fastness to dry cleaning.

[32] DIN 54 023: Bestimmung der Lösungsmittelechtheit von Färbungen und Drucken. ISO 105-X05-1987: Textiles – Tests for colour fastness – Part X05: Colour fastness to organic solvents.

[33] DIN 54 028: Bestimmung der Säureechtheit von Färbungen und Drucken. ISO 105-E05-1989: Textiles – Tests for colour fastness – Part E05: Colour fastness to spotting: Acid.

[34] DIN 54 030: Bestimmung der Alkaliechtheit von Färbungen und Drucken. ISO 105-

E06-1989: Textiles – Tests for colour fastness – Part E06: Colour fastness to spotting: Alkali.

[35] DIN 54 031: Bestimmung der Sodakochechtheit von Färbungen und Drucken. ISO 105-X06-1987: Textiles – Tests for colour fastness – Part X06: Colour fastness to soda boiling.

[36] DIN 54 033 bis 54 037: Bestimmung der Peroxid-, Hypochlorit-, Chlorit-Bleichechtheit (leichte bzw. schwere Beanspruchung). ISO 105-N-1978: Textiles – Tests for colour fastness – Part N: Colour fastness to bleaching agencies.

[37] DIN 54 060: Bestimmung der Trockenhitzeplissier- und Trockenhitzefixierechtheit von Färbungen und Drucken. ISO 105-P-1978: Textiles – Tests for colour fastness – Part P: Colour fastness to heat treatments.

[38] F. Gläser, XII. Congres FATIPEC, Garmisch, 1974, Congress book 363–370.

[39] W. McDowell, J. Soc. Dyers Colour. 88 (1972) 212–216.

[40] DIN 53 775: Prüfung von Pigmenten in PVC weich.

[41] Kumis and Roteman, J. Polym. Sci. 55 (1961) 699.

[42] W. Herbst and K. Merkle, Farbe + Lack 75 (1969) 1137–1157.

[43] DIN 55 943: Farbmittel; Begriffe. ISO 4618-1-1984 and ISO 4618-2-1984: Paints and varnishes – Vocabulary – Part 1: General terms; Part 2: Terminology relating to initial defects and to undesirable changes in films during ageing.

[44] DIN 55 945: Anstrichstoffe und ähnliche Beschichtungsstoffe; Begriffe. ISO 4618-4-1984: Paints and varnishes – Vocabulary – Part 4: Further general terms and terminology relating to changes in films and raw materials.

[45] DIN 53 405: Bestimmung der Wanderung von Weichmachern. ISO 177-1988: Plastics-Determination of migration of plasticizers.

[46] DIN 53 991: Bestimmung des Ausblutens.

[47] DIN 53 415: Bestimmung des Ausblutens von Farbmitteln. ISO 183-1976: Plastics – Qualitative evaluation of the bleeding of colorants.

[48] DIN 53 236, in connection with DIN 6174.

[49] J. Richter, Plastverarbeiter 18 (1968) 933–935.

[50] F. Memmel, Plastverarbeiter 21 (1971) 325–328.

[51] H.G. Völz, G. Kämpf and H.G. Fitzky, X. Congr. FATIPEC, Montreux, 1970, Congress book 107–112.

[52] G. Kämpf, W. Papenroth and R. Holm, XI. Congr. FATIPEC, Florenz, 1972, Congress book 569–574.

[53] DIN 53 159: Bestimmung des Kreidungsgrades von Anstrichen und ähnlichen Beschichtungen nach Kempf.

[54] DIN 53 223: Bestimmung des Kreidungsgrades nach der Klebebandmethode. ISO 4628-6-1984: Paints and varnishes – Evaluation of degradation of paint coatings – Designation of intensity, quantity and size of common types of defect – Part 6: Designation of degree of chalking.

[55] U. Johnsen and K.-H. Moos, Angew. Makromol. Chem. 74 (1978) 1–15.

[56] W. Woebcken and E. Seus, Kunststoffe 57 (1967) 637–644, 719–723.

[57] H. Rumpf, Chem.-Ing.Techn. 37 (1965) 187.

[58] H. Pahlke, Farbe + Lack 72 (1966) 623–630, 747–758.

[59] L. Gall and U. Kaluza, DEFAZET Dtsch. Farben Z. 29 (1975) 102–115.

[60] A.W. Neumann and P.J. Sell, Z. Phys. Chem. 187 (1969) 227.

[61] G. Hellwig and A.W. Neumann, Farbe + Lack 73 (1967) 823.

[62] J. Schröder and B. Honigmann, Farbe + Lack 87 (1981) 176–180.
J. Schröder, Farbe + Lack 91 (1985) 11–17.

[63] P.J. Sell and D. Renzow, Progr. Org. Coat. 3 (1975) 323–348; Farbe + Lack 83 (1977) 265–269.

[64] K.L. Wolf: Physik u. Chemie der Grenzflächen, Bd. I, Springer Verlag, Berlin, 1957.

[65] T.C. Patton: Paint Flow and Pigment Dispersion, Interscience Publishers, New York 1964.

[66] E.W. Washburn, Phys. Rev., 17 (1921) 273.

[67] G.D. Parfitt, J. Oil Colour Chem. Assoc. 50 (1967) 826.

[68] W. Herbst, DEFAZET Dtsch. Farben Z. 26 (1972) 519–532, 571–576.

[69] W. Herbst, Farbe + Lack 76 (1970) 1190–1208.

[70] W. Herbst, Farbe + Lack 77 (1971) 1072–1080, 1197–1203.

[71] W. Herbst, Progr. Org. Coat. I (1972/73) 267–331.

[72] P. Quednau, Polym. Paint Colour 18 (1982) 515–520.

[73] U. Kaluza, DEFAZET Dtsch. Farben Z. 33 (1979) 355–359, 399–411.

[74] B.V. Deryaguin and L.D. Landau, Acta Physicochim. URSS 14 (1941) 633.

[75] E.J.W. Verwey and J.Th.G. Overbeck: "Theory of the Stability of Lyophobic Colloids", Elsevier, Amsterdam, 1948.

[76] D.H. Napper, J. Colloid Interface Sci. 58 (1977) 390.

[77] V.T. Crowl, J. Oil Colour Chem. Assoc. 50 (1967) 1023.

[78] E.J. Clayfield and E.C. Lumb, J. Colloid and Interface Sci. 22 (1966) 269.

[79] A. Topham, Progr. Org. Coat. 5 (1977) 237.

[80] W.K. Asbeck and M. van Loo, Ind. Eng. Chem. 41 (1949) 1470.

[81] W.K. Asbeck, G.A. Scherer and M. van Loo, Ind. Eng. Chem. 47 (1955) 1472–1476.

[82] Literature references see O. Kolár, B. Svoboda, F. Hájek and J. Korinský, Farbe + Lack 75 (1969) 1039.

[83] K. Merkle and W. Herbst, Farbe + Lack 77 (1971) 214–223.

[84] W. Herbst and K. Merkle, Farbe + Lack 78 (1978) 25–34.

[85] DIN 53 203: Beurteilung der Körnigkeit mit dem Grindometer. ISO 1524-1983: Paints and varnishes – Determination of fineness of grind.

[86] DIN 53 238 T 20: Bestimmung der Änderung der Körnigkeit (Grindometer). ISO 8721-2-1986: Dispersion characteristics of pigments and extenders – Methods of assessment – Part 2: Assessment of dispersion characteristics from the change in fineness of grind.

[87] DIN 53 238 T 10-12: Prüfung des Dispergierverhaltens. ISO 8780-1-1986: Dispersion characteristics of pigments and extenders – Methods of dispersion – Part 1: Dispersion using an oscillatory shaking machine. Part 4: Dispersion using an automatic miller. Part 5: Dispersion using a triple roll mill.

[88] DIN 53 238 T 30-33: Prüfmedien.

[89] A. Klaeren and H.G. Völz, Farbe + Lack 81 (1975) 709.

[90] D.v. Pigenot, VII. Congr. FATIPEC 1964, Congress book 249.

[91] U. Zorll, Farbe + Lack 80 (1974) 17.

[92] O.J. Schmitz, R. Kroker and P. Pluhar, Farbe + Lack 79 (1973) 733.

[93] DIN 53 238 T 22: Bestimmung der Farbstärkeentwicklung und der Dispergierhärte.

[94] DIN 53 238 T 23: Bestimmung der Glanzentwicklung. ISO 8781-1-1987: Dispersion characteristics of pigments and extenders – Methods of assessment – Part 3: Assessment of dispersion characteristics from the change in gloss.

[95] DIN 67 530: Reflektometer als Hilfsmittel zur Glanzbeurteilung an ebenen Aufstrich- und Kunststoff-Oberflächen.

[96]   A.S. Gomm, G. Hull and J.L. Moilliet, J. Oil Colour Chem. Assoc. 51 (1968) 143–160.

[97]   J.W. White, Am. Paint Coat. J. 61 (1976) 52–56.

[98]   DIN 54 003: Bestimmung der Lichtechtheit von Drucken und Färbungen mit Tageslicht. ISO 105-B01-1989: Textiles – Tests for colour fastness – Part B01: Colour fastness to light: Daylight.

[99]   DIN 16 525: Prüfen von Drucken und Druckfarben des graphischen Gewerbes; Lichtechtheit.

[100] H. Schmelzer, XVII. Congr. FATIPEC, Lugano, 1984, Congress book Vol. 1, 499–509.

[101] J. Boxhammer, Original Hanau Quarzlampen GmbH, Hanau, VDI-Seminar "Alterung von Kunststoffen", 1977.

[102] DIN 53 387: Kurzprüfung der Lichtechtheit. ISO 4892-1981: Plastics – Methods to exposure to laboratory light sources.

[103] DIN 1249: Fensterglas.

[104] DIN 53 236: Meß- und Auswertbedingungen zur Bestimmung von Farbunterschieden bei Anstrichen, ähnlichen Beschichtungen und Kunststoffen. ISO 7724-2-1984: Paints and varnishes – Colorimetry – Part 2: Colour measurement.

[105] DIN 53 387: Kurzprüfung der Wetterbeständigkeit (Simulation der Freibewitterung durch gefilterte Xenonbogen-Strahlung und Beregnung). ISO 4892-1981: Methods to exposure to laboratory light sources.

[106] G. Kämpf, W. Papenroth and R. Holm, Farbe + Lack 79 (1973) 9–21.

[107] E. Baier, Farbe + Lack 83 (1977) 599–610.

[108] W. Herbst and O. Hafner, Farbe + Lack 82 (1976) 394–411.

[109] DIN 53 772: Bestimmung der Hitzebeständigkeit durch Spritzgießen.

[110] DIN 24 450: Maschinen zum Verarbeiten von Kunststoffen.

[111] DIN 53 774: Prüfung von Farbmitteln in PVC hart.

[112] DIN 53 775: Prüfung von Pigmenten in PVC weich.

[113] P. Hauser, M. Hermann and B. Honigmann, Farbe + Lack 76 (1970) 545–50 and 77 (1971) 1097–1106.

[114] O. Fuchs, DEFAZET Dtsch. Farben Z. 22 (1968) 548 und 23 (1969) 17, 57, 111.

[115] U. Zorll, Farbe + Lack 86 (1980) 301–307.

[116] U. Kaluza: "Physikalisch-chemische Grundlagen der Pigmentverarbeitung für Lacke und Druckfarben", BASF Aktiengesellschaft 1979.

[117] H. Pahlke, X. Congr. FATIPEC, Montreux, 1970, Congress book 549–554.

[118] G.R. South, Am. Ink Maker 32 (1968) 35–37.

[119] DIN 16 515/1: Farbbegriffe im Graphischen Gewerbe.

[120] A.C. Zettlemoyer, R.F. Scarr and W.D. Schaeffer, Proc. TAGA, 9A (1957) 75, Int. Bull. Print Allied Trades 80 (1958) 88–96.

[121] W. Herbst, Farbe + Lack 75 (1969) 431–436.

[122] J. Schurz and T. Kashmoula, Farbe + Lack 82 (1976) 895–901.

[123] Manufactures: Anton Paar KG, Graz.

[124] DIN 53 019: Messung von Viskositäten und Fließkurven mit Rotationsviskosimetern mit Standardgeometrie. ISO 8961-1987: Plastics – Polymer dispersions – Definition and determination of properties.

[125] T.C. Patton, J. Paint Technol. 38 (1966) 656–666.

[126] Snell-Hilton: Encyclopedia of Industrial Chemical Analysis, Vol. 3, John Wiley and Sons, 1966, 408–463.

[127] DIN 53 018/1: Messung der dynamischen Viskosität newtonscher Flüssigkeiten mit Rotationsviskosimetern; Grundlagen.

[128] DIN 53 018/2: Fehlerquellen und Korrekturen bei Zylinder-Rotationsviskosimetern.
[129] DIN 53 229: Bestimmung der Viskosität bei hoher Schergeschwindigkeit. ISO 2884-1974: Paints and varnishes – Determination of viscosity at a high rate of shear.
[130] DIN 53 222: Bestimmung der Viskosität mit dem Fallstabviskosimeter.
[131] J. Schurz: Viskositätsmessungen an Hochpolymeren, Verlag Berliner Union Kohlhammer, Stuttgart, 1972.
[132] Otto Plajer, Plastverarbeiter 29 (1978) 169–175, 249–252, 311–314, 376–378.
[133] H. Schmelzer, Farbe + Lack 79 (1973) 1066–1071.

1.7 Particle Size Distribution and Application Properties of Pigmented Media

The application properties of a pigment-vehicle system depends largely on the particle size distribution of the pigment. This seemingly straightforward correlation would make it easy to predict the application properties of an entire pigment-vehicle system from pigment powder data if it was not for the fact that the particle size of a pigment in its medium is not normally the same as that in the powder (Sec. 1.5.2). It is therefore difficult to gain more than qualitative information about the final properties, unless the particle size distribution is determined directly on the dispersed pigment in its medium of application.

1.7.1 Tinctorial Strength

Fig. 60 reflects the relation between pigment particle size and the ability of a pigment-vehicle system to absorb visible electromagnetic radiation, i.e., to the tinctorial strength of a pigment in its application medium (Sec. 1.6.1.1). The ability of a given pigment to absorb light (its tinctorial strength) increases with decreasing particle diameter and accordingly increased surface area, until it approaches the point at which the particles are entirely translucent to incident light. Particle size reduction beyond this point does not improve the tinctorial strength of a pigmented system [1]. The tinctorial strength increases within the above-mentioned range of particle sizes, but only if the self-scattering of the pigment, for instance in white reductions with TiO_2, can be neglected. Unusual optical characteristics may also be found in transparent systems such as printed layers and may change the relation between tinctorial strength and particle size distribution.

To study the effect of changes in the particle size or particle size distribution on the optical properties of a TiO_2-reduced pigment-vehicle system, a series of aqueous Pigment Yellow 1 pastes in emulsion paints were compared [2,3]. Ultracentri-

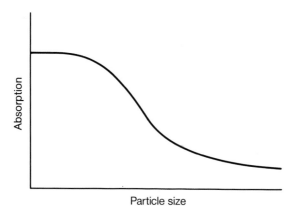

Absorption

Particle size

Fig. 60: Relation between particle size and absorption (tinctorial strength) of a pigment.

fugation techniques were used to separate the pigment pastes into three fractions according to the particle size. Fig. 61 shows the electron photomicrographs of the individual fractions; the respective particle size distributions as determined by ultrasedimentation are outlined in Fig. 62, in which the median for each curve (Sec. 1.5.2) is indicated by an arrow. Each fraction was then blended with an appropriate TiO_2 dispersion and the resulting samples matched against the original batches. At equal reduction ratio, the tinctorial strength of 1% colorations of the medium sized particle fraction (paste 2) matches that of 0.7% colorations of the batch with a fine particle size (paste 3) and the tinctorial strength of 2.1% colorations of the product with a coarse particle size (paste 1). The results can be interpreted quantitatively: the tinctorial strength of the three fractions in the order of increasing particle size thus corresponds to a pigment/tinctorial strength ratio of 100:145:300.

There are a number of specific cases in which, independent of the application, the tinctorial strength of a product is even more sensitive to the particle size distribution.

A comparative study of three letterpress proof prints containing Pigment Yellow 13 also illustrates the correlation between particle size distribution and the tinctorial strength of prints. The electron photomicrographs of the three samples in Fig. 63 (p. 125) correspond to the particle size distribution histograms in Fig. 64 (p. 126). All three pigments were dispersed under equal conditions in a sheet offset vehicle using a three-roll mill; and printed 1 µm thick with a proofing printer. Colorimetric evaluation shows that compared to sample III, which contains the variety with a coarse particle size, sample II shows a 25% increase in strength; while the print with a fine particle size exceeds the standard with a coarse particle size by as much as 36%.

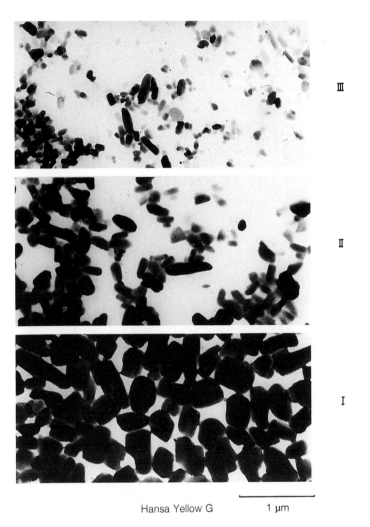

Ⅲ

Ⅱ

Ⅰ

Hansa Yellow G   1 µm

Fig. 61: Electron micrographs of samples I through III containing Pigment Yellow 1.

1.7.2   Hue

The shade of a product is another parameter which is influenced by the particle size (Sec. 1.6.1.2). This is easy to see by comparing the reflection curves of two Pigment Yellow 83 white reductions in an alkyd-melamine resin system. The electron photomicrographs in Fig. 65 and the particle size distribution histograms in Fig. 66 represent these samples. At equal pigment concentration, curve 1 of the two remission curves in Fig. 67 reflects the behavior of the pigment with the smaller particle

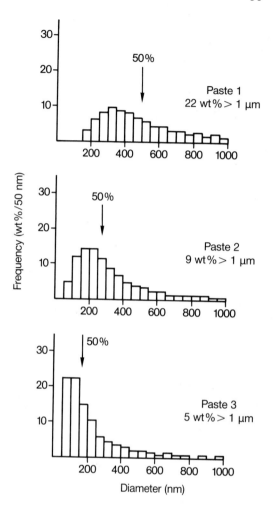

Fig. 62: Particle size distributions of the aqueous preparation of Pigment Yellow 1 samples I to III.

size, and curve 2 that of the type with the larger particle size. In the wavelength region between about 520 and 560 nm, curve 2 undergoes a sideways shift, indicating a distinct change of shade towards a more reddish yellow.

In general, the following rule applies: increasing particle size causes yellow pigments to undergo a red shift. Orange pigments turn more reddish; yellowish reds become more bluish; bluish reds turn more yellowish; brown shades undergo a red shift; violet pigments, such as Dioxazine Violet, develop a more bluish hue; and blues, such as phthalocyanine pigments, show more of a reddish shade. The influence of the particle size on the shade is especially evident in white reductions.

Paints behave similarly. Fig. 68 illustrates the color shifts which are observed in Pigment Yellow 1 pastes containing different particle size distributions (Sec. 1.7.1). The colors are identified according to their location on the DIN Color Chart, in which each color is defined by two values, x and y [4]. Increasing amounts of pig-

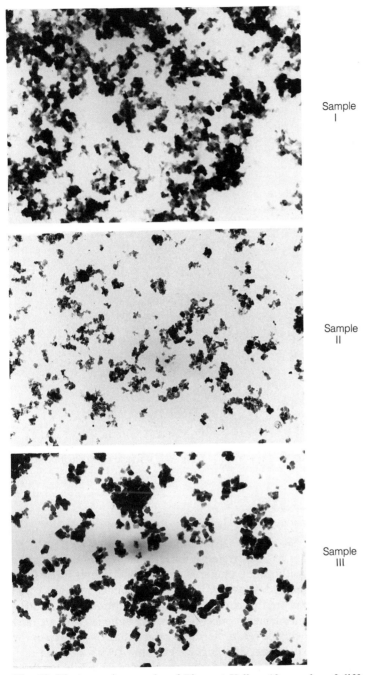

Sample
I

Sample
II

Sample
III

Fig. 63: Electron micrographs of Pigment Yellow 13 samples of different particle sizes.

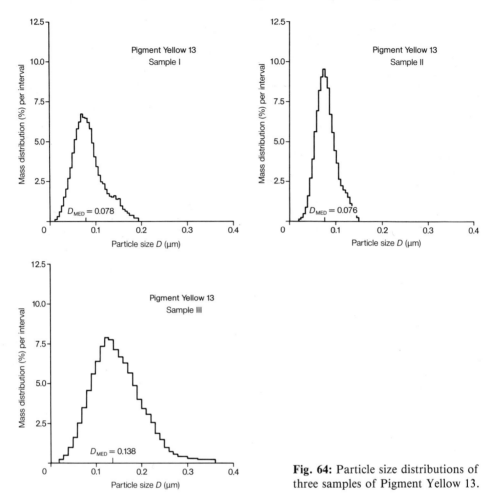

Fig. 64: Particle size distributions of three samples of Pigment Yellow 13.

ment paste were added to the sample of a pure white dispersion, referred to by location W. Likewise, the shades of the differently sized Pigment Red 3 (Toluidine Red) dispersions were plotted within the same area of the Color Chart [2]. All locations along one of the numbered lines originating in point C indicate equal shade; while the concentric rings are lines of equal color saturation. It is not difficult to recognize that at constant pigment concentration, the varieties with a coarser particle size of the two pigments differ in their tinctorial behavior relative to the respective type with a fine particle size. The Pigment Yellow 1 with a coarse particle size sample tends more towards the reddish end of the scale, while its Toluidine Red counterpart shifts clearly toward the bluish region. Similar color shifts are observed if the amount of pigment in the dispersion is increased, a tendency which is most pronounced in the respective samples with a fine particle size.

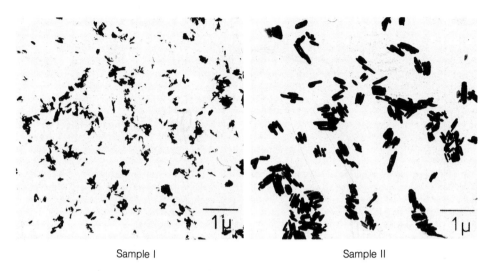

Sample I   Sample II

Fig. 65: Electron micrographs of two samples of Pigment Yellow 83 of different particle sizes.

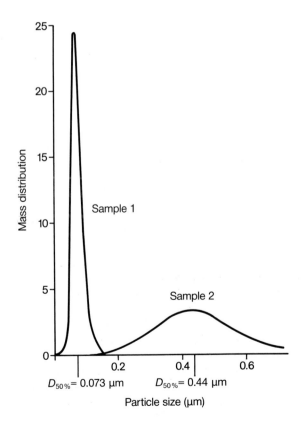

$D_{50\%}= 0.073$ μm $D_{50\%}= 0.44$ μm

Particle size (μm)

Fig. 66: Particle size distributions of two P.Y.83 grades.

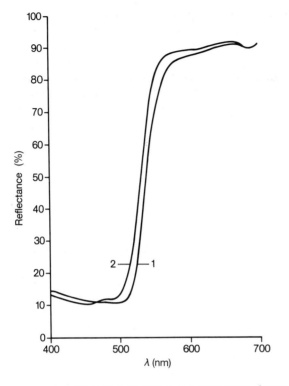

Fig. 67: Influence of the particle size of P.Y.83 on the shade of alkyd-melamine resin baking enamels (white reductions) demonstrated by means of reflectance curves.

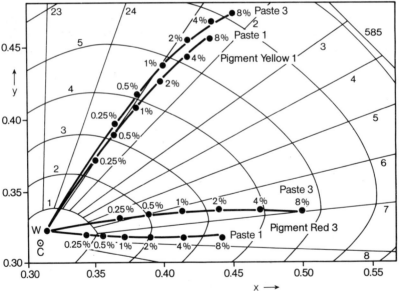

Fig. 68: Color locations of aqueous Pigment Yellow 1 and Pigment Red 3 pastes of different particle sizes. Section of the xy-chromaticity diagram, including the mid point, source C (DIN 6164).

It has often been observed that the coloristic properties of an organic pigment are a function not only of the size of particles but also of their shape. This is due to the anisotropy of the optical properties in different crystallographic directions within the crystal forms of a pigment. In 1974 [5, 6], it was demonstrated that of the equally sized but differently shaped particles of beta copper phthalocyanine blue, the almost completely cubic, i.e., more or less isometric form produces greenish blue shades, while acicular forms are responsible for reddish blue hues. The optical behavior of ordered pigment particles in systems has been reported in the literature [7, 8].

1.7.3   Hiding Power, Transparency

The hiding power of a pigment — a feature that is controlled by parameters such as the absorption coefficient of the pigment, the relative refractive indices of pigment and vehicle, the degree of pigment dispersion, the thickness of the layer, and other factors — is also largely determined by the scattering coefficient of the pigment. Like absorption, this coefficient is in turn a function of the particle size. The curve in Fig. 69, plotting the scattering coefficient versus the particle size, goes through a maximum. Particles above this critical size exceed the scattering maximum only to lose opacity, a phenomenon known as ultratransparency [9]. It is interesting to note that the scattering maximum is frequently observed at a particle size which provides comparatively poor absorption and inadequate tinctorial strength.

A pigment which absorbs a large portion of light will increase the hiding power of its application system despite insufficient scattering. In highly concentrated formulations, colored organic pigments with very high absorption coefficients and low scattering coefficients are referred to as hiding, even if the resulting full shade tends to appear somewhat dark. The excellent hiding power of copper phthalocyanine blue, for instance, is a result of a very large absorption coefficient, i.e., high tinctorial strength [10].

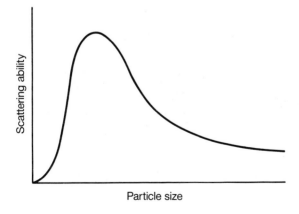

Fig. 69: Relation between particle size and scattering ability of a pigment.

However, there is no exact quantitative method which makes it possible to calculate the critical particle size that will produce an optimum in hiding power in a colored pigment formulation. Although the basic theory behind this phenomenon has been treated extensively, it is still most advantageous for practical R & D purposes to experimentally determine the particle size that affords a maximum in hiding power. A more approximate than quantitative rule has been established by H.H. Weber, which makes it possible to estimate the particle size that will afford a maximum in scattering:

$$D_{max} = \frac{\lambda}{2,1 \ (n_P - n_B)}$$

in which λ is the wavelength of the scattered light, and n_P and n_B denote the refractive indices of pigment and vehicle. The refractive indices of organic pigments range between 1.3 and 2.5, dependent on pigment and wavelength, while most of the commonly used vehicles have an index somewhere near 1.5. Assuming, for example, that $n = 2$, then Weber's rule predicts that the scattering maximum for wavelengths between 400 and 700 nm will be realized by particles of 0.4 to 0.7 µm.

There are several techniques which determine the hiding power of a pigmented medium for practical application. Colored systems are typically tested by measuring their ability to entirely cover up a standard black and white checkerboard substrate. Parameters used as variables are the pigment concentration and/or the thickness of the layer (pigment concentration per area unit), either one or both of which are adjusted until the ΔE^*_{ab} between the colors above the black and the white substrates equals 1 [11] (Sec. 1.6.3.1).

Using Pigment Orange 34 in a long-oil alkyd resin paint as an example, this technique clearly reveals the effect of the particle size distribution on the hiding

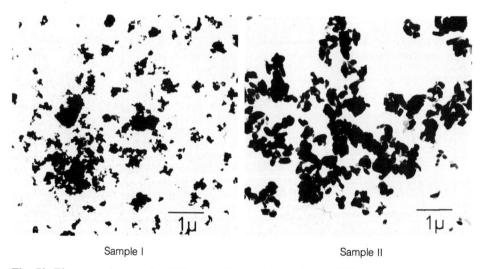

Sample I   Sample II

Fig. 70: Electron micrograph of Pigment Orange 34 samples of different particle sizes.

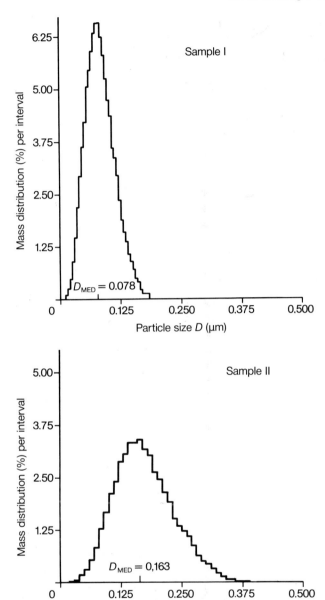

Mass distribution (%) per interval

$D_{MED} = 0.078$

Particle size D (µm)

Sample II

Mass distribution (%) per interval

$D_{MED} = 0,163$

Particle size D (µm)

Fig. 71: Particle size distributions of Pigment Orange 34 samples I and II.

power of the film. The two electron photomicrographs in Fig. 70 reflect the structure of a transparent type with a fine particle size on the one hand and a considerably more opaque type of the same pigment with a very coarse grain on the other hand. These correspond to the two different particle size distributions shown in Fig. 71. It is remarkable to see that two equal amounts of the same compound,

Fig. 72: Pigment Orange 34 samples of different particle sizes (above: small, below: large) in a long-oil alkyd resin system. Pigment concentration in the system: 6%; thickness of the wet film: 75 μm.

incorporated into the same vehicle and applied at equal thickness, give an entirely different impression above a checkerboard surface (Fig. 72).

If the hiding power of a system is particle-size dependent, then the same is necessarily true for the transparency of the system. The criteria for satisfactory transparency in a pigment-vehicle system depend on the applicaton purpose for which it is targeted. The printing industry, for instance, requires inks which do not appreciably cover black substrates and therefore uses formulations which scatter very little light. Printing inks are accordingly tested by applying standard layers of ink (1 to 2 μm) at certain pigment levels, i.e., at a certain pigment surface concentration, to black test paper and evaluating the results visually or colorimetrically.

It is possible to colorimetrically determine the ability of a pigmented layer to scatter light above a black substrate; i.e., its transparency (Sec. 1.6.1.3) by finding the normal value Y or the distance ΔE^*_{ab} on the DIN 6174 Color Chart. This method may also be used to compare increasingly thick layers. To quantitatively describe the transparency of a system, the so-called Transparency Number T has been introduced. It is defined as the inverse of the distance ΔE^*_{ab} between colors printed on top of a black surface and ideal black for a pigmented medium that has been applied in a layer of thickness h [12].

$$T = \frac{h}{\Delta E^{*}_{ab}}$$

However, the transparency number is not only a function of pigment and vehicle; it usually changes with the thickness of the layer (e.g., the pigment surface concentration). The transparency number of a sample which is applied at a thickness somewhere in the interval between two limits h_1 and h_2 is calculated as follows:

$$T = \frac{h}{\frac{h - h_1}{h_2 - h_1}(\Delta E_2 - \Delta E_1) + \Delta E_1}$$

It is not difficult, for instance, to find the respective transparencies of the two Pigment Orange 34 samples which differ in their particle size distribution. In letterpress proof prints, printed 1.5 µm thick, 15% pigment formulations of both yield the following results:

	Transparency Number T
Pigment sample 1	0.135
Pigment sample 2	0.045

1.7.4 Lightfastness and Weatherfastness

The lightfastness and weatherfastness of an organic pigment are largely dependent on the particle size, which is not surprising in view of the fact that it is the interaction of pigment particles with absorbed radiation that determines the response of a system to light. Very little is known about the chemistry which is responsible for the destruction of organic pigment crystals through incident light. The radiation which ultimately cleaves the pigment structure only penetrates the pigment crystals by an average of 0.03 to 0.07 µm, which explains why in larger particles it is only the uppermost layers, those in the vicinity of the surface, that are affected. The effect of light thus destroys one layer after another, beginning with the outer one. More centrally located pigment particles in such large units are safe until the outer layers are destroyed, therefore large particles resist light much longer than small ones.

A few examples will exemplify the relation between the particle size distribution of a pigment and its lightfastness. Fig. 73 shows histograms of the respective particle size distributions of three aqueous pastes containing Pigment Orange 5 [13], determined by the Marshal ultrasedimentation method. Particle sizing is designed such that every two out of these three samples agree in one size-related aspect. Although the medians of samples 1 and 2 are almost equal, their size distributions have an entirely different shape. Pastes 1 and 3, on the other hand, show a similarly narrow distribution, but their medians are dissimilar. White emulsion paint was

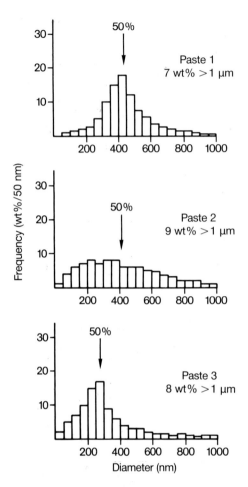

Fig. 73: Particle size distributions of three Pigment Orange 5 samples.

added to all three samples to afford equal depth of shade, and the samples were exposed to light. After 1500 hours, the tinctorial strength of sample 1 was 72% of its original value; while samples 2 and 3 only retained 46% and 39%, respectively, of their tinctorial strength (Fig. 74).

It is interesting to see the difference in weatherfastness between two pigments as related to the particle size. Two types of Pigment Yellow 151 were incorporated in an alkyd/melamine baking enamel and examined by electron microscopy. The particle images are reflected by the electron photomicrographs and the particle size distributions as shown in Fig. 75. The coatings were exposed in a Xenotest 1200 type accelerated weathering cabinet [14]; cycles were 17/3 minutes (Sec. 1.6.6.2) (Fig. 76, page 137). Fig. 76 demonstrates that the type with the larger particle size shows significantly better weatherfastness than the type with the smaller particle size.

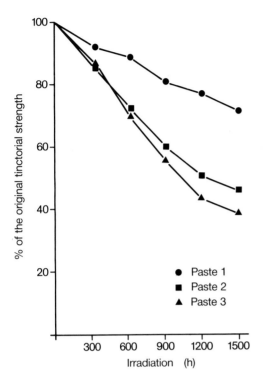

Fig. 74: Relative tinctorial strength of irradiated dispersion layers containing Pigment Orange 5 of different particle sizes (white reductions).

1.7.5 Dispersibility

All dispersion conditions being equal, the dispersibility of a given organic pigment in a particular vehicle is largely controlled by its particle size distribution. However, the mechanisms underlying the relation between particle size and dispersibility are somewhat complex and not easy to study. Although the problem has been discussed in the literature, most publications fail to provide detailed information as to the degree of agglomeration, surface structure, and pigment preparation. Some studies simplify the problem by standardizing the pigments through uniform finishing procedures in order to attain equal dispersibility and equal or similar surface structure [15, 16].

The dispersibility of a pigment depends on the number of surfaces and edges that can, through attractive surface forces, unite the particles into aggregates and agglomerates. These surface forces are a function of the particle size and size distribution. Theoretically, the larger the particles, the lower the surface forces, and the easier it is to disperse the agglomerates [17]. A wide particle size distribution considerably increases the number of points of adhesion, allowing small particles to "glue" the large units together [18]. Simultaneously, the small particles fill in or partially occupy the gaps and pores. The inner surfaces of the agglomerates thus

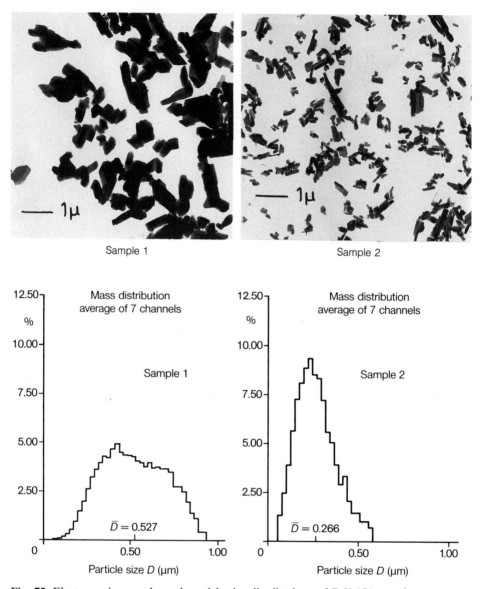

Fig. 75: Electron micrographs and particle size distributions of P.Y.151 samples.

become less accessible to the surrounding medium, and so reduce the wetting rate (Sec. 1.6.5). In all particle size ranges, pigments with a comparatively narrow particle size distribution are therefore easier to disperse than corresponding types with a wider range of sizes.

The particle size of a pigment, however, has an impact not only on those steps within the dispersion process in which pigment agglomerates are broken down. It

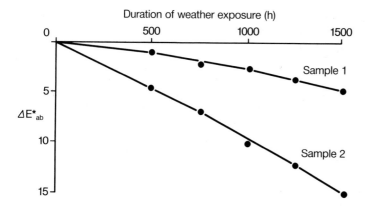

Fig. 76: Effect of the particle size of P.Y.151 in alkyd-melamine resin systems on the weatherfastness. Organic pigment: $TiO_2 = 1:3$.

also affects retroactive processes, such as reagglomeration, flocculation, and rubout effects [19]. Reagglomeration (Sec. 1.5) occurs if the pigment particles are not covered adequately by the binder. Flocculates (Sec. 1.5) are loose associations of particles which separate easily through low shear such as stirring. In contrast to agglomerates, flocculates are adequately wetted and the space inside them is filled with vehicle molecules. Although they are only loose units, flocculates adversely affect the coloristic properties of a pigment-vehicle system by decreasing its tinctorial strength and gloss. The tendency to flocculate is a property not only of a pigment but of an entire system. Since it is a result of interaction between the surface

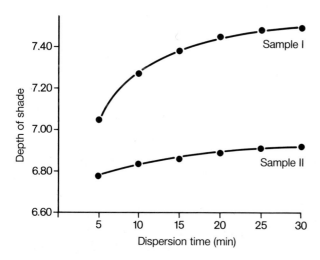

Fig. 77: P.O.34 in a long-oil alkyd resin system. Influence of the particle size on the dispersibility. Dispersion unit: paint shaker, organic pigment: $TiO_2 = 1:5$. Sample 1: fine particle pigment, sample 2: coarse particle pigment.

of the pigment particles and the surrounding medium, this phenomenon is particle-size dependent. Flocculation can be inhibited or even avoided by adding traces of antiflocculation agents to the application medium [20]. Likewise, these agents prevent pigment floating or flooding; and a variety of agents are available to cope with other particle-size dependent failures that may be observed during the processing of a low-viscosity system [21]. Moreover, the particle size is also responsible for specking, seeding, and similar phenomena which are frequently observed in paints.

Fig. 77, which plots the depth of shade of two P.O.34 samples versus the time needed to disperse the pigment, represents the dispersibility curve of two types with different particle size distributions. The two samples of Pigment Orange 34 are characterized in Figs. 70 and 71.

1.7.6   Gloss

The gloss of a pigmented system is a function not only of the degree of pigment dispersion but also frequently of the particle size distribution (Sec. 1.6.5). This is a common phenomenon in very thin applications, such as in offset printing, which produces layers less than 1 μm thick.

The effect of the particle size on the gloss of a coating is readily reflected through the haze which is frequently observed in air-dried alkyd systems. This is a surface abnormality which develops with increasing drying time as a result of the contraction of the film volume. It is particularly pronounced in Pigment Red 3 types, where it is known as the Toluidine Red haze. Bluish toluidine red pigments,

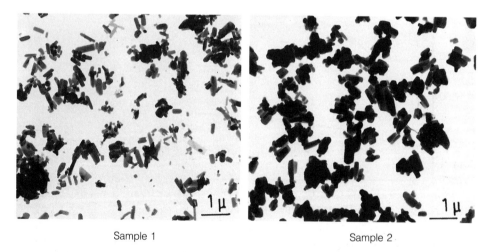

Sample 1   Sample 2

Fig. 78: Electron micrographs of Pigment Red 3 samples of different particle sizes.

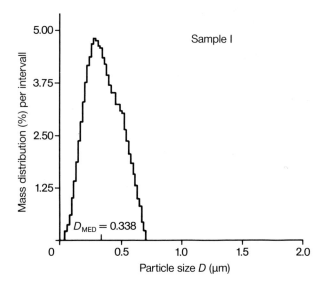

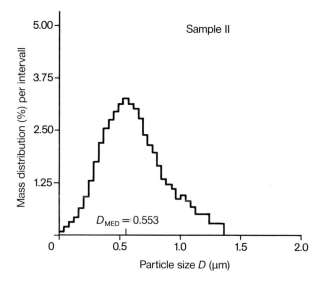

Fig. 79: Particle size distributions of P.R.3 Samples I and II.

which on the average contain more coarse particles, are more prone to turn hazy than the yellowish types with finer particle sizes [22]. This phenomenon is illustrated in Figs. 78–82.

As an example, two samples of Toluidine Red with different particle size distributions (Figs. 78 and 79) were dispersed in a long-oil alkyd system using an oscillating shaking machine (Paint Shaker) [23]. The paints were applied to a polished glass surface and the gloss values monitored to detect any developing haze. After drying for 16 days, both specimens were evaluated for surface gloss. The hazy type

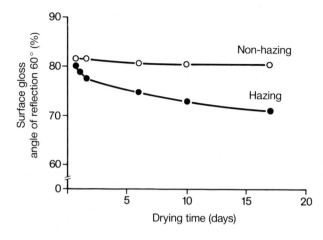

Fig. 80: Influence of the gloss of air-dried alkyd resin systems containing Pigment Red 3 of different particle sizes with the drying time. Pigment content of the systems: 8%.

with a coarser particle size loses more than twice as much gloss as the type with a fine particle size, which is imperceptibly hazy (Fig. 80). Scanning electron micrographs show the different surfaces of the two coatings after 1 day (Fig. 81) and after 8 days (Fig. 82).

Considerable shearing forces will break down the particles in the variety with a coarser particle size. It is interesting to note that the tendency of a pigmented system to become hazy decreases as the dispersion time increases, accompanied by improved gloss, color shift towards more yellowish shades, enhanced color strength in white reductions, and increased viscosity [22].

Similar results have been reported for other types of products with coarse particle sizes, such as Pigment Yellow 1.

1.7.7   Solvent and Migration Fastness

The solubility of any crystalline material decreases as the particle size increases; the correlation is quantitatively described by the Ostwald equation [24]:

$$\ln \frac{c}{c_\infty} = \frac{2\sigma V}{r \cdot R T}$$

in which: c = solubility of crystals with radius r, c_∞ = solubility of very large crystals, σ = surface tension, V = molecular volume of the crystal, R = gas constant, and T = Kelvin temperature. For several reasons, it is very difficult to quantitatively determine the solubility of an organic pigment as related to the particle size. F. Gläser [25] found that a particle diameter of 0.3 μm represents somewhat of a threshold value. The solubility of pigment crystals above this size essentially equals that of very large particles (c_∞) and is therefore largely constant; in moving towards finer particle sizes, however, the solubility increases appreciably (Table 7).

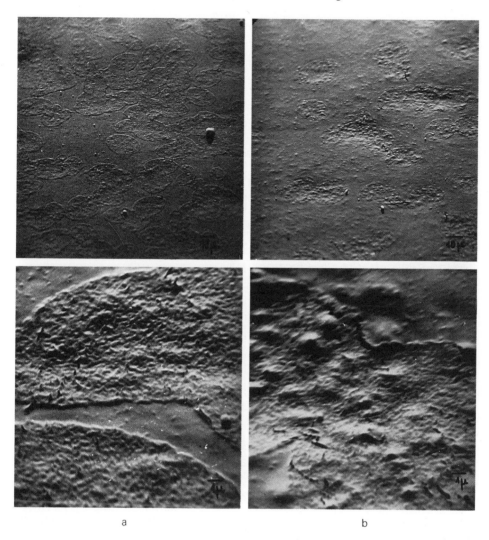

a   b

Fig. 81: Scanning electron micrographs of long-oil alkyd resin systems containing P.R.3, (a) nonhazing, (b) hazing. Drying time: 1 day.

Table 7: Solubility as a function of the particle size.

r (μm)	10	3	1	0.3	0.1	0.03	0.01	0.003
α (r)	1.004	1.013	1.03	1.14	1.5	3.78	54.6	83 300

$\alpha(r)$ is the factor by which the solubility of a particle with radius r increases compared to the solubility of a very large crystal.

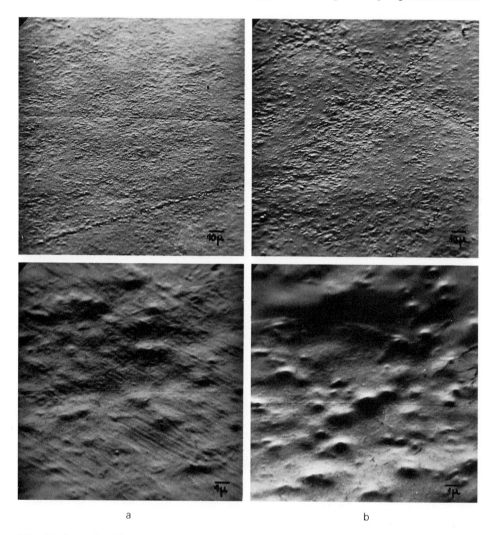

a   b

Fig. 82: Scanning electron micrographs of long-oil alkyd resin systems containing P.R.3, (a) non-hazing, (b) hazing. Drying time: 8 days.

The kinetics of the dissolution process obeys the Noyes and Whitney equation [26], revised by Nernst and Brunner:

$$\frac{\mathrm{d}C}{\mathrm{d}t} = \frac{D}{\delta} \cdot O(C_\infty - C)$$

in which C_∞ = concentration of the saturated, evenly stirred solution, C = the respective concentration t minutes after the material begins to dissolve, O = surface area, D = diffusion constant, and δ = thickness of the interface. According to this equation, the dissolution rate is proportional to the surface area O of the particles and to the solubility of the pigment. The dissolution rate decreases more and more as the concentration of the solution approaches saturation, i.e., the limit of solubility.

Pigments are supplied with solubility tables determined by the manufacturer. The results are reported with respect to the test method described in Sec. 1.6.2.1. Although pigment dissolution is not normally very advanced in these tests, many samples distinctly bleed, irrespective of the particle size.

Table 8 lists the effect of a number of solvents on three chemically different pigment pairs as a function of the particle size; sample 1 is the type with the finer particle size. The specimens are rated according to the extent of pigment bleeding into the solvent. Grading is from 1 to 5, with 1 referring to considerable coloration of the solvent, while a completely colorfast sample is rated as 5. Solvent coloration is referred to relative to the Gray Scale for evaluation of bleeding [27].

Table 8: Solvent stability as a function of the particle size.

	Specific surface area (m^2/g)	Ethanol	Toluene	Butyl acetate	Acetone	Dibutyl phthalate
P.Y. 1						
Sample 1	43	3	1	1	1	1
Sample 2	16	3–4	3	3	3	3
P.Y. 12						
Sample 1	86	5	2	3–4	3	4
Sample 2	34	5	3	4	4	4–5
P.R. 53:1						
Sample 1	43	3	4	4	3–4	4–5
Sample 2	16	3	4–5	4–5	4	4

Another facet that is affected by the particle size of a pigment is its tendency to recrystallize, a direct result of poor solvent fastness (Sec. 1.6.5). This is illustrated by three samples which differ in terms of particle size distribution (Sec. 1.7.3). The samples were dispersed in an offset vehicle, divided into batches, stored at different temperatures for 8 hours, and finally printed on a black-coated test paper under standardized conditions. The transparency of each specimen shows the effect of the particle size on the occurrence of recrystallization during storage (Fig. 83) (Sec. 1.6.5).

The tendency of a pigment to migrate is less sensitive to changes in the particle size (Sec. 1.6.3.1, 1.6.3.2).

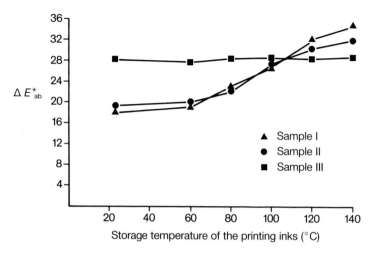

Fig. 83: Effect of the particle size of Pigment Yellow 13 on its tendency to recrystallize in a sheet offset printing ink. The samples were stored for eight hours at different temperatures, printed in equally thick layers over a black background, and evaluated in relation to standard black. The particle size increases from Sample 1 to Sample 3.

1.7.8   Flow

Most of the flow behavior and the rheological demands of a pigmented system are controlled essentially by the particle size of the dispersed pigment.

The correlation between rheological data and the most frequent particle diameter D_{mf} has been studied on a dispersion of the gamma modification of Pigment Violet 19 and various types of copper phthalocyanine blue in an offset vehicle on the basis of linseed oil [15]. The results show that while the ultimate viscosity η_∞ (Sec. 1.6.8.1) is independent of the particle size, the yield value increases with the cube of the particle diameter.

The considerable influence of the pigment particle size in controlling the flow behavior and the rheology of a system is a useful asset to the manufacturer who tries to produce highly opaque paints. Comparatively large particles lead to an optimum in hiding power and consequently provide a relatively small surface area to be wetted. The resulting products exhibit good flow behavior and are not usually very viscous. Larger amounts of such pigments can be worked into an application medium without affecting the rheology or the gloss of the system. To give a quantitative example, the following table lists the respective amounts of Pigment Yellow 74 types which, incorporated in a long-oil alkyd resin system, will afford essentially equal flow curves; the specimens differ in terms of particle size distribution and therefore vary in their opacity:

- 8.8 wt% pigment sample 1 (highly opaque type with coarse particle size),
- 5.6 wt% pigment sample 2 (standard of comparison),
- 4.0 wt% pigment sample 3 (transparent, tinctorially strong).

If sample 2 is used as a standard with a hiding power of 100, then the relative opacity of the type with a coarse particle size equals 275, while the transparent variety with a fine particle size has a relative hiding power of 30.

References for Section 1.7

[1] L. Gall: "Farbmetrik auf dem Pigmentgebiet", BASF AG, Ludwigshafen 1970.
[2] M.A. Maikowski, Melliand Textilber. (1970) 574–578.
[3] M.A. Maikowski, Ber. Bunsenges. Phys. Chem. 71 (1967) 313–326.
[4] DIN 6164: DIN-Farbenkarte.
[5] P. Hauser, D. Horn, and R. Sappok, XIII. Congr. FATIPEC, Juan les Pins, 1974, Congress book 191.
[6] R. Sappok, J. Oil Colour Assoc. 61 (1978) 299.
[7] F. Gläser, DEFAZET Dtsch. Farben Z. 32 (1978) 338–342.
[8] U. Zorll, Farbe + Lack 72 (1966) 733–742.
[9] W. Herbst, J. Paint Technol. 45 (1973) 39–50.
[10] L. Gall, Farbe + Lack 72 (1966) 955–965.
[11] DIN 55 987: Bestimmung eines Deckvermögenswertes pigmentierter Medien. ISO 6504-5-1987: Paints and varnishes – Determination of hiding power – Part 5: Colour difference method for dark coloured paints.
[12] H.G. Völz, DEFAZET Dtsch. Farben Z. 30 (1976) 392–454. DIN 55 988: Bestimmung der Transparenz (Lasur) von pigmentierten und unpigmentierten Systemen; Farbmetrisches Verfahren.
[13] M. Maikowski, Farbe + Lack 77 (1971) 640–647.
[14] Manufacturer: Original Hanau Quarzlampen GmbH, Hanau.
[15] P. Hauser, H. Herrmann and B. Honigmann, Farbe + Lack 76 (1970) 545–550; 77 (1971) 1097–1106.
[16] W. Herbst and K. Merkle, Farbe + Lack 74 (1968) 1072–1089, 1174–1179.
[17] W.C. Carr, J. Oil Colour Chem. Assoc. 50 (1967) 1115–1155.
[18] L. Gall and U. Kaluza: DEFAZET Dtsch. Farben Z. 29 (1975) 102–115.
[19] U. Kaluza, DEFAZET Dtsch. Farben Z. 33 (1979) 355–359, 399–411.
[20] e.g. V.T. Crowl, J. Oil Colour Chem. Assoc. 55 (1972) 388.
[21] F. Haselmeyer, Farbe + Lack 74 (1968).
[22] W. Herbst and K. Merkle, Farbe + Lack 75 (1969) 1137–1157.
[23] DIN 53 238 Teil 10: Dispergieren in niedrigviskosen Medien mit einer Schüttelmaschine. ISO 8780-1-1986: Dispersion characteristics of pigments and extenders – Methods of dispersion – Part 1: Dispersion using an oscillatory shaking machine.
[24] W. Ostwald, Z. Phys. Chem. 34 (1900) 295.
[25] F. Gläser, XIII. Congr. FATIPEC, Juan les Pins, 1976, Congress Book 239–243.
[26] A. Noyes and W. Whitney, Z. Phys. Chem. 23 (1897) 289.
[27] DIN 54 002: Herstellung und Handhabung des Graumaßstabes zur Bewertung des Ausblutens. ISO 105-A03-1987: Textiles – Test for colour fastness – Part A03: Grey scale for assessing staining.

1.8 Areas of Application for Organic Pigments

The considerable increase in demand for colored pigments over the last three decades is in response to a number of external factors, primarily of a psychological nature. Advertising effectiveness and product promotion have become increasingly dependent on the public attention value of colored materials. On the other hand, both industry and the private sector show an ever increasing tendency to color products for identification and differentiation purposes.

Organic pigments are used to color a variety of media. It is useful to distinguish between three primary fields of application: the coatings and paints industry, the printing inks industry, and the plastics and fibers industry. Besides, organic pigments are used for special purposes, for instance in office materials and in the mass coloration of paper.

In discussing commercially significant organic pigments and their performance in use, it seems appropriate to define the limitations placed on pigment use by the different fields of application. A given pigment may perform well in one application but poorly in another, and the standards that are acceptable to the pigment user depend on the scope of the product. The chemistry and physics of the medium to be colored play as significant a role as the processing methods and operational conditions. Ultimately, the choice of pigment and binder system for a specific application almost invariably represents a compromise between optimum technical and bearable economical considerations.

It is useful in this context to focus on the performance of entire pigmented systems rather than to reflect upon the properties of individual components. The properties of a pigment powder, for instance, bear very little resemblance to its ultimate property in use.

As an example, a given pigment, incorporated in one medium, may match grade 8 on the Blue Scale for lightfastness, which is an excellent value; while a change of application medium may render the same pigment sensitive enough to reach only step 2. Similar observations apply to other technical properties such as weatherfastness, migration fastness, and solvent fastness, along with certain coloristic properties, such as tinctorial strength, hue, gloss, gloss retention, and transparency, resp. opacity.

The technical requirements for pigments are subject to constant change because of the development and introduction of new vehicle systems fostered by the proliferation of materials to be colored. The recurring necessity to replace traditional pigments that are not usable with the newer vehicle systems is a permanent incentive for the development of new products.

However, instead of exploring new pigments, it is often preferable to optimize the properties of one of the already available compounds. This explains why a given type of pigment is often marketed in more than one physical form to provide a variety of coloristic and application properties, each of which is designed for a specific vehicle system or condition of use.

Where this approach fails, new pigments have to be developed. A large number of patents have been issued in recent years describing compounds to be used as pigments; novel classes of pigments have been proposed. Only very few, however, have entered the market, and even less have turned into a commercial success. This is why R & D laboratories today have shifted their attention preferably towards optimizing the properties of known pigment structures for special purposes. Moreover, it should be noted that economical and environmental concerns remain significant issues in the development of synthetic novelties.

This chapter lists the primary areas in which organic pigments are employed, along with the respective requirements for pigment use. It is occasionally useful to include information as to whether and in which way special demands can be fulfilled.

1.8.1   Printing Inks

The graphic industry ranks among the leading pigment consuming industries; principally organic pigments. The 1989 trade sales volume of diarylide yellow pigments to be used in printing inks alone is estimated at 22 to 25000 metric tonnes. A preference for certain printing processes, which determines the demand for a given type of pigment, is a result of economical and environmental considerations and may vary from one year to another. Regional differences and preferences add to the complexity of pigment use. At the present time, web offset and publication gravure printing are the techniques that dominate the market.

The technology of printing has in recent years expanded drastically; printing processes in particular have been accelerated considerably. It is therefore necessary to constantly adjust both printing inks and pigments to the needs of a new technology; high printing speeds can only be maintained by using an ink that performs well and is sufficiently strong. Since printing inks, especially those designed for web offset, are typically pastes, formulated at high pigment concentrations, there is a certain emphasis on good rheology and high tinctorial strength; however, transparency and gloss are often equally important.

Certain areas of pigment application in the printing ink industry challenge the performance of a pigment more severely than others. To be useful in inks for food cans, for instance, the inside coatings must be capable of being sterilized with steam without degradation. For bank notes, currency, and cheques, special security properties are needed.

1.8.1.1   Offset/Letterpress Printing

These printing processes are particularly sensitive to the degree of pigment dispersion. Therefore additional shearing forces between the cylinders of the printing presses are applied to convert the original pigment formulation to an adequately dispersed ink [1]. Incomplete breakdown of agglomerates may present problems in the thin layers of 0.8 to 1.1 µm as produced by offset techniques, affecting the tinctorial strength and the gloss. This is because agglomerates which protrude from the surface of the film scatter much of the incident light. Figures 84 and 85 show electron micrographs of printed layers. The pigment in this case was dispersed in an offset vehicle at different temperatures, using a three-roll mill (Sec. 1.6.5.3). The samples for the electron microscopic images in Fig. 84 were produced with an ultramicrotome, and the images reflect the cross section of printed layers on paper; Fig. 85 shows the corresponding surface images. The photographs clearly show that increasing the dispersion temperature improves the distribution of pigment throughout the vehicle, produces a smooth surface, and thus enhances the gloss.

A pigment which is not dispersed thoroughly in its medium will also affect the transparency of the resulting printed layer, which is a particularly sensitive issue in multicolor printing. As a result, offset techniques in particular depend on good pigment dispersion; and offset equipment, pigment manufacture, and resin preparation are designed to cope with this problem. Not only have three-roll mills and other dispersion units been improved continuously in recent years, but new technologies have been developed, resulting in specialized mechanically agitated ball mills with a high speed shaft and other machinery. Process automation and high throughput make printing ink manufacture more economical. Modern dispersion techniques have been developed along with the new dispersion equipment, presenting a severe challenge to pigment performance. Modern ball mills, for instance, require pigments that are easy to disperse and stable to recrystallization, because even efficient cooling does not entirely protect the printing inks from being exposed to temperatures of 70 to 90 °C and more during dispersion.

Pigments such as Pigment Yellow 12 and 13 type diarylide yellow pigments, used primarily in offset printing inks, dissolve to a certain extent in mineral oils under these conditions and may therefore recrystallize. In practice, a pigment may be required to resist such high temperatures for up to several hours as the pigment is incorporated into the vehicle, especially if high speed impellers are used. In large scale production, there is usually some time lag in between the two steps of incorporating the pigment with a dissolver and dispersing it using a bead mill. This may give rise to problems, since the tendency of a pigment to recrystallize increases with the temperature and duration of pigment exposure, resulting in loss of tinctorial strength and transparency, often to a considerable extent. Moreover, recrystallization also affects the gloss of a product; and yellow pigments tend to turn more reddish and become somewhat dull. A printing ink which suffers from extensive pigment recrystallization will perform like an insufficiently dispersed system. For these reasons, special-purpose products have been developed and are commercially available which satisfy the demands of offset dispersion.

It is not only the dispersion equipment, however, which defines the dispersibility standards for organic pigments that are targeted for offset and letterpress application. The composition of the binder also plays a major role. Moreover, ever improving printing techniques are instrumental in creating severe application conditions. Web offset inks, for instance, have been improved further by reducing the amount of easily wetting alkyd resins and vegetable oils and by increasing the content of hard resins and mineral oil. This shortens the time and decreases the temperature needed to dry the printed layers. Gel varnishes, for example, which are

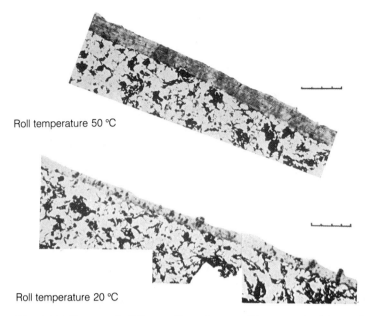

Roll temperature 50 °C

Roll temperature 20 °C

Fig. 84: Influence of different dispersing conditions on the degree of dispersion of Pigment Yellow 17 in a sheet offset varnish and on the gloss of prints which were produced from this system.
Images of ultramicrotome-cut thin sections of the printed layers taken with a transmission electron microscope.

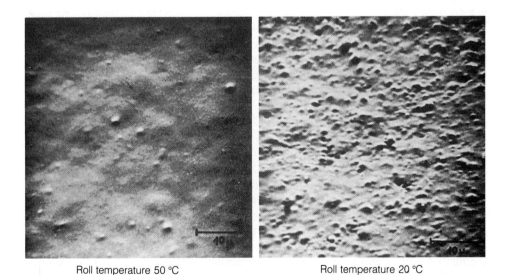

Roll temperature 50 °C   Roll temperature 20 °C

Fig. 85: Images of the surfaces of the prints taken with a scanning electron microscope.

added to printing inks in order to confer more advantageous printing properties on the host system, wet a pigment with difficulty. Gel varnishes are obtained by reacting varnishes which contain large amounts of hard resin and a small portion of vegetable oil with a metal chelate.

Special-purpose pigment preparations, especially those containing various resins, are manufactured in an attempt to provide systems which are easy to disperse. There is a disadvantage in the fact that certain resins — especially those contained in the pigment in large concentration — may adversely affect the evaporation of solvent, i.e., the drying process. It is not possible to use surfactants, which also slightly disperse a pigment in an offset vehicle, because these otherwise useful agents may disturb the equilibrium between water and ink on the offset printing plate.

Considering the effect and the impact of water on offset printing, it is not necessary to enlarge on the fact that pigments that are incorporated in such inks must be insoluble in water. This, however, is a requirement which is not always completely satisfied. A number of pigment lakes, such as the Lake Red C types (Pigment Red 53:1) are not completely insoluble in water. Insufficient fastness to water affects the print through a phenomenon known as "toning". The term refers to the effect of making the water receptive nonimage areas of the offset plate absorb some color and produce an image in print.

Multicolor Printing

Pigments which are designed for multicolor printing have to satisfy special coloristic requirements, which are defined by standardized color scales for letterpress and offset printing. Manufacturers of photographic film have developed useful color scales, which used to be very important and are still referred to today. The most well known is the Kodak Scale. This scale was designed to aid in the selection of color filters used in the processing of color prints to match the Kodak Color Control Standards. The range of shades is comparatively wide. To match these shades, for instance a yellow hue, with organic pigments in a printing system, a larger number of different pigments over a relatively wide concentration range may have to be used. In print, the Kodak scale produces warm colors, a fact which is particularly appreciated in connection with art prints. It is for this reason that the Kodak scale enjoys continued importance.

In contrast to the color scale used by the film industry, there are other scales which were first developed on a national and later also on a European basis. Conditions to be met in color reproduction are referred to in color co-ordinates. Various standards for three- and four-color printing in letterpress and offset processes have been developed, among others, to provide standardized inks for photoengraving and photolithography as well as for the printing industry. These products, printed on standardized art paper under standardized conditions, therefore produce standardized results. According to the European Color Scale, the thickness of a freshly applied reference film must be between 0.8 and 1.1 μm. Compared with the DIN scale [2], the European Color Scale allows stronger reproduction in the range of red blends (yellow and magenta). Magenta, for instance, is less blue on this scale. Colorimetric tolerances are given. The European Standard, in contrast to the DIN scale, does not strictly define the sequence in which the colors have to be printed for an offset print [3]. To satisfy European letterpress standards, however, inks must be printed in the sequence (black)-yellow-magenta-cyan [4].

The standard yellow on the European Scale is accessible through a range of azo pigments, such as Pigment Yellow 13, 126, or 127. Quite a number of yellow pigments, whose shades are outside the tolerance limits, may be adjusted coloristically by appropriate shading. The ultimate shade of a pigment in application is largely defined by the pigment concentration in the ink and by the degree of pigment dispersion. Standard magenta may be approached, for instance, by Pigment Red 57:1, P.R. 184, or P.R. 185 type azo pigments; while standard cyan is produced by using the β-modification of phthalocyanine blue (Pigment Blue 15:3 and 15:4).

However, not all pigments that satisfy the coloristic standards are eligible for use in multicolor printing. The requirements of different areas of application present a considerable challenge to pigmented materials in terms of fastness to overspraying and calandering, lightfastness, thermostability, migration fastness, transparency/opacity, and rheology. The number of applicable pigments is also considerably restricted by economic considerations, which is especially true for yellow pigments.

Metal Deco Printing

Metal deco printing can be considered a specialty area of offset printing. Traditional printing inks, which are processed at temperatures of 140 °C and more, must be quite thermally stable to retain their color value during application. Moreover, printed metal sheets which are to be used in food cans must be capable of being sterilized in the presence of food without degradation. Tests are typically carried out in water at 120 °C and 2 bar. The pigment is usually expected to be fast to overspraying with clear "silver lacquer" (Sec. 1.6.2.1). In recent years, however, metal deco inks have been developed which are less demanding in terms of pigment heat stability. Like in offset printing, metal deco inks are applied in thicknesses of only about 1 μm, which makes it necessary to use inks containing pigments with high tinctorial strength at high concentrations. Pigments targeted for metal deco printing therefore require good rheology, i.e., adequate flow behavior in oil-based inks.

Ultraviolet-Cured Printing Inks

Ultraviolet-cured printing inks are used to some extent in metal deco printing and are sold primarily for ecological reasons. The demand for such inks, which harden through exposure to UV radiation, varies considerably according to the region. At typical pigment surface concentrations, organic pigments do not significantly affect the drying and hardening of these printing inks. Since UV light causes scattering and absorption by the pigment, some effect is to be expected, but there are no exceptional specifications for pigments which are targeted for this area of application.

1.8.1.2   Gravure Printing

There are two types of gravure printing: publication gravure and special or packaging gravure. Publication printing is used for magazines, magazine supplements, mail-order catalogs, and other areas. Packaging gravure inks are used on packaging materials such as paper or plastic films. Two kinds of use requirements for pigments arise from the two types of gravure, which use printing inks of different composition.

Publication Gravure Printing

Inks (and therefore pigments) which are designed for use in publication gravure printing must be characterized by extremely good rheology. Printing speeds of almost 45000 revolutions per minute of the printing cylinder (which equals 14 m/s on paper) require inks which are able to almost entirely fill a 40 µm deep cell within a fraction of a second. These inks are then transferred immediately onto paper. The performance of the pigment, especially its physical characteristics, largely determine the flow properties of the resulting printing ink. Flow properties are based on very complex phenomena (Sec. 1.6.8.4) and are far from being entirely understood. The phenomenon of thixotropy in these extremely thin layers, for instance, has not yet been explained on a microbasis. Besides, the requirements at different stages of ink application vary and may be contradictory. Although a structurally viscous or thixotropic printing ink makes it difficult to fill and empty the tiny cells in an engraved cylinder, these are exactly the properties that will make an ink perform well on paper and prevents it from penetrating too deep into the paper. Clearly, the optimum performance is realized if properties like structural viscosity or thixotropy develop immediately after the ink is applied onto paper: the printed dots have sharp edges and the resulting print is clear. The importance of ink rheology can hardly be overestimated, since it also determines the ease with which a printed layer will dry: solvent removal, either by penetration into the paper or by evaporation, is largely a function of the flow behavior of the pigmented ink.

Pigments used for applicable printing inks are also expected to be nonabrasive. A printing ink or ink components which cause mechanical abrasion to the engraved printing cylinder may prematurely damage the printing cylinder, destroying or scratching the etched cells in the printing surface and thus adversely affecting pictorial reproduction. Some pigments consist of crystals that are hard enough to cause abrasion; typical products which are prone to mechanically attack printing cylinders are often pretested before they are sold and nonabrasive grades labeled accordingly.

Publication gravure printing inks are commonly toluene-based; in some countries, mixtures of toluene and aliphatic hydrocarbons are also used. Ready-made printing inks normally contain solvents in excess of 60% solvent and between 4 and 10% pigment. The solvent content makes it necessary to employ pigments with sufficient fastness to solvents; i.e., pigments which do not recrystallize too much under the conditions of dispersion and processing.

Pigments are selected for their tinctorial strength, which must be very high. Weaker types clearly have the disadvantage of requiring relatively high pigment concentrations in order to afford equal color value at a given size of the tiny cells in the printing cylinder. The result of using a tinctorially less strong pigment would be a highly viscous ink which behaves poorly in print. For this reason, monoazo yellow pigments are practically no longer used for publication gravure printing inks.

Although Pigment Yellow 12 types show less of a tendency to recrystallize in toluene and aliphatic hydrocarbons, this effect is still quite pronounced. Using pigments with relatively high specific surface areas at the high concentrations that are required to afford a high color value produces very viscous printing inks. To eliminate these problems, diarylide yellow pigments are used almost exclusively in amine-treated form (Sec. 2.4.1.4). The resulting printing inks are not only much less viscous, but the pigments are also easily dispersible and exhibit high tinctorial strength. Fig. 86 shows the shear stress–shear rate curve for amine-treated versus traditional types of Pigment Yellow 12 at equal concentrations in toluene-based publication gravure inks. Amine treatment, which leads to excellent rheology and high tinctorial strength, has made it possible to develop yellow pigment types which satisfy the specifications of technologically advanced printing techniques, including modern high-speed printing.

However, amine-treated pigments also have their problems. The treatment process affords dark red compounds which are easily soluble in toluene and appreciably enhance the tinctorial strength of the resulting printing ink (Sec. 2.4.1.4). The same pigment formulation, however, will be considerably less soluble and thus weaker in printing inks based on aliphatic hydrocarbons or in media which contain

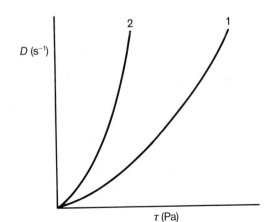

Fig. 86: Rheograms of publication gravure printing inks containing 15% pigment.
Curve 1: Pigment Yellow 12
Curve 2: Pigment Yellow 12, amine treated.

blends of toluene and aliphatic hydrocarbons. As the ink is printed on to the gravure paper, the dissolved colorant, together with the solvent, penetrates the paper. An ink which fails to remain on the surface of the substrate will lose some of its optical effectivity. The ink penetrates through the sheet, i.e., produces a stain on the back of the sheet. An ink that penetrates through the paper is particularly inconvenient since users frequently employ low cost, thin, and light papers in an effort to prevent increasing postal expenses in many countries as far as possible. The demand for less penetrating types of diarylide yellow pigments is satisfied by commercially available special amine-treated pigments which are presently dominating the European market.

The results of amine-treated ruby red and blue pigments, which are used in three and four color printing, i.e., Pigment Red 57:1 and Pigment Blue 15:3 type pigments, do not match the effect of the corresponding types of diarylide yellow pigments. Amine-treated red and blue pigments are therefore of no commercial value. Besides, these pigments are not quite as problematic in the nonamine-treated form as are the amine-free diarylide yellow pigments.

Packaging Gravure Printing

Inks behave very differently depending on their application, which makes it necessary to formulate products with respect to the substrate. The list of common solvents ranges from alcohols and glycol ethers to acetates and other esters to ketones, such as methylethylketone and methylisobutylketone. Aromatic hydrocarbons, especially toluene, are also often used. It is very common to employ a blend of two or more of these solvents to formulate printing inks. Pigments that are used in such inks must be fast to the solvents which are part of the ink system. Moreover, plasticizers such as dibutyl phthalate, dioctyl phthalate, and epoxy plasticizers, which are contained in substrates such as soft PVC, and which are components of printing inks such as nitrocellulose inks, present a challenge to pigment stability. A pigment that satisfies these requirements will show less of a tendency to migrate or recrystallize as the ink is produced, stored, and shipped.

Recrystallization is a particular problem in inks which are formulated to provide transparency in prints on aluminum film, for example. Pigments are usually selected for complete fastness to bleeding in an effort to prevent successive printed layers from interacting with each other — a measure which is especially crucial if a white layer is printed last. The problem receives an added degree of complexity in intermediate layers of composite films, which commonly contain pigments that are not sufficiently fast to the respective laminating adhesive (Sec. 1.6.2.3). Pigments which are to be printed onto packaging films must be particularly fast to a variety of chemical agents and physical influences. Standardized tests have been developed to examine prints for fastness to acid, sodium hydroxide solution, soap (Sec. 1.6.2.2), detergents, cheese, fats, paraffin, wax, and spices. The product is selected with respect to its intended use (Sec. 1.6.2.3).

To facilitate the printing process, pigments which are targeted for use in packaging gravure inks should match the good flow behavior of publication gravure inks. The flow properties of a pigmented system are largely determined by the pigment itself; it is a question not only of the particle size but also of the agents that are being used to prepare the surface of the pigment particles. The advantage of optimizing particle sizes to provide good rheology is frequently compromised by simultaneously deteriorating coloristic properties and performance in application. Such changes may sometimes be prevented by suitable pigment preparation or by predispersing a pigment in nitrocellulose resins (NC chips). Nevertheless, it is advisable to always consider a pigment in the context of binder and solvent. The viscosity of a product may thus be adjusted so as to match certain standards by adding appropriate solvents or certain resins to the formulation. However, before making any attempt to change the flow properties of a pigmented system, it is useful to remember that it is necessary for the solvent to evaporate from the printed ink film. In practice, any ink which fails to dry properly after having been applied to the substrate is not acceptable.

Decorative Printing

The conditions to be met in decorative printing are particularly demanding. Decorative papers for gravure and flexographic printing are used to produce laminated plastic sheets. Pigment selection is a matter not only of the type of resin (melamine or polyester), but also a function of the processing method.

Polyester sheets require pigments which are fast to certain solvents. A pigment that is used to color polyester dissolved in styrene must therefore be completely fast to both polyester and styrene monomer. In order to tolerate the diallyl phthalate process, which involves dissolving polyester in acetone, a pigment should be completely fast to acetone.

Pigments that are intended for use in inks for melamine resin sheets are chosen primarily with respect to the particular process characteristics. The list of possible methods includes the high-pressure process with or without an overlay paper or with chipboard coating; or by the short-phase technique, which also makes it possible to manufacture polymer films. Working in the absence of overlay paper, i.e., without covering the printed decorative paper with a transparent resin-impregnated paper, makes it necessary for a pigment to be fast to scumming. Scumming is related to plate-out (Sec. 1.6.4.1); the term refers to the deposition of pigment on the impression cylinder or plate, which in turn affects the surface structure of the following articles. The complex nature of the scumming phenomenon makes it necessary to select pigments on an empirical basis only. Fastness to binders and to the solvents that are typically contained in inks to be printed on decorative paper is less of a concern, as experience shows that useful pigments can also be expected to be compatible with the printing ink. When considering the final product, however, excellent lightfastness is a priority issue.

1.8.1.3   Flexo and Screen Printing and Other Printing Processes

The technical requirements for pigments incorporated into flexo, screen, and other inks depend on the desired fastness properties, such as fastness to light, to solvents, and to bleeding, as well as on the processing conditions, dispersion techniques, etc. The limitations imposed on pigment use by these methods are not more demanding than in offset/letterpress application (Sec. 1.8.1.1) and gravure printing (Sec. 1.8.1.2).

1.8.2   Coatings

The requirements for pigments which are targeted for use in paints depend primarily on the final coated product, but also take into account the paint composition and color as well as manufacture and process characteristics. In discussing these requirements, a classification system is used based on pigment related characteristics and the drying process. Systems which do not pose a particular challenge to the fastness of a pigment are therefore of no concern in this context. This is also true for the majority of the physically drying paints containing nitro and nitro-combination lacquers.

1.8.2.1   Oxidatively Drying Paints

Heading the list are air-drying medium and long-oil alkyd resin systems, which command a good share of the market. They are broad in scope; areas of application range from architectural paints to industrial paints, such as special-purpose automotive repair finishes. These systems contain solvents such as aliphatic (white spirit) and aromatic hydrocarbons, turpentine, and higher alcohols, in which most organic pigments are almost insoluble under common processing conditions. Consequently, there is no limit to the choice of pigment with respect to solubility. Lightfastness and weatherfastness are the primary criteria for pigment selection besides coloristic parameters.

1.8.2.2   Oven Drying Systems

Oven drying systems often require highly specialized equipment and production units for application and drying. Such systems are mainly processed in industrial operations. At the present time, the majority of industrial coatings are oven drying (i.e., heat set) systems.

Besides the solvents which are found in air-drying paints, most industrial finishes also contain solvents which may be more or less aggressive to organic pigments, such as glycols and glycol ethers, including ethyl and butyl glycol; esters like ethyl and butyl acetate; ketones, such as acetone, methylethylketone, and methylisobutylketone; and finally chlorinated hydrocarbons and nitroparaffins.

Conditions to be met in oven drying enamels depend also on the composition of the binder. Paint systems containing melamine formaldehyde or urea formaldehyde resins, for instance, harden by polycondensation with other resins, such as epoxy resins, short-oil alkyd or acrylic resins at elevated temperatures. Baking is carried out at temperatures between 100 and almost 200 °C and may last from a few minutes to more than an hour. A general trend towards energy conservation has in recent years shifted public attention towards binders which require low baking temperatures.

Two-component finishes begin to harden immediately upon mixing the components, which makes it necessary to consider the respective pot life. Components may react at room temperature; but commercial scale operations are usually accelerated by working at temperatures up to 120 °C. This is true for polyesters containing hydroxy groups, cross-linked with isocyanate, and acrylic resins designed for automotive refinishes. High temperatures are an equally common curing method throughout the wood coatings and furniture industry, which uses partially pigment-free unsaturated polyester lacquers and acid hardening alkyd/melamine or urea systems, respectively.

In recent years, considerable efforts have been made to restrict or completely eliminate solvent emission during oven drying. Intensive research has focused on developing paint systems which contain very little solvent or none at all. A variety of approaches are possible; systems that have found their way to pilot testing tend to present a special challenge in terms of pigment selection:

- Materials known as high-solids and medium-solids systems have a higher solids content and a consequently lower amount of solvent in application, compared with conventional baked enamels [5]. The binders which are typically used in these systems have a lower molecular weight and are somewhat less efficient wetting agents, but they present no particular problems with regard to pigment dispersion and application.

- Nonaqueous-dispersion systems (NAD systems) consist of a blend of dissolved and dispersed binders. The higher solid content of these resin systems—and the correspondingly smaller solvent content—is achieved by partially dispersing the binder. This is of some consequence to pigment dispersion, which is usually carried out in dissolved resin. Where the resin content is not high enough to sufficiently wet the surface of the pigment particles and to stabilize the dispersion, flocculation may occur, especially in pigments with a high specific surface area and at a high pigment concentration. These effects may be avoided or reduced by methods such as increasing the dissolved resin content in the NAD system.

- Water-reducible systems are usually waterborne but also contain some portion of organic solvents. The high surface tension of water has a considerable effect on the wettability of a pigment. Highly polar pigments are generally wetted easily by aqueous media and thus disperse readily. Pigments which lack polarity, on the other hand, may be difficult to disperse; but it is possible to improve the dispersibility of a nonpolar pigment through chemical modification. Polar functions, introduced into the pigment structure, or incorporation of specific pigment additives can be highly advantageous.

Most of the commercially available organic pigments show a behavior some-where in between these two extreme cases. Possible dispersion problems, which may become evident through lack of gloss, can be avoided by suitably formulat-ing the mill base and by adapting the processing conditions [6]. Water-reducible systems have been discussed extensively in the literature, and a large number of patents have been issued [7–10].

- At present, systems cured by ultraviolet light or electron beams [11–13] are used primarily in wood coatings, which contain very little or no pigment; they are employed to a lesser extent in paper coating and other applications. In contrast to the printing field, the comparatively thick layers in which paint systems are typically applied pose somewhat of a challenge to the drying process, especially since pigments absorb UV radiation.

- The requirements for a pigment used in a powder coating [14, 15] are for var-ious reasons extremely stringent. The shade of a coating, for instance, frequent-ly changes as the material is cured; the effect may be considerable, depending on the temperature and the time [6]. Problems may be avoided or reduced by suitably adjusting the curing agent and the pigment. The color change has in some cases been shown to result from a reaction between pigment and hardener (Sec. 1.6.7).
  Plate-out is another one of the phenomena which effect color change in a powder coating (Sec. 1.6.4.1). Plate-out is largely due to insufficient wetting of the surfaces of solid particles within the coating. It is possible to reduce or to avoid this effect by selecting suitable pigments and/or by using certain harden-ers which are only effective at elevated temperatures.

Pigments which are designed to color paints are selected partly on the basis of their properties and performance with respect to the applied coating; this includes lightfastness, weatherfastness, and fastness to certain chemicals. On the other hand, pigments must also tolerate paint synthesis and processing as well as applica-tion of the pigmented medium. The rheology of a system is one of the primary issues, since good flow behavior facilitates dispersion and application. The list also includes fastness to the respective solvents and resulting properties, such as fastness to overspraying, tendency to bloom, and recrystallization stability. Solubility re-lated fastnesses are controlled by the particle size distribution and concentration of the pigment and mainly by the processing temperature (Sec. 1.6.3.2). The composi-tion of the paint, the choice of solvents, and the duration of heat exposure are also important.

These requirements considerably restrict the selection of pigments for oven dry-ing systems. The complex nature of the problem makes it necessary to submit any pigment which is basically applicable to a pilot experiment under the exact pro-cessing conditions.

The worldwide trend towards lead-free automotive finishes has given rise to special requirements for organic pigments in this area. Lead chromate pigments, for instance, are not sufficiently fast to high sulfur dioxide concentrations and are affected by materials containing sulfuric acid, such as soot or moist dead leaves, to

which automotive finishes are frequently exposed. This is especially true in industrial and urban areas. Fig. 87 shows a paint containing Chrome Yellow, which was
subjected to a test as described in Section 1.6.2.2: treatment with 1 N sulfuric acid
for one hour at 70 °C. The distinct color change on the circular test area is accompanied by a considerable gloss reduction. In the corresponding lead chromate-free
system, it is only the binder that is affected, which reduces the gloss; the color
value on the other hand is retained.

Chrome Yellow and Molybdate Orange Red pigments may be replaced by other
pigments to cover the spectral range from greenish yellow to bluish red. Such pigments must be opaque enough to completely hide the substrate in 40 μm thick
layers in which automotive finishes are typically applied. It is therefore necessary
for suitable organic pigments to provide considerable hiding power. With a number
of organic pigments, the particle size may be optimized (Sec. 1.7.3) and the pigment concentration in the lacquer simultaneously increased to satisfy the opacity
specifications. The very small specific surface areas of such pigments compared to
conventional types make it possible to appreciably increase the pigment concentration without affecting the flow properties, which are vital to ensure easy processing
(Sec. 1.7.8). However, products with an optimized particle size, i.e., organic pigments with high hiding power rarely reach the hiding power of corresponding
Chrome Yellow or Molybdate Orange/Red full shade pigments, even if the pigment concentration in the paint is high [16]. This is principally a result of the higher refractive indices of the inorganic pigments (Sec. 1.7.3) compared to their organic counterparts.

Commonly, blends of organic pigments with suitable inorganic pigments are
used, such as nickel titanium yellow, chrome titanium yellow, bismuth–molybdenum–vanadium–oxide, or iron oxide.

Formulations which do not contain lead chromate are becoming increasingly
important, not only for automotive finishes, but also for architectural paints and
other areas of paint application.

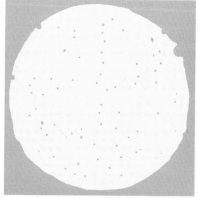

Fig. 87: Systems containing organic pigments (left) and Chrome Yellow (right). Determination of resistance to acids with 1 N sulfuric acid.

The application conditions for pigments in metallic finishes regarding hiding power and lustre, on the other hand, vary considerably (Sec. 1.7.3). Pigments which are targeted for use in such systems must be highly transparent. The base coat of a typical European two-coat metallic finish usually contains a combination of oil-free polyester, cellulose acetobutyrate, and melamine resin. The clear top coat commonly consists of acrylic resin and melamine resin, frequently containing UV absorbants to improve the lightfastness and weatherfastness of the entire system.

Depending on the composition of the mill base and the dispersion and application conditions, the dispersion of an organic pigment in the base coat frequently affords coloristially different coatings. Good dispersion is essential in order to produce the desired high transparency; intensive shear, however, is required to realize this end with pigments with typically fine particle sizes. Aluminum pastes which are dispersed in a binder solution (for metallic finishes), on the other hand, are adversely affected by high shear, because shear changes the structure of the aluminum flakes. Even as the aluminum pastes are manufactured, the particle size distribution is kept as narrow as possible; the smallest particles, which create a grey, dull visual impression, are removed. Applying intensive shear as the paste is incorporated into the base coat seriously compromizes the coloristic advantage of fine grain separation. The same applies to the use of pearlescent pigments.

It should be noted, however, that the dispersion process is of general importance for the application of a pigment in industrial paints. Mill base formulation, various additives, dispersion unit, and dispersion conditions, type of application and operating conditions, and a number of other parameters (Sec. 1.6.5) determine the degree of pigment dispersion in the finished paint. Neglecting the cause-effect relationships and interactions will disturb further processing, an effect which is easily revealed in laboratory experiments. Rub-out phenomena (Sec. 1.7.5) are among the more common of these effects.

A spraying-pouring test has been devised to facilitate comparisons between application conditions. Films made of the same paint, applied by spraying or pouring, are compared with respect to the gloss in full shade and the tinctorial strength of the reduction. It is possible to try and eliminate differences by adjusting the mill base formulation and the dispersion conditions. Specialized additives for paints are available for systems which do not respond to these attempts. These agents are very useful in avoiding or eliminating floating in pigment blends. The theoretical background of floating and other reasons to explain rub-out effects are discussed in the literature [17, 18, 19].

A wide variety of dispersion equipment is employed by the paint industry to produce oven drying paints. All have advantages and disadvantages for pigment application, which have to be weighted in each case. A unit for a particular purpose is selected on the basis of the tendency of the pigment to recrystallize and the degree of pigment dispersion at optimized mill base formulation. Equally important are energy considerations and product output, that is, the amount of paint produced in a certain time. The list also includes the type of operation, i.e., by continuous or batch process, and cleaning a unit before color change. It is important to note that not all types of mills are suitable for pigments which are difficult to disperse.

There are a number of specialty areas which pose a particular challenge to suitable candidates for pigmentation. One example is the long-term heat stability in specialty paint systems for radiators. A number of azo and polycyclic pigments have been tested for color stability by exposing the coated metal sheets to 180 °C for 1000 hours and subsequent irradiation in an accelerated weathering cabinet. Compared to unheated specimens, the lightfastness of the thermally treated samples is inferior by an average of one step on the Blue Scale.

Another specialty area is coil coating, which involves coating metal coils by continuous operation. Modern roller systems afford speeds of up to 200 m/min. Most coils are made of cold-rolled and surface treated steel, aluminum, or alloys of the latter with manganese or magnesium. Coating systems are based on alkyd or acrylic resins, oil-free polyester, silicone-modified polyester or acrylic resin, poly(vinylidene fluoride), or poly(vinyl fluoride). Recently, water-reducible systems, mainly based on acrylic resins, have been developed for aluminum as well as for steel coils [20–23]. Drying is carried out by continuous operation in gas- or oil-heated multichamber ovens.

The layers are heated to 280 °C or higher for a few seconds up to several minutes. The coated metal coils are then processed, i.e., deformed, profiled, or stamped. The resulting coated products are broad in scope; they are employed in a variety of areas from packaging and the vehicle industry to household items and building materials.

The conditions to be met by pigments which are targeted for this area of application vary considerably, depending on the purposes and the coating system. For products which are sufficiently heat resistant it is important to consider the solvent fastness and recrystallization stability of the pigment in relation to the respective coating system, its composition, and processing history. Organic pigments rarely satisfy the demand for extreme weatherfastness over a period of several years, which is crucial, for instance, in applications such as sheet metal panels for wall covering. Polycyclic pigments and some very fast azo pigments are commonly selected where the requirements are less stringent. A pigment which adversely affects the mechanical properties of a coating, such as its hardness or scratch fastness, or extreme elasticity are not acceptable in practice.

1.8.2.3 Emulsion Paints

Emulsion paints are based on aqueous synthetic resin dispersions, which afford a lacquer-like paint film. The resin dispersions which are commonly used by the paint industry contain water as the carrier phase. A large number of such dispersions are available, based on different resins such as poly(vinyl acetate), which may be employed as a copolymer with vinyl chloride, maleic dibutyl ester, ethylene, acrylic acid esters, polyacrylic resin, and copolymers of the latter with various monomers, as well as styrene–butadiene or poly(vinyl propionate). These dispersions are highly balanced systems, prepared by adding emulsifiers to the inner or outer dispersion phase.

The dispersions are stabilized by agents such as cellulose derivatives, poly(vinyl alcohol), starch, or gelatine, which more or less swell in the presence of water and act as protective colloids. Using pigment powders may give rise to problems, because increasing the solid surface area within the dispersion will almost invariably disrupt the equilibrium between the two phases. Therefore it is most advantageous to employ pigment preparations which contain surface active agents that do not noticeably affect the dispersion equilibrium. Preventing the viscosity from increasing and coagulating the entire paint system is critical to the success of a system. Moreover, these pigment preparations sometimes lose tinctorial strength as they are stored, due to wetting, dispersion, and stability deficiencies. Less solvent resistant pigments may also tend to recrystallize, a trend which is frequently enhanced by surface active agents. Loss of tinctorial strength, a phenomenon which is commonly observed in pigmented paints, may be caused by a number of reasons. Inadequate wetting of the particle surfaces will affect the tinctorial strength of a pigmented system, and every care must be taken to ensure adequate wetting in the presence of water-immiscible solvents, large amounts of antifoam, or later addition of plasticizers [24].

Depending on the area of application, pigments in such paints must satisfy a number of additional specifications. Exterior house paints, for instance, should not only exhibit excellent weatherfastness but also tolerate lime and concrete (Sec. 1.6.2.2).

Apart from pigment preparations which have been developed specifically for aqueous emulsion paints, there are also products known as multipurpose tinting pastes that show outstanding fastness. They have the advantage of being useful not only in emulsion paints but also in solvent containing architectural paints.

Multipurpose tinting pastes usually contain pigment and hydrophilic solvents, sometimes some amount of water, and suitable wetting agents, which define the equilibrium between hydrophilic and lipophilic character. Conditions to be met in each case depend on the type of paste and on the method by which it is produced and also on the intended use of the coating or paint.

1.8.3   Plastics

The large number of different plastics on the market is responsible for the variety of requirements for organic pigments which lend color to polymers. Within the scope of this book, it is possible only to discuss the most important types of plastics.

Pigments are selected with regard to the plastics to be colored and the conditions to be met by the final product; among the major determinants of pigment use are the processing conditions.

Almost all plastics contain additives by the time they are processed. Various types of stabilizers are used in order to render a material stable to heat and light; the list also includes plasticizers, lubricants, wetting agents, flame retardants, fillers, extenders, foaming agents, antistatic agents, fungicides, substances to improve the impact fastness, optical brighteners, UV absorbants, and, last but not least, colorants. Any one of these additives may influence the coloristic properties of the colored plastic to a more or less appreciable extent, and additives may act quite differently. Adding even large amounts of barium sulfate to PVC, for instance, only slightly reduces the transparency of the product; while traces of antimony trioxide, a fire retardant which forms antimony trichloride, have a considerable impact on the transparency. Similar observations are made with solid lead stabilizers which, when worked into PVC, act as strong brighteners; this, however, is not true for other common agents, such as tin/sulfur or barium/cadmium stabilizers.

A number of plastics, such as polyethylene, are delivered in the form of ready-made blends which already contain additives. In other cases, such as with PVC, it is up to the user to add suitable agents. A wide range of different units are available to premix the plastics and the additives and/or pigment; the type of equipment and the operating conditions determine the coloristic outcome. The color of the product is thus defined; it is not possible to change the color later by subsequent processing.

A battery of different premixing units are available, ranging from simple gravity mixers, such as rotary mixers, tumbling mixers, dry-blend mixers, and pebble mills, to agitated ball mills and high-speed mixers, including fluid mixers, mixing rolls, and kneaders for materials such as natural and synthetic rubber. Mixing is followed by comminution unless the process affords a mixture in liquid or powder form. Thermoplastics, for instance, may be granulated or cooled so that the melt turns solid. Several options are available.

A variety of different techniques are employed to process plastics; and processing conditions vary considerably between methods. Heading the list are operations such as injection molding and extrusion of profiles, hollow articles and blown films, calandering, coating, sintering, dipping, foaming, and varnishing.

Plastics processing, no matter by which technique, almost always presents somewhat of a challenge to suitably dispersed pigments; candidates are often rejected for insufficient dispersibility. Pigments are usually dispersed in polymers by applying shear (Sec. 1.6.5). Wetting the surface of the pigment particles by polymer molecules is only of minor importance. The shearing forces that are necessary to satisfactorily break down pigment agglomerates in thermoplastic materials, such as PVC or polyolefins, are only realized if the polymer is not too plasticized (soft)—a demand which is rarely satisfied under common processing conditions.

It is primarily for economical reasons that the primary parameters which govern pigment dispersion in plastics have in recent years generated an increasing number of problems. Calandering temperatures and speed in the processing of plasticized PVC, for instance, have increased so much that the time span over which shearing forces are applied has shortened considerably. The disadvantage is that the shearing forces in a heavily plasticized polymer are no longer high enough to satisfactorily break down the agglomerates of conventional organic pigments. Particular problems are encountered in thin films, in which holes may appear through insufficient dispersion. Another example is spin dyeing, where failure to disperse properly may cause the filaments to break.

Therefore pigment powders are only used in plastics if thick walled products are manufactured, such as hollow articles, injection molded articles, or for sheet extrusion. They have the advantage of being broad enough in scope to be used in almost all commercially available plastic materials. Apart from a lesser degree of dispersion, which may show in the form of filler specks or color striation, pigment powders present somewhat of a problem, partially because they make it difficult to clean the processing units (such as mixers, metering devices, and pneumatic conveyors). Where conventional pigment powders are employed, irrespective of these problems, it is advisable to pretreat the pigment in an additional processing step in order to achieve good dispersion before it is worked into the plastic.

Problems may be avoided by coloring plastics with pigment preparations or pigments which feature a comparatively narrow particle size distribution.

Pigment preparations have the advantage of being predispersed by the manufacturer or by companies who specialize in this area. A suitable carrier material is used which either closely resembles the plastic to be colored or is at least compatible with it. Such preparations are produced using kneaders, three-roll mills, extruders, or other high shear dispersion equipment.

The pigment concentration of a given preparation depends on the shade and on the desired pigment preparation-to-polymer ratio. The maximum pigment concentration is defined by the absorption capacity of the carrier material.

A pigment which is predispersed in a pigment preparation has the advantage of being equally distributed throughout the vehicle, which facilitates color reproduction under the normal processing conditions for the respective polymer. No additional effort is therefore needed in terms of time, energy, and equipment to distribute the pigment preparation equally throughout the polymer and independently of processing temperature.

Pigment preparations are available in solid form as granulates or powders and also in paste form. They usually contain between 10 and 80% pigment. Each product is designed for very few plastics only, depending on the carrier material. Universally applicable pigment preparations play no role on the market since they require pigments which are both expensive and equally broad in scope.

To ensure good distribution of the pigment preparation throughout the polymer, it is often necessary to use a carrier material which flows more easily and is more easy to distribute than the polymer to be colored. There are different ways to achieve this end, for example by employing a carrier with a low average molecular weight, a suitable mixed polymer, or using a high plasticizer content.

In recent years, there has been an increasing trend in favor of premixed pigment concentrates. These are pigment preparations which contain two or more pigments and which afford a defined shade if they are worked into a certain defined amount of polymer. Suitable color-matched pigment mixtures in powder form are available for the same purpose. Such powder mixtures used to be produced only by means of edge roll mills, which have since been replaced by high-speed mixers and specialized pigment mills. It is customary also to add trace amounts of surface active agents.

The dispersibility of an organic pigment in a given polymer depends on a variety of factors. This may give rise to problems if mixtures of such pigments are used. It is important to monitor the conditions very closely, since even minor temperature changes, which have a plasticizing effect on the polymer and affect the shear forces, may make it difficult to reproduce a desired shade. Pigment blends and color concentrates are also referred to as preprogrammed colorants.

Plastics are commonly marketed and processed in powder or granulated form. Certain techniques have proven to be advantageous in coloring polymers with pigments or with pigment preparations [25]. Polymer powders, for instance, are frequently premixed with pigment powders in fluid mixers. The pigments are then dispersed in the course of the plastic processing. PVC, for instance, requires kneaders or roll mills, while polyolefins are processed by means of extruders or screw-type injection molding machines. The latter disperse a pigment through shearing forces between screw and cylinder surface [26]. The degree of dispersion which is realized in such machines satisfies the specifications for thick-walled products but not the more stringent requirements for other articles such as thin films.

Granulated plastics are best colored by granulated pigment preparations or preparations in paste form, such as pastes made of pigment/plasticizer combinations. They are used to advantage where automatic dosage is employed. The necessary premixing of the polymer with the pigment preparation is usually achieved by using slow running mixers. Pigment powders tend to adhere to the surface of the granulate through electrostatic forces. The adhesion is frequently enhanced by adding trace amounts of an adhesion promoter, i.e., a wetting agent, to the granulate. The adhesion promoter, which covers the granulate as a fine microfilm, must be applied prior to pigment addition.

Granulated pigment preparations are only marginally suitable candidates to color polymer powders. No sufficient pigment distribution is achieved by this coloration method, and separation may occur, especially if pneumatic conveyors are used [27]. More information on the dispersion of organic pigments in plastics is found in Chapter 1.6.5.

Pigments which are targeted for use in plastics are often selected for their solvent fastness and insolubility in the polymer and its components, especially under the processing conditions. The phenomenon of migration, for instance, which includes the effects of blooming and bleeding, is a result of complete or partial pigment dissolution in the plastic at the processing temperature (Sec. 1.6.3).

Recrystallization is also an effect that is created through partial solubility of a pigment in the plastic. Like in other media, a partially dissolved pigment will change the transparency or the hiding power of a transparent product and also affect the depth of shade in white reductions. Insufficient recrystallization stability shows in the manufacture and processing of pigment/plasticizer pastes, as well as in various polymers at elevated processing temperatures.

High processing temperatures restrict the number of applicable pigments for a variety of plastics (Sec. 1.6.7). One of the important parameters is the duration of heat exposure. This is why it is important to consider the thermal history of any regenerated material being used. It is also important to consider the relation between heat stability and pigment concentration and also the ratio of colored to white pigment concentration. The required heat stability is a function of the type of plastic.

Some organic pigments create shrinkage in thick-walled, large-surfaced, non-symmetrical extrusion products such as bottle crates by acting as nucleation agents to partially crystalline polymers (Sec. 1.6.4.3).

A large number of organic pigments satisfy the lightfastness conditions for pigments which are to be used in plastics. The lightfastness of a colored system is normally limited by the lightfastness of the plastic.

The fastness of a pigment to light may vary considerably, depending on the plastic carrier. It is possible for a pigment to bleach considerably in one polymer, while, when incorporated in another polymer, it is not even affected by light. This is why the lightfastness of a pigment in a plastic is always mentioned with reference to the entire pigmented medium and not to individual components.

Lead stabilizers or flame retardant antimony trioxide scatter so much light in PVC that they influence the lightfastness of the plastic system as much as titanium dioxide does in white reductions. The comparative values given in Chapters 2 and 3 of this book always refer to colorations of equal standard depth of shade in accordance with DIN 53 235/2.

Industrial standards have been developed to determine the fastness of colored pigmented plastics to daylight exposure and to xenon arc lamps by using accelerated test methods. Similar normative tests exist for weatherfastness determinations (Sec. 1.6.6).

A number of polymer systems are somewhat sensitive to the compatibility of a pigment with the polymer to be colored. In other cases, the chemical and/or mechanical properties of the plastic material must be prevented from deteriorating as a result of pigment incorporation. These requirements are specific to certain polymers and are discussed in the respective chapters.

Pigments designed for plastics articles, such as films, which are used as packaging materials for food or cosmetics, must not only be migration resistant and extraction proof, but also physiologically safe. Purity regulations differ between countries.

Only very few pigments satisfy all these specifications and additional ones to suit certain specialty purposes without being extremely costly. Generally, more stringent fastness requirements are met only by more expensive pigments. Like in other areas, pigment selection for a particular coloration purpose in the plastics field usually represents a compromize between fastness requirements and economical considerations. It should be noted in this context that inorganic pigments also play a considerable role in the coloration of plastics. Technical and economical considerations determine whether organic or inorganic pigments are used to advantage in each individual case.

Organic pigments perform satisfactorily where transparent colorations are desired and high tinctorial strength is concerned, especially in thin-walled articles such as films or spin dyed fibers. Multicolor printing on films is another one of the areas where organic pigments are commonly used profitably. If, on the other hand, a very light shade is required, along with high hiding power and considerable lightfastness and weatherfastness, inorganic pigments are preferred.

Inorganic and high grade organic pigments are frequently combined to advantage. Using an excess of the usually weaker inorganic pigment makes it possible to afford the desired hiding power; while a smaller amount of the organic pigment will mainly improve the tinctorial strength of the blend and enhance the brilliance of the product.

Whenever plastics are pigmented, it is important to also consider the optical properties of the carrier. The individual absorption of each plastics material, for instance, frequently accounts for a more or less pronounced yellowing. Crystalline polymers scatter light as much as so-called blends, which are mixtures of various different plastics components with different refractive indices. Orientation effects are frequently observed in such plastics blends. Fillers not only absorb and scatter light but also resemble the plastic in their optical properties. As foam products are colored, the highly lightening effect of the bubbles must be taken into consideration.

The following sections discuss the primary types of plastics and the most important properties with regard to pigment incorporation, as well as the processing conditions. Conditions to be met by suitable pigments are listed and applicable test methods are described.

1.8.3.1 Poly(Vinyl Chloride)

Vinyl chloride has been known for more than a century. It was first polymerized in 1912 to form poly(vinyl chloride), known as PVC. Industrial scale production started in 1927, and PVC continues to exhibit the broadest and most useful range of properties of any of the plastics which are available today. A variety of polymerization techniques and methods to adequately process the monomer facilitate the production of a wide range of VC homo- and copolymers. The polymer can, for instance, be molded by using any of the available processing units at elevated temperatures, as long as its thermal sensitivity is taken into account. PVC may be glued, bent, heat-sealed, printed on, and thermoformed.

VC may be polymerized by a variety of techniques, known as mass or bulk, suspension, solution, or emulsion methods, by which a range of different products may be obtained. The choice of polymerization technique largely defines the criteria for pigment selection, because a number of organic pigments are quite sensitive coloristically to the polymerization method.

In contrast to mass or suspension polymerization products, emulsion PVC (PVC-E) is made in the presence of emulsifiers, which remain more or less within the plastic product. While PVC-E usually also contains sodium carbonate, which lends a milky appearance to the resulting plastic article, emulsifiers such as alkylaryl sulfonate considerably reduce the tinctorial strength of the product upon pigmentation. This effect is most noticeable in PVC-E as opposed to mass or suspension polymerized PVC.

At equal pigment concentrations and analogous processing conditions, the tinctorial strength of a PVC-E product may be as much as 50% below that of mass or suspension made types, especially if less easily dispersible polycyclic pigments are used. The K value, i.e., the average molecular weight of the PVC type, has no effect on this pigment behavior. The drop in tinctorial strength is less noticeable if higher shearing forces are applied to disperse a pigment. This effect is used successfully by processing systems at low temperatures by means of roll mills. The difference in tinctorial strength between PVC-E and other types of PVC accordingly loses significance as the plasticizer content decreases; but it is still noticeable even in rigid PVC at low processing temperatures and long dispersion times, i.e., high shearing forces.

In order to ascertain whether it is the emulsifier which affects the tinctorial strength of a system, pigments were dispersed in mass and suspension-made PVC samples in the presence of emulsifiers which are typically used only in PVC-E. The results were the same as in the first experiment, which confirms this assumption.

PVC-E emulsifiers, even in traces of only 0.005%, have a considerable impact on the tinctorial strength of a pigment, which is much more noticeable if they are used during pigment preparation. Being extremely active tinctorial strength determinants, these agents even affect pigment preparations such as pigment/plasticizer pastes despite the fact that such systems only contain predispersed pigment.

Loss of tinctorial strength, however, is not as problematic with all products. Color changes are not as obvious, for instance, in PVC-E which is processed in the absence of plasticizer. Moreover, very dark pigmented products (frequently containing inorganic pigments) make it more difficult to discern differences in tinctorial strength compared to other types of PVC. Therefore a drop in tinctorial strength is not always considered a disadvantage.

It should be noted at this point that it is very difficult either in the laboratory or in industrial scale operations to attain exact and reproducible PVC coloration, unless the working conditions are precisely defined and specialized processing units are used.

The question of whether a given pigment is suitable for use in PVC, as in other plastics, is a function of its dispersibility. It is becoming increasingly more difficult to adequately disperse a pigment under processing conditions which are subject to economical considerations. This problem presents an added degree of complexity at high processing temperatures. Calandering, for instance, is frequently carried out at temperatures in excess of 200°C, which improves the output efficiency and reduces the dispersion time.

In order to test pigments for dispersibility in plasticized PVC under standardized conditions, a national German industrial standard was developed in 1975 [28]. Cold rolling was carried out by means of a standardized laboratory scale roll mill; however, this technique has for various reasons failed to gain industrial recognition. In the meantime, pigment manufacturers and users, in carrying out comparative studies, have worked out new test conditions and test materials. The essence of these studies is expected to develop into an international industrial standard (ISO), in which a standardized PVC blend is pigmented under defined conditions at 160°C (moderate shearing forces) and at 130 or 75°C (higher shearing forces).

The γ-modification of Pigment Violet 19 may serve as an example to illustrate the influence of increasing temperatures on pigment processing by means of a roll mill; the tinctorial strength-time curve in PVC is examined for the temperature range between 130°C and 180°C at a plasticizer content of 33% dioctyl phthalate. The two temperature extremes result in a difference in strength of almost 60% (Fig. 88); the milling time is 8 minutes in both cases.

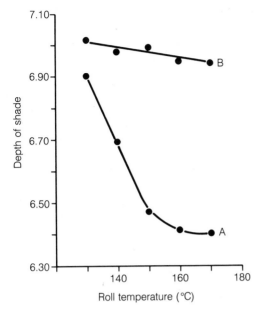

Fig. 88: Pigment Violet 19, γ-modification, in plasticized PVC. Influence of the processing temperature on the depth of shade.
Pigment content: 0.1%; TiO$_2$: 0.5%
Curve A: pigment powder
Curve B: pigment preparation

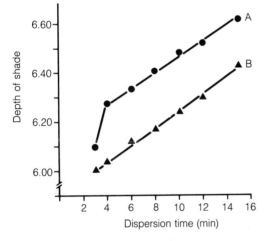

Fig. 89: Influence of the dispersion time and of the dispersion temperature in a roll mill on the depth of shade of P.V.19, β-modification, in plasticized PVC.
Organic pigment: 0.1%; TiO$_2$: 0.5%
Curve A: 130°C
Curve B: 160°C

The effect of the milling time is demonstrated by means of an analogous system, which contains the β-modification of Pigment Violet 19 (Fig. 89). It is easy to see that even a 15 minute effort does not suffice to completely disperse the pigment.

Reducing the plasticizer content at a given temperature improves the degree of pigment dispersion by enhancing the viscosity of the plasticized polymer. However, roll mills are inadequate to disperse a pigment completely, even in plasticizer-free PVC, under the given dispersion conditions (see also Sec. 1.6.5).

On the other hand, only a minor temperature effect is observed in a pigment preparation which also contains the β-modification of Pigment Violet 19, incorporated in a vinyl chloride/vinyl acetate copolymer as a carrier (curve B in Fig. 88). The preparation contains the predispersed pigment, which makes it possible to achieve uniform pigment distribution throughout the equally plasticized polymer as the pigment is thermally worked into the carrier material after the latter has been plasticized. However, differences in tinctorial strength are colorimetrically obvious even in this case, but the deviations approach the tolerance limits of the methods used. Such preparations are equally suited for use in flexible and in rigid PVC.

The results of these studies are universally applicable. Although there are other pigments which are much easier to disperse, conventional types on the whole tend to be inferior to comparable pigment preparations. At equal pigment concentration, incorporating a pigment preparation into a polymer will nearly always improve the tinctorial strength of the product, which reflects a higher degree of pigment dispersion.

Specific problems arise with pigment/plasticizer pastes, which are common colorants for PVC. If the plasticizer content in a PVC-plasticizer mixture exceeds a certain value, the polymer passes from a plastic to a brittle state, due to a miscibility gap in the PVC/plasticizer system. This miscibility gap is plasticizer-specific; with dioctyl phthalate, for instance, it occurs in the concentration range between 5 and 18%. Therefore such pastes are either only conditionally suitable to color rigid PVC or not at all.

Dispersion units such as three-roll mills are quite adequate to disperse organic pigments in plasticizers. It should be noted, however, that the output efficiency is not always as desired. Pigment/plasticizer pastes, which are typically less tacky than systems such as letterpress or offset inks, respond best to treatment in agitated ball mills.

Plasticizer pastes usually contain between about 20 and 35% pigment. Incorporating a pigment in a plasticizer is frequently facilitated by adding special-purpose dispersion agents. These additives make it easier for the medium to wet the pigment particles, facilitate attempts to increase the pigment concentration in a paste, and shorten the time needed for dispersion. Fewer passes on a three-roll mill are therefore necessary to achieve satisfactory pigment dispersion. Because pigments may vary appreciably in their dispersion behavior and differences between pigments may be considerable, some caution should be observed as pigment/plasticizer pastes are applied. A pigment blend which is designed to lend a specifically defined color to a product should never be dispersed as such. It is, on the contrary, advisable to disperse the pigments separately in different batches of paste and then to homogeneously blend the ready-made pastes containing one pigment each. Pigments which tend to recrystallize in their plasticizer paste will present problems during manufacture, storage, and processing. PVC pastes (plastisols) are colored most elegantly using pigment/plasticizer pastes; although pigment powders are equally suitable for such pastes. This is primarily due to the fact that plasticizers wet pigment particles quite readily (Sec. 1.6.5).

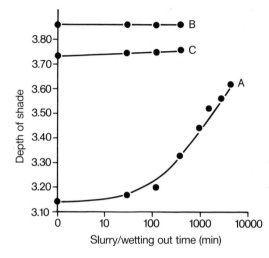

Fig. 90: Influence of the dispersing conditions (stirring, slurry/suspension time, temperature) on the degree of dispersion (depth of shade) of P.Y.19, γ-modification, in a heavy PVC paste. Organic pigment: $TiO_2 = 1:50$
(A) pigment powder; suspended at 20 °C
(B) pigment preparation; suspended at 20 °C
(C) pigment-plasticizer paste (obtained: 1 pass three-roll mill, 15 °C), wetted out at 20 °C.

This is illustrated by dispersing a type of Pigment Violet 19 (Fig. 90), which is difficult to disperse, in a plastisol containing 39% dioctyl phthalate [29]. The pigment is incorporated in the plastisol under slow agitation by hand, and the pigmented system is then stored at room temperature. After some days, the depth of shade almost matches that of the reference sample consisting either of a pigment/plasticizer paste dispersed on a three-roll mill, or a pigment preparation based on a vinyl chloride/vinyl acetate copolymer. The technique of dispersing a paste is referred to as wetting out, an operation in which a pigment is often dispersed almost entirely through interaction between pigment surface and plasticizer.

In making use of the good wetting behavior of DOP, attempts have been made to improve the dispersion behavior of pigments in PVC by preparing the materials with 20% DOP. However, surface treated types fall short of the expectations.

In order to reduce the viscosity of PVC plastisols, the pastes are often diluted with solvents such as volatile aliphatic or aromatic hydrocarbons, mainly mineral spirits, dodecylbenzene, or glycols. The solvents do not have a gelating effect and may be removed by vaporization, followed by gelation, in order to avoid formation of fine cracks or bubbles. Before diluting a paste, it is important to ensure that all pigments are sufficiently fast to the solvents at the respective processing temperature.

In practice, PVC plastisols are commonly manufactured by means of high-speed planetary mixers which can be evacuated. It is best to feed the plasticizer into the unit before stirring in the pigment or pigment pastes, and finally to admix the other solid components in portions. Where plasticized PVC mixtures are to be colored, on the other hand, it is necessary to first incorporate the pigments in the system before adding plasticizer. Pigment/plasticizer pastes, however, may be admixed simultaneously with a plasticizer.

PVC may be processed by any of the methods which are recommended for thermoplastic polymers. The requirements to be met by suitable organic pigments in terms of heat stability and migration fastness vary accordingly.

A number of international industrial standards (ISO) have been developed to determine the heat stability and migration fastness of a pigment in a PVC system. Standards regarding both the composition and manufacture of master batches of rigid and plasticized PVC [30] and PVC plastisols [31] have been developed, standardized methods to prepare test samples have been defined [32], and there are regulations regarding the performance and evaluation of tests.

Continuous roll tests are employed to assess the heat stability of pigments in rigid and plasticized PVC. According to the standard procedure, the test system is subjected to continuous stress under defined conditions at 190 to 195 °C by means of a laboratory scale two-roll mill. At set intervals of 10, 20, and 30 minutes, samples are taken from the center of the sheet, injected, and visually and colorimetrically compared to an untreated specimen for color change.

Tests to assess the heat stability of a pigment in a PVC paste (plastisol) are carried out in a thermostatted oven. Specific test samples are tested either at 180 °C for 30 minutes or at 200 °C for 8 minutes, followed by colorimetric or visual colormatch comparison with a reference sample which has been stored at room temperature [33]. The determination method for heat stability during injection molding is also described in a standard [34].

From a purely practical standpoint, a certain temperature-time limit has evolved which often decides whether or not an organic pigment is thermally stable enough to be applied in PVC. Pigments, incorporated in well stabilized PVC, are expected to tolerate exposure at 200 °C for 5 minutes. It should be noted, however, that the color change of PVC white reductions containing a high percentage of white pigment is a familiar phenomenon in otherwise satisfactory PVC systems. This is caused by the limited heat stability of the used pigment.

There are processing techniques which are thermally so demanding that pigment selection must be carried out on a more differentiated basis. This is true for emulsion foam floor covering and wall covering made of PVC with a particular history. In these cases, the PVC plastisol is coated on a backing, such as jute fiber, felt, or paper, covered with a gravure printed film, and topcoated with a transparent PVC layer. Incorporated in this topcoat is a foaming agent, usually azo dicarbonamide, which decomposes in the presence of zinc oxide. As the system gelates, a foam builds up inside the PVC layer. Prior gelation is carried out at temperatures of about 150 to 210 °C, while the gelation process itself requires exposure to 220 °C or more for about 5 minutes. Inhibitors, such as certain acids or benzotriazol, which may migrate from the printed film into the PVC layer, help control the extent of foam formation and define the surface structure. Obviously, pigments which are designed for this purpose must be extremely heat stable. Diarylide pigments, such as Pigment Yellow 83, decompose at processing temperatures above 200 °C to form monoazo colorants and other decomposition products and are therefore unsuitable for such applications.

Detailed standards have also been developed in various countries to test for fastness to bleeding [35] (details see Sec. 1.6.3, 1.7.7). Plate-out is a familiar phenomenon which frequently occurs during PVC coloration and processing, especially as the polymer is calandered and rolled (Sec. 1.6.4.1).

Chalking is a disintegration of the plastics surface which especially in PVC appears similiar to the effect of plate-out (Sec. 1.6.4.2).

Pigments which lend color to articles that are designed for permanent outdoor use, such as garden furniture, outdoor siding, and profiles for blinds and shutters, must be completely weatherfast. Only very few organic pigments tolerate long-term weathering; moreover, the response of a pigmented system to weathering is largely determined by the type of PVC stabilization.

In pigmenting the various different types of PVC, care should be taken to ensure that the pigments are compatible not only with the polymer but also with all additives, a requirement which is no great challenge to most organic pigments. Pigment lakes (Sec. 2.7.1), on the other hand, may present problems as they hydrolyze in emulsion PVC which contains emulsifier, forming the metal free colorant compounds. Hydrolysis results in color change and affects the fastness properties of the pigmented system. Azomethine metal complex pigments exchange the metal with tin stabilizers, altering the shade of the material (Sec. 1.6.7, 2.10.4). Manganese laked pigments may present problems in the presence of epoxy compounds. Pigment preparations based on epoxy/soy bean oil, for instance, frequently replace dioctyl phthalate pigment pastes in the automotive industry to color plastic parts (roofs) made of PVC.

Similarly, pigments are not expected to affect or at least not noticeably affect the physical and mechanical properties of a polymer. The rheology of a PVC plastisol or PVC melt, for instance, may easily suffer during processing operations through the influence of an unsuitable pigment.

Colored PVC which is used in cable insulations is particularly susceptible to changes in the electrical resistance. It is not the organic pigment itself that affects the electrical resistance but oxyethylated additives which are used during the manufacture especially of azo pigments. Contrary to a frequently quoted opinion, the electrolyte content in a typical azo pigment lake is insufficient to affect the dielectrical properties of PVC*.

A few pigment manufacturers supply special-purpose products with defined and tested dielectrical properties.

The shades and color designations for cables and insulated wires have been defined by standards in various European countries [37].

---

* Plasticized PVC itself may be used in cable coverings up to about 20 kV DC or for alternating current at low frequencies. It is unsuitable for the insulation of high voltage cables, such as long-distance cables and main power lines. The material property which is particularly relevant to electrical insulation is the dielectric loss of a polymer $\varepsilon \times \tan \delta$, in which ε is the dielectric constant of the polymer and $\tan \delta$ stands for the dielectric loss factor [36]. This is the energy that is dissipated in a plastic, which is a nonideal insulator containing oscillating dipoles that transform electric energy into heat as the polarization changes. The $\tan \delta$ of plasticized PVC is about 0.1, but only 0.0001 for polyethylene. On the other hand, PVC has the advantage of being less flammable than PE and being somewhat of a fire retardant.

1.8.3.2   Polyolefins

Polyolefins are manufactured and used in much greater volume than all other plastics. Depending on the starting materials (monomers) and the density (processing temperature), polyolefins may be classified according to three main categories:

– Low-density polyethylene (LDPE), formerly known as high pressure polyethylene,
– High-density polyethylene (HDPE), formerly known as low pressure polyethylene,
– Polypropylene (PP).

The respective processing temperatures, which determine the criteria for pigment selection, are as follows:

$$
\begin{array}{ll}
\text{LDPE} & 160-260\,°C \\
\text{HDPE} & 180-300\,°C \\
\text{PP} & 220-300\,°C
\end{array}
$$

At any given temperature, the flow property of a system is defined by the density and the molecular weight of the plastic and characterized by the melt flow index. Being partially crystalline, polyolefins scatter very little light and therefore tend to appear lighter on coloration, depending on the processing temperature. Another type of lightening effect is observed in stretched and foamed polyolefins. Polyolefins are processed primarily by injection molding and extrusion techniques.

Ethylene/propylene products reign supreme among the copolymers. They are elastomers. Plastics containing about 20% or more propylene perform like natural rubber and can be cured by peroxide cross-linking. They are more fast to chemical and to ageing than other types of natural rubber.

Polyolefins are produced in a variety of forms: HDPE and PP are produced as powders, while LDPE emerges from the melt preferably in the form of lenticular granules. All types, however, are supplied primarily as granules. As a rule, any thermoplastic transformation of a polymer powder to a granulate is carried out in the presence of additives. This is also partly true for pigments.

Since polyolefins have a low glass transition temperature (Sec. 1.6.3), it is not surprising to note that pigments which are partially soluble in polyolefins tend to migrate. Like in other media, this trend is concentration and temperature dependent. However, the type of polyolefin, especially its density and molecular weight, influences the tendency of a pigment to migrate in LDPE to a much larger extent than in other plastics. As a rule, pigments migrate more easily in LDPE than in HDPE or in PP. HDPE and PP, for instance, may safely be colored by organic pigments which migrate in LDPE.

The trend among pigments to migrate in LDPE, which includes both bleeding and blooming, increases with increasing melt flow index and decreasing molecular weight of the polymer, respectively. Additives such as lubricants or antistatic agents may also play a role.

A number of organic pigments cause distortion in certain types of polyolefins, especially in HDPE. Pigments act as nucleating agents in such partially crystalline plastics; i.e., they promote crystallization, which creates stress within the plastic product (Sec. 1.6.4.3). These pigments also enhance the shrinkage of polyolefins, particularly in the direction of the flow.

The effect of heat ageing on a polyolefin, as on modified polystyrene, is considerably influenced by particular pigment lakes. Heavy metal ions, especially copper, manganese, and iron adversely affect the thermostability of polyolefins, while sulfide-containing inorganic pigments have a distinctly improving effect. The response of a system to additives or impurities is tested by measuring the loss of tensile strength after exposing the polymer to heat. The polyolefins are stored at temperatures close to the melting point of the crystallites and the pigmented and pigment-free test samples compared for brittleness.

The lightfastness of pigments in a polyolefin, like that of other polymers, paints, or printing inks is always measured for an entire pigmented system, including the polymer with all its additives. This aspect gains an added degree of importance since it has become increasingly important in recent years to blend plastics, especially PP and HDPE, with sterically hindered amines, known as HALS stabilizers (hindered amine light stabilizers) so as to protect the plastics material against light and especially weather. However, some caution should be exercized in combining pigments and stabilizers, because the effectivity of these agents is inferior in the presence of certain pigments.

The specific requirements for organic pigments in terms of heat stability result from the temperature levels at which individual polyolefins are processed. Standards have been developed to test the heat stability of pigments in polyolefins [34] (see also Sec. 1.6.7).

The dispersibility of an organic pigment which is to be incorporated in a polyolefin is of particular importance. This is especially true for the coloration of extrusion films and for HDPE or PP ribbon made from stretched blown film or from sheeted extrusion film, as well as for the coating or melt spin dyeing.

As a result of the high processing temperatures and the consequently high degree of softening, only limited shearing forces are available to disperse pigments in polyolefins. Pigment dispersion is therefore far from adequate. Therefore pigment preparations, usually in combination with polyolefins as a carrier, are frequently used to color polyolefins. Not only is the tinctorial strength improved, but the safety of the operations is also superior. Poor dispersion causes filler specks, holes in films, and other faults, which are avoided by using pigment preparations. Recently, there has been a considerable trend towards using color concentrates, which are easy to apply in metered amounts and which facilitate the production of exactly defined shades (Sec. 1.8.3).

Pigment powders continue to be used in thick-walled articles, such as extruded sheets, or hollow objects, or in injection-molded products; although color concentrates are beginning to be more important even in these areas.

There are a number of mixing and processing techniques which have been used successfully to color granulated or powdered polyolefin with pigments or pigment preparations. Powders, for instance, are processed primarily with high-speed mixers, while granulated types respond better to slower mixers.

Pigment/polyolefin paste preparations containing between 20 and 70% pigment are also used. They are manufactured by means of three-roll mills, agitated ball mills, or dissolvers or similar equipment. Such preparations are employed primarily to produce bottles, injection-molded articles, or extrusion sheets. These preparations have the advantage of being applicable in volumetric doses and may be pre-mixed in slow mixers or gravitation mixers to uniformly adhere to the plastic granulate.

Details are discussed individually in the literature [27, 38].

1.8.3.3   Polystyrene, Styrene-Copolymers, Poly(Methyl Methacrylate)

Polystyrene (PS), a highly rigid and surface-hardened thermoplastic, is glass clear and almost colorless. Its typical slight yellow tinge is easy to compensate for by adding transparent blue colorants to adjust the color. Polystyrene softens between 80 and 100 °C. It is processed between about 170 and 280 °C, up to a maximum of 300 °C, without color change, by any of the methods which are recommended for thermoplastics. The list of products includes extrusion made sheets, profiles, and films, which are often foamed.

The mechanical properties of PS may be improved drastically by copolymerizing the monomer with one or more out of a variety of rubber-like materials (graft polymers). Impact resistant PS, for instance, contains 5 to 25% natural rubber, which is not dissolved but dispersed in PS. This and other dispersed additives, due to a difference in refractive indices between plastic and filler, scatter much of the incident light and therefore afford opaque products; the degree of opacity in each individual case depends on the amount of rubber added. Impact resistant PS types are processed at 170 to 260 °C. Copolymers with acrylonitrile and butadiene have high impact strength and excellent fastness to aging. The inherent color of these ABS polymers is somewhat more yellowish than polystyrene; their high opacity is attributed to extensive light scattering. In recent years, transparent types of ABS have been marketed; these are processed at about 210 to 250 °C.

Poly(methyl methacrylate) (PMMA) is extremely stable to aging and to weathering; it is hard and glass clear.

The transparency of a plastic, with polystyrene and styrene/acrylonitrile copolymer (SAN) being most transparent, to impact resistant PS, down to highly opaque ABS, clearly has some influence on the coloration of a product. The coloristic properties of a colorant depend on the plastic in which it is incorporated. In discussing the coloration of polymers whose glass transition temperature is far above room temperature, special rules have to be observed (Sec. 1.6.3). It is rarely possible for dissolved molecules to migrate, thus these polymers may be colored with pigments and also soluble dyes. Some of these dyes are even acceptably lightfast and afford brilliant shades, especially in combination with opaque inorganic or organic pigments.

The requirements regarding heat stability are stringent for pigments which are used to color these plastic materials. Few organic pigments tolerate the high end of the processing temperature range between 280 and 300 °C. A number of types, however, withstand moderate temperatures between 220 and 260 °C.

Depending on the temperature, a large number of organic pigments dissolve more or less in this class of plastics. The color almost invariably changes as the pigment dissolves, frequently accompanied by a change in the fastness properties, especially in the response of the system to light.

Completely dissolved pigments should be referred to as dyes and be tested as such. This concerns features such as the extraction properties in the finished article. In PS, SAN, and other transparent plastics with a high glass transition temperature, they afford transparent, glass clear colorations.

Pigments which at low concentrations dissolve reasonably well at the temperature at which their medium is processed, are frequently employed to advantage at higher concentrations. This phenomenon is readily explained by considering the influence of the dissolved versus the undissolved portion of pigment. The coloristic properties of the polymer will be enhanced if the undissolved pigment particles dominate over the dissolved portion in determining the coloristics of the system, often resulting in high brilliance. According to the laws of physical chemistry, organic pigments (and dyes) dissolve at a rate which is largely dependent not only on the pigment particle size but also on the available time. PS and its copolymers are normally processed as pellets, which may be colored by colorants in powder, granulated, or paste form. Pigment preparations for PS are also becoming increasingly important.

Pigments and pigment blends in powder form are incorporated in their medium by means of slow mixers. Organic pigments are often insufficiently wetted by molten PS or its copolymers and accordingly difficult to disperse. It is possible, however, to facilitate dispersion by initially applying an adhesion agent, i.e., a wetting agent, in concentrations up to 0.3% relative to the granulated plastic. PS, being a brittle and hard material, may cause metal abrasion after a certain mixing time, which gives a dull effect to otherwise clean, brilliant shades; fluorescent colorants may even lose their fluorescence.

The advantage of using pigment preparations in paste form, which afford easy color matching by blending the corresponding pastes, is compromised by the fact that a higher content of liquid component may affect the mechanical properties of the plastic and lead to stress corrosion cracking. Similarly, the applicability of pastes is restricted by physiological considerations. Used as carriers, for instance, they contain paraffin oil.

In exterior exposure, PS yellows somewhat, due to UV radiation. In order to shield the plastic from degradation in UV light, it is also supplied in combination with UV absorbents. This prolongs the lifetime of the products by a factor of three to five. Grades which contain UV absorbents are slightly yellowish, a fault which may be corrected by adding transparent blue colorants such as soluble dyes or Ultramarine Blue [38].

1.8.3.4   Polyurethane

Polyurethane (PUR) has one of the broadest and most useful ranges of properties of any of the plastics yet developed. A wide variety of starting materials and the fact that PUR is compatible with almost all of the known processing methods which are used throughout the plastics industry make it possible to manufacture a wide range of products. The above-mentioned considerations largely apply also to the coloration of PUR.

Thermoplastic PUR, for instance, is colored by the same organic pigments which are also used for plasticized PVC; moreover, migration plays the same role in both cases.

This is particularly important in the coloration of artificial leather made from PUR, which is usually carried out at high pigment concentrations.

PUR, like PVC, may be processed in combination with plasticizers. There are also thermoplastic types of PUR that can be used to soften PVC, either by themselves or in combination with other plasticizers. They have the advantage of being more oil and abrasion resistant and fast to plasticizer migration.

Pigment preparations are also used to color plasticized PUR. Carrier materials range from vinyl chloride/vinyl acetate copolymers like those used for PVC to low molecular weight polyethylene, and finally PUR itself.

The coating of textiles with two and one component PUR has increasingly attracted attention among manufacturers of artificial leather. Depending on the solvent system, the types of PUR may be classified according to whether aromatic (solvents: dimethylformamide or tetrahydrofuran with methylethylketone, toluene, etc.) or aliphatic types (solvents: frequently isopropanol/toluene blends) are used. Every care must be taken to ensure almost complete insolubility of the organic pigments in the respective solvents. Not many organic pigments are fast enough especially to dimethylformamide to be applicable. Even quinacridone pigments dissolve to an appreciable extent in pure dimethylformamide.

Azo pigment lakes are altogether unsuitable for application in PUR because they have the added disadvantage of bleeding in the presence of moisture.

Pigment powders are dispersed in a portion of the PUR solution which is to be colored by means of ball mills, sand mills, or other closed dispersion units. The pigment concentration is typically between about 20 and 40%. The resulting pigment paste is then homogeneously mixed with the PUR solution; it is critical to the success of the operation to avoid flocculation. Pigment preparations may also be used; care should be taken, however, to ensure that the carrier material of the preparation dissolves in the solvent system and does not noticeably affect the so-called grip of the artificial leather product.

Organic pigments represent a frequent choice for foam articles made from PUR; they are becoming increasingly important in foam products which form integral parts of automobiles and furniture. Pigments which are used in PUR foam must as a rule be extremely heat resistant. To prevent failure during PUR foaming by the high pressure technique, in which isocyanate and polyol are sprayed through narrow jets, pigments must be completely dispersed to avoid clogging of the dyes. Pigment preparations are quite common in this field. The carrier material of a pigment preparation is typically involved in the reaction of the isocyanate during PUR formation, or it may cooperate in the formation of foam. More detailed information concerning carrier materials for such preparations, for instance the hydroxy number, is provided by the manufacturers.

1.8.3.5 Polyamide, Polycarbonate, Polyester, Cellulose Derivatives

Among the large number of other commercially available plastics, polyamide (PA), polycarbonate (PC), poly(ethylene terephthalate) (PETP), and cellulose derivatives maintain an important position. Pigmentation of these polymers largely follows the rules which are described above. The problem of selecting extremely heat-resistant pigments for use in polyamide for injection molding and extrusion purposes gains an added degree of complexity through the slightly basic and reducing character of the polymer melt. A wide range of organic pigments is available for coloring the various plastics, depending on the individual use requirements for each purpose. Economic considerations make it increasingly necessary to compromise between price and performance.

1.8.3.6 Elastomers

Elastomers are characterized by rubber-like elastic behavior. They soften at temperatures below room temperature. Elastomers which are not cured, whose molecular chains are therefore not cross-linked, are plastic and deformable. Once cured (within a certain temperature range), they show elastic deformation.

Vulcanization turns raw natural rubber into rubber. A large number of synthetic types of rubber and elastomers are commercially available. They have much improved special properties compared to rubber, such as considerably enhanced elasticity, heat stability, fastness to cold, and fastness to weathering and oxidation, abrasion fastness, fastness to various chemicals, oils, etc.

Pigments which are used to color rubber blends must meet a series of requirements, especially regarding the absence of so-called rubber poison. Even minor amounts of copper or manganese, for instance, have a considerable adverse effect not only on the vulcanization of natural rubber, but they also accelerate the aging of the rubber. Suitable pigments are therefore expected to contain a total of less than 0.01% of both these heavy metals. Copper phthalocyanine blue pigments are somewhat of an exception in that slightly higher values are accepted for metals which are not part of the complex; but the total content of "free" heavy metals should in no case exceed 0.015%. Pigments which are targeted for this application are frequently sold only after having been tested for their heavy metal content.

Moreover, pigments which are used to color rubber should also be fairly heat resistant. The heat stability is determined in five samples with different pigment concentrations, varying from 0.01 to 1%, with a tenfold amount of chalk. The specimens are arranged in a row and vulcanized together at 140°C for 15 minutes.

The results are evaluated in terms of color-change comparison with the respective untreated reference samples.

Migration fastness is another common requirement for pigments which are targeted for use in rubber. Tests are carried out for heat stability by using five samples at different pigment concentrations. Fastness to bleeding is determined by bringing nonvulcanized samples in defined contact with a white rubber sheet of standard composition and then steam vulcanizing the specimens at 140°C without pressure. Samples are commonly half covered with a wet cotton cloth to see whether the "wrapper", the natural rubber, or both are colored by a bleeding pigment.

Testing for fastness to blooming is carried out by rubbing several differently pigmented samples with a white cloth immediately after manufacture, after 6 months of storage at room temperature, and possibly also by an accelerated technique, by which samples are exposed to 70°C for 24 hours.

The weatherfastness of a pigment which is to be used in natural rubber is practically of no concern, since rubber itself is quite sensitive to weather. A few high grade synthetic elastomers which are very weatherfast, on the other hand, require correspondingly selected pigments. The demand for pigment fastness to certain chemicals depends on each individual application.

Rubber blends are colored with pigment powders and increasingly with granulated pigment preparations, which especially in the rubber industry have come to be known as master batches. Latex, which is an aqueous dispersion of synthetic caoutchouc, and its polymers are colored successfully with aqueous pigment preparations.

1.8.3.7 Thermosetting Plastics

Thermosetting plastics are made by cross-linking (hardening) reactive, linear, or branched macromolecules by polycondensation, polymerization, or polyaddition. Thermosetting plastics are therefore processed only once, using heat and pressure, to form semifinished goods or finished articles. The process is not reversible, formation cannot be undone. Among the best known thermosetting plastics are formaldehyde derivatives with phenol, resorcin etc. (phenoplasts, phenolic resins), urea, aniline, melamine, and similar compounds (aminoplasts).

The dark color of phenolic resins restricts the coloration of these materials, which are generally processed at high filler content (up to 80%).

The processing of thermosetting plastics depends on the chemical composition and the structure of the polymer: the choice of techniques includes compression molding, transfer molding, injection molding, and extrusion. Compression temperatures range from about 150 to 190°C, which defines the corresponding requirements for pigments. Dyes are also frequently used to color thermosetting plastics.

In order to ensure homogeneous coloration, colorants are worked into the resins before they are cross-linked. This may happen in the resin melt, for instance in a kneader at about 90°C, or in a dissolved or liquid resin using a method which is known as the liquid resin technique. Ball mills are commonly used to color prewetted molding powders which are still to be hardened.

Cast resins based on epoxy resins, methacrylate, or unsaturated polyester are also considered thermosetting plastics. Epoxy resins are cured by using amines or phthalic anhydride. Organic pigments do not affect resin hardening; the presence of water, however, inhibits it. Pigments thus must be largely dry, which as a rule may be safely relied on.

Cast resins made of unsaturated polyester or methacrylate are normally hardened with organic peroxides. Polyester cast resins are dissolved in monostyrene and polymerized at up to 200 °C. Especially thick-walled articles are frequently exposed to heat for a longer period of time, which defines and limits the choice of suitable organic pigments.

Methacrylate resins are made from monomeric methyl methacrylate, which is polymerized at a considerably lower temperature. The thermostability of pigments in this medium is therefore of little concern.

Application in areas such as carbody production typically requires excellently lightfast and weatherfast pigments. However, these fastness properties are considerably influenced by the type and amount of peroxide catalyst. As a consequence, suitable pigments should be fast to peroxides. At the same time, however, they are not expected to influence the hardening process in any direction, i.e., neither to accelerate it nor to slow it down. It is known, however, that organic pigments may respond very differently to light and weather, depending on the hardening technique and on the type and amount of peroxide used [39]. Diarylide yellow pigments, for instance, such as Pigment Yellow 17 and 83, have no effect under certain conditions; while copper phthalocyanine green considerably inhibits the hardening process, and copper phthalocyanine blue prevents it altogether. The conditions may be changed drastically by exchanging the hardener; using a different organic peroxide, for instance, reverses the conditions, so that copper phthalocyanine green now has a slightly accelerating effect.

Coloring unsaturated polyester and methacrylate resins is usually performed with pigment/plasticizer (DOP) pastes. These do not noticeably affect the primary mechanical properties of the finished article. To a lesser extent, pigments are also dispersed directly in a portion of the monomer.

1.8.3.8 Spin Dyeing

The technique of spin dyeing chemical fibers may be located somewhere between the textiles and the plastics area. In contrast to textile coloration, the material which is to be extruded is colored before the fiber is made. The requirements to be met by pigments are therefore similar to those which apply to the coloration of plastics. Likewise, heat stability is the foremost concern in connection with pigment selection for spin dyeing. In addition, the pigment in this application must be insoluble in the solvents used.

The demands regarding pigment dispersibility are particularly high. Every care must be taken to ensure that the size of remaining pigment agglomerates does not exceed 2 to 3 µm [40, 41]. Larger particles adversely affect the tensile strength of the fiber and frequently cause failure through breakage, especially as the fiber is stretched. In practice, however, pigment powders rarely afford such a high degree of dispersion, and there is no guarantee for the quality of dispersion. It is therefore almost unavoidable to replace pigments by pigment preparations in spin dyeing processes. Clearly, a carrier material should be chosen which does not present any processing problems.

Three different methods are available for spin dyeing [42]:

Melt spinning. This technique is used with thermoplastic material, such as polyester, polyamide, or polypropylene. The polymer is melted in the extruder and then pressed through a spinnerette. It solidifies by cooling as it falls vertically through a shaft to the bottom of the extruder. Pigments are therefore expected to exhibit excellent heat stability. Regarding organic pigments, the spinning temperatures may vary over a wide range, depending on the melting point of the polymer (Table 9).

Table 9: Melting points and spinning temperatures of various polymers.

	Melting point (°C)	Spinning temperature (°C)
Polyester	255	285
PA 6 (6-polyamide)	220	250 – 280
PA 66 (nylon)	245	275
Polypropylene	175	240 – 300

As a rule, pigment preparations for this application are based on a carrier material which is identical or similar to the polymer which is to be extruded.

Wet spinning. This technique is characterized by spinning a filtered viscous polymer mass, dissolved in a suitable solvent, into contact with a precipitation or coagulation bath. Polyacrylonitrile, poly(vinyl acetate), cellulose acetate, and other materials are processed by this method. Thermal requirements for pigments are less stringent than for melt spinning; but pigments are expected to be fast to the solvents and chemicals used.

Dry spinning. The polymer, dissolved in a suitable solvent and filtered, is pressed through spinnerettes and, in an oxygen-free atmosphere, pulled by vacuum through a heated shaft, where the polymer solidifies as the solvent evaporates. The requirements of this process regarding the heat stability of pigments are much less stringent than in melt spinning. However, like in wet spinning, pigments must be fast to the solvents used. Polyacrylonitrile, triacetate, and polycarbonate are processed by this method.

Newly developed coloration techniques have recently been introduced for a number of thermoplastics such as polyester [43, 44, 45]. Instead of using pigments, dyes are employed which dissolve in the polymer melt and are sufficiently heat stable and sublimation-fast to tolerate the processing conditions. These new techniques have certain advantages over traditional methods based on organic pigments; they avoid fiber failure through breakage during stretching, plugging of the filter, and other faults.

Polyacrylonitrile (PAC)

Polyacrylonitrile, which decomposes already at 220 °C, which is below its melting point (about 290 °C), cannot be processed by melt extrusion. Dry and wet spinning techniques are therefore used, and the list of appropriate solvents includes dimethylacetamide, dimethylformamide, dimethylsulfoxide, and aqueous solutions of inorganic salts. Besides being insoluble in the appropriate solvents, pigments which are to be used in specialty applications such as window blinds, awnings, and tents, should meet the respective demands regarding lightfastness and weatherfastness. Less lightfast pigments are acceptable for use in clothing, decorative and home textiles, i.e., upholstery, curtains, and carpeting. Polyacrylonitrile fiber is commonly used for these purposes and is quite adequate: it is by far the most weatherfast of all synthetic and natural fibers. Pigment preparations are also supplied for the coloration of PAC. Specialized cationic dyes have recently been introduced and are becoming increasingly important.

Polyolefins (PO)

In practice, polyolefin fibers are colored only through mass coloration. Traditional methods used by the textiles industry are unsatisfactory. Severe processing conditions make it necessary to select pigments with excellent heat stability which tolerate temperatures between 240 and 300 °C. In recent years, types of polymers with lower melting points have been introduced, which use pigment preparations based on PE or PP as carrier materials. In order to ensure easy homogenization, the carrier may be adapted to the extrusion polymer which is to be colored [46, 47].

Polyester (PETP)

The thermal requirements for pigments which are targeted for PETP melt extrusion are particularly severe. However, it is important to consider the individual conditions at the various stages of polymer coloration. Pigments, for instance, which are added during the so-called condensation process in a glycol dispersion prior to transesterification or condensation in the autoclave, are exposed to temperatures between 240 and 290 °C for 5 to 6 hours [45]. These harsh conditions are only tolerated by very few polycyclic pigments, primarily by representatives of the quinacridone, copper phthalocyanine, naphthalenetetracarboxylic acid, and perylene tetracarboxylic acid series.

The choice of pigments which are to be added to a ready-made polyester is much less restricted. At this stage, a pigment will only be exposed to heat for about 20 to 30 minutes; although the temperature will be equally high. This is made possible by mixing the granulated polyester with a pigment concentrate or with a pigment preparation, or by transferring the molten pigment concentrate to the melting zone of the spin extruder, for instance via a side-screw extruder. Clearly, the carrier material of the pigment preparation is equally subject to compatibility restrictions. In the case of polyester, the novel coloration technologies are carried out to advantage with dyes which are melt-soluble, sufficiently heat resistant, and sublimation, proof during spinning.

Polyamide (PA)

Pigments which are targeted for polyamide spin dyeing, apart from being extremely heat stable, must also be chemically fast to the highly reducing medium of the PA melt. Spinning temperatures are between 250 and 290 °C, depending on the type of polyamide. As with polyester extrusion, only very few polycyclic pigments are suitable for this purpose. There are no organic yellow pigments which satisfy the specifications; inorganic cadmium pigments are therefore used instead to produce yellow shades. Pigment preparations are also available for this purpose.

Viscose

Viscose, the alkaline solution of sodium cellulose xanthate, is produced by treating cellulose with sodium hydroxide solution and carbon disulfide. Viscose is colored in the form of cellulose xanthate prior to extrusion. Aqueous pigment preparations in paste form are employed. Apart from the usual fastness requirements, pigments are expected to be fast to strong acids, alkali, and reducing agents. Besides, pigments which affect the so-called maturation of the viscose or the coagulation and regeneration in the spin bath or in one of the after-treatment baths are unacceptable. However, a considerable number of pigments satisfy these specifications [48].

1.8.4   Other Areas of Application

Organic pigments are also used to color a variety of materials outside the above-mentioned areas of printing inks (Sec. 1.8.3.1), paints (Sec. 1.8.3.2), and plastics and spin dyeing (Sec. 1.8.3.3). Pigment selection depends on the individual manufacturing and processing conditions, which define the demands placed on a pigment regarding solvent fastness and the performance of the ultimate product, especially its fastness to light.

Areas of application include: wood coloration [49], paper mass coloration [50] and paper surface coating in the lime press [51], the office articles and artists' colors sector; pigments are used in colored pencils, crayons, and writing and pastel chalks or in water colors, as well as in cosmetics, especially soap [52].

Printing on textiles is one of the major areas of pigment application and is usually considered separately from the graphics industry. Requirements to be met by pigments for this purpose depend especially on the expected performance of the final article, the printed textile (Sec. 1.6.2.4).

References for Section 1.8

[1] A. Rosenberg and O.-J. Schmitz, Farbe + Lack 90 (1984) 362–370.
[2] DIN 16 508: Farbskala für den Buchdruck außerhalb der europäischen Farbskala; Normdruckfarben und Druckreihenfolge.
   DIN 16 509: Farbskala für den Offsetdruck; Normfarben.
[3] DIN 16 539 (CIE 12-66): Europäische Farbskala für den Offsetdruck.
[4] DIN 16 538: Europäische Farbskala für den Buchdruck.
[5] A. Goldschmidt, Farbe + Lack 84 (1978) 675–680.
[6] W. Herbst and O. Hafner, Farbe + Lack 82 (1976) 393–411.
[7] H.L. Beeferman and J. Water-Bone, Coatings 1 (1978) 4.
[8] E. Levine and J. Kuzma, J. Coat. Technol. 51 (1979) 35–48.
[9] R. Dhein, H.U. Pohl and J. Schoeps, Farbe + Lack 84 (1978) 680.
[10] W. Burckhard and H.J. Luthardt, J. Oil Colour Chem. Assoc. 62 (1979) 375–385.
[11] C.B. Rybny et al., J. Paint Technol., 46 No. 596 (1974) 60–69.
[12] T.A. Du Plessis and De Hollain, J. Oil Colour Chem. Assoc. 62 (1979) 239–245.
[13] P.S. Pappas, 6th. Int. Conf. Org. Coat. Sci. Technol. Proc. Athens, 1980, 587–597.
[14] W. Brushwell, Farbe + Lack 85 (1979) 662–665.
[15] B.D. Meyer, 5th Int. Conf. Org. Coat. Sci. Technol. Proc. Athens, 1979, 177–206.
[16] H. Schäfer and G. Wallisch, J. Oil Colour Chem. Assoc. 64 (1981) 405–414.
[17] U. Kaluza, DEFAZET Dtsch. Farben Z. 33 (1979) 399–411.
[18] F. Haselmeyer, Farbe + Lack 74 (1968) 662–668.
[19] P. Kresse, DEFAZET Dtsch. Farben Z. 24 (1970) 521–526.
[20] E.J. Percy and F. Nouwens, J. Oil Colour Chem. Assoc. 62 (1979) 392–400.
[21] William Brushwell, Farbe + Lack 80 (1974) 639–642 and 85 (1979) 1035–1038.
[22] D.S. Newton, J. Oil Colour Chem. Assoc. 56 (1973) 566–575.
[23] E.V. Schmid, Farbe + Lack 85 (1979) 744–748.
[24] R. Gutbrod, Farbe + Lack 69 (1963) 889–892.
[25] J. Richter, Das Einfärben von Kunststoffen und Kunstfasern in der Masse, Hoechst, SD 119.
[26] H.-J. Lenz, Ber. Techn. Akad. Wuppertal, Heft 8, 1972.
[27] N. Nix, Farbmittel von Hoechst für die Kunststoffindustrie, SD 435[1].
[28] DIN 53 775 T 7: Bestimmung der Dispergierhärte durch Kaltwalzen.
[29] W. Herbst, DEFAZET Dtsch. Farben Z. 26 (1972) 519–532, 571–576.
[30] DIN 53 774 T 1 also DIN 53 775 T 1: Prüfung von Pigmenten in PVC hart bzw. PVC weich; Zusammensetzung und Herstellen der Grundmischungen.
[31] DIN 53 773: Prüfung von Farbmitteln in Polyvinylchlorid-Pasten (Plastisolen).
[32] DIN 53 774 T 2 also DIN 53 775 T 2: Herstellen der Probekörper.

[33] DIN 53 775 T 6: Bestimmung der Hitzebeständigkeit im Wärmeschrank.

[34] DIN 53 772: Bestimmung der Hitzebeständigkeit durch Spritzgießen.

[35] DIN 53 775 T 3: Bestimmung des Ausblutens.

[36] DIN 53 483: Bestimmung der dielektrischen Eigenschaften.

[37] DIN 47 002 / BS 6746: Farben und Farbkurzzeichen für Kabel und isolierte Leitungen.

[38] H.-J. Lenz, Kunststoffe 66 (1976) 683–687.

[39] F. Kuckelsberg, Kunstst. Rundsch. (1972) 654–660.

[40] W.K.J. Teige, 33rd Annu. Tech. Conf. Soc. Plast. Eng. Atlanta, 1975, 57–59.

[41] W.K.J. Teige, Chemiefasern/Textilind. 33/85 (1983) 636–642.

[42] E. Welfers, Chemiefasern/Textilind. 26/78 (1976) 1079–1086; 27/79 (1977) 42–59.

[43] DE-AS 2 608 481 (Hoechst AG) 1976.

[44] F. Steinlin and M.J. Wampetich, Melliand Textilber. 61 (6/1980) 509–513.

[45] W.K.J. Teige, Chemiefasern/Textilind. 33/85 (1983) 127–131.

[46] H.-J. Sohn, Chemiefasern/Textilind. 32/84 (1982) 712–717.

[47] H.-J. Sohn, Chemiefasern/Textilind. 34/86 (1984) 827–829.

[48] W.K.J. Teige, Chemiefasern/Textilind. 31/83 (1981) 632–636.

[49] J. Steier, Das Färben von Holz. Hoechst AG, 6.76, SD 441.

[50] Farbstoffe und Pigmentpräparationen für die Papierindustrie. Hoe 3023.

[51] Papieroberflächenfärbung in der Leimpresse. Techn. Rat aus Hoechst/Pigment, 30 (1975).

[52] Das Färben von Seifen. Techn. Rat aus Hoechst 21 (1969).

2 Azo Pigments

Azo pigments carry an azo function ($-N=N-$) between two sp^2-hybridized C atoms. They are structurally based on the general formula

$$Ar-N \diagdown_{\diagdown N-R}$$

in which Ar is an aromatic or a heteroaromatic moiety and R is either an aromatic unit or represents the function

$$\longrightarrow C \diagdown^{\diagup C=O \;\; (R^1)}_{\diagdown C-OH \;\; (R^2)}$$

An alkyl or an aryl group is attached in the R^1 position; most commercially important pigments are derivatives with CH_3 as R^1. R^2 mainly represents $-HN-Ar$, in which Ar is an aromatic or a heteroaromatic moiety.

According to the demand for insolubility in water and the commonly used organic solvents, substituents attached to the alkyl, aryl, or heteroaryl groups should not contain sulfo (SO_3H) groups or long-chain alkyl groups which render the colorant soluble.

Monoazo and disazo pigments contain one and two azo functions, respectively. Compounds with more than two azo groups (tris, tetra,..., polyazos), however, have failed to gain commercial recognition as pigments.

The first important event in the history of azo pigments was the discovery of the diazotization reaction by P. Gries in 1858. In 1875, Caro and Witt synthesized chrysoidine, the first azo dye, through a reaction sequence of diazotization and coupling (Sec. 2.2); a technique which continues to be used today.

Nomenclature

Due to the complexity of their chemical names, azo pigments are rarely referred to by IUPAC or Chemical Abstracts nomenclature. Practical considerations make it more convenient to classify these compounds according to the nature of the azo coupling reaction which leads to pigment formation (Sec. 2.2.2). This system defines the constitution of a pigment by the starting materials for the coupling reaction, the diazo component D and the coupling component C; and by the direction of the coupling ($\rightarrow$). The formation of a monoazo pigment is thus represented by

$$D \rightarrow C$$

while disazo formation obeys the equation

$$C \leftarrow D \rightarrow C \quad \text{or} \quad D \rightarrow C \leftarrow D$$

This principle is equally accepted in the technical literature.

Although azo pigments exhibit a wide range of colors covering essentially the entire visible spectrum, the blue and green representatives have no commercial significance. These shades are made available almost exclusively by phthalocyanine, triarylcarbonium, and indanthrone pigments (Sec. 3.1, 3.8, and 3.6.3.2). Among the hues produced by azo pigments are all shades of yellow, orange, red, bordeaux, carmine, and brown. Within these color ranges almost all possible hues combined with the demanded properties can be provided.

The reaction sequence of diazotization and coupling is the basic reaction of the azo pigment industry. Alternative routes to the formation of azo groups are sometimes used in the production of azo dyes, but they are less commonly employed for azo pigments. Economical production methods make azo pigments by far the largest fraction of organic pigments on the market today. Not only are the starting materials easily accessible (Sec. 2.1), but the azo group is formed in a coupling reaction which is easily performed on a commercial scale, usually in an aqueous medium.

Azo pigments have become considerably more important in recent years as new products have been developed to meet increasingly stringent industrial requirements. Their properties in application, including lightfastness and weatherfastness, resistance to solvents, migration, and heat has improved appreciably. The proliferation of azo compounds has raised the fastness standards of azo pigments to a previously unknown level.

2.1   Starting Materials, Synthesis

Azo pigments are typically formed by a reaction sequence of diazotization and coupling, involving a primary aromatic amine, which is referred to as a diazo component, and a nucleophilic aromatic or aliphatic compound with active methylene groups as a coupling component [1, 2, 3].

2.1.1   Diazo Components

The diazotization reaction typically involves an aromatic amine, such as a mono-, di-, or trisubstituted aniline, as a diazo component. Coupling components are also frequently based on aniline or its derivatives. The following examples are thus important diazo components:

Substituted aniline derivatives

Aminobenzamides and aminobenzanilides, aminobenzosulfonamides

Another family of technically important diazo components for pigment formation includes a series of aromatic diamino compounds, such as 3,3'-dichlorobenzidine, 3,3'-dimethoxybenzidine (o-dianisidine), and 2,2',5,5'-tetrachlorobenzidine:

Aromatic diamines

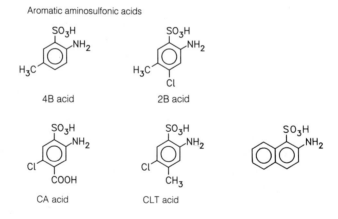

3,3'-Dichlorobenzidine

o-Toluidine

o-Dianisidine

2,2',5,5'-Tetrachlorobenzidine

Polycyclic amines, such as α-aminoanthraquinone, and heterocyclic amines, such as aminophthalimide, aminobenzazoles, or aminoquinazolines are less interesting coupling components for azo pigments.

Aromatic aminosulfonic acids, which play a major role in connection with pigment lakes, are produced by sulfonating the corresponding nitro compound and then reducing it to an aminosulfonic acid. An alternative technique, known as baking process, involves exposing an amine/dihydrosulfate to a temperature of 200 to 300 °C in order to effect rearrangement to p-aminosulfonic acid. Ortho-sulfonation prevails if the para position is occupied. In contrast to sulfonation techniques with sulfuric acid, this method avoids wastewater contamination with sulfuric acid.

Aromatic aminosulfonic acids

4B acid

2B acid

CA acid

CLT acid

The synthesis of an aromatic amine generally starts from the corresponding nitro compound. Nitration and subsequent reduction are key reactions in the synthesis of intermediates for azo pigments.

Aromatic nitro compounds are obtained by nitrating appropriately substituted benzene derivatives with nitric acid. This reagent can be employed in a more or less concentrated form and is often used in combination with concentrated sulfuric acid as so-called mixed acid. This medium traps water and also has the advantage of serving as a "diluting agent"; in other words, as an agent to stabilize the temperature–time curve. It is also a very effective solvent for the nitro compound. Aromatic compounds with free amino functions can be nitrated in the presence of a considerable excess of sulfuric acid (formation of amino sulfates).

Additional substituents may be introduced into the aromatic nucleus by halogenation, oxidation, or nucleophilic replacement. The aromatic halogen is "activated" by electronegative substituents (electron acceptors) in the ortho and para positions. $-NO_2$, $-COOH$, or $-COO$ alkyl groups are effective electron acceptors. These convert the chlorine atom into a leaving group, which can be replaced by a suitable nucleophilic group, such as CH_3O-, C_2H_5O-, or $-NH_2$.

Reduction Methods

There are a few methods of reducing nitro functions to amino groups which are of considerable technical importance:

a) Catalytic Hydrogenation

Catalytic hydrogenation with molecular hydrogen is by far the most favored option. It is carried out at temperatures between 20 and 120 °C and at pressures between 10 and 100 bar in 2 to 10 m^3 tanks, which are pressure and shock resistant. Catalysts typically contain nickel (Raney-nickel), which may in some cases be replaced by a precious metal, such as palladium.

b) Reduction with Iron

The traditional technique of reducing nitro compounds with iron powder in dilute acid (Béchamps–Brimmeyr reduction) continues to be used for nitro compounds that are adversely affected by the catalytic reduction method with hydrogen. The list of examples includes aromatic nitro compounds carrying halogen substituents, especially if these are attached in ortho or para position to the nitro group. The solution containing only a small amount of acid (such as acetic acid) is almost neutral and allows iron to precipitate as Fe_3O_4.

c) Reduction with Zinc in an Alkaline Medium

4,4′-Diaminodiphenyl derivatives are obtained from appropriately substituted nitrobenzenes by alkaline reduction with zinc powder/sodium hydroxide solution, which affords hydrazobenzene. Subsequent acid-catalyzed rearrangement with hydrochloric acid yields the hydrochloride of the target diamine:

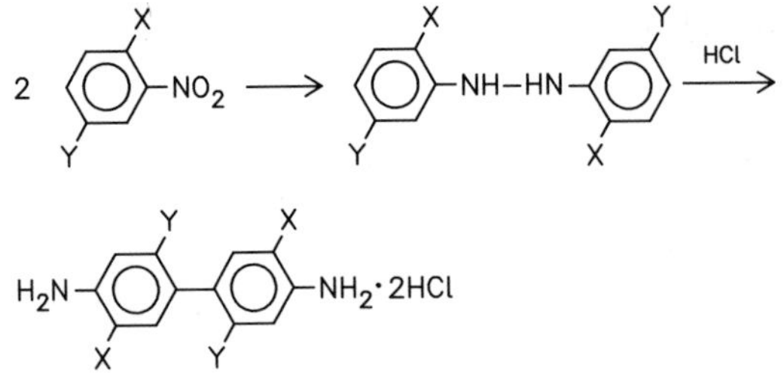

X: Cl, CH$_3$, OCH$_3$, OC$_2$H$_5$
Y: H, Cl

d) Reduction with Sodium Hydrogen Sulfide or Sodium Sulfide

Selective reduction (e.g., partial reduction of one of two nitro groups) is carried out with an alkali sulfide, such as sodium hydrogen sulfide NaHS ("sodium sulfhydrate") or sodium sulfide Na$_2$S, in an aqueous or alcoholic solution. Azo groups are not affected by this method. The reaction converts sodium hydrogen sulfide or sodium sulfide mainly to sodium thiosulfate.

Example:

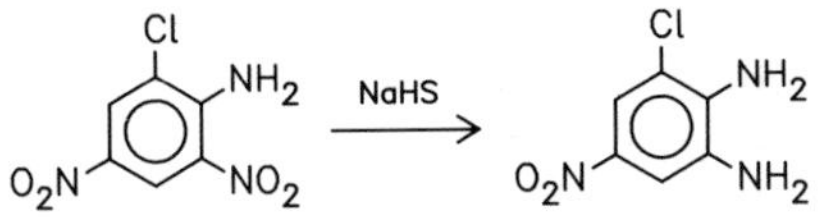

2.1.2   Coupling Compounds

The technically most significant groups of coupling compounds are:

a) Compounds containing activated methylene groups of the type:

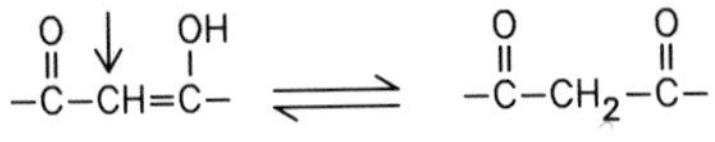

especially acetoacetic arylides

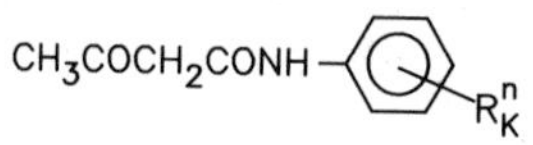

b) 2-Hydroxynaphthalene and its 3-carboxylic acid derivatives:

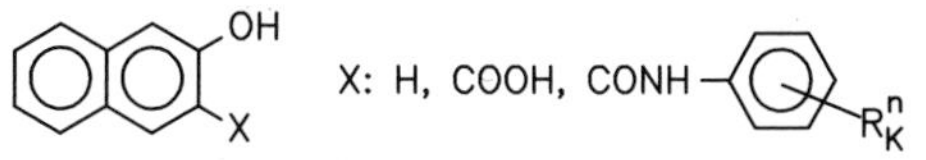

X: H, COOH, CONH

R_K: CH₃, OCH₃, OC₂H₅, NO₂, Cl

or 5-membered or 6-membered heterocycle, condensed to the phenyl ring

n = 0-3

c) Pyrazolone derivatives

Supplement to a) Compounds based on the general structure

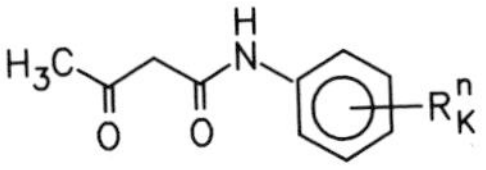

are produced by reacting acetoacetic ester or diketene with aromatic or heteroaro-

matic amines based on the structure H_2N —.

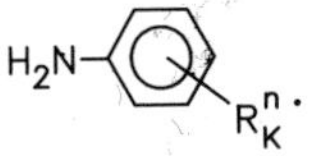

The reaction may proceed in water, acetic acid, or any other organic solvent or mixture that is inert to diketene. This group also includes bifunctional coupling components of the bisacetoacetic diaminodiphenyl type:

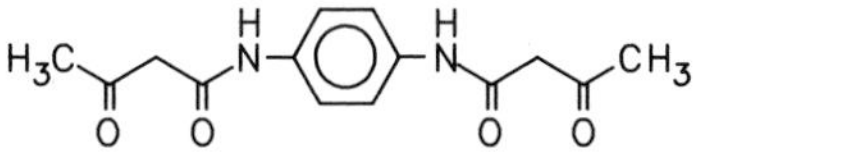

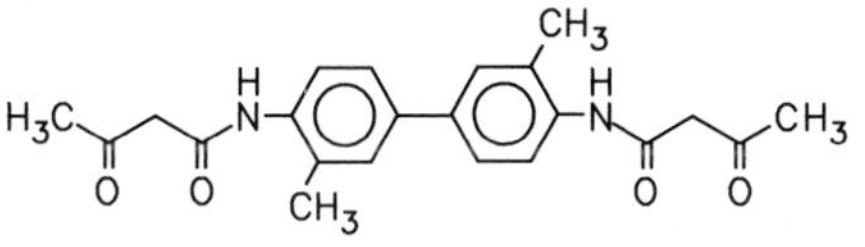

Naphtol AS–G*)

Supplement to b) 2-Hydroxynaphthalene (β-naphthol) is obtained from naphtha-lene. The reaction sequence includes sulfonation of the starting material and subse-quent alkaline baking of the intermediate sodium naphthalene-2-sulfonate with so-dium hydroxide at 300 to 320°C for 6 to 8 hours. After the reaction mixture has

---

* A few "Naphthol Yellow pigments" which are coupling components of the acetoacetic arylide series are also considered members of the "Naphthol AS" series (see footnote p. 281). These are internationally known as "Naphtol AS-G", "Naphtol AS-IRG", etc.. Naphtol AS-IRG, for instance, is acetoacetic-2,5-methoxy-4-chloroanilide (see also Table 10).

been allowed to cool, it is dissolved in water, sodium sulfite is removed by filtration, and the basic solution is neutralized with sulfuric acid. The liquid crude naphthol begins to precipitate at about pH 8. It is separated from the aqueous solution and purified by vacuum distillation:

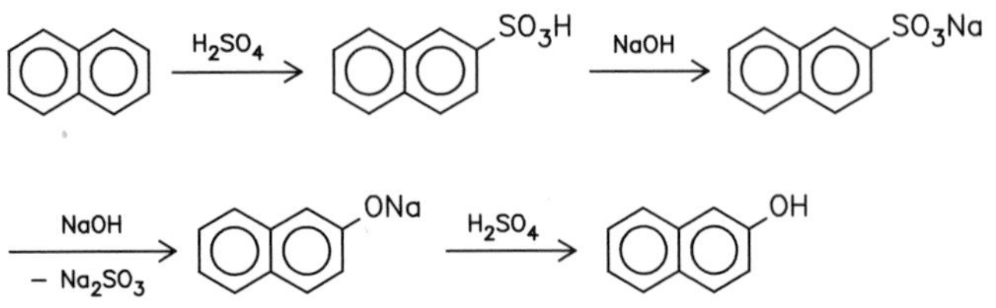

2-Hydroxy-3-naphthoic acid ("BONA" or "BON") is prepared by heating the sodium salt of 2-hydroxynaphthalene with carbon dioxide in a pressure chamber at 240 to 250°C at a pressure of 15 bar (Kolbe synthesis). The reaction mixture is continuously agitated. Remaining 2-naphthol is separated and recycled:

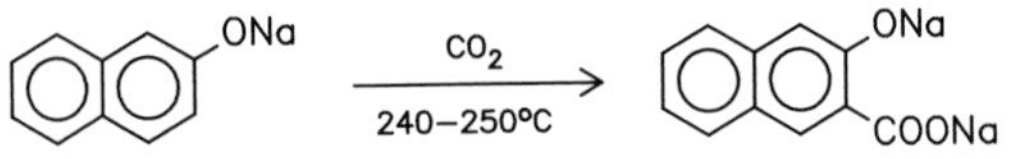

The formation of the technically important 2′-hydroxy-3′-naphthoylanilines (Naphthol AS derivatives) is accomplished primarily by a condensation reaction between 2-hydroxy-3-naphthoic acid and an aromatic amine in the presence of phosphorus trichloride at 70 to 80°C. Appropriate reaction media are organic solvents, such as toluene or xylene. In stoichiometric terms, one mole of 2-hydroxy-3-naphthoic acid reacts with 0.4 to 0.5 moles of phosphorus trichloride. The solution is allowed to cool to room temperature, then neutralized with a sodium carbonate solution, and the Naphthol AS derivative is isolated by filtration. Mechanistically, the reaction is thought to proceed via the phosphoazo compound (**11**):

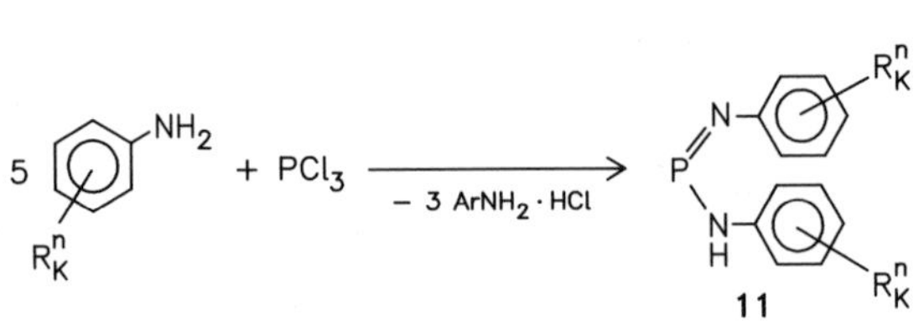

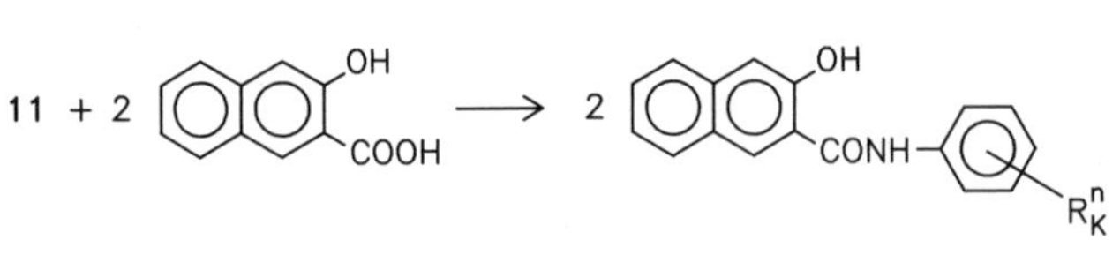

An alternative is to react 2-hydroxy-3-naphthoic acid with thionyl chloride to form the naphthoyl chloride. Condensation with the aromatic amine is then typically carried out in the presence of a tertiary organic base. Table 10 lists the currently most significant Naphtol AS derivatives.

Tab. 10: Important Naphtol AS derivatives* as coupling components for azo pigments.

Naphtol AS Derivatives	C.I. Azoic Coupl. Comp. No., C.I. Constitution No.	R^2	R^3	R^4	R^5
Naphtol AS	2, 37505	H	H	H	H
Naphtol AS-D	18, 37520	CH_3	H	H	H
Naphtol AS-OL	20, 37530	OCH_3	H	H	H
Naphtol AS-PH	14, 37558	OC_2H_5	H	H	H
Naphtol AS-BS	17, 37515	H	NO_2	H	H
Naphtol AS-E	10, 37510	H	H	Cl	H
Naphtol AS-RL	11, 37535	H	H	OCH_3	H
Naphtol AS-VL	30, 37559	H	H	OC_2H_5	H
Naphtol AS-MX	29, 37527	CH_3	H	CH_3	H
Naphtol AS-KB	21, 37526	CH_3	H	H	Cl
Naphtol AS-CA	34, 37531	OCH_3	H	H	Cl
Naphtol AS-BG	19, 37545	OCH_3	H	H	OCH_3
Naphtol AS-ITR	12, 37550	OCH_3	H	OCH_3	Cl
Naphtol AS-LC	23, 37555	OCH_3	H	Cl	OCH_3

* Naphtol AS is a trade name of Hoechst AG, Germany. Other ranges are assigned Naphthol AS.

Likewise, heterocyclic coupling components are derived from acetoacetic ester, diketene, or 2-hydroxy-3-naphthoic acid and a heterocyclic amine:

Supplement to c) Pyrazolone derivatives. Derivatives of pyrazolone-(5) are important heterocyclic coupling components. They are based on the following general
structure:

R: CH_3, $COOCH_3$, $COOC_2H_5$
R': H, CH_3

The pyrazolone ring system is accessible through ring closure by the condensation
of 1,3-diketo compounds with hydrazine derivatives.

Methods of synthesizing pyrazolone focus especially on ring closure by reaction
of acetoacetic ester with phenyl or p-tolylhydrazine, affording water and ethanol as
by-products.

R' : H, CH_3

2.1.3   Important Intermediates

The list of important diazo components for the diazotization reaction includes aniline and a number of aniline derivatives, diamino diphenyl derivatives, and aromatic aminosulfonic acids. Important coupling components are acetoacetic arylides, pyrazolones, β-naphthol, 2-hydroxy-3-naphthoic acid, and their aniline derivatives.

The following list includes starting materials which have found their way to
large-scale production and application in the azo pigment industry.

Important intermediates for azo pigments are:

aniline
p-toluidine
m-xylidine
2,5-dichloroaniline
4-chloro-2-nitroaniline
4-methyl-2-nitroaniline
4-chloro-2,5-dimethoxyaniline
3,3′-dichlorobenzidine
4-aminotoluene-3-sulfonic acid
2-chloro-5-aminotoluene-4-sulfonic acid
acetoacetanilide
acetoacet-o-chloroanilide
acetoacet-m-xylidide
acetoacet-4-chloro-2,5-dimethoxyanilide
β-naphthol

Other common intermediates for azo pigment production are 2,4-dinitroaniline, acetoacet-o-anisidide, acetoacet-o-toluidide, phenyl and p-tolyl-methylpyrazolone, 2-hydroxy-3-naphthoic acid, Naphtol AS and its derivatives, and 2-chloro-4-amino-toluene-5-sulfonic acid.

References for Section 2.1

[1] N.N. Woroshzow: Grundlage der Synthese von Zwischenprodukten und Farbstoffen, 4. Ed. Akademie-Verlag, Berlin, 1966.
[2] H.R. Schweizer: Künstliche organische Farbstoffe und ihre Zwischenprodukte, Springer-Verlag, Berlin, 1964.
[3] Winnacker: Küchler Chemische Technologie, Vol. 6 (4th edition), 143–310 (1982).

2.2   Synthesis of Azo Pigments

The production of azo pigments relies almost exclusively on the azo coupling reaction [1] to afford the azo group. Diazotization of an aromatic amine yields a diazonium compound, which subsequently reacts with a coupling component ("coupling").

Generally speaking, the coupling reaction links an aromatic amine to a nucleophilic partner RH (coupling component); the amine is treated with a nitrosyl source XNO to form an azo compound. This sequence is expressed by the following equation:

$$Ar-NH_2 + XNO + RH \xrightarrow[-H_2O]{} Ar-N{\underset{N-R}{\diagdown}} + HX$$

Ar: aromatic or heteroaromatic group
R: coupling component residue
X: Cl, Br, NO_2, HSO_4.

As a result of the stabilization of the coplanar structure, azo groups exhibit cis-trans isomerism. Azo compounds are generally written in the trans configuration, which is the more stable one. However, X-ray structure analyses show azo pigments exclusively in their hydrazone form (see pp. 216, 273).

The reaction proceeds via two main steps: a diazotization reaction, which is the basis of diazonium salt formation, followed by azo coupling, which is responsible for the formation of the azo compound.

2.2.1   Diazotization

Diazotization is the reaction of a primary aromatic amine with a nitrosating agent, such as sodium nitrite; or, to a lesser extent, with nitrosylsulfuric acid $NOSO_4H$, nitrous gases, or organic nitrites in an aqueous acidic solution at a temperature between 0 and 5 °C, converting the amine to its diazonium salt:

$$ArNH_2 + 2\ HY + NaNO_2 \longrightarrow ArN\overset{\oplus}{\equiv}N\ Y^{\ominus} + 2\ H_2O + NaY$$

The diazotization reaction was discovered in 1858 by Peter Griess. Following a suggestion by Kolbe, he treated picramic acid (2-amino-4,6-dinitrophenol) with nitrous gases in an ethanol solution in an attempt to replace the amino function by an OH group. Kolbe, who had been the first to carry out this substitution on p-aminobenzoic acid, worked at elevated temperature, so that the formation of the intermediate diazonium compound escaped his attention. Griess not only worked at low temperature, but diazopicramic acid also has the advantage of being relatively stable, which facilitated its detection as an intermediate. Griess described it as a "diazo" compound, erroneously assuming that he had substituted two hydrogen atoms on the benzene ring by nitrogen.

On an industrial scale, diazotization reactions are carried out by dissolving the aromatic amine in hydrochloric or sulfuric acid. Despite the fact that 2 equivalents of acid per equivalent of amino group should theoretically suffice, as much as 2.5 to 3 equivalents per amino function are actually required to ensure complete diazonium salt formation. One equivalent of an aqueous sodium nitrite solution is added to the resulting mixture at 0 to 5 °C. The exothermic nature of the reaction, combined with the heat sensitivity of most diazonium salts, makes it necessary to provide cooling, usually by direct addition of ice.

Amines of low basicity require a higher acid concentration in order to avoid the formation of diazoamino compounds, which do not couple:

$$ArN{\equiv}N\ Y^{\ominus} + H_2N-Ar' \longrightarrow Ar-N{\diagdown}_{N-NH-Ar'}\ + HY$$

Diazoamino compounds occur only in the trans configuration. Their formation may be prevented by adding a small excess of nitrite during or after diazotization.

Nitrosylsulfuric acid is employed for amines of very low basicity; e.g., those with more than one electronegative function. Compounds such as di- and trinitroanilines, halogenated nitroanilines, and tetrahalogenanilines will not react under more moderate conditions. If necessary, these amines can also be diazotized after dissolving them in a mixture of glacial acetic acid and concentrated hydrochloric acid.

Aromatic diamines can be diazotized twice to produce bisdiazonium compounds (bisdiazotization).

2.2.1.1   Diazotization Mechanism

Contrary to earlier assumptions, the aryl ammonium ion is not able to undergo diazotization since the crucial step in the diazotization mechanism is the electrophilic nitrosation of the free amino group in the free-base primary aromatic amine **12**:

$$ArNH_2 + XNO \xrightarrow[-HX]{} Ar\overset{H}{N}-N{=}O \longrightarrow Ar-N{\diagdown}_{N-OH}$$

12   **13**

$$X: Cl,\ Br,\ NO_2,\ HSO_4$$

Formation of the diazonium ion **14** proceeds via the diazo hydroxide **13**:

$$Ar-N{\diagdown}_{N-OH} \underset{OH^{\ominus}}{\overset{H^{\oplus}}{\rightleftharpoons}} Ar-N{\equiv}\overset{\oplus}{N} + H_2O$$

13   **14**

The diazotization reaction requires an excess of acid; formation of the active nitrosating agent XNO proceeds via the underlying equilibrium:

$$NaNO_2 \xrightarrow{H^{\oplus}} HNO_2 \underset{}{\overset{H^{\oplus}}{\rightleftharpoons}} H_2NO_2^{\oplus}$$

$$H_2NO_2^{\oplus} + X^{\ominus} \longrightarrow XNO + H_2O$$

2.2.1.2   Methods of Diazotization

The following methods are currently used to produce azo pigments on an industrial scale. The choice depends on the basicity and on the solubility of the individual product.

Direct Diazotization

An aqueous sodium nitrite solution is added to a cold solution or suspension of the primary aromatic amine in an excess of hydrochloric or sulfuric acid. A temperature of 0 to 5 °C is maintained by adding ice directly to the reaction mixture.

Indirect Diazotization Method

This method is particularly useful for aromatic aminocarboxylic and aminosulfonic acids, which are often only sparingly soluble in dilute acid. The amino compound is dissolved in water or in weak alkali and combined with a stoichiometric amount of sodium nitrite, upon which the resulting solution is poured into a mixture of acid and ice. Alternatively, the process may be reversed by pouring the acid into the amine–nitrite mixture.

Diazotization of Weakly Basic Amines

Amines of very low basicity undergo diazotization after being dissolved in concentrated sulfuric acid. The nitrosating agent is provided by nitrosylsulfuric acid, which is either purchased commercially or easily prepared by dissolving solid sodium nitrite in concentrated sulfuric acid.

Sulfuric acid as a reaction medium may be replaced by glacial acetic acid or by a mixture of glacial acetic acid/nitrosylsulfuric acid. In the former case, half-concentrated hydrochloric acid is added, and diazotization proceeds in an aqueous sodium nitrite solution. The combination of glacial acetic acid/nitrosylsulfuric acid is a particularly useful medium for the bisdiazotization of 1,2-, 1,3-, or 1,4-diaminobenzenes (phenylenediamines).

Diazotization in Organic Solvents

Amines which are considerably or entirely insoluble in water are dissolved in glacial acetic acid or in other organic media, possibly mixed with water; e.g., alcohols or aprotic solvents. Addition of acid is followed by a typical diazotization reaction in an aqueous sodium nitrite solution. Other possible sources of the nitrosating species NO include nitrosylsulfuric acid, nitrosylchloride, alkylnitrite, and nitrous gases.

The outcome of a diazotization reaction is largely controlled by the temperature, the pH, and the concentration of the medium in which the reaction is carried out. Reactions involving sparingly soluble amines are also dependent on physical parameters, such as distribution and particle size, as well as on the possible presence of emulsifiers or dispersion agents.

A new diazotization technique is described in connection with the different coupling methods (Sec. 2.2.2.1).

In aqueous media, most diazonium salts are only stable at low temperature. Heating frequently leads to decomposition; as a result, nitrogen and the corresponding phenol are formed. The stability of a diazonium compound is a function of the substitution pattern of the aromatic ring system. Electronegative (electron withdrawing) substituents (electron acceptors), such as halogens or nitro functions, render their host structure more sensitive to decomposition than do electron donors such as $-CH_3$, $-OCH_3$, $-OC_2H_5$. Some amines can even be diazotized up to a temperature of $50\,°C$. Other factors that affect the stability of a diazonium compound are visible light or heavy metal ions.

On an industrial scale, diazotization reactions are chiefly carried out in cast iron kettles that are lined with brick or rubber as a protection against acid. Wooden vats also continue to be used.

In most cases, diazonium salts are unstable in the dry state and are sensitive to heat and impact. Since isolation is not necessary for azo pigment production, the diazonium compound is coupled with the coupling component as it is formed in solution or suspension.

2.2.2 Coupling

A coupling reaction is an electrophilic substitution of the diazonium compound with a nucleophilic partner (coupling component RH):

$$Ar-N{\equiv}N^{\oplus} \ Y^{\ominus} + RH \longrightarrow Ar-N{\underset{N-R}{\searrow}} + HY$$

$$Y: Cl, HSO_4$$

Considering the electrophilic nature of the reaction mechanism, suitable coupling components for the synthesis of azo pigments should carry a nucleophilic center on the aromatic ring system. Examples include naphthols and enolizable compounds with reactive methylene functions. Naphthols enter the reaction as naphtholates; compounds with reactive methylene groups participate as enolates.

The free acid which is produced in a coupling reaction according to the above equation makes it necessary to add bases or buffers to the reaction mixture in order to maintain a constant pH and to optimize the outcome of the coupling process. Highly basic solutions do not permit coupling because they shift the underlying equilibrium and thus convert the diazonium compound to a trans ("anti") diazotate ion, which does not couple:

$$Ar-N\overset{\oplus}{\equiv}N \;\overset{OH^{\ominus}}{\rightleftarrows}\; Ar-N\overset{}{\underset{N-OH}{\leqslant}} \;\overset{OH^{\ominus}}{\rightleftarrows}\; Ar-N\overset{}{\underset{N-O^{\ominus}}{\leqslant}} + H_2O$$

Phenols, naphthols, and enols therefore couple best in the weakly acidic to weakly basic pH range. In order to avoid the necessity of permanent pH correction, buffers such as sodium acetate, sodium phosphate, magnesium oxide, calcium carbonate, sodium or potassium hydrogen carbonate, or sodium or potassium carbonate are employed, unless a dilute (3 to 6%) sodium hydroxide solution is constantly added during the coupling reaction. Substituents showing $-I/-R$ effects (electron acceptors) on the aromatic ring system of the diazo compound generally increase the reactivity. The same substituents, however, have an adverse effect on the coupling reaction if they are located in the aromatic (anilide) moiety of the coupling component. The reverse is also true; namely, $+I/+R$ groups (electron donors) in this part of the molecule enhance the reaction.

This principle is exemplified by the following sequence of substituted anilines used as diazo compounds, whose coupling energy decreases with an increasingly electron donating substitution pattern on the aniline skeleton:

polynitroanilines > nitrochloroanilines > nitroanilines > chloroanilines > anilinesulfonic acids > aniline > anisidines > aminophenols

As mentioned above, the nature of the coupling mechanism, which is an electrophilic substitution reaction, makes the carbon atom with the highest electron density the most likely coupling location. This explains why hydroxy or amino functions (although the latter do not play a major role in pigment chemistry) direct the attack of the diazonium ion exclusively to the ortho or para position of an aromatic system. Blocking both these positions by substituents other than hydrogen precludes the reaction entirely or effects expulsion of one of the substituents. There is no other option; the meta position never participates in a coupling reaction.

Naphthalene derivatives as coupling components generally couple better than their benzene analogs; the latter only play a minor role in pigment chemistry.

The pH of the reaction medium is not the only parameter to determine the outcome of an azo coupling reaction. As mentioned above, most diazonium salts decompose at elevated temperature:

$$Ar-N\overset{\oplus}{\equiv}N \;\; Cl^{\ominus} + H_2O \longrightarrow ArOH + N_2 + HCl$$

This undesirable effect largely compromises the advantage that increased temperature has in accelerating a coupling reaction. Rather than risking decomposition of the diazonium compound by elevating the reaction temperature, it is therefore much more useful to increase the pH or the concentrations of the reactants in order to enhance the rate of a coupling reaction.

A number of bases, with pyridine heading the list, act as proton acceptors in the electrophilic coupling reaction. Their contribution is particularly useful if voluminous substituents exist in ortho or peri position relative to the coupling location of the intermediate; they may also facilitate coupling with diazonium ions of low electrophilicity (e.g., diazophenols).

Coupling components for azo pigments are generally almost insoluble in water itself, but phenols and enols dissolve rapidly as phenolates and enolates in aqueous alkali solutions:

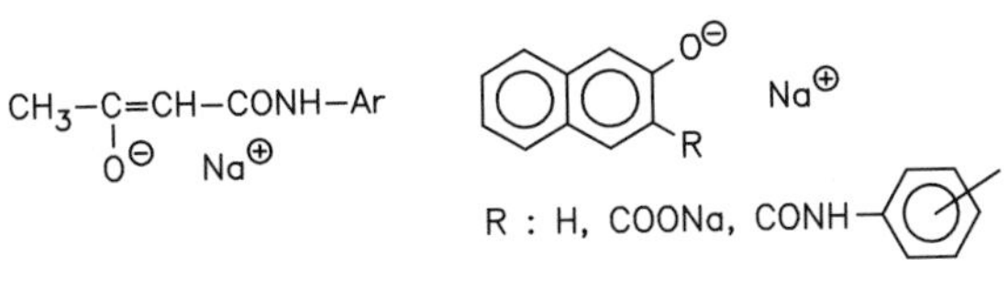

$$CH_3-C=CH-CONH-Ar$$

R : H, COONa, CONH

The remaining difficulty of working with a solution that is too basic for a coupling reaction can be avoided by carefully adding acid and thus precipitating the coupling component from the alkaline solution. A dispersing agent may be useful to produce an aqueous suspension with fine particle sizes, which couples easily with the diazonium compound.

2.2.2.1 Coupling Techniques

The following methods play a role in the commercial production of azo pigments:

Direct Coupling

The coupling component is dissolved in an alkaline solution and, after adding a clarifying agent and possibly charcoal to the solution, it is filtered through a sparkler filter or a filter press.

The solution is then transferred into the coupling vessel equipped with a mechanical stirrer and, possibly in the presence of a surfactant, precipitated with acetic acid, hydrochloric acid, or phosphoric acid. The coupling component may also be precipitated "indirectly"; i.e., the appropriate mixture of acid and emulsifier is filled into the kettle first and the alkaline solution of the coupling component is then added gradually to the clear solution by gravity flow. The clarified solution of the diazonium compound is then introduced into or onto this coupling suspension.

Precipitating the coupling component with acetic acid or phosphoric acid often automatically provides the buffer that is necessary to maintain a certain pH throughout the coupling reaction. Otherwise, buffers such as sodium acetate, sodium phosphate, or calcium carbonate ("chalk coupling") must be added.

Indirect Coupling

The clarified acidic diazonium salt solution is first transferred to a coupling vat equipped with a mechanical agitator, and the clarified alkaline coupling component solution is then added above or under the surface of the diazonium salt solution. Constant agitation is essential.

"Pendulum" Technique

If a coupling reaction is carried out in the absence of a buffer system, a constant pH is maintained throughout the coupling process by continuously adding dilute sodium hydroxide solution ("pendulum solution") to the reaction mixture.

Organic Solvents as Coupling Media

Starting materials which are only sparingly soluble in water may require solvents that are either partially or entirely organic. Diazotization can either be carried out as usual with an aqueous sodium nitrite solution, or alternatively with nitrosylsulfuric acid or an organic nitrite. Appropriate solvents must be stable to the reactants. Examples include aromatic hydrocarbons, chlorohydrocarbons, glycol ethers, nitriles, esters, and dipolar aprotic solvents, such as dimethyl formamide, dimethyl sulfone, tetramethylene sulfone, tetramethyl urea, and N-methylpyrrolidone.

A special type of azo bridge formation is observed in aprotic polar media (solvents with a dielectricity constant below 15). In these solvents, the reaction proceeds via an aprotic diazotization–coupling mechanism [2], which unites both diazotization and coupling in one step. The process involves adding a volatile alkyl nitrite to a slightly acidic suspension of a combined solution of diazo and coupling component. Organic solvents have the double advantage of being almost completely accessible to recycling and of producing only very small amounts of contaminants which affect the wastewater.

Coupling with "Masked" Diazonium Compound

Coupling reactions in organic solvents are occasionally carried out with "masked" diazonium compounds; e.g., with special diazonium moieties which are incorporated into a larger organic structure [3], for instance in a diazoamino compound (**15**) or a benzotriazinone (**16**):

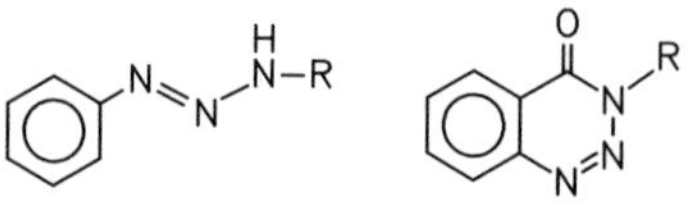

15   16

R: alkyl, aryl

The diazonium compound itself is liberated by adding a strong organic acid, such as halogenated acetic acid, after the coupling component has been introduced into the reaction mixture.

Azo pigments are difficult to purify, because they emerge from the manufacturing process as almost insoluble substances. The purity of the starting materials is therefore of major concern to the producer. However, contamination is not the only factor to influence the outcome of pigment synthesis; the following factors have a much greater impact on pigment production than on the synthesis of dyes:

- coupling technique, e.g., sequence and rate of reactant addition,
- concentrations of the reactants,
- temperature of the reaction mixture,
- choice of organic solvent added,
- technical and operational parameters: shape and size of the reaction vats and the agitator, as well as the speed of agitation.

2.2.3   Finishing

Azo dyes differ greatly from azo pigments in the form in which they emerge from the manufacturing process. Dyes are precipitated from the (aqueous) solution by salts. Little labor is involved in washing, drying, and finally standardizing a dye formulation: auxiliary agents and solid diluents are added to the product, which is now ready to be sold. Azo pigments, on the other hand, emerge from the synthesis in the form of extremely small, insoluble particles (primary crystallites), which require aftertreatment or finishing. Physical properties, such as crystal shape, crystal size, and crystal quality, as well as particle size distribution, must be optimized to reach the desired quality. The properties of the primary crystals obviously depend on the pigment itself and can therefore be gauged by adjusting the parameters of the coupling reaction, such as temperature and pH.

More or less extensive finishing is generally necessary to prepare a crude azo pigment for technical application. Washing and drying the crude pigment presscake directly on formation has a detrimental effect on the product. The primary particles may associate to form agglomerates and aggregates. As a consequence, the resulting hard particles will be difficult to disperse, affording a hard grained pigment with poor tinctorial strength. At this stage, it is not possible to convert the material to a useful pigment by milling.

Combining the physical parameters of the crystals to the advantage of pigment performance is a prerequisite to developing optimum application properties. Thermal treatment is the most important process on the way to this target.

Heating the crude pigment suspension or the salt-free pigment presscake in water and/or organic solvents improves the quality of the crystals. Presscakes are prewashed with water in order to remove salt, isolated, and mixed with water and/or solvent. This finishing process reduces the portion of extremely fine particles,which are the primary source of agglomeration, so that the particle size distribution narrows. Particularly insoluble pigments are finished in an organic medium, such as alcohol, glacial acetic acid, chlorobenzene, o-dichlorobenzene, pyridine, or dimethylformamide at 80 to 150 °C. Extensive thermal treatment appreciably enlarges the particle size.

The particle size distribution thus shifts towards larger sizes, which not only improves the rheological properties but frequently also effects an increase in opacity.

It is also possible to optimize the application properties of a pigment by adding appropriate auxiliaries with different chemical structures to the reaction medium in order to influence the surface structure of the resulting pigment particles. This process is known as surface treatment. Rosin or other resins, for instance, added to the reaction mixture during or immediately after coupling, will inhibit crystal growth, resulting in transparent pigments with fine particle sizes. Features such as these are often required in pigment application.

Preparation with aliphatic amines, on the other hand, may promote side reactions, converting portions of a pigment to compounds that are somewhat soluble in toluene. Toluene is the most important solvent for publication gravure printing inks. This preparative method reduces the viscosity of the printing ink. The pigment is thus partially converted to a soluble azomethine (Schiff's base), which is formed by reaction between the acetoacetic arylide and an aliphatic amine:

R′: long-chain aliphatic residue

Dispersion is the most important prerequisite for the technical application of a pigment. Pigment preparation typically comprises the following steps: Azo pigment synthesis – drying (aggregation and agglomeration) – milling – combining the pigment with its application medium (dispersion). This sequence is often uneconomical, and it is useful to find ways to shorten it.

The last step – pigment dispersion – may be facilitated by adding appropriate agents to the reaction mixture before agglomeration can occur. These agents can either be chemically identical or be part of the medium of application. There are polymer dispersions, for instance, that may even be added during or after coupling. Pigment preparations which are produced via this route can often be distributed in the application medium (i.e., plastics) without extensive dispersion.

Finishing not only improves the application properties of a pigment, such as hue, tinctorial strength, brilliance, transparency/hiding power, dispersibility, and flow behavior, but also considerably enhances its lightfastness and weatherfastness and its solvent and migration resistance.

2.2.4   Filtration, Drying, Milling

Following manufacture and possibly further processing, the pigment is separated from the suspension and dried. Both processes can be carried out either continuously or by batch operation, depending on the tonnage of the product.

Plate-and-frame filter presses are particularly suited to batch operations. Modern frames are made of polymers (mainly polypropylene), which have replaced wood as a building material. Large amounts of products, on the other hand, are filtered routinely via continuous-belt filters or rotary drum-type filters. Following the filtration step, inorganic salts are removed by washing with water.

Before being dried, the resulting pigment presscake may be flushed in a paste mixer. Since drying operations are traditionally slow, lasting from about 10 hours to as long as 2 days, it is advantageous to increase the surface area within a pigment presscake in order to accelerate the drying process. This can be effected by mechanically chopping the presscake on the drying trays or by granulating the material by means of an extruder.

Batch drying is usually carried out in a steam-heated drying oven with a circulating air stream. Depending on the heat sensitivity of the pigment, vacuum-operated drying ovens may also be used for batch processes. Continuous-drying operations, on the other hand, are carried out with belt dryers or spray dryers. Belt dryers typically operate on the principle of hot air being blown through a tunnel in counter-direction relative to a running metal belt, which transports the pigment presscake. Parameters such as temperature and duration are variable.

In a spray dryer, the aqueous pigment paste passes through a rotating disk or nozzle into a cone-shaped spray chamber fed with hot air. The dried pigment powder trickles through a grate at the bottom.

A variety of mills are available for the pulverizing of pigment particles. In order to ensure optimum properties, the best type of mill for a given pigment is determined by a pilot experiment. Great care should be taken to avoid excessive pulverization, which might lead to reagglomeration of the primary particles and thus has a detrimental effect on the application properties of a pigment.

Before being milled on an industrial scale, each pigment has to pass a test concerning its sensitivity to dust explosion. There are standardized milling regulations for each hazardous class.

2.2.5   Azo Pigment Synthesis by Continuous Operation

A number of publications describe continuous techniques used to manufacture azo colorants, especially azo pigments (selection see [4]).

The widespread interest in continuous methods is readily explained by the expected advantages:

– product standardization through uniform reaction conditions,
– increased productivity compared to batch processes,
– improved methods of process and quality control.

In contrast to azo dyes, whose coloristic properties are almost exlusively defined by their chemical structure, the property of an azo pigment depends largely on the physical characteristics of its particles. The continuous process of pigment synthesis is therefore designed to afford a product which already satisfies the commercial specifications. In other words, in a continuous operation, the ultimate performance of the material is determined by the coupling process. Converting a traditional batch technique to a continuous operation adds an extra degree of complexity: in the form of aqueous suspensions, diazo compounds sometimes and coupling components generally react under coupling conditions. It is therefore often difficult to maintain standardized conditions throughout such a heterogeneous reaction.

The following factors determine the outcome of azo pigment production by continuous process:

– coupling rate,
– pH,
– temperature,
– concentration of the diazonium compound,
– impurities,
– form in which the coupling component precipitates,
– surfactants,
– rate of crystal nucleus formation,
– rate of crystal growth.

The first five of these parameters apply to all continuous azo-coupling reactions, while the last four pertain to azo pigments only.

The overall process can be related to two operations: diazotization and coupling. Out of the vast number of patents concerning continuous methods of azo pigment formation, the following examples have been chosen.

Diazotization by Continuous Technique

The most important side reaction during diazotization may lead to the formation of the diazoamino compound (Sec. 2.2.1). The diazonium compound may combine with unreacted amine to form the side product (Sec. 2.2.1). This effect is particularly prominent if it arises from a change of concentration. Insufficient nitrite prevents the free amine from immediately converting to the diazonium salt. It is possible to reduce this undesirable side reaction by maintaining an excess of nitrite during the diazotization reaction and thus enhancing the formation of diazonium compound over the side product. This provides an important method of process control. Obviously, the amount of formed diazoamino compound is largely determined by the reactivity of the diazonium salt. Weakly basic amines afford the side product much more readily than do more reactive amines (those with electron donating substituents).

Diazotization by continuous process may, for instance, be carried out as follows [5]:

The three components (the aqueous amine suspension or its salt in aqueous mineral acid, a nitrite solution, and mineral acid) are transferred simultaneously into a diazotization vessel. For adequate process control, a portion of the reaction mixture is diverted between the storage tank and the diazotization vessel. It enters the diazotization vessel after remaining in an analyzer only for about 1/3 to 1/4 of the time required for the main stream in the diazotization vessel. The analyzer is responsible for maintaining constant reactant concentrations throughout the coupling process. Any change to the excess of nitrite initially provided in the storage tank is continuously recorded and controlled through variations of the redox potential or the polarization voltage (polarography) within the analyzer. The azo coupling process is thus operating at continuously constant concentrations. The following flow chart (Fig. 91) outlines the principle of this particular example of indirect diazotization by continuous process.

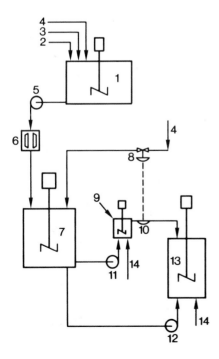

Fig. 91: Diagram showing a continuous diazotization process.

A storage tank (1) is used to combine the amine (2) with water (3) and nitrite (4). This make-up vessel (1) should contain approximately 90% of the theoretically required amount of nitrite. The milled amine suspension is transported into the reaction vessel (7) by a pump (5). Milling the amine in a roll mill, a carborundum mill, or a ball mill (6) in order to reduce the particle size prior to the reaction is particularly essential for species with coarse particle sizes. Additional nitrite then enters the reaction vessel (7) through a valve (8). A 2% excess of nitrite is maintained throughout the reaction (7). An analyzer (9) is responsible for precise nitrite addition. Electrochemical data continuously collected by a redox potentiometer, a voltmeter, or a polarograph (10) connected to the analyzer (9) indicate the nitrite concentration in the amine suspension or solution leaving the reaction vessel (7) at any given time. In case of demand, the analyzer transfers additional nitrite through the valve (8) into the reaction vessel (7). As soon as the desired excess is reestablished within the monitor (10), the valve (8) interrupts the nitrite transfer. The reaction mixture is typically represented by only a small portion of the entire mass flow (11) that is diverted via the analyzer. The main portion of the amine suspension leaving the reacton vessel (7) is pumped (12) directly into the diazotization vat (13). To ensure a quantitative diazotization procedure, mineral acid (14) is transferred into the analyzer (9) and the diazotization vat (13).

Coupling by Continuous Process

There are basically two variations to the continuous-coupling technique [6]:
- Reaction in a homogeneous medium, which means the coupling component is dissolved.
- Reaction in a heterogeneous medium, using a suspension of the coupling component, which is obtained by precipitation.

The kinetics of the coupling mechanism include a number of sometimes very fast and competitive side reactions. The following steps, for instance, proceed simultaneously as a separately prepared diazonium salt solution is combined with an initially dissolved coupling component:
- the coupling reaction itself,
- decomposition of the diazonium salt,
- precipitation of the originally dissolved coupling component under the reaction conditions.

The manufacturer, whose main concern is to maintain a prevalence of the coupling reaction over the two others, i.e., a sufficiently fast coupling rate, achieves his end by combining the reactants at the exact moment of interaction (turbulence). Suitable equipment, such as a mixing nozzle or a static mixing tube, is therefore indispensable for a continuous operation; but it is on the other hand quite successful in helping to avoid side reactions.

The particle size of the resulting pigment can only be influenced to a limited extent by adjusting the reaction parameters, because the decisive factor is the ratio of the rate of formation of the crystal nucleus to the rate of crystal growth.

Another variety of the continuous-coupling technique operates by transporting the coupling component suspension as a laminar flow upwards inside a vertical reaction tube. Portions of the diazonium compound, dissolved in an acidic aqueous medium, are added through appropriately located inlets in the walls of the reaction tube. The concentration of the added solution decreases as the reaction mixture flows upward and is designed to synchronize the uppermost inlet for the diazonium salt solution with the stoichiometric end point of the coupling reaction.

Measurement

Reliable measuring techniques [7] and appropriate process control are the basic elements of successful azo pigment synthesis by the continuous process. These are the parameters which are responsible for maintaining constant reaction conditions: flow rate, pH, temperature, and the concentrations of the reactants, before and after the point of mixing itself.

Potentiometric methods (e.g., with Pt/Hg_2Cl_2 or Au/Hg_2Cl_2 electrodes) are particularly useful to monitor both the diazotization and the coupling reaction. Polarography (determining the polarization voltage) indicates changes of the nitrite content during diazotization. It is also possible to measure the amount of nitrous gases escaping into the air above the agitated and therefore constantly renewed surface of the liquid [8].

On a cost/performance basis, large-scale pigment manufacture by the continuous operation as compared to batch operation remains uneconomical as far as technical considerations are concerned. The difficulty of attaining starting materials with a standardized quality and maintaining constant coupling rates adds to the complexity of process control (e.g., stability of the potentiometric electrode systems). These are clearly factors in favor of batch techniques, which remain the rule throughout the azo pigment industry.

The dominant role of the batch operation over the continuous process throughout the pigment industry is somewhat in contrast to the patent literature, which includes numerous proposals for complete azo pigment manufacture by the continuous process as well as descriptions of the partial steps, such as the diazotization or the coupling reaction.

2.2.6 Production Units for Azo Pigment Manufacture by Batch Operation

A typical production unit consists of an acid-proof diazotization kettle ("diazotizer"), a dissolution vessel to dissolve the coupling component in its medium, and a reaction vessel with an agitator, in which the coupling reaction is carried out. Typical vessel capacities are 20 to 80 m^3, corresponding to batches of 0.5 to 2.5 t of azo pigment.

Clarifying filters or clarifying presses are installed between diazotization kettle and dissolution tank and the reaction vessel. The crude pigment slurry from the coupling vessel is filtered in a filter press; and a pressure vessel equipped with an agitator (for thermal aftertreatment) connected to a filter press completes the processing unit for the synthesis.

Drying and milling the resulting wet pigment presscakes, possibly preceded by extrusion or granulation, finally affords the desired pigment powder. Drying may be carried out either by a continuous process on a conveyor belt or as a batch operation in a convection oven.

The flow chart in Fig. 92 outlines the typical sequence of azo pigment synthesis, including essential equipment:

The coupling component, usually in an alkaline solution, is clarified by mechanical and adsorptive methods in clarifying filters or presses and then charged into a "coupling vessel". Acids, and possibly also surfactants, are added in order to precipitate the material. In the "diazo kettle", the aromatic amine is dissolved in aqueous acid and then diazotized with aqueous sodium nitrite. After the coupling component slurry has been clarified by filtration, the diazonium salt solution is slowly transferred above or below the surface of the coupling component. It is also possible to reverse the sequence in which the reactants are combined, or to add them simultaneously to a buffer solution in the coupling vessel. Further treatment, such as precipitation (metal salt formation), metal complexation, or even thermal aftertreatment may be carried out advantageously in the same vessel.

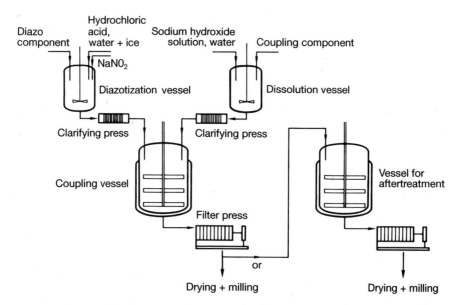

Fig. 92: Diagram showing the equipment which is used to manufacture an azo pigment.

The solid azo pigment is then separated by filtration, washed, and dried immediately. It is also possible to reslurry the pigment presscake in another agitation vessel in order to prepare it for thermal aftertreatment. Milling follows drying.

Depending on the technical requirements such as corrosion resistance, pressure and temperature stability, industrial scale azo pigment synthesis is carried out in appropriate equipment. Suitable materials include cast iron, stainless steel, steel lined with rubber, acid-proof brick, enamel, synthetic resins supported by glass fiber, and wood.

The most widely used material for diazotization, dissolution, and the coupling-vessels is rubber-lined steel.

The rubber lining resists temperatures up to 100 °C for some time (about 1 h) without suffering damage; however, contact with organic solvents must be avoided. Recent developments include coupling vessels made of synthetic resins on a fiber glass structure, which have the multiple advantage of being very light, low-cost materials which for repair are easily accessible. Moreover, these vessels are fast to hydrochloric acid and resist temperatures up to 100 °C. Traditional wooden vats also continue to be used, which is not as surprising as it seems, considering the fact that they are corrosion resistant in aqueous media and are economically installed and repaired. Stainless steel is the material of choice for reactions in neutral or alkaline media or in organic solvents, as well as in autoclaves. It should be noted, however, that the advantage of stainless steel is compromised by the fact that it will corrode if the vessel or any other part of the production unit is exposed to mineral acid. Thus, autoclaves made of this material cannot be used for reactions in slightly acidic media or salt solutions; chloride ions are particularly destructive. Such operations are therefore carried out in enamel lined pressure vessels, which in turn are not resistant to alkali, but easy to clean. Stainless steel autoclaves lined with rubber or brick are also suitable for acidic reaction media; agitators, heating coils, and thermometer jackets are made of a nickel alloy (e.g., Hastelloy) with a varying content of Mo, Cr, Mn, Cu, Si, Fe, and C. Brick lined vessels are primarily used in large-scale production, where the cost of installation and repair exceeds the advantage of enamel lining.

Reaction mixtures are transferred from one piece of equipment to another by pumping, particularly if the production unit includes wooden vats or vessels made of synthetic resins, which may not be pressurized. Steel vessels, on the other hand, allow transport by air or nitrogen pressure.

References for Section 2.2

[1] H. Zollinger: Chemie der Azofarbstoffe, Basel, Birkhäuser, 1958.
[2] A.C. Rochat and E. Stocker, XII. Congr. FATIPEC, Garmisch 1974, Congress book, p. 371.
[3] e.g. DE-OS 1 644 119 (Ciba-Geigy) 1966; DE-OS 1 644 127 (Ciba-Geigy) 1966.
[4] H. Nakaten, Chimia 15 (1961) 156–163; D. Patterson, Ber. Bunsenges. Phys. Chem. 71 (1967) 270–276.

[5] EP-PS 1236 (Bayer) 1977; EP-PS 3656 (ICI) 1978.
[6] EP-PS 10219 (Hoechst AG) 1978.
[7] DE-OS 2635536 (Ciba-Geigy) 1975; DE-OS 2635778 (Ciba-Geigy) 1975.
[8] EP-PS 58296 (Hoechst AG) 1981.

2.3   Monoazo Yellow and Orange Pigments

The synthetic route to monoazo yellow pigments involves the coupling of a diazo-tized substituted aniline with a coupling component containing an active methylene moiety in a linear structure.

Yellow monoazo pigments were discovered by Meister Lucius & Brüning in Germany (today: Hoechst AG) in 1909 and entered the market in 1910 under the trade name of "Hansa Yellows".

Azo pigments were essentially unknown until around the turn of the century, when coupling reactions were first carried out with β-naphthol as a coupling com-ponent. However, since the range of available colors was somewhat restricted at the time, it was not possible to manufacture shades of yellow beyond that of Dini-troaniline Orange (2,4-dinitroaniline→β-naphthol). The discovery of acetoacetic

arylides (N-acetoacetanilide CH_3COCH_2CONH—⬡R_K) as coupling components has

therefore considerably broadened the scope of organic pigments in general [1]. The synthetic pathway to such compounds had originally been suggested by the obser-vation that 1,3-diketo compounds couple with diazonium salts to form yellow dyes, which was published as early as 1897. At present, acetoacetic arylides are not the only 1,3-diketo compounds to be used in azo pigment synthesis. Introducing the pyrazolone-(5)-ring skeleton provides a convenient method of synthesizing struc-tures in which the acetoacetyl aniline forms a heterocycle:

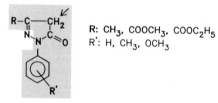

R: CH_3, $COOCH_3$, $COOC_2H_5$
R': H, CH_3, OCH_3

In 1884, H.J. Ziegler first used pyrazolones as coupling components. In an attempt to find a new dye by synthesizing a colored osazone from phenyl hydrazine-4-sulfonic acid and dioxo tartaric acid, he obtained yellow tartrazine by condensa-tion:

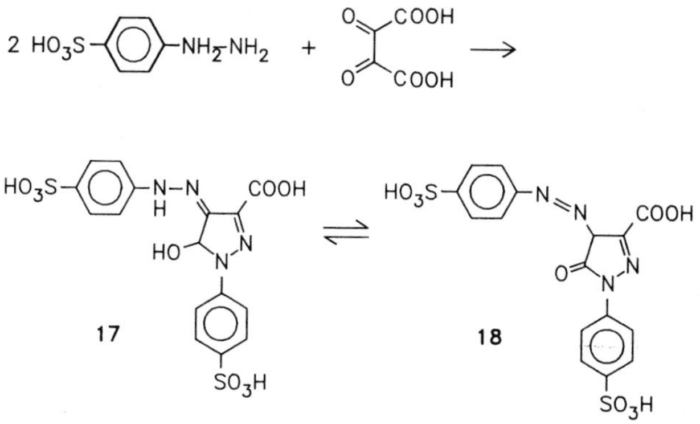

$$2 \text{ HO}_3\text{S} - \langle \bigcirc \rangle - \text{NH}_{\overline{2}}\text{NH}_2 \quad + \quad \text{(structure)} \quad \longrightarrow$$

17   ⇌   18

Ziegler not only confirmed the chemical structure of his product but also described the tautomeric structures of the hydrazone **(17)** and the azo form **(18)**. He then proceeded to prepare tartrazine by coupling diazotized p-sulfanilic acid with 1-sulfophenyl-3-carboxy pyrazolone-(5). The discovery of the pyrazolone skeleton as a coupling component, which improved the versatility of the method, led to the production of a large number of monoazo pigments and disazo pigments. However, the production of pyrazolone-based monoazo pigments (monoazo orange pigments) has declined largely in favor of the corresponding disazo pigments.

The list ends here; other heterocycles have failed to gain significance as coupling components for diazonium salts. The only exception is 2,4-hydroxyquinoline, which is used as a starting material for a copper complex pigment (Sec. 2.10.1.1).

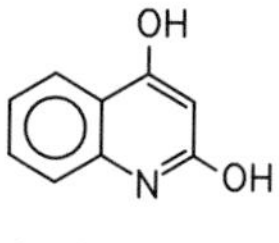

2,4-Dihydroxyquinoline

2.3.1 Chemistry, Manufacture

2.3.1.1 Non-laked Monoazo Yellow and Orange Pigments

Most non-laked monoazo yellow and orange pigments are derivatives of the following structure:

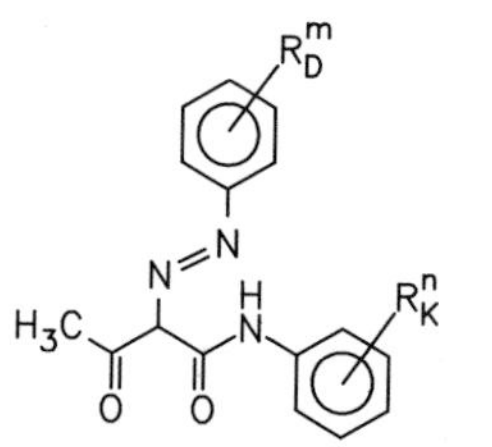

R_D: substituents of the diazo component
R_K: substituents of the coupling component
m, n: between 1 and 3

R_D and R_K represent substituents such as CH_3, OCH_3, OC_2H_5, Cl, Br, NO_2, CF_3; in important azo pigments, R_D and R_K are identical. In some cases, the coupling component may also be a pyrazolone derivative:

R: CH_3, $COOCH_3$, $COOC_2H_5$
R': H, CH_3

The customary method of preparation involves diazotizing a substituted aniline with an aqueous sodium nitrite solution at low temperature (0 to 5 °C). Coupling with acetoacetic arylide is carried out in a weakly acidic solution (pH 4 to pH 5). The resulting pigment suspension is heated shortly to 70 to 80 °C and then filtered. Ionic (salt) impurities are removed by washing the presscake with water, after which the product is dried at 60 to 80 °C. Controlling the particle size is essential. Powders with fine particle sizes are produced by adding appropriate agents, such as dispersing or emulsifying agents, to the reaction mixture before or during coupling. Coarse particle sizes (monoazo yellow or orange pigments), on the other hand, are attained by thermal aftertreatment. Heating either the crude pigment suspension or the isolated and washed pigment presscake to a temperature above 80 °C, possibly under pressure, affords the desired particle size.

The diazo component is typically a substituted aromatic amine which carries either an NO_2 group or, less frequently, OCH_3, Cl, or CH_3 in the ortho position relative to the amino function. A number of recent studies propose methods to improve the solvent and migration resistance of monoazo yellow and orange pigments, which is normally somewhat poor. Introducing carbonamide or sulfonamide groups into the diazo compound and/or the coupling compound provides a convenient method of modifying the chemical structure. The results were as follows:

R: CH_3, C_2H_5, C_6H_5

$-NHCOR'$ R': CH_3, C_6H_5

Only a few of these species, however, have gained commercial significance.

A more traditional approach consists of introducing sulfonic acid substituents into yellow monoazo compounds. These can simply be precipitated, i.e., converted to lakes by salt formation, especially with calcium, and thus afford the more solvent and migration resistant monoazo yellow pigment lakes.

The chemical structure of non-laked monoazo yellow pigments has recently been reexamined with respect to tautomerism. It seems that the common habit of referring to non-laked monoazo yellow pigments in the 2-oxoazo form probably fails to reflect the true structure of the pigment crystals [2]. Three dimensional single crystal X-ray diffraction analysis indicates that monoazo yellow pigments prefer the 2-oxohydrazone form. This is demonstrated by Pigment Yellow 6, 11670:

Traditional assumption:

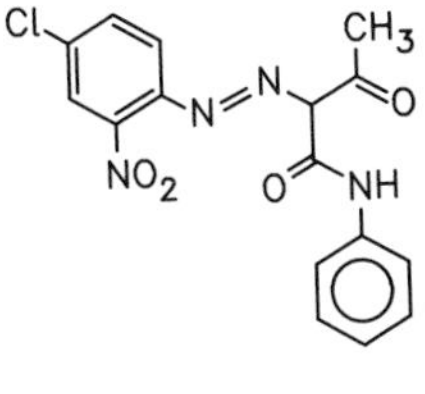

Actual structure:

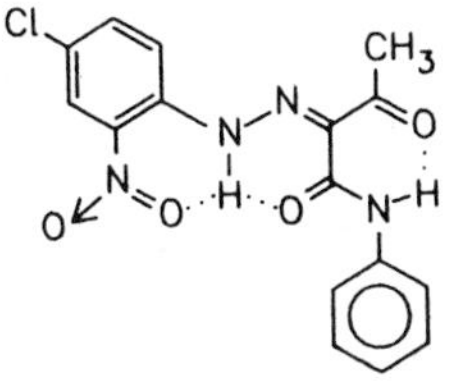

A maximum of possible intramolecular hydrogen bonds renders the entire molecule, with the exception of the aniline ring, almost completely planar. The hydrogen atom of the hydrazone group bridges with two oxygen atoms simultaneously, leading to a "forked" hydrogen brigdge. Among all possible conformations, this is the energetically most favorable one.

A more recent publication [3] indicates that although a number of other monoazo yellow pigments also prefer the 2-oxohydrazone form, perfect planarity is not always ensured. Compound **19**, for instance, with the structure (R=H)

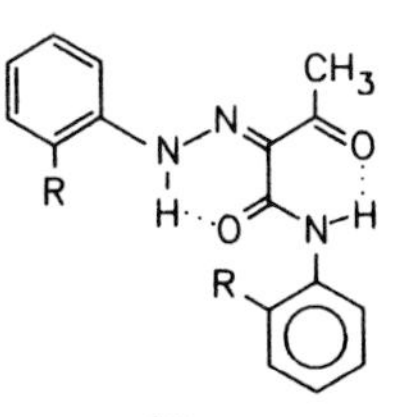

19

is completely planar; introducing a substituent R=OCH$_3$, however, increases the angle between the two terminal phenyl groups by more than 37°.

2.3.1.2   Monoazo Yellow Pigment Lakes

Monoazo yellow pigments lakes are basically synthesized by introducing acidic groups into the diazo or coupling component and by precipitating the product as an insoluble salt. In practice, however, only those species which carry a sulfonic acid function on the diazo component are commercially important. The list includes derivatives of the following structure:

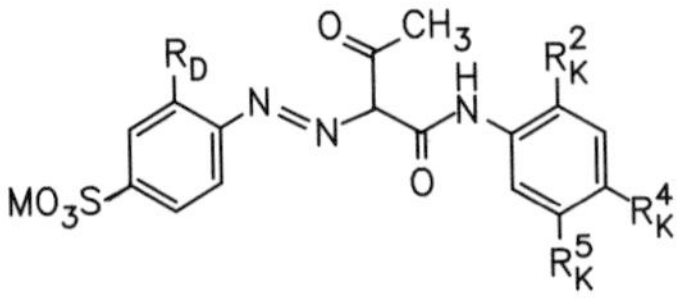

R_D: NO$_2$, Cl, CH$_3$
R_K^2, R_K^4, R_K^5: H, CH$_3$, Cl, OCH$_3$
M: Na, $\frac{Ca}{2}$, $\frac{Ba}{2}$

A number of commercially available yellow monoazo pigment lakes are based on a pyrazolone sulfonic acid derivative as a coupling component. An example is the aluminum tartrazine lake, listed in the Colour Index as Pigment Yellow 100, 19140:1.

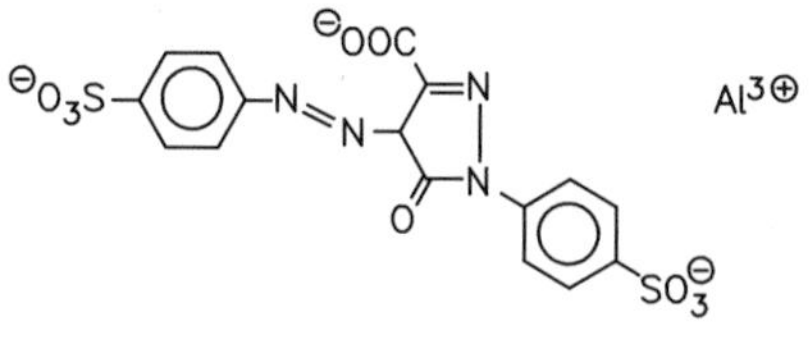

Recent patents [4,5] also describe precipitated yellow pyrazolone monoazo pigment lakes based on the general structure

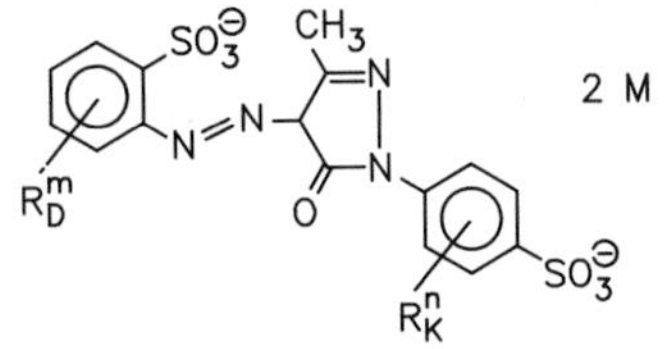

M: Na$^{\oplus}$, K$^{\oplus}$, $\frac{Ca^{\oplus\oplus}}{2}$

2.3.2   Properties

2.3.2.1   Non-laked Monoazo Yellow and Monoazo Orange Pigments

Monoazo yellow and orange pigments provide a range of colors from intensely greenish to very reddish yellow or yellowish orange shades. Monoazo pigments based on pyrazolone as the coupling component are reddish yellow; commercially, however, they have largely been displaced by other products. Most of the industrially significant representatives of this class of pigments exhibit a tinctorial strength which is only about half of that of the diarylide yellow pigments covering the same range of colors (Sec. 2.4.1.2); there is only one monoazo yellow pigment which is strong enough to compete with comparable diarylide yellow pigments.

Monoazo yellow pigments exhibit good lightfastness and durability in full shades and close to deep shades. In order to optimize the pigment performance for a certain target application, physical parameters, such as particle size or crystallinity, may be adjusted accordingly. It should be mentioned, however, that such techniques do not open up new areas of use.

The fact that monoazo yellow and orange pigments are not completely insoluble in organic solvents frequently affects their performance in certain vehicles, resulting in poor migration resistance, i.e., a trend towards bleeding and blooming, as well as a tendency to recrystallize. This obviously restricts the use of such pigments in a variety of areas and in a number of application materials.

Although it is sometimes possible to considerably improve the fastness properties of a monoazo yellow pigment by introducing carbonamide groups or sulfonamide functions into the molecule, the applicability of a pigment is not likely to extend basically beyond its original scope.

Although monoazo yellow pigments are low-cost materials, this advantage is compromised by their limited performance in application, which is a function of their chemical constitution.

2.3.2.2   Monoazo Yellow Pigment Lakes

Introducing acidic substituents into the basic structure of a typical monoazo yellow pigment makes it possible to convert the material into a lake by salt formation. This improves the application properties of a pigment compared to its non-laked counterpart. Such pigments exhibit particularly good migration resistance and heat stability, making them useful colorants for plastics.

2.3.3 Application

The largest fraction of monoazo yellow pigments today are used throughout the air drying paints and emulsion paints industry. Recommended types with large pigment particles show only a slight tendency to recrystallize in most application media used by the paint industry; this slight recrystallization is generally tolerated. Monoazo yellow pigments, especially Pigment Yellow 1 and Pigment Yellow 3, are used routinely in air drying paints for their excellent lightfastness and durability in full shades. However, the application performance of these types deteriorates upon white reduction with TiO_2. The lightfastness of a particular type of P.Y.1, for instance, which in full shades equals step 7–8 on the Blue Scale, decreases to step 5–6 upon 1:5 reduction with TiO_2 and will be reduced to as little as 4–5 at 1:60 TiO_2. Excellent hiding power and clean, brilliant full and medium shades make some of these pigments suitable to replace Chrome Yellow in chromium free pigmentation. Good rheology makes it possible to prepare comparatively highly pigmented formulations without affecting the flow behavior of the paint. Although this class of organic pigments does not typically lack opacity, a high monoazo yellow pigment concentration will enhance the hiding power of the material in question even more.

Monoazo yellow and orange pigments are easily dispersed in most media. Aided by a dissolver, a number of types can even be worked into long-chain alkyd resin systems.

Poor solvent resistance and a tendency to bloom considerably restrict the use of monoazo yellow pigments. As the fastness requirements for pigments in industrial paints have become increasingly stringent, monoazo yellow pigments are only used in special cases. The ultimate decision is made on the basis of the concentration limits beyond which blooming occurs in a given system (Sec. 1.6.3.1). Such coatings are stable to blooming, but they still bleed.

As mentioned above, regular monoazo yellow pigments may be substituted with sulfonamide groups in order to improve their fastness to organic solvents. Products such as Pigment Yellow 97 thus become eligible for application in stoving enamels. These products do not bloom under the typical processing conditions and may even resist bleeding if they are cured at moderate temperature.

Strong migration tendencies practically exclude monoazo yellow and orange pigments from being used for the mass coloration of plastics: they bleed and bloom considerably in most polymer systems. Again, Pigment Yellow 97 is an exception which can be used under certain conditions in PVC pastes. A few selected types are to a limited extent applied in urea–formaldehyde resins, also known as thermosetting plastics.

The printing inks field is the primary user of certain monoazo yellow pigments, especially in areas for which the routinely used diarylide yellow pigments are not sufficiently lightfast; such as for posters and packaging and partially for wallpaper. In the field of printing inks today they are primarily used for the packaging of printing inks.

However, problems may arise through the solvents which are commonly used to formulate specialty packaging gravure inks (Sec. 1.8.1.2). Depending on the processing conditions and especially on the method of dispersion, pigments may recrystallize in these media. Tinctorially strong, transparent types are particularly sensitive to recrystallization, which not only reduces the tinctorial strength but also opacifies the ink. Using inks based on water or water/alcohol will circumvent the problem. Their low thermostability and migration fastness prevent these pigments from being useful for metal decoprinting.

In the printing inks field, monoazo yellow and orange pigments rank next to the much stronger and solvent resistant diarylide yellow pigments. Monoazo yellow and orange pigments, however, reign supreme where lightfastness is a concern. This is also true for other media.

The paint, printing ink, and plastics industries are not the only users; monoazo yellow and orange pigments have gained recognition in a host of applications. A number of types—sometimes sold as pigment preparations—are used in the office article sector, where they lend color to felt-tip pen inks, drawing inks, colored pencils, wax crayons, watercolors, etc. They are also used as colorants for wood stains, plywood, veneered wood, shoe polish, floor polish, fertilizers, match heads, and in the cosmetics industry, where they are used, for instance, to color soaps. These pigments also lend themselves to textile printing and for paper spread-coating and mass coloration of paper.

2.3.4   Commercially Available Monoazo Yellow and Orange Pigments

General

The oldest monoazo yellow pigment is Pigment Yellow 1, which continues to be produced and shipped on a large scale. Its discovery and introduction in 1910 was followed by the development of a variety of related pigments of this type, not all of which have turned into a commercial success. A few, such as pyrazolone-based Pigment Yellow 10, are still produced by some regional companies and are targeted for smaller niche applications.

Tables 11 and 12 list the currently available non-laked monoazo yellow and orange pigments. Most of them carry a nitro substituent in the diazonium component, usually in ortho position relative to the azo bridge. The more migration-resistant Pigment Yellow 97 lacks a nitro group in its molecule. Five monoazo yellow pigments are based on 2-nitro-4-chloroaniline as a diazo component.

Acetoacetic-o-anisidide, a frequently used coupling component for monoazo yellow pigments, is generally considered one of the most important intermediates for the production of monoazo and disazo yellow pigments worldwide.

The currently available laked monoazo yellow pigments are based on 2-nitroaniline-4-sulfonic acid as a diazo component and are mostly sold as calcium lakes (Table 13).

Tab. 11: Non-laked monoazo yellow and orange pigments.

C.I. Name	C.I. Constitution No.	R_D^2	R_D^4	R_D^5	R_K^2	R_K^4	R_K^5	Shade
P.Y.1	11680	NO_2	CH_3	H	H	H	H	yellow
P.Y.2	11730	NO_2	Cl	H	CH_3	CH_3	H	reddish yellow
P.Y.3	11710	NO_2	Cl	H	Cl	H	H	very greenish yellow
P.Y.5	11660	NO_2	H	H	H	H	H	very greenish yellow
P.Y.6	11670	NO_2	Cl	H	H	H	H	yellow
P.Y.49	11765	CH_3	Cl	H	OCH_3	Cl	OCH_3	greenish yellow
P.Y.65	11740	NO_2	OCH_3	H	OCH_3	H	H	reddish yellow
P.Y.73	11738	NO_2	Cl	H	OCH_3	H	H	yellow
P.Y.74	11741	OCH_3	NO_2	H	OCH_3	H	H	greenish yellow
P.Y.75	11770	NO_2	Cl	H	H	OC_2H_5	H	reddish yellow
P.Y.97	11767	OCH_3	SO_2NH-C_6H_5	OCH_3	OCH_3	Cl	OCH_3	yellow
P.Y.98	11727	NO_2	Cl	H	CH_3	Cl	H	greenish yellow
P.Y.111	11745	OCH_3	NO_2	H	OCH_3	H	Cl	greenish yellow
P.Y.116	11790	Cl	$CONH_2$	H	H	$NHCOCH_3$	H	yellow [6]
P.Y.130	–	–	–	–	–	–	–	yellow
P.O.1	11725	NO_2	OCH_3	H	CH_3	H	H	very reddish yellow

Tab. 12: Non-laked monoazo yellow and orange pigments.
Diazo or coupling components which either deviate from the general structure shown in
Table 11 or which are unknown.

DC = diazo component, CC = coupling component, PMP :

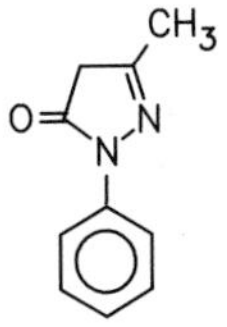

C.I. Name	C.I. Constitution No.	DC	CC	Shade
P.Y.10	12710	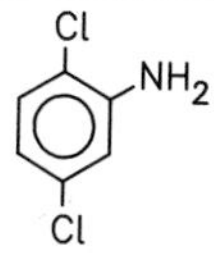	PMP	reddish yellow
P.Y.60	12705	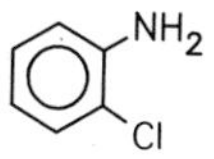	PMP	reddish yellow
P.Y.165	–	–	–	reddish yellow
P.Y.167	11737	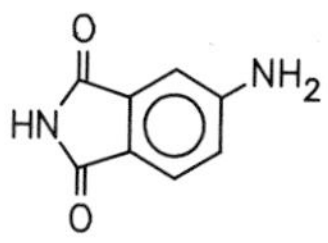	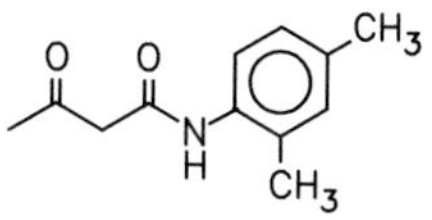	greenish yellow
P.O.6	12730	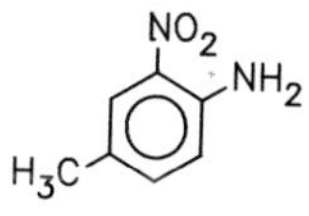	PMP	orange

Tab. 13: Monoazo yellow pigment lakes.

C.I. Name	C.I. Constitution No.	R_K^2	R_K^4	M	Shade
P.Y.61	13880	H	H	Ca	greenish yellow
P.Y.62:1	13940:1	CH$_3$	H	Ca	yellow
P.Y.133	–	–	–	Sr	yellow
P.Y.168	13960	Cl	H	Ca	greenish yellow
P.Y.169	13955	H	OCH$_3$	Ca	reddish yellow

Structure either deviating from above-mentioned structure or unknown:

P.Y.100	19140:1		greenish yellow
P.Y.183	18792		reddish yellow
P.Y.190	–	–	yellow
P.Y.191	18795		reddish yellow

Individual Pigments

Non-laked Pigments

Pigment Yellow 1

This compound has entered pigment history as "Hansa Yellow G". The commercial varieties are products with comparatively coarse particle sizes and specific surface areas of approximately 8 to 30 m^2/g. P.Y.1 provides good hiding power, it is used primarily in air drying paints as well as in packaging and textile printing.

Decreasing the particle size produces more transparent, more greenish, and cleaner pigment types; the tinctorial strength frequently improves. This advantageous trend is compromised by a considerable decrease in lightfastness and resistance to organic solvents. In a number of application media, the tendency to recrystallize is also enhanced (Sec. 1.7.7). Pigment Yellow 1 types with very fine particle sizes lose the advantage of superior lightfastness compared to diarylide yellow pigments with comparable shades, whose solvent resistance is much better and whose tinctorial strength is almost twice as high as that of P.Y.1.

Although inferior fastness to organic solvents, involving unsatisfactory migration properties, excludes P.Y.1 from important areas of application, such as baking enamels, there are certain conditions under which it may be applied in such media. The user must in this case observe a certain concentration limit beyond which the pigment may bloom. The pigment is stable up to 140°C.

For decades, P.Y.1 represented the standard source of yellow for all printing purposes. Today, however, P.Y.1 has been replaced in this field by the stronger diarylide yellow pigments.

The tinctorial strength of a printing ink is measured in terms of the pigment concentration in a standardized ink film that will produce a certain depth of shade in letterpress proofprints. Depending on the type, inks made from P.Y.1 must contain about 8 to 11 wt% pigment in order to produce 1/3 SD films; while inks made from the similarly shaded diarylide yellow pigment P.Y.14 require only between 4 and 6% pigment. The lightfastness of such a printed P.Y.1 sample equals approximately step 5 on the Blue Scale, while the corresponding P.Y.14 specimen only reaches step 3. The corresponding values for 1/1 SD prints are 6–7 and 3 on the Blue Scale, respectively.

P.Y.1 prints are comparatively sensitive to most common organic solvents, such as esters, ketones, and aromatic hydrocarbons, but completely fast to alcohols and aliphatic hydrocarbons. The prints are also resistant to soap, alkali, and acid.

In the printing ink industry, P.Y.1 is used especially to replace diarylide yellow pigments in areas where the latter are not sufficiently lightfast to meet the technical standards. This is especially true for the packaging printing industry, which particularly appreciates the good hiding power of most types of P.Y.1. Since they are sensitive to organic solvents, such grades are mainly used in alcohol-based nitrocellulose (NC) printing inks and in waterborne inks. Poor migration resistance and heat stability largely precludes the use of P.Y.1 in plastics.

In general, P.Y.1 is used extensively in a variety of applications, including those generally mentioned in connection with monoazo yellow pigments. As a result of its good lightfastness, this pigment is particularly interesting for textile printing; however, its fastness to dry-cleaning solvents and to dry heat setting (fixation) is poor. P.Y.1 is easily dispersed in most media, irrespective of the dispersion equipment.

Pigment Yellow 2

This pigment has little impact on the market today and is only occasionally found in printing inks or in office articles, for instance in colored pencils. Its shade is more reddish and its tinctorial strength superior to that of P.Y.1. The low specific surface area of the types which are still commercially available makes for good hiding in print. Solvent resistance and other fastness properties equal those of P.Y.1; P.Y.2 is only slightly less lightfast than P.Y.1.

Pigment Yellow 3

P.Y.3 grades produce a pure, much greener yellow than the Pigment Yellow 1 types. Pigment Yellow 3 can easily be combined with blue pigments to produce shades of green; it may also be used to modify the hue of a green pigment, such as Copper Phthalocyanine Green. The commercial types exhibit low specific surface areas and confer good hiding power on paints, coatings, and prints.

P.Y.3 is even less stable to most organic solvents than P.Y.1; it migrates in baking enamels and is applied in this context only rarely and at a pigment concentration below the concentration limit above which blooming occurs (Sec. 1.6.3.1). As a comparison, the concentration limit beyond which P.Y.3 blooms in a certain urea/alkyd resin paint at a temperature of 120 °C (30 min) is 1%; at 140 °C, the limit increases to 2.5%. Higher temperatures always cause blooming, irrespective of the concentration. Any pigment that is to be applied in a baking enamel should be submitted to a test run in the specific material. There are commercially available grades which recrystallize considerably less in alkyd resin paints.

P.Y.3 is completely water, acid, and alkali resistant. Alkali fastness is a major prerequisite for a number of applications, especially in aqueous media.

The pigment exhibits excellent lightfastness and durability; its performance in this respect is superior to that of P.Y.1, especially if it is reduced with titanium dioxide. Compared to P.Y.98, which exhibits a similar hue but is much stronger, P.Y.3 grades have the advantage of being more reasonable in terms of cost but less fast to solvents and migration. In air drying paints, for instance, medium white reductions of P.Y.3 are somewhat less lightfast but more durable than comparable P.Y.98 grades.

P.Y.3 and P.Y.1 are used in similar applications. The paint industry employs these pigments in air drying, cellulose nitrate, polyurethane, acid hardening, and similar paints. P.Y.3 may lend color to emulsion paints, and its full and medium shades even comply with the industrial standards for exterior house paints. In the printing inks field, the pigment is primarily used in a variety of packaging inks. P.Y.3 also plays a role in textile printing, office articles, artists' paints, and a number of special purposes. It is, however, not used in plastics; a certain exception is the coloration of unsaturated polyester resins, whose hardening process by peroxide is not influenced by P.Y.3 and in which the pigment exhibits good lightfastness.

Pigment Yellow 5

Over the last few decades, commercial recognition of Pigment Yellow 5 has decreased considerably. Two crystal modifications of the pigment are available [7]; both show practically identical shades. P.Y.5 is a greenish yellow pigment, applied mainly in printing inks and occasionally also in air drying systems. Although its hue is considerably redder than that of P.Y.3, its application properties parallel that of P.Y.3. It is slightly less lightfast and in this respect similar to the more reddish P.Y.1.

Pigment Yellow 6

This representative is only of limited regional importance. Its shade is a medium yellow, which places it in a competitive position with a number of similar colorants of the same class, which it also resembles in its fastness properties.

Pigment Yellow 10

P.Y.10 provides clean, reddish yellow hues. Its very good lightfastness makes it almost as fast as P.Y.97. However, its solvent fastness is poor and in this respect inferior to most other monoazo yellow pigments; it shares this deficiency with P.O.1. Its migration resistance is also poor and considerably less satisfactory than that of the slightly greener P.Y.65. In the face of increasingly stringent application requirements, P.Y.10 has in recent years almost entirely vanished from the European market. Originally, it was used in special printing inks and in air drying paints. In the USA, it is applied primarily in traffic marking paints.

Pigment Yellow 49

This very lightfast pigment has a clean greenish yellow shade and is used for viscose spin dyeing and for the mass coloration of viscose foils, sponges, etc. It is commercially available in the form of aqueous pigment preparations, in which the pigment is predispersed. In 1/1 to 1/3 SD, the lightfastness equals step 8, in 1/12 SD it reaches a value of 7–8 on the Blue Scale. The pigment performs excellently on textiles, but its fastness to wet and dry crocking is not always satisfactory.

Pigment Yellow 60

This colorant is produced in the USA, where it also plays a minor role on the market. Its shade is a reddish yellow. P.Y.60 is used in trade sales paints and in emulsion paints. Poor resistance to organic solvents, lack of overcoating fastnesss, and a lightfastness that deteriorates rapidly with the degree of reduction with TiO_2 are reasons for the limited commercial interest in this pigment. Crystal data are published by A. Whitaker [8].

Pigment Yellow 65

Pigment Yellow 65 provides reddish yellow shades. Although its importance as a colorant on the European and Asian market has diminished over the last years, there is still some demand for it in the USA. The types which are still commercially available consist of pigments with comparatively coarse particle sizes and with specific surface areas of approximately 6 to 20 m^2/g. The pigment thus provides excellent hiding power. There is some resemblance between Pigment Yellow 65 and P.Y.74 types with high hiding power in that they perform equally well in terms of solvent resistance, recrystallization fastness in solvent containing media, lightfastness, and durability. Tinctorially, however, P.Y.65 is considerably inferior, which precludes its application in printing inks. Opaque varieties of the slightly more greenish diarylide yellow pigment P.Y.83, for instance, are more than three times as strong as P.Y.65, yet provide similar lightfastness. Media in which P.Y.65 grades are applied include air drying and emulsion paints. Combinations of the positional isomers P.Y.65 and 74 are also commercially available. Crystal data were reported by A. Whitaker [9].

Pigment Yellow 73

Depending on particle size distribution and the application medium, this pigment more or less bears some resemblance to P.Y.1 as far as shade and tinctorial strength are concerned. At equal tinctorial strength, however, P.Y.73 is more lightfast and durable. Compared to P.Y.1, it is also considerably more insoluble in various organic solvents, such as aliphatic and aromatic hydrocarbons, and thus also less prone to recrystallize. This facilitates its incorporation into solvent containing media, such as air drying systems and emulsion paints, as well as in nitrocellulose systems etc. throughout the printing inks field. Compared to P.Y.1, Pigment Yellow 73 grades generally afford more transparent prints and are also somewhat redder than the P.Y.74 types with fine grains, which provide a tinctorial strength which is about twice as high. P.Y.73 representatives are more lightfast and durable than the tinctorially similar disazo yellow pigments, which are also very stable to solvents.

Pigment Yellow 74

P.Y.74, a commercial pigment of considerable significance, is used primarily in the printing ink and paint industries. Its greenish yellow shades are somewhere between those of P.Y.3 and P.Y.1. The pigment is considerably stronger than and superior to all comparable monoazo yellow pigments. P.Y.74 grades with very different particle sizes are commercially available. Varieties with fine particle sizes, i.e., pigments with specific surface areas between approximately 30 and 70 m^2/g, are used primarily in the printing ink industry. They provide brilliant prints, which are brighter and more transparent than those of other monoazo yellow pigments. In terms of tinctorial strength, these P.Y.74 varieties reach the level of the similarly colored disazo yellow pigments, such as the slightly redder P.Y.12. The pigment concentration that is necessary to produce a 1/3 SD P.Y.12 proofprint is 4.5%, compared to 4.2% in the case of P.Y.74. In letterpress application, P.Y.74 has a yellow hue, which corresponds to the DIN 16 508 standard for letterpress; in offset printing, it matches not only the DIN 16 509 standard in the yellow section of the DIN Scale for process color printing (Sec. 1.8.1.1), but also the standard yellow on the Kodak scale. On the European Scale (Sec. 1.8.1.1), P.Y.74 is somewhat more greenish than the standard yellow. Its shade can be adjusted with suitable reddish yellow pigments.

The lightfastness of types with fine particle sizes exceeds that of coloristically similar disazo yellow pigments (P.Y.12) by 2 to 3 steps on the Blue Scale. This explains why types with fine particle sizes are preferred where excellent lightfastness is required, such as in packaging printing. P.Y.74 is not different from other yellow pigments of this class in that it performs poorly in solvents. The prints are not very suitable for calandering and are not colorfast to sterilization. As typical monoazo yellow pigments, P.Y.74 specimens are fast to alkali, acid, and soap. Other aspects of their fastness in print, such as butter resistance and fastness to the solvent mixture described in DIN 16 524 (Sec. 1.6.2.1), are not convincing. However, like other pigments of their class, P.Y.74 varieties with comparatively fine particle sizes are also used in the paint industry, where they lend color to air drying and emulsion paints. Excellent tinctorial strength, especially in pastel shades, i.e., in white reductions, make P.Y.74 more advantageous on a cost/performance basis than the somewhat redder P.Y.1. A strong tendency to bloom, however, precludes the use of P.Y.74 in baking enamels.

For some years, the paint industry in particular has focussed increasingly on P.Y.74 types with very low specific surface areas between about 12 and 20 m^2/g and correspondingly very large particle sizes. Such materials exhibit good hiding power and are particularly interesting as replacements for Chrome Yellows in air drying systems. As a result of favorable pigment rheology, P.Y.74 concentrations in paints can be increased to even improve the hiding power without affecting the rheology, i.e., the flow properties, the ease of processing, or the brilliance of the product (Sec. 1.7.8). Compared to transparent types with fine particle sizes, the more opaque varieties are less brilliant, more reddish, and more lightfast and durable. They are also faster to various organic solvents; but their tendency to migrate is similar.

Pigment Yellow 75

Although it is sold in the USA, this pigment has failed to gain commercial importance. Its hue is a reddish yellow, much redder than that of P.Y.74 but greener than that of P.Y.65. In full shade and in white reductions, it resembles the shade of the pigment which is produced by mixed coupling between the two, or plain mixtures of the positional isomers, which are also marketed in the USA. P.Y.75 is somewhat more lightfast than the above-mentioned mixture. Its lightfastness in full shade (5%) corresponds to step 7–8 in an air-dried alkyd resin paint, in reductions (1:5 TiO_2), it matches step 7 on the Blue Scale. Its tinctorial strength is poor, and it is used mainly in traffic marking paints.

Pigment Yellow 97

Pigment Yellow 97 provides a medium yellow shade with properties that are somewhat of an exception compared to other monoazo yellow pigments. Its hue is similar to that of P.Y.1. P.Y.97 is considerably more resistant to most organic solvents than the other representatives of this group, its migration fastness properties are also superior. In baking enamels, for instance, the pigment resists blooming up to about 180°C. However, its fastness to overpainting in such coatings is not always sufficient. In a white baking enamel on alkyd-melamine resin basis, P.Y.97 does not bleed after 30 minutes at 120°C, but very slight bleeding is observed at 160°C. Its high quality in terms of fastness to clear lacquer overcoatings and sterilization (Sec. 1.6.2.3), which are important aspects in the printing inks field, is very unusual for a monoazo yellow pigment.

In contrast to most other representatives of its class, P.Y.97 is also much more heat resistant. In a performance test, in which prints formulated at 5% pigment concentration are applied to sheet metal covered with white paint, most monoazo yellow pigments clearly change color after 30 minutes exposure to a temperature up to 140°C (Sec. 1.6.7). P.Y.97, however, being very heat stable, retains its color for 30 minutes at 180°C or for 10 minutes at 200°C.

P.Y.97 is very lightfast and durable; in this respect it excels over P.Y.1, especially in white reductions. In a variety of systems (air drying and baking paints), P.Y.97 exhibits a lightfastness wich corresponds to step 6–7 on the Blue Scale, even if it is reduced by as much as 1:140 TiO_2.

Its tinctorial strength is average and comparable to that of other monoazo yellow pigments, except P.Y.74. P.Y.97 disperses easily in all application media.

P.Y.97 is used in a variety of fields. Even in pastel shades, it is used in industrial finishes; while its full shades lend color to automobile refinishes. In emulsion paints, both its medium and full shades are suited to exterior application. The printing ink industry uses P.Y.97 in high grade printing products, especially where excellent fastness is required, such as in stable posters, etc. It lends itself without difficulty to all printing techniques. However, lack of fastness to monostyrene and acetone and therefore a certain tendency to bleed in these media precludes its application in deco printing inks, i.e., for decorative laminates.

P.Y.97 can also be applied in plastics. In rigid PVC, for instance, its lightfastness is excellent, both in transparent and in opaque shades with TiO_2. Depending on its concentration, the lightfastness reaches values between steps 6–7 and 8 on the Blue Scale. In tests designed to determine the color fastness in rigid PVC films containing 0.5% pigment, coconut fat, for instance, is not stained. In plasticized PVC, however, the pigment tends to migrate at low concentrations. P.Y.97 exhibits good thermal stability in polyolefins; in 1/3 and 1/25 SD HDPE foils, for instance, it is stable up to 240 °C for 5 minutes. In HDPE extrusion the pigment does not affect shrinkage. P.Y.97 performs similarly in polystyrene and PE as far as thermal stability is concerned. At common coloration concentrations, the pigment dissolves almost completely at temperatures above 200 °C. The lightfastness of transparent polystyrene colorations is excellent (step 7–8 on the Blue Scale); its opaque shades show good lightfastness (0.1% pigment/0.5% TiO_2) (step 5). In ABS, the lightfastness is superior even in reduction with TiO_2 (step 7 on the Blue Scale). If polymethacrylate is used as a medium, the pigment changes color only after 5 minutes at 280 °C; in transparent and opaque colorations; the color change is 1 CIELAB unit in the system of color difference. The lightfastness of transparent colorations in such media corresponds to step 8 on the Blue Scale.

P.Y.97 is also used to lend color to cast epoxy resins and to unsaturated polyester resins; it considerably accelerates the hardening process in the latter (Sec. 1.8.3.7).

Basically, P.Y.97 can also be applied in other areas, such as in pigmented water-based paints. However, a variety of more economical organic pigments covering the same range of hues, such as P.Y.1, frequently compete with P.Y.97 in this area.

Pigment Yellow 98

At present, this pigment is being removed from the market worldwide because one of the intermediates for its production is no longer available. P.Y.98 resembles P.Y.3 in shade, i.e., it provides clean greenish shades of yellow. P.Y.98, like P.Y.3, is a particularly suitable pigment for clean green mixtures. No perfect replacement for P.Y.98 is commercially available. A mixture of P.Y.3 and P.Y.111 has recently been introduced to the market which approaches the coloristic, application, and fastness properties of P.Y.98 in printing inks. However, this mixture is not quite as lightfast as P.Y.98. In the paint field, P.Y.3 can replace P.Y.98 in many areas of application. Compared to P.Y.98, however, P.Y.3 is much less resistant to organic solvents. The interval in which blooming occurs in baking enamels is consequently much wider for P.Y.3, which is why this pigment is unsuitable for such applications.

Pigment Yellow 111

P.Y.111 has been commercially available since the 1970s. It is much redder and somewhat duller than P.Y.98, but considerably greener than P.Y.74. In printing inks it is tinctorially stronger than the more greenish P.Y.98 and less lightfast than the latter; the difference between printing inks containing the same amount of P.Y.74 and 98, respectively, equals one step on the Blue Scale. Prints that exhibit the same depth of shade are even more different in their lightfastness. Packaging gravure and flexographic inks, some of which are based on alcohol-soluble nitrocellulose, are much less subject to differences in tinctorial strength than are letterpress or offset inks. The commercially available version of P.Y.111 is considerably more opaque than P.Y.98 and 74 types with fine particle sizes.

Pigment Yellow 116

Pigment Yellow 116 is a product of somewhat less commercial importance. Its shade is a medium to reddish yellow. The commercial types with their low specific surface area (about 15 to 18 m²/g) exhibit good hiding power.

Areas of application include coatings, printing inks, and plastics. Compared to other representatives of its class, P.Y.116 is exceptionally resistant to common organic solvents. Its fastness to overpainting in baking systems, however, is not satisfactory at baking temperatures which are typically around 150°C. The pigment is acid and alkali resistant and thermally stable up to 180°C. It is very lightfast; in a medium-oil alkyd resin paint, its full shades (5% pigment) are equal to step 7–8 on the Blue Scale, while in 1/3 SD formulations it equals step 7 on the Blue Scale. P.Y.116 is also very weatherfast. It is applied especially in combination with inorganic yellow pigments and also in baking enamels that are baked at high temperature, such as 200°C (30 minutes); and in automobile repair finishes. The pigment is also found in emulsion paints; exterior use is possible if the demands are not too high.

P.Y.116 generally exhibits good fastness properties in printing inks. In letterpress and offset application, for instance, it resists a number of organic solvents, such as the standard DIN 16 524/1 solvent mixture (Sec. 1.6.2.1), paraffins, butter, soap, alkali, and acid. Its thermostability is good up to 180°C. The pigment is also recommended for metal deco printing. The lightfastness of 1/1 to 1/3 SD letterpress proofprints is equal to step 6 on the Blue Scale. Its good solvent resistance qualifies P.Y.116 as a colorant for packaging gravure inks. The pigment can, for instance, be applied in mixed polymers based on vinyl chloride/vinyl acetate for PVC films.

P.Y.116 is also used in plastics. In plasticized PVC, it shows little tendency to bleed and is thermally stable up to 180°C. The lightfastness of transparent PVC colorations (0.1% pigment) equals step 7–8; in 1/3 SD (with 5% TiO₂), it corresponds to step 6 on the Blue Scale. Insufficient heat resistance limits the application of P.Y.116 in polyolefins, polystyrene, and other polymers which are processed at high temperature.

Pigment Yellow 130

This pigment is not marketed in Europe and enjoys only limited commercial importance in other parts of the world. As one of the less resistant representatives of its class, its lightfastness and resistance to organic solvents are particularly deficient.

Pigment Yellow 165

Its exact chemical constitution has not yet been published. The pigment enjoys limited commercial success, mainly in the Japanese market. Its hue is reddish yellow, its lightfastness is almost as good as the much greener and tinctorially stronger P.Y.97. P.Y.165 is recommended for use in paints.

Pigment Yellow 167

The constitution of this pigment differs from that of common monoazo yellow pigments, since it is synthesized from aminophthalic imide as a diazonium compound.

P.Y.167 enjoys only regional importance and is recommended for application in paints. The currently available type provides medium hiding power. In full shade and in similar deep shades, it shows a medium yellow; white reductions are greenish yellow and very clean. Its fastness properties correspond to those of other monoazo yellow pigments; P.Y.167 is not alkali resistant.

Pigment Orange 1

P.O.1 has lost most of its commercial significance. Its shade is a very reddish yellow; the Color Index classifies it as a yellowish orange. Resistance to organic solvents is very low. P.O.1 dissolves more easily in common organic solvents than do other monoazo orange pigments, and it is also less lightfast.

P.O.1 is not entirely acid resistant; acids change its color to a more reddish shade. The pigment has not been able to meet the increasingly stringent requirements of the printing ink and paint industry in recent years. It is now limited in scope to air drying paints and has largely been replaced by mixtures of P.O.5 or others with other monoazo yellow pigments.

Pigment Orange 6

P.O.6 has also suffered considerable loss of interest throughout the European market and is rarely sold today. It provides a very reddish yellow shade. Its lightfastness is very good and almost reaches the level of that of P.Y.1.

Monoazo Yellow Pigment Lakes

Pigment Yellow 61

This calcium pigment lake is of little commercial value. It is used to lend color to plastics and to spin dyeing of polypropylene. Its shade is a greenish yellow, and 1/3 SD formulations in polyolefin are thermally stable up to 250 °C. Its tinctorial strength, however, is not satisfactory. 1/3 SD HDPE colorations (1% TiO$_2$), for instance, require 0.7% pigment. The more costly but coloristically identical azo condensation pigment P.Y.94 requires only 0.44% pigment in order to afford the same depth of shade, and only 0.13% of the somewhat redder diarylide yellow pigment P.Y.17 suffice for this purpose. In partially crystalline polymers, P.Y.61 has a considerable influence on the shrinkage during polymer extrusion. The pigment provides good lightfastness; in HDPE, transparent and opaque 1/3 SD colorations are equal to step 5–6 on the Blue Scale. Its lightfastness is even better in rigid and in plasticized PVC, but the pigment also exhibits very poor tinctorial strength in these media. 1/3 SD colorations require a 3.6% pigment concentration. Its migration resistance in plasticized PVC is good.

P.Y.61 is also used for industrial and trade sales paints. Its full shade and close to deep shades, frequently in combination with inorganic yellow pigments (especially Nickel Titanium Yellow), exhibit good to excellent lightfastness and durability. However, the pigment is technically unsuitable for exterior application. Its hiding power and its fastness to overpainting are good. However, P.Y.61 is of little overall importance to the paint industry.

Pigment Yellow 62:1

There is equally little commercial interest in the calcium pigment lake P.Y.62:1. Its use is in the plastics industry, where it produces colors in the medium to reddish yellow range. Although its solvent resistance is only average, P.Y.62:1 shows better fastness to plasticizers, such as dioctyl phthalate and dioctyl adipate, and consequently exhibits good bleeding fastness in plasticized PVC. It is also thermally stable. Transparent colorations and white reductions up to 1/3 SD equal step 7 on the Blue Scale for lightfastness. At 1/25 SD, the lightfastness drops to step 5–6.

P.Y.62:1 exhibits comparably low tinctorial strength. A 0.5% pigment concentration in HDPE is necessary to produce 1/3 SD colorations (containing 1% TiO$_2$), compared to the 0.17% pigment concentration required to afford equal depth of shade with the somewhat more reddish diarylide yellow pigment P.Y.13.

P.Y.62:1 is thermally stable up to 250 °C. It has a considerable effect on the shrinkage of HDPE and other partially crystalline polymers. The pigment is an equally suitable colorant for polystyrene and polyurethane and lends color to polypropylene spin dyeing products with minimal application requirements.

Pigment Yellow 100

This pigment is of very little technical importance, except in the USA, where it is sometimes used to lend color to NC-based flexographic inks. P.Y.100 provides greenish yellow shades of little alkali and soap fastness. It also performs poorly in organic solvents and even water and exhibits little fastness to overcoating. Its tinctorial strength is inferior to that of other organic pigments, and it provides poor lightfastness.

As long as it complies with certain purity requirements, the pigment is approved as a colorant for food and cosmetics in the USA under FD&C Yellow 5 and in Europe under E102.

Pigment Yellow 133

This pigment is of regional commercial importance in East Asia. It is a strontium lake of a monoazo yellow pigment, whose exact chemical composition has not yet been published. P.Y.133 provides greenish to medium yellow shades and is used in plastics and for the spin dyeing of polyolefins. Its tinctorial strength is poor. 1/3 SD HDPE colorations (1% TiO_2), for instance, require 0.85% pigment. In this medium, the pigment is thermally stable up to 260 °C.

Pigment Yellow 168

P.Y.168 is a calcium pigment lake and chemically closely related to the P.Y.61 and 62:1 lakes. It has been commercially available for several years, but has as yet gained only regional importance. It provides a clean, somewhat greenish yellow shade somewhere between the shades of P.Y.1 and 3. P.Y.168 is used in paints and plastics.

The pigment exhibits good fastness to aliphatic and aromatic hydrocarbons; but it shows only limited resistance to alcohols, esters, and ketones. It is largely acid and alkali resistant. P.Y.168 is not fast to overpainting. It is used in inexpensive industrial finishes wherever the application requirements, especially as far as lightfastness and durability are concerned, are not high.

P.Y.168 shows good migration resistance in plasticized PVC, but its tinctorial strength is relatively poor. In transparent colorations, its lightfastness (0.1% pigment) is equal to step 6 on the Blue Scale; at 1/3 SD it corresponds to step 5. In HDPE, the pigment affects the shrinkage of extrusion products, which is typical of azo pigment lakes of this class. P.Y.168 is recommended especially for use in LDPE.

Pigment Yellow 169

P.Y.169 is another calcium monoazo yellow pigment lake, which has been available for several years. Its hue is a reddish yellow. Its properties and application are similar to those of the much greener P.Y.168.

Pigment Yellow 183

P.Y.183 entered the market a few years ago and is considered a specialty product for the coloration of plastics.

This monoazo yellow calcium pigment lake affords reddish, somewhat dull yellow shades with poor tinctorial strength. 1/3 SD HDPE colorations (1% TiO_2) require 0.32% pigment. In such media, the pigment is thermally stable up to 300 °C. Polymer shrinkage is practically unaffected. The pigment provides very good lightfastness; 1/3 SD colorations equal step 7–8 on the Blue Scale. P.Y.183 is also recommended for other high temperature processed plastics. In ABS, for example, it is thermally stable up to 300 °C.

P.Y.183 is bleed resistant in plasticized PVC, where its poor tinctorial strength is of some disadvantage. 1/3 SD samples (5% TiO_2) require 1.64% pigment. Such systems provide a lightfastness which is equal to step 6 on the Blue Scale; 1/3 SD transparent colorations correspond to step 6–7 on the Blue Scale.

Pigment Yellow 190

The monoazo calcium pigment lake P.Y.190, which was only recently introduced to the market, is considered a specialty product for the coloration of plastics.

The commercially available type is tinctorially very weak. 0.66% pigment are required to produce 1/3 SD HDPE colorations (1% TiO_2). P.Y.190 provides medium shades of yellow. In HDPE, the pigment resists heat up to 300 °C.

In polyamide, P.Y.190 is heat stable up to 270 °C. In this medium, the pigment has a somewhat redder shade and demonstrates noticeably higher tinctorial strength than in HDPE. 0.3% of the commercially available type are sufficient to produce 1/3 SD colorations.

This commercial grade is slightly soluble in water, which may cause problems during extrusion.

Pigment Yellow 191

The monoazo calcium pigment lake P.Y.191 was only recently introduced to the market. It produces reddish shades of yellow, covering the same range of shades as the diarylide yellow pigment P.Y.83. P.Y.191, however, exhibits distinctly less tinctorial strength than P.Y.83. 0.32% P.Y.191 are needed to produce 1/3 SD colorations (1% TiO_2) in HDPE. These colorations are heat stable up to 300 °C, while 1/3 SD colorations without TiO_2 resist heat up to 290 °C. The pigment does not affect the shrinkage of the polymer. P.Y.191 shows good lightfastness: 1/3 SD colorations containing 1% TiO_2 equal step 6–7 on the Blue Scale, while 1/3 SD colorations without TiO_2 equal step 8. In media such as polystyrene and ABS, P.Y.191 shows similar values in terms of tinctorial strength, heat stability, and lightfastness. In polycarbonate, 1/3 SD colorations (containing 1% TiO_2) are heat stable up to 330 °C; such colorations equal step 4–5 on the Blue Scale for lightfastness. P.Y.191

is migration resistant in plasticized PVC; up to a concentration limit of 0.005%, the pigment can be used at 180 °C and in rigid PVC up to a concentration limit of 0.005% at 200 °C.

P.Y.191 displays excellent solvent fastness in aliphatic and in aromatic hydrocarbons, as well as in the commonly used plasticizers. The pigment is almost completely fast to alcohols and esters but not to water, ketones and methylglycol.

References for Section 2.3

[1] DRP 257 488 (Meister Lucius & Brüning) 1909.
[2] K. Hunger, E.F. Paulus and D. Weber, Farbe + Lack 88 (1982) 453–58.
[3] E.F. Paulus, W. Rieper and D. Wagner, Z. Kristallogr. 165 (1983) 137–149.
[4] DE-OS 2 616 981 (BASF) 1976.
[5] EP 37972 (BASF) 1981.
[6] NPIRI Raw Materials Data Handbook, Vol. 4, Pigments, 1983, 4–234, B.L. Kaul, JOC-CA 1987, 349–354.
[7] A. Whitaker, J. Soc. Dyers Col. 10 (1985) 21–24.
[8] A. Whitaker, J. Soc. Dyers Col. 104 (1988) 225–226.
[9] A. Whitaker, J. Soc. Dyers Col. 102 (1988) 136–137.

2.4 Disazo Yellow Pigments

In the industrial context, the term disazo yellow pigments is used exclusively to refer to pigments which possess two azo groups and contain the diaminodiphenyl skeleton according to the general formula:

in which X = Cl, OCH$_3$, or CH$_3$, and Y = H, Cl, or the diaminophenylene moiety

with U, V = H, CH$_3$, OCH$_3$, or Cl, although the latter is less common. The term is somewhat misleading, because it does not relate to all compounds which possess two azo groups; azo condensation products, for instance, are not included (Sec. 2.9). This book follows the generally accepted definition.

Structurally, there are two basic types of disazo pigments, depending on whether the bifunctional element is introduced through the diazonium compound or through the coupling component. A bifunctional diazo component results in products of the type

CC : coupling component group

Acetoacetarylides and 1-aryl-pyrazolone-5 are the two most common coupling components. The former affords diarylide yellow pigments (Sec. 2.4.1), while the latter produces disazopyrazolone pigments (Sec. 2.4.3).

The general structure of a typical disazo pigment based on a bifunctional coupling component is

n can assume the values 1 or 2, and the substituents on the aromatic ring are accordingly U or V for n = 1, and X and Y for n = 2. R_D^2, R_D^4, and R_D^5 can be H, Cl, CH_3, OCH_3, OC_2H_5, or $COOCH_3$. These products are known as bisacetoacetarylide pigments (Sec. 2.4.2).

Diarylide yellow pigments reign supreme among disazo pigments, followed by the commercially much less important disazopyrazolone pigments and the bisacetoacetarylide pigments. Disazo pigments provide colors ranging from greenish yellow to reddish orange. Compared to monoazo yellow pigments, whose molecular weight is typically only about half of that of a disazo pigment, the latter are considerably more solvent and migration resistant.

2.4.1   Diarylide Yellow Pigments

Diarylide yellow pigments were first patented as early as 1911 in the name of the company Griesheim–Elektron. However, the new invention was not utilized for some time because monoazo yellow pigments, which had just entered the market under the name of Hansa Yellow, were faster to light than diarylide yellow pigments. It took as long as 25 years for diarylide yellow pigments to be appreciated. Their commercial value began to be rediscovered when they were used in Germany to lend color to rubber products: monoazo yellow pigments, which show a tendency to bleed, proved unsatisfactory for this purpose.

The printing ink industry, currently the most important area of application for diarylide yellow pigments, first introduced these products around 1938 in the USA.

The chemically most simple diarylide yellow pigment is made from bisdiazotized 3,3'-dichlorobenzidine and acetoacetanilide. This pigment exhibits high tinctorial strength, a property which is particularly useful for process printing inks, for which other diarylide yellow pigments were later developed.

It was not until after the Second World War that Europe followed this trend. During the early fifties, the European printing ink industry experienced a change in pigments from monoazo yellow pigments to diarylide yellow pigments. In the course of this event, the German market added other diarylide yellow pigments with improved lightfastness and migration properties (Pigment Yellow 83, 81, and 113). As a result, diarylide yellow pigments are by far the largest fraction of organic yellow pigments in the market today.

2.4.1.1   Chemistry, Manufacture

Commercially significant diarylide yellow pigments have the following chemical structure:

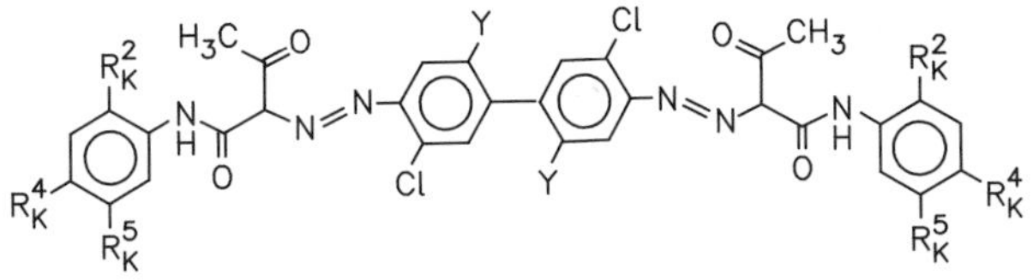

with Y $=$ H or Cl and R_K^2, R_K^4, and R_K^5 $=$ H, Cl, CH_3, OCH_3, or OC_2H_5.
3,3'-dichlorobenzidine (Y:H) is the preferred bisdiazo component. It is synthesized by alkaline zinc reduction (Sec. 2.1.1) or by catalytic reduction of o-nitrochlorobenzene with subsequent rearrangement of the resulting 2,2'-dichlorohydrazobenzene with dilute hydrochloric acid.

In order to manufacture a diarylide yellow pigment, the diamine, dissolved in hydrochloric acid or sulfuric acid, is bisdiazotized with an aqueous sodium nitrite solution. The resulting diazonium compound is subsequently coupled onto two equivalents of acetoacetarylide. Since a material with a very fine particle size is needed for a fast and complete coupling reaction, the coupling component is prepared by dissolution in an alkaline medium and reprecipitation with an acid, such as acetic acid or hydrochloric acid. Dichlorobenzidine cannot be diazotized stepwise. A lack of sodium nitrite does not produce the monodiazonium compound, but an excess of 3,3'-dichlorodiaminodiphenyl remains.

Surfactants, dispersing and coupling agents are not the only additives used in pigment preparation. Depending on the application, the list includes resins, aliphatic amines, and other compounds. Surfactants are primarily used as dispersing agents, while resins are responsible for the very fine particle size consistency of the pigments (Sec. 2.2.3).

It has recently been discovered [1] that the presence of an excess amount of resin (rosin) during the production of diarylide yellow pigments of the Pigment Yellow 12 type affords an additional crystal modification, which can be identified by X-ray diffraction spectroscopy.

Additives may exert not only a physical but also a chemical influence on the coupling suspension: long-chain aliphatic or cycloaliphatic amines RNH_2, which react partially with the pigment molecules, will enhace the effect of other additives. The carbonyl function of the acetyl group is converted to a ketimine (azomethine) or, in the presence of the enol form, it reacts to form an alkylammonium enolate [2].

Pigment preparations containing such amines are widely used in publication gravure inks (see P.Y.12, p. 248).

The coupling reaction may be followed by more or less intense thermal processing of the crude aqueous pigment suspension. This finishing is sometimes carried out in the presence of organic solvents, especially to develop high hiding power. Solvent treatment is most economical if the solvent can easily be separated from the aqueous medium by distillation, in order to be recycled. An aqueous isobutanol suspension, for instance, may undergo steam distillation to afford a solvent-free pigment suspension, which can then be passed through filter presses for separation.

In order to extend the range of available products, mixed couplings are carried out with combinations of two or three different acetoacetarylide coupling components. The most common ones are acetoacetanilide, acetoacet-2-methylanilide, -2-methoxy anilide, -2-chloroanilide, -4-methoxyanilide, -2,4-dimethylanilide, and acetoacet-2,5-dimethyl-4-chloroanilide. It is also possible to carry out mixed couplings with two bisazo compounds. Combining 3,3'-dichlorobenzidine and 3,3'-dimethoxybenzidine, for instance, affords products which cover a wide range of colors and combine a variety of different properties [3].

The performance of a pigment which is produced by mixed coupling does not always equal the sum of the characteristics of the individually coupled products. A new crystal modification, for instance, may lead to an unpredictable deviation in the application properties. On a cost-performance basis, the best approach is to add to the reaction mixture traces of a costly coupling component which is known to confer excellent properties on the product, in an attempt to promote a more favorable crystal modification. The resulting mixed product is expected to assume the modification of the trace coupling component and thus to improve its commercial performance without representing a financial strain [4].

2.4.1.2 Properties

Diarylide yellow pigments range in shade from very greenish to extremely reddish yellow. The more greenish types almost exclusively contain 2,2',5,5'-tetrachlorobenzidine in place of the otherwise more common 3,3'-dichlorobenzidine.

Diarylide yellow pigments are colorants with high tinctorial strength. Their tinctorial strength is typically about twice as high, compared to monoazo yellow pigments which cover the same range of colors (Sec. 2.3). In letterpress application, for instance, the strength of commercially available types of Pigment Yellow 12 is more than twice as high as that of transparent P.Y.1 varieties. In air drying alkyd systems, the intensely reddish P.Y.83 is three times as strong in white reductions as the similarly colored monoazo yellow pigment P.Y.65. Structurally, it is the larger conjugated system that is the primary color strength determinant in diarylide yellow molecules. Physical considerations, especially regarding the particle size distribution, also play a major role.

Most commercial varieties of diarylide yellow pigments are materials with comparatively fine pigment particles and specific surface areas between 50 and 90 m^2/g. The specific demands of the printing industry often require surface-coated pigments with resins or other substances.

The specific surface areas mentioned in this context are frequently minimum values; pigment preparation tends to increase the surface area, so that the actual value may be much higher. A resinated P.Y.83 type, which is washed repeatedly with petrol ether, for instance, will slowly increase the measured specific surface area from 70 m^2/g to more than 100 m^2/g (Sec. 1.5.1). It should be noted that considerable amounts of resin remain on the pigment surface, even after repeated washing.

Monoazo yellow pigments, on the other hand, are products with larger particle sizes and with much smaller specific surface areas. P.Y.74 is the only species of which there are types with specific surface areas approaching those of diarylide yellow pigments (Sec. 2.3.4).

The pigment industry produces primarily the more transparent diarylide yellow pigment types. This has its advantages, particularly in the field of printing inks, since yellow is printed as the last color in three or four color printing (Sec. 1.8.1.1). Highly transparent varieties are almost exclusively resinated and are often easy to disperse.

Commercial diarylide yellow varieties are generally formulated with hard resins, primarily with rosin or its derivatives. These resins are easily used for the common types of printing ink systems, in which they dissolve more or less entirely during dispersion. Resins for pigment preparation may also be used advantageously in other application media, as long as the concentration is not too high. Polymers such as polyolefins are the only media in which a resin may decompose during processing if the degradation temperature is exceeded. Moreover, decomposition residue may adhere to parts of the production unit and present processing problems, owing to surface interaction.

A number of diarylide yellow pigments are also available in the form of very opaque products with small specific surface areas. These are used to an appreciable extent in Chrome Yellow free paints, in packaging printing inks, and other media. Fastness to light and weather are much improved in opaque varieties, compared to the corresponding transparent types; their rheological characteristics are superior. Opaque pigments are not resinated.

Diarylide yellow pigments are typically surface treated to produce taylor-made products for specific applications, especially for printing purposes. Types that are treated with various amines are often very easily dispersible. However, they have their disadvantages in some applications. In offset inks, for instance, amine-treated types generally present printing problems, owing to a disturbed equilibrium between pigment and water. Use in nitrocellulose chips, i.e., preparation on nitrocellulose basis, is not recommended because amines reduce the self-ignition temperature of nitrocellulose, which may lead to deflagration during chip production. Publication gravure printing inks, however, rely almost exclusively on diarylide yellow pigments prepared with amines to produce yellow shades.

In practice, diarylide yellow pigments are frequently selected for their excellent fastness to a variety of organic solvents, in which they perform much better than monoazo yellow pigments. Properties such as migration or recrystallization resistance are thus distinctly superior. A number of diarylide yellow pigments, such as P.Y.81 or 83, exhibit excellent fastness to overcoating (Sec. 1.6.3.2). Characteristics such as migration resistance in plastics, particularly in plasticized PVC, as well as fastness to acid and alkali, are equally superior.

A variety of diarylide yellow pigments are very fast to light and weather – although monoazo yellow and orange pigments are superior in this respect.

2.4.1.3   Application

The bulk of diarylide yellow pigments are targeted for use in printing inks, for which they are selected on the basis of their excellent tinctorial strength and versatile transparency, which can be optimized according to customer specification. The printing ink industry requires good or at least sufficient fastness to the commonly used solvents. Recrystallization can be minimized by maintaining an appropriate processing temperature and by keeping the dispersion temperature low, so that products with very fine particle sizes do actually provide highly transparent prints. This is a primary consideration in connection with modern processing technology. In agitated ball mills, for instance, which are used to prepare inks for web offset printing presses, a medium based on mineral oil is used which is exposed to a temperature of at least 70 to 90 °C for several hours. Even the most effective cooling will not sufficiently reduce this temperature (Sec. 1.8.1.1).

In this type of application, there is no alternative to diarylide yellow pigments. It is not yet clear whether preparation agents such as hard resins, used in highly transparent types, enhance the recrysallization resistance of the product. The resin content of highly transparent varieties of diarylide yellow pigments, especially of the P.Y.12 and 13 grades, may vary considerably; this is also true for P.Y.127, derived from the above-mentioned pigments. The printing ink formulations are typically adjusted accordingly.

The effect of resin addition is a primary concern in offset printing. Good dispersibility is needed to guarantee easy incorporation of a diarylide yellow pigment into its offset vehicle by low shear equipment, such as agitated ball mills. During the dispersion process, the resin dissolves more or less in the ink, which exposes the surfaces of the previously resin-coated pigment particles. The units are distributed throughout the medium, i.e., dispersed. Test results may be distorted by partially undissolved or by swollen resin, which falsely indicates insufficient pigment dispersion. A variety of test methods, including microscopic examination of a printing ink, may fall prey to this error. The printed sample itself is usually exempt from such effects: tinctorial strength or gloss typically remain unaffected.

Good resistance to recrystallization also explains the frequent choice of diarylide yellow pigments as colorants in solvent containing gravure inks for packaging printing. The packaging industry does not generally require high transparency, as in offset printing inks; although transparency may be conferred on a product by applying nitrocellulose chips in order to preclude recrystallization during processing and to make for a glossy appearance. With P.Y.83, however, high transparency is usually expected, even for packaging purposes: it produces brilliant shades of gold on aluminum foil. Aluminum printing generally requires pigments with fine particle sizes and high transparency. This applies also to other diarylide yellow pigments, such as P.Y.17, and to a variety of other organic pigments.

Diarylide yellow pigments exhibit good heat stability in printing inks; they withstand temperatures of 180 to 200 °C. The fastness of such prints thus far exceeds that of some of the monoazo yellow pigments. Diarylide yellow pigments have stimulated widespread interest as colorants for metal deco printing purposes. They are generally resistant to clear lacquer overcoating and to sterilization; fastness to alkali, acid, and water are excellent, which is also true for a number of special application fastness properties (Sec. 1.6.2).

A number of diarylide yellow pigments meet the lightfastness requirements for wallpaper printing. A certain tendency to migrate precludes, for instance, the more lightfast monoazo yellow pigments from use in PVC wallpaper.

Compared to the demand for printing inks, there is somewhat less of a demand for diarylide yellow pigments throughout the paint industry. In terms of lightfastness and weatherfastness, they frequently fail to meet the requirements for exterior application. Apart from P.Y.83 with its good properties, the paint market is dominated by monoazo yellow pigments, which are used in air drying paints; other suitable products are benzimidazolone, isoindolinone, flavanthrone, and anthrapyrimidine yellow pigments, which lend themselves to use in baking enamels.

It is their moderate lightfastness compared to monoazo yellow pigments that restricts the use of most diarylide yellow pigments in emulsion paints, with the exception of P.Y.83.

In contrast, diarylide yellow pigments are used widely throughout the plastics field. This is particularly true for P.Y.13, 17, 81, 83, and 113. The German edition of this book reported that the heat stability (Sec. 1.6.7; 1.8.3) of these pigments, for instance in polyolefins, was up to 200 to 270 °C for 5 minutes, depending on the depth of shade. Results that have been published in the meantime [5], however, require rectification of these numbers.

According to these new studies, thermal decomposition may occur during the processing or application of these pigments in polymers above 200 °C. The same is true for Pigment Orange 13, 34, and Pigment Red 38 types. The decomposition products are monoazo colorants and aromatic amines. The formation of such colorants had at earlier times escaped attention because they afford a range of shades which almost equals that of the respective basic pigment. Comparative coloristic studies, for instance, concerning such pigmented, thermally treated versus untreated polymer materials do not show pigment decomposition or, respectively, formation of a colorant.

Analytical identification of monoazo colorants and the other decomposition products requires effective (analytical) methods of concentration, which today is made possible by high performance liquid chromatography (HPLC). Prior to HPLC analysis, the pigmented medium was extracted for 20 hours with toluene in a soxhlet extractor. These analytical methods also showed that above 240 °C, especially after prolonged exposure of the pigmented polymer material to heat, dichlorobenzidine (DCB) is also formed.

These results show that the described types of diarylide yellow pigments are not recommended for use in polymers which are processed at temperatures above 200 °C. This is true not only for their use in plastics, such as polypropylene and polystyrene, but also for applications such as metal deco prints, which are baked at temperatures above 200 °C, or powder coatings, which are processed above 200 °C.

The assumption is that thermal decomposition starts as soon as the diarylide yellow pigment begins to dissolve in the polymer. This means that the decomposition process proceeds via the dissolved state of the pigment. This assumption, if true, also explains why these pigment powders are heat resistant up to temperatures as high as 340 °C, as shown by differential thermoanalysis (DTA).

Blooming is not normally observed, but may occur under certain conditions. Depending on the processing temperature and the choice and amount of used plasticizer, a pigment concentration below 0.05% in plasticized PVC may have unacceptable results. Consideration must therefore be given to the respective threshold concentrations. Under certain conditions, a number of diarylide yellow pigments are very resistant to bleeding. These species are used to an appreciable extent in various PVC products, and their lightfastness also qualifies them for a variety of applications throughout the plastics industry. These pigments exhibit high tinctorial strength in plastics, and favorable cost-performance considerations and generally good fastness properties broaden their scope within the plastics sector. Diarylide yellow pigments lend color to rigid and plasticized PVC and polyolefins, polyurethane foam, rubber, other elastomers, and various cast resins. Diarylide yellow pigments are sometimes also used in spin dyeing.

High tinctorial strength and good fastness properties make diarylide yellow pigments important products in areas other than the printing inks, paints, and plastics fields. Out of the numerous applications, only a few options should be mentioned. Cleaning agents of various kinds, solvent stains, and office articles employ diarylide yellow pigments to lend color to pencils, watercolors, chalks, or artists' paints, provided the pigment is sufficiently lightfast. Diarylide yellow pigments are also used to an appreciable extent in textile printing inks.

2.4.1.4 Commercially Available Diarylide Yellow Pigments

General

Diarylide yellow pigments entered the market in 1935, when P.Y.13 was introduced by IG Farben in Germany. Sold as Vulcan–Echtgelb GR, this pigment was designed to lend color to natural rubber articles. P.Y.12, which was produced and offered a few years later in the USA, replaced the P.Y.1 type monoazo yellow pigments, which up until then had dominated the yellow pigment market in most parts of the world. In subsequent years, methods for diarylide yellow pigment synthesis advanced steadily and resulted in a broad range of products, many of which have since then either vanished from the market altogether or are now only regionally important. This is true for diarylide yellow pigments based on 3,3'-dimethoxy, 3,3'-dimethyl, or 2,2'-dimethoxy-5,5'-dichlorobenzidine as a bisdiazo component instead of the most commonly used 3,3'-dichlorobenzidine. P.O.15, 16, and 44 are the only species which have maintained a somewhat important position in the American and Japanese market.

Table 14 lists the commercially significant diarylide yellow pigments. It should be noted that the acetoacetarylides in the aromatic moiety possess mostly methoxy or methyl substituents.

Individual Pigments

Pigment Yellow 12

P.Y.12 is a commercially important product; it comprises a large portion of the worldwide pigment production. The unsubstituted skeleton of the acetoacetanilide as coupling component makes P.Y.12 chemically the most simple representative of diarylide yellow pigments. The fastness properties of P.Y.12 are somewhat lower than those of the substituted derivatives. This is particularly true for the lightfastness. 1/1 to 1/3 SD letterpress proof prints exhibit a lightfastness around step 3 on

Tab. 14: Commercially available diarylide yellow pigments.

C.I. Name	C.I. Constitution No.	X	Y	R^2_K	R^4_K	R^5_K	Shade
P.Y.12	21090	Cl	H	H	H	H	yellow
P.Y.13	21100	Cl	H	CH₃	CH₃	H	yellow
P.Y.14	21095	Cl	H	CH₃	H	H	yellow
P.Y.17	21105	Cl	H	OCH₃	H	H	greenish yellow
P.Y.55	21096	Cl	H	H	CH₃	H	reddish yellow
P.Y.63	21091	Cl	H	Cl	H	H	yellow
P.Y.81	21127	Cl	Cl	CH₃	CH₃	H	very greenish yellow
P.Y.83	21108	Cl	H	OCH₃	Cl	OCH₃	reddish yellow
P.Y.87	21107:1	Cl	H	OCH₃	H	OCH₃	reddish yellow
P.Y.90	–	–	–	H	H	H	reddish yellow
P.Y.106	–	Cl	H	CH₃	CH₃	H	greenish yellow
				OCH₃	H	H	
P.Y.113	21126	Cl	Cl	CH₃	Cl	H	very greenish yellow
P.Y.114	21092	Cl	H	H	H	H	reddish yellow
				H	CH₃	H	
P.Y.121	–	Cl	H	–	–	–	yellow
P.Y.124	21107	Cl	H	OCH₃	OCH₃	H	yellow
P.Y.126	21101	Cl	H	H	H	H	yellow
				H	OCH₃	H	
P.Y.127	21102	Cl	H	CH₃	CH₃	H	yellow
				OCH₃	H	H	
P.Y.136	–	Cl	H	–	–	–	yellow
P.Y.152	21111	Cl	H	H	OC₂H₅	H	reddish yellow
P.Y.170	21104	Cl	H	H	OCH₃	H	yellowish orange
P.Y.171	21106	Cl	H	CH₃	Cl	H	yellow
P.Y.172	21109	Cl	H	OCH₃	H	Cl	yellow
P.Y.174	21098	Cl	H	CH₃	CH₃	H	yellow
				CH₃	H	H	
P.Y.176	21103	Cl	H	CH₃	CH₃	H	yellow
				OCH₃	Cl	OCH₃	
P.Y.188	21094	Cl	H	CH₃	CH₃	H	yellow
				H	H	H	
P.O.15	21130	CH₃	H	H	H	H	yellowish orange
P.O.16	21160	OCH₃	H	H	H	H	yellowish orange
P.O.44	–	OCH₃	H	H	Cl	H	reddish orange

the Blue Scale; in 1/5 SD, the lightfastness decreases to about step 2. At equal depth of shade, P.Y.12 is thus 1 to 2 steps less fast to light than other diarylide yellow pigments, such as P.Y.13, 83, 127, or 176, which cover the medium to reddish yellow range. The only exception is P.Y.14, whose lightfastness is about the same.

At equal pigment concentration, P.Y.12 is as much as 2 to 3 steps less lightfast than its counterparts. In other words, in order to achieve the same depth of shade, the tinctorially stronger diarylide yellow pigments in the medium to reddish yellow range require much lower pigment concentrations than a comparative P.Y.12 sample, which has to be more highly pigmented. Since the lightfastness of a pigment in an application medium increases with its concentration, differences between pigments become more obvious as the pigment level increases. Printed in equally thick layers (1 µm) and provided the particle sizes of the two samples are similar, a 1/1 SD P.Y.12 letterpress proof print requires a 10% pigment concentration, while a P.Y.13 sample will afford the same depth of shade if it is formulated at only about 7%.

P.Y.12 has a medium yellow hue, which is used to a large extent in letterpress and offset printing inks as the yellow component in three and four color printing. To adjust P.Y.12 to the standard yellow of the European standard (Sec. 1.8.1.1), the hue may be shaded with traces of a redder component, such as one of the orange pigments P.O.13 or 34. No such shading is necessary for P.Y.12 to match the yellow on the Kodak scale.

The transparency of P.Y.12 can be optimized to meet a variety of application requirements. The product line ranges from fairly opaque versions, which are targeted for packaging and newspaper printing inks, to semitransparent and highly transparent, typically resinated types. As the pigment particle size is reduced for the more transparent versions, the hue shifts increasingly towards more clean, greenish shades; fine pigment particles also render the resulting products more sensitive to light. In this respect, P.Y.12 types are satisfactory only in areas where fastness to light is of less concern. High gloss has its advantage for types that are to be used in the uppermost and very thin layer of a three or four color print. Common printing techniques (Sec. 1.8.1.1) for letterpress and offset application produce roughly 1 µm ink films. This requirement explains why good dispersibility and optimized ink formulation are of major importance to the manufacturer.

While in Europe P.Y.13 and its derivatives are preferred to P.Y.12 for offset printing inks, the reverse is true in the USA, where P.Y.12 is almost exclusively used. In America, most oil-based inks are manufactured from pigment presscake with a fine particle size, which is transfered directly into the flush paste and then converted to the printing ink, avoiding agglomeration and, to a large extent recrystallization (Sec. 1.6.5.8). This technique has become practically obsolete in Europe, considering the easy dispersibility of pigment powders for process colors, i.e., colors for three or four color printing in letterpress and offset application.

Compared to other diarylide yellow pigments, P.Y.12 is only moderately fast to organic solvents. Its tendency to recrystallize is particularly detrimental to modern agitated ball mills which are used to disperse pigments in web offset printing vehicles. Under these processing conditions, highly transparent P.Y.12 types are inferior to the corresponding P.Y.13 types; improved reddish varieties of P.Y.12 with distinctly improved recrystallization stability have recently been introduced.

P.Y.12 is a typical diarylide yellow pigment in that its prints are fast to clear lacquer overcoating, which is used to protect tin prints against scratching and scrubbing. They are also fast to sterilization (Sec. 1.6.2.3), which is one of the major requirements of an ink to be printed on food cans.

Offset and letterpress inks are usually formulated at high pigment concentrations; most contain more than 15% pigment. This is a particular challenge to highly transparent products with very fine particle sizes. These depend on good rheology, which determines the flow behavior of a material in a high-speed printing press.

Yellow shades for publication gravure printing inks are produced almost exclusively by specialty types of P.Y.12. These types have the additional advantage of providing extra depth of shade. A side reaction takes place between the aliphatic amines that are used to prepare the pigment, in which a portion of the pigment reacts to form ketimines (Sec. 2.4.1.1). The reaction products dissolve in the media which are commonly used to formulate publication gravure printing inks, shifting the color of the ink to a dark red. Toluene is more effective than benzene in this respect. The result is a seemingly deeper color in print. At equal pigment concentration, printing inks containing such specialty pigments also exhibit much better flow behavior than inks made from traditional types.

Recently, amine-treated pigments have been developed which show less of a tendency to penetrate low-cost gravure paper (Sec. 1.8.1.2). In certain circumstances, the storage stability of a gravure printing ink is a major concern. This is particularly true for the transportation of such inks in summertime, when products are exposed to high temperature and agitation. The resins that are commonly used to formulate these printing inks are frequently unable to prevent a pigment from changing its coloristic properties in an ink. Ketimine formation is reversed and the previously dissolved compound precipitates as a crystalline pigment. The result is a shift towards a greener shade, reduced tinctorial strength, and viscosity increase, due to the formation of additional pigment surface [4]. The storage stability of a given publication gravure printing ink may be improved considerably by choosing appropriate resins for the product. It should be noted that in contrast to calcium resinates, the commonly used zinc resinates shorten the shelf life of an ink. However, the underlying mechanisms are not yet entirely known.

P.Y.12 types are used in packaging gravure and flexographic inks for their reasonable price, but they can also be utilized in aqueous printing inks, provided they are sufficiently fast to light. They cannot be applied as decorative printing inks to laminated paper.

The paint industry shows little interest in P.Y.12, since it is not sufficiently fast to overcoating for use in baking enamels. In air drying paints, lightfastness of P.Y.12 in white reduction (with 1:5 TiO_2) only equals step 2 on the Blue Scale.

There is also little demand for P.Y.12 as a colorant for polymers. Although a certain tendency to migrate precludes its use in plasticized PVC, rigid PVC somewhat enhances its performance. Transparent varieties become sufficiently fast to light to equal step 6 on the Blue Scale; opaque types afford values between 2 and 5, depending on the pigment concentration and amount of TiO_2 or depth of shade. Good heat resistance also qualifies P.Y.12 for use in polyurethane foam; aromatic polyurethanes themselves are not very fast to light.

Like modified P.Y.12 (P.Y.126), P.Y.12 is selected to lend color to a variety of specialty products, such as cleaning agents or office articles. Where better lightfastness is required, P.Y.12 is replaced by less intense and less solvent resistant monoazo yellow pigments, especially by P.Y.1 types, which provide a similar hue.

Pigment Yellow 13

P.Y.13 also represents a frequent choice throughout the printing ink industry, especially in offset application. Its hue corresponds to the standard yellow on the European Scale for three and four color printing, and it also matches the yellow on the Kodak Scale (Sec. 1.8.1.1). High transparency is a major asset to a yellow pigment which is to be used as the uppermost layer in the multicolor printing sequence. P.Y.13 types, as well as their derivatives P.Y.127 and 176, are therefore surface coated with hard resins. At comparable specific surface area and particle size distribution and consequently similar transparency, prints containing P.Y.13 or its modifications are up to 25% stronger than those formulated with P.Y.12 or 126.

Although P.Y.13 types, especially the resinated, highly transparent versions, are generally easy to disperse, problems may arise with low shear equipment like agitated ball mills. Incompletely dispersed pigment may appear to recrystallize during dispersion, affecting both the transparency and the tinctorial strength of the product.

P.Y.13 is not only more resistant to solvents, but also shows less of a tendency to recrystallize than P.Y.12. This makes it much more advantageous to incorporate such pigments in offset vehicles containing large amounts of mineral oil by using agitated ball mills like the ones mentioned above (Sec. 1.8.1.1). P.Y.127 and 176 are even more recrystallization resistant than P.Y.13.

At equal depth of shade, the lightfastness of P.Y.13 exceeds that of similarly transparent types of P.Y.12 by one to two steps on the Blue Scale.

P.Y.13 is supplied as a wide selection of different types, ranging from highly transparent systems to semitransparent and semiopaque versions to highly opaque varieties.

P.Y.13 and its chemically modified derivatives, due to their higher solvent fastness compared to p.Y.12, are used in much greater volume in packaging gravure inks. P.Y.13 is also fast to protective clear lacquer overcoatings and may be sterilized and calandered.

P.Y.13 is not used in publication gravure printing inks based on toluene or toluene-benzene mixtures; amine-treated grades are not commercially available at present.

The paint industry shows only limited interest in either P.Y.13 or 12. Although the redder varieties of P.Y.13 are more lightfast than the P.Y.12 types by a few steps on the Blue Scale, they do not reach the lightfastness of Hansa Yellow type monoazo pigments. P.Y.13 is not fast to overpainting.

P.Y.13 is used to a more appreciable extent in plastics, where it meets average standards. Depending on the choice and the amount of plasticizer and processing temperature, a P.Y.13 content below 0.05% in plasticized PVC may cause some blooming (Sec. 1.6.3.1). Fastness to bleeding in plasticized PVC is much better than for P.Y.12; regarding bleed resistance, P.Y.13 is in fact more comparable to the more greenish P.Y.17. In 1/3 SD, which corresponds to a pigment concentration of about 0.3% with 5% TiO_2, the lightfastness (daylight) of P.Y.13 in plasticized PVC is equal to step 6–7 on the Blue Scale; the pigment behaves similarly in rigid PVC, in which it does not migrate. In cable insulations made of plasticized PVC, P.Y.13 is used to match the standards according to DIN 47 002, RAL 1021 and RAL 2003 (Sec. 1.6.1.3); it is also used in PVC-based floor covering.

As a result of the recently discovered thermal decomposition of diarylide yellow pigments (Sec. 2.4.1.3), the use of P.Y.13 in HDPE must be limited to 200 °C, even though the pigment was said to survive a 5 minute exposure to 200 to 260 °C (depending on the depth of shade and the commercial grade). Since P.Y.13 is a product with a high tinctorial strength, a pigment concentration of about 0.12% in HDPE will produce a 1/3 SD sample (1% TiO_2 content). The shrinkage of plastics extrusion products is only affected if the processing temperature is low. In these materials, P.Y.13 is as lightfast as in PVC. P.Y.13 is used to an appreciable extent – usually in the form of pigment preparations – in rubber and other elastomers and in aromatic polyurethane foams. The spin dyeing market employs P.Y.13 to color viscose rayon and synthetic wool.

Pigment Yellow 14

Although P.Y.14 is less important than P.Y.12 and 13, the packaging and textiles printing ink industries use it in large volume. P.Y.14 is somewhat greener than P.Y.12 and considerably more so in comparison with P.Y.13. A number of the P.Y.14 types are appreciably greener than the standard yellow on the European Scale. P.Y.14 is not only weaker than comparable P.Y.13 varieties with similar physical characteristics, such as specific surface area; but it is also less lightfast by 1 to 2 steps on the Blue Scale. Its resistance to solvents is also comparatively poor. This somewhat limits its use for process inks in offset and letterpress application to special cases, which is equally true for P.Y.14 blends with reddish pigments. Types with fine particle sizes, which match highly transparent versions of P.Y.12 and 13, are not available in Europe.

P.Y.14 is more resistant to solvents than P.Y.12. In contrast to P.Y.12, it is resistant to paraffin, which makes it a useful product for packaging inks. This is one of the reasons why P.Y.14 is used so much more widely throughout the USA, where it is produced in considerably larger volumes than P.Y.13. According to the US International Trade Commission, the 1988 production figures were: 2335 t P.Y.14, compared to only 358 t P.Y.13, but 7506 t P.Y.12. Where P.Y.12 is not fast enough to meet the requirements, P.Y.14 is usually the second choice in the USA; while most other countries replace P.Y.12 by P.Y.13.

Suitable P.Y.14 preparation with amines affords specialty types for publication gravure inks. Compared with corresponding varieties of P.Y.12, P.Y.14 specialty types provide a clean, particularly greenish hue; they are also coloristically some-what weaker. However, the graphics industry currently prefers reddish shades for publication gravure printing inks, which explains why P.Y.14 specialty types cur-rently have no value. However, untreated P.Y.14 is of regional importance as a colorant in publication gravure printing inks. The resulting inks exhibit poor rheo-logy.

P.Y.14 is not utilized by the paint industry, and the interest of the plastics area varies according to the region. The pigment is more important in the USA, where it is used in polyolefins. It is thermally stable up to 200 °C, which makes it a useful colorant for elastomers, especially rubber. Whenever P.Y.14 is worked into plasti-cized PVC, there is a certain concentration threshold beyond which the pigment will bloom. P.Y.14 is also used for spin dyeing viscose rayon and viscose cellulose, and it is an especially important product for the mass coloration of viscose sponges. It is not particularly lightfast in these applications: in 1/1 to 1/9 SD, it equals only step 4 to step 5–6 on the Blue Scale. The important textile fastnesses are good.

Pigment Yellow 17

Used primarily in the printing ink field, this pigment is considerably greener than P.Y.14 and much more greenish than P.Y.12. At equal depth of shade, P.Y.17 is more lightfast than P.Y.14 by 1 to 2 steps on the Blue Scale. P.Y.17 is tinctorially weaker than P.Y.14. Printed in a standard layer (1 μm) on a letterpress proof printer, a P.Y.17 ink must be formulated at about 7.5% pigment to afford 1/3 SD prints; while an equally deep P.Y.14 ink only requires 3.7% pigment. In comparing the lightfastness of prints containing P.Y.17 and 13, respectively, it should be re-membered that accelerated exposure to light with high pressure xenon lamps de-stroys P.Y.17 distinctly faster than P.Y.13. The reverse is true for daylight expo-sure, to which P.Y.17 is more resistant than P.Y.13. P.Y.17 is also more resistant to most organic solvents than P.Y.14.

Most commercially available P.Y.17 types are highly transparent, often resi-nated pigments, which have gained recognition in packaging inks. Combination with equally transparent, very reddish varieties of P.Y.83 affords a range of inter-mediate shades which exhibit good transparency and are very lightfast.

P.Y.17 is suitable for all printing techniques. It is somewhat more greenish than the standard yellow for offset and letterpress application on the European Scale. Shading with traces of P.Y.83, however, alleviates the problem. P.Y.17 is also ap-plied in a variety of packaging printing inks. The choice ranges from nitrocellulose and polyamide based inks to vinyl chloride/vinyl acetate mixed polymers on a ke-tone/ester basis for PVC or aluminum foil. In ester-based NC colors and in other media, the pigment, particularly in its fine particle size, highly transparent varie-ties, tends to exhibit poor flow properties and is highly viscous. It may also thick-en, but this is rare. Publication gravure printing inks do not normally contain P.Y.17 but are mostly based on specialty types of P.Y.12 and 14. Reliable heat stability at temperatures up to 200 °C makes P.Y.17 suitable for metal deco print-ing.

P.Y.17 is only rarely employed in paints. It is used in interior linings of cans and other products for its high transparency. In air drying paints, its more opaque versions (organic pigment : TiO_2 = 1:5) equal step 5 on the Blue Scale for lightfastness; full shades correspond to step 7. P.Y.17 is not completely fast to overpainting when employed in baking enamels.

The plastics industry, however, uses P.Y.17 extensively, although problems may occur in plasticized PVC, where blooming is observed if pigment concentrations are low. P.Y.17 is almost as lightfast as the somewhat redder P.Y.13 (step 6–7 at 1/3 SD). Likewise, both pigments are almost equally lightfast in rigid PVC. In polymers, as in other materials, P.Y.17 is also considerably weaker than P.Y.13.

P.Y.17 may be used for mass coloration and also to print PVC film. For these purposes, P.Y.17 is frequently prepared on a VC/VAc (vinyl chloride/vinyl acetate) mixed polymer basis. Good dispersibility in plastics makes these preparations suitable even for thin films. The dielectrical properties of P.Y.17 allow its application in PVC cable insulations.

P.Y.17 is also frequently used in polyolefins, sometimes in the form of pigment preparations. Its heat stability in these media was said to be about 220 to 240°C, but must now, as a result of the recently detected thermal decomposition of diarylide yellow pigments in plastics, be limited to 200°C. This tendency to decompose excludes P.Y.17 from recommendation for use in polystyrene, in which the pigment largely dissolves under the processing conditions. The same is true for ABS.

Aromatic polyurethane foams are equally suitable media for P.Y.17, which is extremely lightfast in a variety of cast resins. In 3 mm thick methylmethacrylate resins containing 0.025% pigment, P.Y.17 equals step 7–8 on the Blue Scale for lightfastness. Similar values are found for unsaturated polyester resins, whose hardening is not affected by the pigment (Sec. 1.8.3.5). In heat setting plastics such as melamine or urea/formaldehyde types, however, P.Y.17 is one to two steps on the Blue Scale less lightfast, which is true both for transparent and opaque versions. P.Y.17 is also used for the coloration of rubber.

The textiles printing industry has an appreciable interest in P.Y.17 and applies it in the form of pigment preparations. Where its fastness properties satisfy the specifications and where the use requirements are not too demanding, the pigment is also utilized for spin dyeing purposes. Manufacturer recommendations include media such as polyacrylonitrile and cellulose acetate fibers, on which 1/3 SD pigment prints exhibit a lightfastness which is equal to step 5 on the Blue Scale.

Pigment Yellow 55

P.Y.55 is less important than some other diarylide yellow pigments, especially the P.Y.12, 13, and 83 types. It affords a very reddish yellow, which is not quite as red as P.Y.83 or 114.

The main area of application for P.Y.55 is in the printing ink industry, where it is used particularly in specialty products for packaging inks. A variety of types are available with different specific surface areas to satisfy customer specifications regarding transparency or opacity. The flow properties are considerably different, especially in oil-based and aqueous systems, which are the most common media for P.Y.55. According to expectation, types with lower specific surface areas are redder, more opaque, show less tinctorial strength, and are less viscous than types with fine particle sizes. The pigment is occasionally highly resinated in order to enhance the transparency of the product. Most P.Y.55 types are distinctly weaker than comparable diarylide yellow pigments covering the reddish and medium yellow range. In 1/3 SD letterpress proof prints, P.Y.55 equals step 4 on the Blue Scale: it is as lightfast as the chemically and tinctorially similar diarylide yellow pigment P.Y.114.

The paints market shows only limited interest in P.Y.55, although the pigment is occasionally used to lend color to low quality industrial paints. In 1/3 SD white reductions, the pigment matches the lightfastness of P.Y.13 samples; its full shades are more lightfast. Addition of TiO_2 decreases its resistance to light.

P.Y.55 is suitable for application in polymers such as PVC and rubber. In plasticized PVC, it exhibits a certain tendency to bleed; its tinctorial strength is good to average. 0.7% pigment, incorporated in plasticized PVC, will produce a 1/3 SD sample (5% TiO_2); while only 0.28% Pigment Yellow 83 is sufficient to afford the same depth of shade. The lightfastness equals step 7 on the Blue Scale. P.Y.55 is also used throughout the textiles printing ink industry.

Pigment Yellow 63

P.Y.63 continues to be manufactured in Japan, where it enjoys only limited regional importance. It provides a greenish to medium yellow hue with a lightfastness that is much inferior to that of other diarylide yellow pigments. Under standardized conditions, 1/1 SD letterpress proof prints equal only step 2 on the Blue Scale. Fastness to a variety of organic solvents that are common in printing inks is good, and the pigment shows no tendency to recrystallize. Preparation with amines considerably shifts the hue of publication gravure printing inks towards greener shades, which is also observed with P.Y.14.

Pigment Yellow 81

This pigment affords a very greenish yellow. Its hue is similar to that of the monoazo yellow pigment P.Y.3, but white reductions are much stronger exhibiting considerably improved solvent and migration resistance. Although P.Y.81 shows satisfactory lightfastness, it is not quite as lightfast as P.Y.3. P.Y.81 resembles P.Y.113 as far as shade and fastness properties are concerned, although P.Y.113 is more resistant to solvents and also more migration resistant.

P.Y.81 is found in a variety of media. It is equally suited to all printing purposes, including textile printing. Good resistance to many organic solvents facilitates its use in a variety of solvent-containing printing inks, such as mixed polymer paints on ketone/ester basis for PVC. Its heat resistance qualifies the pigment for application in metal deco printing inks. The prints are resistant to clear lacquer coatings, sterilization, and calandering. In 1/3 SD, the lightfastness is equal to step 5-6 on the Blue Scale. Comparative values are: P.Y.3: step 6-7, P.Y.113: step 5-6.

P.Y.81 is recognized throughout the paint industry for its fastness to overcoating and very good solvent fastness in industrial application. Most types exhibit good hiding power. In an alkyd-melamine paint, the pigment equals step 7-8 on the Blue Scale in full shades and step 6-7 in white reduction (1:4 TiO_2) for lightfastness. In paints, P.Y.81 provides a tinctorial strength which is similar to that of P.Y.3 of comparable shade, but it is not quite as lightfast and weatherfast.

At low pigment concentrations in plasticized PVC, the pigment may bloom, depending on the processing conditions and formulation of the polymer. In rigid PVC, however, P.Y.81 performs well, irrespective of its concentration. Its daylight fastness in 1/3 SD is equal to step 7 on the Blue Scale, in which it equals the distinctly stronger P.Y.113. At exposure to light from a xenon lamp, P.Y.113 is much more lightfast than P.Y.81, especially in transparent types. The difference is sometimes more than one step on the Blue Scale. P.Y.81 is very resistant to bleeding, provided the appropriate concentration limits are observed. Its heat resistance makes it a useful colorant for polyolefins. P.Y.81 was said to be heat stable up to 260 °C for 5 minutes, depending on the depth of shade. The pigment was thus much more heat resistant than other diarylide yellow pigments. As a consequence of new insight into the thermal decomposition of diarylide yellow pigments in the presence of plastics (Sec. 2.4.1.3), the heat stability of P.Y.81 must now be limited to 200 °C. The pigment hardly affects the shrinkage of preformed HDPE articles and those made from similar, partially crystalline plastics. It is also useful in other plastics and for spin dyeing of materials such as secondary acetate.

Pigment Yellow 83

P.Y.83 possesses excellent fastness properties, which make it almost universally applicable. It provides a reddish yellow hue, which is considerably more reddish than that of P.Y.13 and at the same time very strong.

P.Y.83 can be used for all printing techniques and purposes. The printing ink industry often prefers highly transparent, mostly resinated types. Printing such types on aluminum foil or on metal sheets produces brilliant shades of gold. Combining P.Y.83 with transparent versions of the greenish P.Y.17 affords transparent prints with good lightfastness, which cover the range of intermediate shades. Both pigments are similar in that their varieties with fine pigment particles afford transparent 1/3 SD prints whose lightfastness equals step 5 on the Blue Scale. For comparison, P.Y.12 equals step 2; P.Y.13 reaches step 3-4.

P.Y.83 is the standard pigment within the reddish yellow range. The similarly shaded monoazo yellow pigment P.Y.10 (see pp. 223, 227) is only about half as strong as P.Y.83, but more lightfast (about 1.5 steps on the Blue Scale). At the same time, it is less transparent and considerably less solvent resistant, which may cause significant recrystallization problems during pigment incorporation into the medium of application.

P.Y.83 shows good to very good resistance to most solvents which are typically found in application media. Recrystallization is therefore rare under common processing conditions, even in highly transparent types. Resistance to clear lacquers, calandering, and sterilization are consequently excellent.

P.Y.83 is also used to an appreciable extent in plastics. Because of its good solvent resistance, migration is no problem in plasticized PVC even at low pigment levels; it does not bleed or bloom. Its daylight fastness at 1/3 SD equals step 8 on the Blue Scale; at 1/25 SD, it still reaches step 7. The results in rigid PVC are similar. P.Y.83 is also very strong in polyolefins. The pigment concentration required for 1/3 SD HDPE coloration (1% TiO_2) is only 0.08%. A considerable variety of pigment preparations is available for coloring a number of plastics. The choice ranges from pigment formulations on VC/VAc copolymer basis to pigment plasticizer pastes for PVC mass coloration, including application in thin films or cable insulations and printing on PVC films. Certain preparations are recommended for polyolefins or other plastics. The pigment has so far, on the basis of application tests in polyolefins, been said to be heat stable up to 250 °C. As a result of new insight into the thermal decomposition of diarylide yellow pigments in monoazo colorants and aromatic amines (Sec. 2.4.1.3), this heat stability must now be limited to 200 °C.

P.Y.83 shows excellent lightfastness, even in methylmethacrylate or unsaturated polyester cast resins. The pigment does not affect the hardening of the latter.

The high quality of the fastness properties is the basis for frequent pigment use in textile printing. Dry cleaning with perchloroethylene or washing has almost no effect on the color. P.Y.83, sometimes in the form of a preparation, is used for viscose spin dyeing, secondary acetate, and polyacrylonitrile.

In the paint industry, P.Y.83 types with fine particle sizes are intended for use in transparent and metallic finishes and are also generally employed in industrial coatings and emulsion paints, provided the demands on the light stability of the pigment are not too stringent. In full shades, the pigment reaches step 6–7 on the Blue Scale; some darkening is observed as a consequence of eposure to light. Opaque finishes with TiO_2 (1:10) are equal to step 6 and mixtures of (1:125) correspond to step 4–5. P.Y.83 is fast to overcoating in baking enamels.

One of the P.Y.83 types is an extremely opaque form. It provides finishes with good flow properties and makes it easy to increase the pigment concentration, which in turn improves the hiding power. This special opaque type is considerably more weatherfast than its more transparent counterparts, which makes its full shades useful colorants for original automobile finishes (O.E.M.). The product also lends itself to paints for automobile refinishes and tractor and implement finishes. It is found in a variety of high grade industrial finishes.

Good overall fastness and considerable tinctorial strength broaden the scope of P.Y.83 application. The list includes office articles, artists' colors, and solvent-based wood stains, in which the pigment is frequently combined with red pigments and carbon black to produce shades of brown.

Pigment Yellow 87

Although this pigment has recently been offered by a number of manufacturers throughout the USA and Japan, it is not produced in Europe. Its color is a reddish yellow, similar to that of the chemically closely related P.Y.83. Although both provide almost equal strength, P.Y.87 shows very poor fastness to overcoating in baking enamels and plasticized PVC. It is also less lightfast. Under standard conditions, 1/1 SD letterpress proof prints are equal only to step 4 on the Blue Scale, while corresponding P.Y.83 prints reach step 5. The pigment is used to a certain extent in textile printing.

Pigment Yellow 90

P.Y.90 has little commercial impact, it is a regional product. The pigment provides very reddish yellow shades, similar to those of the much lightfaster P.Y.114. 1/1 SD and 1/3 SD letterpress proof prints equal only step 3 on the Blue Scale, while corresponding P.Y.114 specimens reach step 4–5 and 4, respectively.

Pigment Yellow 106

The shade of P.Y.106 is greenish, somewhat redder than P.Y.17, and its tinctorial strength is greater than that of P.Y.17. Standard films of 1/1 SD letterpress proof prints are made with only 12%P.Y.106, while the necessary P.Y.17 content exceeds 25%. The required pigment concentrations for 1/3 SD proof prints are 6% and 7.5%, respectively. At equal SD, the lightfastness of P.Y.106 is inferior to that of P.Y.17; the difference is about 1/2 step on the Blue Scale. Both pigments lend themselves to similar purposes, mainly in the area of printing inks. P.Y.106 is thermally stable up to 200°C (10 minutes); in prints, it is bleed resistant toward clear lacquer coatings and sterilization, which qualifies it for metal deco printing.

Pigment Yellow 113

At present, P.Y.113 is being removed from the market worldwide because one of the intermediates for its production is no longer internationally available. P.Y.113 is a yellow pigment with a very greenish shade. In its coloristic and application properties, the pigment largely resembles P.Y.81, by which it must widely be replaced. In contrast to P.Y.81, however, P.Y.113 is completely migration resistant in plasticized PVC. Even at very low pigment concentrations, P.Y.113 does not bloom in this medium and has no limitation in this respect. Both pigments display approximately equal lightfastness.

Pigment Yellow 114

P.Y.114 affords very reddish yellow shades which closely resemble those of P.Y.83, although the level of the fastness properties of P.Y.114 is lower; it is somewhere between those of P.Y.12 and 13. In print, P.Y.114 is less lightfast than P.Y.83; the difference is 1/2 to 2 steps on the Blue Scale, depending on the depth of shade; but P.Y.114 does not achieve the tinctorial strength of P.Y.83.

P.Y.114 is primarily supplied to the printing ink industry, where it is used especially for packaging inks. The pigment is utilized to produce prints at reasonable cost, especially where exceptional fastness, as provided by P.Y.83, is a minor consideration. Prints made from P.Y.114 are not entirely resistant to a number of organic solvents, including the standard DIN 16 524 solvent mixture, paraffin, and butter; but P.Y.114 prints are soap, alkali, and acid resistant. The fact that the pigment does not withstand a temperature of 140°C and is not stable to sterilization excludes P.Y.114 from use in metal deco printing.

Pigment Yellow 121

The chemical constitution of this product of mixed coupling has not yet been published. The pigment enjoys a certain regional importance and lends itself to a variety of printing inks. It produces a medium yellow shade, which is somewhat less green than the standard yellow on the European Scale CIE 12-66 for process printing (Sec. 1.8.1.1). Commercial types show medium transparency. The pigment is weaker than similarly colored diarylide yellow pigments, such as P.Y.126 or 176. P.Y.121 is fast to a number of organic solvents, which are commonly used in printing inks; the pigment does not recrystallize in these media. The list includes aromatic hydrocarbons, especially toluene, which qualifies the pigment for use in publication gravure printing inks. This is especially true for amine preparations, which shift the color to a much greener shade. A similar color trend is observed with P.Y.14.

Pigment Yellow 124

Its distribution is limited to the USA, where it enjoys little importance. The pigment affords a medium yellow shade.

Pigment Yellow 126

This pigment is roughly as fast as the chemically related P.Y.12. In types with a similar specific surface area, P.Y.126 is somewhat redder. In contrast to P.Y.12, printing inks containing P.Y.126 require little, if any, modification with reddish components to attain the European Standard Yellow (Sec. 1.8.1.1) for process printing. This feature makes P.Y.126 a suitable process color pigment for offset printing. Highly transparent versions are designed for use as the uppermost layer in four color printing.

P.Y.126 is tinctorially stronger than P.Y.12; in order to produce standard films of highly transparent 1/1 SD proof prints, 9% P.Y.126 is needed, as opposed to 10% P.Y.12. P.Y.126 prints are more resistant to light; the difference is one step on the Blue Scale. In 1/1 to 1/25 SD prints, the lightfastness of P.Y.126 and P.Y.12 equals step 4 and 3, respectively. Heat stability up to 200 °C (10 minutes) and complete resistance to clear lacquer coatings and sterilization facilitate its use in metal deco printing. In aqueous or water/alcohol-based printing inks P.Y.126 is also stronger and more lightfast than P.Y.12. Treatment with aliphatic amines leads to ketimine formation, which is also true for P.Y.12 (Sec. 2.4.1.1). The resulting products afford high tinctorial strength and prints with a comparatively reddish shade. Its storage stability and penetration characteristics parallel those of P.Y.12.

P.Y.126 is generally applied wherever P.Y.12 types suit the purpose, such as in office articles and cleaning agents.

Pigment Yellow 127

This pigment performs like the chemically related P.Y.13. Its commercial types are highly transparent, resinated materials. P.Y.127 is principally used in offset printing inks; in Europe it is considered a standard product for this purpose. The pigment provides a yellow shade, which matches the standard yellow for process printing according to CIE 12-66 (Sec. 1.8.1.1). P.Y.127 has a very high tinctorial strength. 1/1 SD letterpress proof prints require 6.5% pigment, 1/3 SD proof prints contain 3.3% pigment. Comparative values for a number of other pigments covering the same range of colors are listed under the respective name.

The fact that P.Y.127 recrystallizes only very slightly facilitates its dispersion in modern agitated ball mills. The pigment is easy to disperse and its heat stability parallels that of P.Y.13. Its fastness to clear lacquer coatings and to sterilization are excellent, which is why the pigment is suitable for metal deco printing.

P.Y.127 is also used to an appreciable extent in packaging gravure printing inks. Highly transparent inks lend gloss especially to nitrocellulose inks. P.Y.127 prints are very fast, in which they largely parallel P.Y.13 prints.

Pigment Yellow 136

This pigment has been commercially available for several years. Chemically, it is a modified version of P.Y.13 and 14. Its coloristic and application properties are comparable to those of P.Y.13 and 14 mixtures. P.Y.136 is supplied to the printing ink industry, primarily for the production of oil-based inks, i.e., offset inks. Both its hue and its tinctorial strength are somewhere between those of P.Y.13 and 14.

Pigment Yellow 152

Although P.Y.152 enjoys little appreciation in Europe, it is used to a considerable extent in the USA. It produces a reddish, somewhat dull yellow shade. Most of the commercial varieties are very opaque versions with good hiding power, which frequently displace Chrome Yellows in paints. P.Y.152 types are more viscous than opaque varieties of the much greener P.Y.83. The pigment is only moderately fast to overcoating: the tendency to bleed is very strong in an alkyd-melamine resin paint at a baking temperature of 120 °C. Its lightfastness is average. White reductions (1:5 TiO_2) equal step 5 on the Blue Scale, while the corresponding P.Y.83 paint reaches step 7–8. In full and deep shades, there is some darkening, as with P.Y.83; the lightfastness of such shades equals step 7. White reductions of P.Y.152 in air drying alkyd resin systems are almost twice as strong as the somewhat more greenish monoazo yellow pigment P.Y.65; full shades are considerably more opaque.

Pigment Yellow 170

P.Y.170 is produced in Japan and is hardly encountered in the European market. Its hue is yellowish orange or very reddish yellow. The commercial product with a coarse particle size exhibits good hiding power and correspondingly little tinctorial strength in white reductions. It is considerably weaker than the more yellowish benzimidazolone pigment P.O.62, which is also far superior in light and weather resistance. P.Y.170 withstands overcoating in baking enamels at commonly used temperatures. Its prints are thermally stable up to 200 °C, are fast to clear lacquer coatings, and may be sterilized.

Pigment Yellow 171

P.Y.171 is found in the Japanese market. Its prints are considerably redder than those made from P.Y.13; besides, its color is off the fastness limits for yellows on the European color scales for offset and letterpress application. P.Y.171 has a dull hue and is weaker than P.Y.13. The pigment is targeted for the coloration of plastics, such as PVC and PE.

Pigment Yellow 172

P.Y.172, like P.Y.171, is produced in Japan. The two pigments are not only coloristically alike but also possess similar application properties.

Pigment Yellow 174

P.Y.174 joined the market a few years ago and bears some resemblance to P.Y.13. The commercially available varieties are very strong, highly resinated products which are correspondingly transparent. They are of great interest to the offset printing industry.

P.Y.174 provides a yellow shade which matches the CIE 12-66 standard yellow for process color printing. The excellent tinctorial strength of the commercial products is accompanied by the high viscosity of the ink.

Pigment Yellow 176

P.Y.176 exhibits application properties which are similar to those of the chemically related P.Y.13. Only highly transparent varieties are available. The hue of P.Y.176 is somewhat redder than that of P.Y.13.

P.Y.176 finds extensive use in offset printing, where it matches the CIE 12-66 standard yellow for process color printing (Sec. 1.8.1.1). It is a particularly strong pigment. Standardized 1/1 SD proof prints are made from inks containing only 6% pigment, while 1/3 SD samples require 3% pigment. P.Y.176 is stronger than P.Y.13 and 127. In certain vehicle systems, however, this advantage can be compromised by increased viscosity and poor flow behavior, respectively. Fastness in prints, including lightfastness, heat stability, and resistance to clear lacquer coatings and sterilization, common organic solvents, and the DIN 16 524 solvent mixture are good to excellent. In this respect, P.Y.176 largely parallels P.Y.13.

Pigment Yellow 188

P.Y.188, which was released onto the market only a few years ago, is closely related to Pigment Yellow 13 in both its chemical and application properties.

The commercially available, highly transparent type is primarily applied in offset printing inks. In this field, P.Y.188 matches the CIE 12-66 standard yellow shade for process printing (Sec. 1.8.1.1). The commercially available type of P.Y.188 is highly resinated and shows very good tinctorial strength. It displays good flow behavior.

Pigment Orange 15

P.O.15 produces yellowish orange shades, which resemble those of P.O.13. Compared to P.O.13, however, P.O.15 has largely lost its importance. Currently, it is only produced in the USA. Although the pigment is less lightfast and less stable to organic solvents than P.O.13, P.O.15 is used to lend color to printing inks and rubber.

Pigment Orange 16

Structurally based on 3,3'-dimethoxybenzidine (o-dianisidine) as a diazo component, P.O.16 is also known as Dianisidine Orange. At present, the pigment only enjoys some importance in the USA and in Japan; according to the US International Trade Commission, 303 t of P.O.16 were produced in the USA in 1988.

Its color is a yellowish orange, which is considerably redder than P.O.13 and 34. Increasingly stringent application requirements regarding lightfastness and durability, solvent and migration resistance present a considerable challenge to P.Y.16 and have largely eliminated the pigment from the market.

The largest area of application for P.Y.16 is in the printing ink industry. The pigment is used particularly to adjust the shades of P.Y.12 type diarylide yellow pigments, to which it is most closely related as far as application performance is concerned. Such blends are employed in low-cost packaging and specialty inks. Resinated types are produced in order to provide higher transparency. There is a certain disadvantage to the poor rheology of a number of P.Y.16 types, which makes it difficult to formulate highly pigmented systems.

P.O.16 is not lightfast and durable enough to be used in paints. In this respect, it performs even more poorly than P.Y.12. Low-cost production, however, makes it a suitable colorant for rubber and for textile printing.

Pigment Orange 44

P.O.44 has lost most of its commercial importance and is at present only applied to a limited extent. This is also true for other pigments whose synthetic route involves 3,3′-dimethoxybenzidine as a diazo component. P.O.44 provides a very reddish orange shade, which is much redder than the color of the β-naphthol pigment P.O.5. Although P.O.44 is more resistant to solvents than P.O.5, the reverse is true for lightfastness. Standardized letterpress proof prints containing P.O.44, for instance, equal step 3 on the Blue Scale, while equally deeply shaded P.O.5 prints equal step 6 on the Blue Scale.

The main area of application for P.O.44 is in the field of textile printing, although it is less lightfast than P.O.5. 1/1 SD prints containing P.O.44 equal step 5–6, as opposed to step 7 for P.O.5 prints; 1/3 SD P.O.44 samples equal step 4–5, again as opposed to step 7; and 1/6 SD P.O.44 prints only reach step 3–4 on the Blue Scale, while P.O.5 still scores as high as step 6–7.

The application fastness of the pigment, however, is more satisfactory. P.O.44 may be dry cleaned and resists dry heat up to 210 °C for 30 seconds; while P.O.5 and also the somewhat redder Naphthol AS pigment P.R.10 is less fast. P.O.44 is also used in PVC coatings.

2.4.2   Bisacetoacetarylide Pigments

The fact that only a few of these pigments enjoy commercial importance eliminates the need to treat each individual species separately.

Bisacetoacetarylide pigments are considered disazo pigments that are obtained from bifunctional coupling components. The latter are obtained by bisacetoacetylation of aromatic diamines, especially of 4,4′-diaminodiphenyl or 1,4-diaminophenyl derivatives with 2 equivalents of diketene or acetoacetic ester:

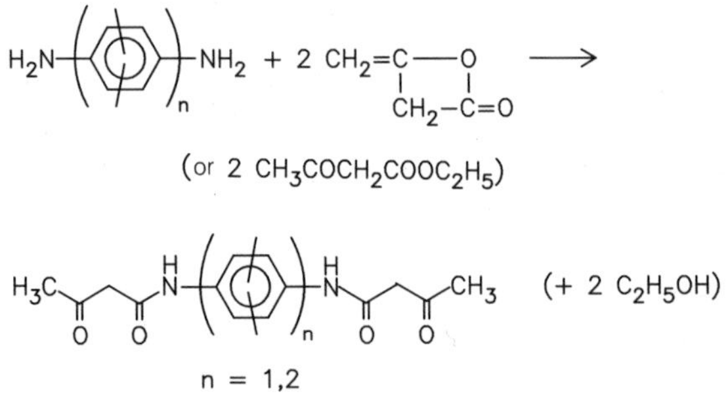

$$H_2N \!-\!\!\left(\!\bigcirc\!\right)_{\!n}\!\!-\! NH_2 \; + \; 2\; CH_2\!=\!C\!-\!O \longrightarrow$$
$$\; CH_2\!-\!C\!=\!O$$

(or 2 $CH_3COCH_2COOC_2H_5$)

$$H_3C \underset{O}{\overset{}{\textstyle\frown}} \underset{O}{\overset{}{\textstyle\frown}} \overset{H}{N}\!-\!\!\left(\!\bigcirc\!\right)_{\!n}\!\!-\!\overset{H}{N} \underset{O}{\overset{}{\textstyle\frown}} \underset{O}{\overset{}{\textstyle\frown}} CH_3 \quad (+\; 2\; C_2H_5OH)$$

n = 1,2

Only common aromatic amines are used as diazonium components. Coupling reactions between aminobenzamides or aminobenzanilides and bisacetoacetylaminophenylenes are described under disazo condensation pigments [see Sec. 2.9]. Bisacetoacetarylide pigments thus have the general chemical structure:

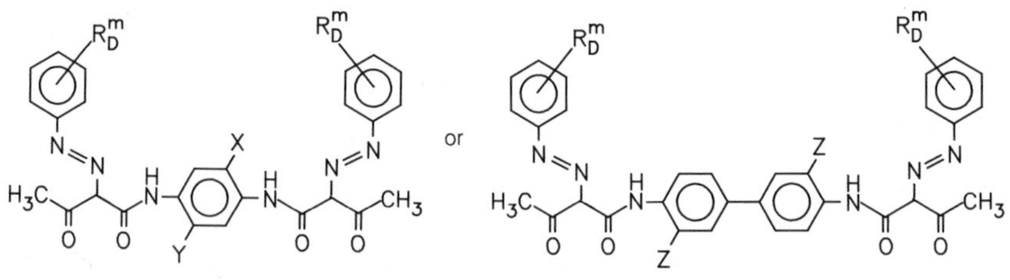

or

with R_D = CH_3, OCH_3, OC_2H_5, Cl, Br, NO_2, $COOCH_3$,
    m = 0 to 3
    X,Y = H, CH_3, OCH_3, Cl, and
    Z = CH_3, OCH_3, Cl.

Bisacetoacetarylide pigments are characterized by good solvent resistance and tinctorial strength; although they are not quite as strong as diarylide yellow pigments.

The large number of patents describing bisacetoacetarylide pigments based on the bifunctional coupling components 1,4-bisacetoacetylaminobenzene and its derivatives [6] is somewhat in contrast to the fact that the products have gained only limited commercial importance.

A commercially available product is Pigment Yellow 155, a yellow pigment which is prepared by coupling the diazonium salt of 1-amino-2,5-dicarbomethoxybenzene onto the bifunctional coupling component 1,4-bisacetoacetylaminobenzene [7]:

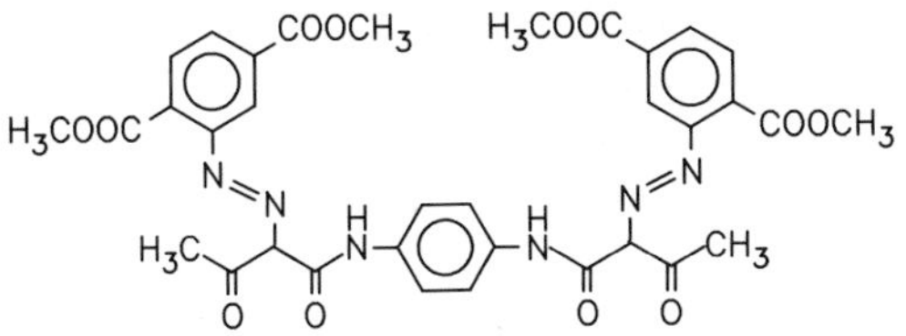

Similarly, bisacetoacetarylide pigments obtained from 4,4′-bisacetoacetylami-
nodiphenyl derivatives as coupling components are of little commercial value. One
exception is Pigment Yellow 16, 20040:

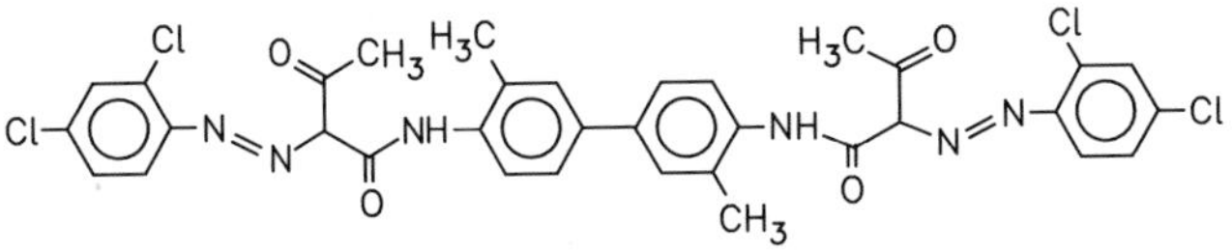

It is obtained by coupling two equivalents of diazotized 2,4-dichloroaniline onto
bisacetoacetylated 3,3-dimethylbenzidine, also known as Naphtol AS-G.

2.4.2.1 Commercially Available Bisacetoacetarylide Pigments and Their Properties

Pigment Yellow 16

P.Y.16 covers the medium yellow range. Increasingly stringent industrial requirements in some areas slowly restrict the use of this originally versatile product.

P.Y.16 shows good to very good resistance to a variety of organic solvents,
including alcohols and esters; however, it exhibits no fastness to some other solvents, such as xylene and other aromatic hydrocarbons. Recrystallization may be a
problem if the pigment is to be processed in systems containing aromatic solvents,
such as some baking enamels and packaging gravure printing inks, resulting in a
considerable shift toward red shades. It has recently been found that P.Y.16 undergoes a similar color shift in the presence of certain solvents in the medium of application, due to a change of crystal modification. This result has been confirmed by
X-ray diffraction studies.

In the paint field P.Y.16 is primarily used in industrial finishes. The similarity
to P.Y.1 is limited to a likeness in shade; but P.Y.16 is fast to overcoating and
shows no tendency to bleed in baking enamels. Although its full shade lightfastness
equals step 7–8 on the Blue Scale, types that are only slightly reduced with TiO_2
(1:5) reach only step 5. Only full shades provide good weatherfastness.

In order to produce chromium free full shade finishes with good hiding power,
special opaque P.Y.16 types are available. Their good rheology makes it possible to
further increase the hiding power by increasing the pigment concentration without
adversely affecting flow, gloss, or other inherent features of the paint. Although

the weatherfastness of such opaque versions with coarse particle sizes is somewhat better than that of traditional types, it is frequently insufficient for long-term exterior application. The opaque varieties are faster to aromatic solvents and solvents other than the standard types, and they recrystallize less readily.

P.Y.16 is supplied as a pigment preparation to the emulsion paint industry. P.Y.16 is suitable for all printing techniques used in the printing ink industry. In some systems, however, it tends to recrystallize. P.Y.16 shows high tinctorial strength and relatively good resistance to light. The pigment is as lightfast as the good lightfast diarylide yellow pigments. 1/1 SD printed samples equal step 5 on the Blue Scale, 1/3 SD prints reach step 4–5. Very opaque types are even faster to light; depending on the depth of shade, these score 1/2 to 1 step higher on the Blue Scale. The prints are resistant to clear lacquer (Sec. 1.6.2.1) and stable to calandering and sterilization. Heat stability at 30 minute exposure up to 200 °C makes P.Y.16 a suitable pigment for metal deco printing. Similarly, it is utilized for textile printing purposes. Application in the plastics field has diminished considerably. P.Y.16 migrates in PVC, which leads to blooming and bleeding. Moreover, its heat resistance in plasticized PVC is unsatisfactory, due to recrystallization. The pigment shows good tinctorial strength in polyolefins. Its heat stability in this medium has been said to equal approximately 230 to 240 °C for 5 minutes, depending on the depth of shade.

Besides its use in cast resin, such as methylmethacrylate mixtures, P.Y.16 has gained recognition as a colorant for felt-tip pens, water colors, and a number of related applications. It lends color to leather, mass and surface paper coloration, and paper pulp.

Pigment Yellow 155

P.Y.155 provides clean, somewhat greenish yellow shades and is characterized by high tinctorial strength, good solvent resistance, and fastness to alkali and acid.

The pigment is recommended for use in paints, plastics, and printing inks. It shows excellent fastness to overcoating in baking enamels: in alkyd-melamine systems, it withstands 30 minutes of exposure to up to 140 °C without bleeding. P.Y.155 is characterized by good lightfastness; its full shades are as fast as P.Y.16. In full shades and in white reductions, P.Y.155 is more weatherfast than P.Y.16; but it is not as weather resistant as the somewhat greener monoazo yellow pigment P.Y.97. A new type consisting of coarser particles, which provides better weatherfastness, has recently been introduced to the market. This type is recommended as a replacement for Chrome Yellow pigments for its higher hiding power.

P.Y.155 is used to an appreciable extent in industrial finishes that are targeted for commercial vehicles, tractors, and farm implements. In paints, it resists temperatures up to 160 °C.

Incorporated in plasticized PVC, P.Y.155 withstands temperatures up to 180 °C, but it shows a certain tendency to bleed under common processing conditions. Its tinctorial strength is good to average. 1/3 SD colorations (5% TiO_2) require 0.7% pigment. In PVC, the ability of both transparent (0.1 %) and opaque

(0.1 % + 0.5 % TiO$_2$) versions to withstand light is equal to step 7–8 on the Blue Scale, while such types are less weather resistant. In HDPE, 1/3 SD P.Y.155 colorations (1% TiO$_2$) may be exposed to temperatures up to 260 °C for 5 minutes without changing color. As in PVC, the tinctorial strength in HDPE is average. Recommendations for application include polypropylene and styrene, but exclude polyesters.

In printing inks, P.Y.155 lends itself to all printing techniques. The prints provide good fastness properties; they are even soap and butter resistant.

2.4.3 Disazopyrazolone Pigments

Like monoazo yellow and diarylide yellow pigments, the first disazopyrazolone pigments were developed as early as 1910; commercial application, however, was delayed by some 20 years. This prolonged lack of interest was caused by a simultaneously ongoing search for high strength, bleed resistant organic pigments to produce orange and yellowish red shades for the coloration of rubber. It was not until the early 1950s that P.O.34 gained recognition as a commercial product.

The preparation of disazopyrazolone pigments is based on a synthetic pathway discovered by Ludwig Knorr in 1883, in a successful attempt to make phenylmethylpyrazolone. It was not until the 1930s that the first pyrazolone pigment, an orange (Pigment Orange 13) became commercially available. Likewise, pyrazolone red pigments were patented at the time; they were released to the market without much delay. Only few disazopyrazolone pigments are still of interest to manufacturers; those that are, however, comprise a large portion of organic pigment production.

2.4.3.1 Chemistry and Manufacture

Industrially produced disazopyrazolone pigments are based on the following general chemical structure:

with X = Cl, OCH$_3$,
R^1 = CH$_3$, COOC$_2$H$_5$,
R^2 = H, CH$_3$.

The originally large number of commercially significant pyrazolone pigments has dwindled considerably, leaving only a few of them which meet modern application standards. Their synthesis parallels that of diarylide yellow pigments: bisdiazotization of a 4,4'-diaminodiphenyl dihydrochloride derivative, primarily 3,3'-dichlorobenzidine dihydrochloride or 3,3'-dimethoxybenzidine dihydrochloride (o-dianisidine), is followed by coupling onto 2 equivalents of the corresponding pyrazolone derivative, which finally affords the crude pigment. Using 3,3'-dichlorobenzidine as the diazo component yields orange pigments, while the increased bathochromic effect of 3,3'-dimethoxybenzidine produces pigments with red shades. Likewise, using 1-aryl-3-carbalkoxypyrazolone-5 as a coupling component affords pigments with red shades. It is possible to modify the physical parameters of a pigment by adding certain agents during or after the coupling reaction, by adjusting the diazotization or coupling technique, or by adapting the aftertreatment of the pigment to the intended use of the product.

Disazopyrazolone pigments, for instance, can be tailor-made for various applications in order to provide features such as high transparency or good hiding power, easy dispersibility, and high tinctorial strength.

2.4.3.2   Properties

Disazopyrazolone pigments range in shade from reddish yellow to orange, red, and maroon. The currently available commercial types, however, come exclusively in orange or red hues. The application properties and fastnesses of these pigments are very versatile. P.O.34, for instance, is as fast as a good diarylide yellow pigment, while P.O.13 is somewhat less fast. This applies to its lightfastness as well as to fastness to solvents and the tendency to migrate. Pigments obtained from 3,3'-dimethoxybenzidine instead of 3,3'-dichlorobenzidine as the bisdiazo component perform much less well as far as fastness to solvents, migration, and light are concerned. They enjoy only restricted commercial significance.

2.4.3.3   Application

Disazopyrazolone pigments are broad in scope. Depending on their physical characteristics, different types are targeted for the printing ink, paint, or plastics industry.

2.4.3.4   Commercially Available Pigments

General

P.O.34 and 13 are the two most frequently used disazopyrazolone pigments obtained by bisdiazotization of 3,3'-dichlorobenzidine. P.O.13 comprises the largest portion of disazopyrazolone pigment production and also reigns supreme among organic orange pigments in general. P.R.38 and 37 have less of an impact on the pigment industry. Both were introduced around the same time as P.O.13. P.R.37 and 41 are obtained with 3,3′-dimethoxybenzidine as bisdiazonium component. Table 15 lists the commercially available disazopyrazolone pigments.

Tab. 15: Commercially available disazopyrazolone pigments.

C.I. Name	C.I. Constitution No.	X	R^1	R^2	Shade
P.O.13	21110	Cl	CH_3	H	yellowish orange
P.O.34	21115	Cl	CH_3	CH_3	yellowish orange
P.R.37	21205	OCH_3	CH_3	CH_3	yellowish red
P.R.38	21120	Cl	$COOC_2H_5$	H	red
P.R.41	21200	OCH_3	H	H	red
P.R.111	–	–	–	–	red

Individual Pigments

Pigment Orange 13

P.O.13, sold in the USA as pyrazolone orange, comes in semitransparent types with specific surface areas between about 35 and 50 m²/g. It is coloristically very similar to P.O.34, but generally somewhat yellower. Occasionally weaker than P.O.34, it is also slightly less fast in many media. Considering its tendency to migrate, incorporation into plasticized PVC is not recommended. The pigment blooms over a large concentration range and bleeds considerably. At concentrations below ca. 0.1%, neither P.O.34 nor 13 are suited for use in rigid PVC.

Application of P.O.13 in polyolefins is limited. It is recommended for use at temperatures up to 200 °C (Sec. 2.4.1.3). This is equally true for polystyrene and other plastic materials which are processed above 200 °C, such as polymethacrylate, in which P.O.13 is used. P.O.13 is one of the pigments which do not affect the extrusion shrinkage of HDPE, but it is nevertheless rarely employed to color such materials. In LDPE, there is some danger of blooming. P.O.13 shows less stability in paints than P.O.34 types of similar particle size. This includes both fastness to overpainting in baking enamels and lightfastness in air drying paints. The volume of trade sales for this purpose is accordingly limited.

The graphics industry, on the other hand, uses P.O.13 to an appreciable extent for packaging printing inks. Its fastness to light is average and corresponds to that of the diarylide yellow pigment P.Y.12, with which it is frequently combined as a shading component. The stability of pigmented prints to a number of organic solvents is excellent or almost perfect. Similarly, the prints are fast to paraffin, butter, and soap. They withstand heat very well and are stable up to 200 °C. P.O.13 thus lends itself to metal deco printing, provided its lightfastness suits the purpose. Likewise, its resistance to clear lacquer coatings and to sterilization are excellent.

Pigment Orange 34

P.O.34 is supplied in a variety of types, which differ considerably in their particle size distributions. Specific surface areas range from 15 m^2/g in highly opaque versions to about 75 m^2/g in transparent types. It is these physical characteristics that determine the coloristic and fastness properties of each type. Even varieties of P.O.34 with fine particle sizes are generally not resinated.

Transparent P.O.34 versions represent the most frequent choice for printing inks. They provide a clean, yellowish orange hue and high tinctorial strength. Standardized 1/1 SD letterpress proof prints require inks formulated at 7.6% pigment. The same is true for corresponding yellow shades produced by the tinctorially strong diarylide yellow pigment P.Y.13. P.O.34 is somewhat redder than the similarly strong P.O.13. At equal depth of shade, prints obtained from P.O.34 tolerate light better than do prints containing P.O.13; the difference is about 1 step on the Blue Scale. The lightfastness of 1/1 and 1/3 SD prints equals step 4 on the Blue Scale, which makes the pigment almost as fast as P.Y.13.

P.O.34 shows good solvent resistance to a number of organic solvents. Its prints are more stable in this respect than those made from P.O.13, which is also true for the standard DIN 16 524 solvent mixture (Sec. 1.6.2.1). In spite of these comparatively good fastness properties, P.O.34 may recrystallize in various printing inks, depending on the processing conditions. P.O.34 prints are fast to paraffin and dioctyl phthalate; likewise, they tolerate clear lacquer coatings and may be sterilized.

Transparent P.O.34 is somewhat sensitive to heat and generally only withstands temperatures up to 100 to 140 °C. Higher sterilization or metal deco printing temperatures may produce a color shift towards a redder orange.

P.O.34 is used for all printing techniques. Packaging printing inks, especially nitrocellulose inks, often use the orange version of the cheaper and more lightfast P.O.5 in areas where fastness to organic solvents is unimportant. The shade of the product may have to be modified accordingly. P.O.34, as do diarylide yellow pigments, exhibits insufficient solvent fastness to be used in decorative printing inks; it is particularly not fast enough to monostyrene and acetone (Sec. 1.8.1.2). The pigment also performs poorly as far as lightfastness is concerned, and it bleeds into solutions of melamine resins, which precludes its use in such fields.

The textiles printing industry, however, uses P.O.34 to an appreciable extent. The pigment provides average lightfastness (at 1/3 SD, it equals step 5–6 on the Blue Scale); its dry cleaning solvent resistance is excellent, and it withstands exposure to dry heat up to 200 °C. P.O.34 performs similarly in connection with spin dyeing of polyacrylonitrile.

Throughout the plastics industry, P.O.34 is used to color plasticized PVC, although a certain tendency to bloom precludes its application at levels below about 0.1%. At higher concentrations, the pigment tends to bleed in plasticized PVC, but it is considerably more stable to light than the weaker P.O.13. At equal depth of shade, 1/3 SD samples equal step 6 on the Blue Scale, while the corresponding P.O.13 sample reaches only step 4. Transparent P.O.34 colorations in rigid PVC are even more resistant to light. Pigment concentrations of less than ca. 0.1% are likewise unsuitable, due to the necessity to avoid blooming. The pigment is also used in vinyl floor coverings and in cable insulations.

P.O.34 is rarely used in polyolefins. In such media, it only withstands exposure to 200 °C, and its opaque colorations show insufficient lightfastness. P.O.34 tends to bloom, especially in extrusion products made of low molecular weight LDPE types. The pigment is, however, recommended for a variety of other media. These range from aromatic polyurethane foams to cast resins of unsaturated polyester, in which the pigment slightly delays the hardening process.

In the paint industry, transparent types are used only to a limited extent. In air drying systems, P.O.34 equals step 6–7 on the Blue Scale in full shades, while opaque colorations with TiO_2 (1:5) only reach step 3 for lightfastness. In baking enamels, the pigment is not fast to overpainting.

Highly opaque versions of P.O.34 with coarse particle sizes and specific surface areas between ca. 15 and 25 m^2/g, however, are gaining recognition within the paint field. Excellent flow properties make it possible to further increase the pigment level and the opacity, in which P.O.34 affords exceptionally good results for an organic pigment. Even at equal pigment concentration, P.O.34 is more opaque than commercially available Molybdate Red pigments which cover the same range of hues. Full shades of this very opaque type are very fast to light and weather, and they have a much better ability to tolerate solvents and migration. This makes P.O.34 attractive as a partial or complete replacement for Molybdate Red in industrial finishes, tractor and agricultural implement finishes, house paints, etc. Its temperature stability is similarly superior. This is also true for application in print, where P.O.34 exhibits an increase in thermal stability from less than 120 °C in transparent prints to 200 °C in specialized opaque varieties. Opaque types are also much more reddish, which changes the DIN color (Sec. 1.6.1.4) of 1/3 SD prints from 4.06 to 5.93. Combinations with other colored pigments, such as iron oxides or titanium mixed oxides, such as nickel titanium yellow and colorations that are reduced with TiO_2 are much less fast.

Pigment Red 37

P.R.37 is a yellowish red pigment which performs poorly in many applications and is largely restricted to the coloration of rubber and plastics.

Lightfastness in rubber, which meets almost all requirements, is accompanied by good curing properties and migration resistance. The resulting pigmented articles are very resistant to water and detergent solutions.

P.R.37 is a very strong pigment in PVC, but deficient in its fastness to light.

1/3 SD colored samples equal step 2–3 on the Blue Scale, while 1/25 SD specimens only reach step 1. In its fastness to light, the pigment thus performs much poorer by several steps on the Blue Scale than diarylide yellow pigments, such as P.Y.13 or P.Y.17. Its fastness to bleeding in plasticized PVC (at 1/3 SD), however, parallels that of these yellows. Pigment preparations of P.R.37 are also available. Due to its good dielectrical characteristics, P.R.37 is frequently selected for use in cable insulations. However, application recommendations exclude polyolefins.

Pigment Red 38

P.R.38 affords a medium red shade. It is used primarily in rubber and plastics. Much faster to a number of organic solvents than P.R.37, rubber colorations are very fast to light and are used under almost any application conditions. P.R.38 is completely resistant to curing and bleeding into natural rubber and fabric backing (Sec. 1.8.3.6). The colored articles are resistant to water, soap, and detergent solutions, and equally fast to a variety of organic solvents, including gasoline. P.R.38 shows high tinctorial strength in PVC, but tends to bloom at low pigment levels. In contrast to P.R.37, low P.R.38 levels in rigid PVC are equally prone to bloom. However, P.R.38 is much more lightfast than P.R.37: in 1/3 SD, it reaches step 6 on the Blue Scale, in which it parallels the properties of P.Y.13. Combinations of the two are frequently used to produce intermediate shades. In PVC, the pigment withstands temperatures up to 180 °C; its good dielectrial properties make it a useful colorant for PVC cable insulations.

In polyethylene, the pigment tolerates temperatures up to 200 °C; such systems are used to make films. Higher processing temperatures carry the risk of inducing blooming and decomposition by heat in LDPE types. Depending on the depth of shade, its lightfastness in HDPE equals between step 3 and step 6 on the Blue Scale. In this medium, it does not affect the shrinkage of the plastic. Especially throughout the USA, P.R.38 is primarily used for paper coatings, mass coloration of paper, artists' colors, crayons, and similar specialized media, and also in specialized printing inks.

Pigment Red 41

P.R.41 is also known as pyrazolone red. It has lost most of its commercial importance in recent years. P.R.41 production is now limited to the USA, where it is mostly employed to lend color to rubber. To a lesser extent, P.R.41 is found in PVC; excellent dielectrical properties make it a suitable candidate for PVC cable insulations. The pigment provides a medium to bluish red; of limited brilliance, it is much bluer than P.R.38. P.R.41 is somewhat less fast than P.R.38, which is also true for its stability to a variety of organic solvents. However, it parallels P.R.38 in its alkali and acid resistance. P.R.41 is very lightfast in rubber; 1% pigment concentrations equal step 6–7 on the Blue Scale, which meets practically any requirement. The pigment also withstands migration, i.e., it does not bleed into natural rubber or fabric backing (Sec. 1.8.3.6). P.R.41 is also completely resistant to curing conditions. Colored rubber articles tolerate boiling water, acid, and soap.

Pigment Red 111

P.R.111 is chemically related to the pyrazolone pigments P.O.34 and P.R.37. Its hue is somewhat bluer than that of its pyrazolone counterparts; it equals that of Signal Red (RAL 3000). The lightfastness of P.R.111 is somewhere between that of P.O.34 and of P.R.37. P.R.111 performs similarly as far as other fastness properties are concerned. The pigment lends itself particularly to the coloration of rubber and PVC; excellent dielectrical properties also render it suitable for cable insulations. Thermally, P.R.111 is not sufficiently stable to be used in polyolefins, styrene, ABS, and similar plastics.

References for Section 2.4

[1] R.B. McKay, Farbe + Lack (1990) 336–339.
[2] W. Herbst and K. Hunger, Progr. in Org. Coatings 6 (1978) 106–270.
[3] T. Kozo et al., Shikizai Kyokai-shi 36 (1963) 16–21.
[4] W. Herbst and K. Merkle, DEFAZET Dtsch. Farben Z. 30 (1976) 486–490.
[5] R. Az, B. Dewald, D. Schnaitmann, Dyes and Pigments 15 (1991) 1–15.
[6] ref. e.g. DE-OS 2 243 955 (Ciba-Geigy) 1972; DE-OS 2 410 240 (Ciba-Geigy) 1973.
[7] DE-AS 2 058 849 (Sandoz) 1969; B. Kaul, Applied marketing information Ltd., Basel, Febr. 1990.

2.5 β-Naphthol Pigments

The same chemistry is involved as with monoazo yellow pigments, except that β-naphthol pigments are obtained by coupling with 2-hydroxynaphthalene (β-naphthol) instead of acetoacetarylides. They have the general chemical structure:

R^2, R^4: H, Cl, NO$_2$, CH$_3$, OCH$_3$, OC$_2$H$_5$

β-Naphthol colorants are among the oldest synthetic dyes known. Likewise, β-naphthol pigments, first manufactured in 1889, were the earliest products of their kind in the pigment industry. Th. and R. Holliday (at Read, Holliday & Sons in the UK) applied for the patent in 1880. The compounds became known as developing dyes in connection with so-called "ice dyeing". They are applied by padding cotton goods through an alkaline β-naphthol solution and drying them evenly. The goods are then processed through tubs containing the cold diazonium compound and acetate buffer, where the color "develops" through coupling. Insolubility of the colorant in water makes for very washfast products. This is the pathway found by Gallois and Ullrich in 1885, who obtained Para Red (Pigment Red 1) by coupling β-naphthol with diazotized 4-nitroaniline. The resulting "pigment" was one of the first successful compounds used in textile dyeing and printing—it is even considered the oldest of all known synthetic organic pigments.

β-Naphthol had been discovered by Schaeffer as early as 1869. It was used initially as a starting material for colorants ("Orange II", Echtrot AV).

o-Nitroaniline Orange followed in 1895 (o-nitroaniline → β-naphthol), and the beginning of the twentieth century saw a steady increase in the number of β-naphthol pigments, including improved products such as Toluidine Red (P.R.3), launched in 1905, chlorinated Para Red (P.R.4), which appeared in 1906, and Dinitroaniline Orange (P.O.5), introduced in 1907. Most of these are still used in a host of applications.

The history of β-naphthol pigments reflects the development of organic pigments in general. First used as developing dyes, the colorant was changed into a pigment by adding an inorganic carrier. As knowledge and expertise progressed and it became clear that such carriers have no effect on the fastness properties of the pigments, these compounds developed into independent products ("para toners"). Likewise, β-naphthol pigment lakes followed the same course of events. To date, two β-naphthol pigments (Toluidine Red and Dinitroaniline Orange) and one β-naphthol pigment lake (Lake Red C) still reign supreme among organic pigments worldwide.

2.5.1   Chemistry, Manufacture

The general formula for β-naphthol pigments is:

In commercial products, R_D primarily stands for CH_3, Cl, NO_2; and m = 1 to 2.

The pigments are obtained by treating the appropriate amine with water/hydrochloric acid to form the amine hydrochloride. In most cases, subsequent diazotization is carried out in the cold (0 to 5 °C) with an aqueous sodium nitrite solution (a). The resulting diazonium salt solution is then transferred onto sodium naphtholate which is dissolved in a sodium hydroxide solution. An acetic acid/sodium acetate buffer maintains a slightly acidic pH throughout the coupling reaction (b). When the coupling process is completed, no diazonium salt may remain in the reaction solution.

Nitrosylsulfuric acid, prepared by dissolving sodium nitrite in concentrated sulfuric acid, is employed for amines of low basicity, whose diazonium salts will hydrolyze in dilute acid. In order to synthesize Pigment Orange 5, for instance, 2,4-dinitroaniline is dissolved in concentrated sulfuric acid and diazotized preferably with nitrosylsulfuric acid. Coupling is carried out with a β-naphthol suspension, produced by acidifying a sodium naphtholate solution.

Comparatively low-cost starting materials and easy synthesis make it possible to manufacture these pigments economically.

The β-naphthol pigment Pigment Red 1 was the first red azo pigment to be submitted to three dimensional X-ray diffraction analysis [1]. Today, three different crystal modifications are known [2].

Later X-ray diffraction studies elucidated the structures of Pigment Red 6 [3] and Pigment Red 3 [4].

	R^2	R^4
P.R.1	H	NO_2
P.R.3	NO_2	CH_3
P.R.6	NO_2	Cl

Members of this family share the following features:
- trans configuration of the substituents at the nitrogen–nitrogen bond,
- prevalence of the o-quinonehydrazone form **20** over the hydroxyazo structure **21**,

- almost planar arrangement of the pigment molecules, especially due to intramolecular hydrogen bonds, partially forked, so that they connect to two oxygen or nitrogen atoms simultaneously,
- absence of intermolecular hydrogen bonds, prevalence of Van der Waals bonds (according to the distances between the atoms of two different pigment molecules within the crystal lattice).

Based on the results of a "direct insight" into the crystal structure of these compounds, it can be assumed that at least all other β-naphthol pigments, including their derivatives, such as Naphthol AS pigments (Sec. 2.6), share these characteristics and adopt the quinonehydrazone structure. It is likely that the name "azo" fails to reflect the actual molecular structure. Although we are fully aware of the underlying misconception, we will nevertheless attempt to avoid confusion and thus adhere to the "traditional" distinction and use the corresponding structure. This is even in spite of the fact that there are similar X-ray analysis results that indicate that the same is true for monoazo yellow pigments (see Sec. 2.3.1.1).

2.5.2 Properties

β-Naphthol pigments come in shades from yellowish orange to bluish red. They are tinctorially weak and in some applications considerably less strong even than monoazo yellow pigments.

Their solvent stability is poor, it approaches that of the above-mentioned yellows (Sec. 2.3.3). Migration characteristics, i.e., fastness to bleeding and blooming, are consequently inferior, although the pigments are generally resistant to alkali and acid.

Most β-naphthol pigments are supplied as opaque products with low specific surface areas, designed to provide good hiding. Large particles, on the other hand, are frequently responsible for loss of gloss in paints and prints. Systems which dry under volume contraction, particularly air drying alkyd resin systems, are sensitive to hazing, a special kind of surface disturbance which also occurs in monoazo yellow pigments with coarse particle sizes (Sec. 1.7.6).

Specialty products with fine particle sizes, high specific surface areas, and enhanced transparency are characterized not only by increased tinctorial strength, but also by inferior stability to solvents and light.

Although standard types resist light well, especially in full shades, they are not as fast as most monoazo yellow pigments. Lightfastness is lost rapidly with decreasing depth of shade and decreasing pigment area concentration, respectively. The commercially available types disperse easily in most media.

2.5.3 Application

The technical applicability of β-naphthol pigments is severely limited by poor fastness to organic solvents and migration.

The more significant members of this family have not lost their impact in the field of paints, which is the main market for β-naphthol pigments.

Pigment Orange 5 and Pigment Red 3 are standard types covering the orange and red range in air drying systems. Pigments used for this purpose are generally characterized by very small specific surface areas between ca. 7 and 20 m^2/g, which makes for products with extremely coarse particle sizes. There is therefore little danger of recrystallization during processing and application. The pigments are used mainly in masstone and similar deep shades. A number of β-naphthol pigments are also incorporated into emulsion paints. In baking enamels, the pigments may bloom considerably over a large concentration range, which practically precludes their use in such systems. In special cases and in concentrations in which there is no danger of blooming, the pigments may be applied in low-temperature curing systems. Application in plastics, however, is severely limited by poor fastness properties, including a tendency to migrate and little fastness to heat. A few special types, which may be incorporated into rigid PVC, are the only exception.

Although β-naphthol pigments, such as P.R.3, have suffered considerable loss of impact in the field of printing inks, this remains an important market. The pigments are preferably used for packaging purposes. They also lend themselves very well to flexo and offset printing. Such prints, like the pigment powders themselves, are fast to water, acids, and bases, but they perform poorly regarding a number of special fastness properties: they are not resistant to butter and paraffin (Sec. 1.6.2.2 and 1.6.2.3). Other deficiencies are poor fastness to clear lacquer, overpainting, and sterilization. β-Naphthol pigment prints do not tolerate heat: they do not withstand exposure to 140 °C. This is another one of the aspects in which this family approximates monoazo yellow pigments.

The grades with coarse particle sizes provide easy dispersion and good rheology. Types with finer particle sizes are tinctorially stronger and somewhat more transparent, and they show less of a tendency to bronze than the products with coarse particle sizes.

β-Naphthol pigments are broad in scope in many fields. Special applications include cleaners and detergents, office articles, and artists' colors, as well as matchhead compositions and fertilizers.

2.5.4   Commercially Available β-Naphthol Pigments

General

Only very few β-naphthol pigments continue to play an important role in today's pigment industry. The list of important products includes Toluidine Red (P.R.3) and Dinitroaniline Orange (P.O.5). Other compounds, such as P.R.6, Parachlor Red, which is the positional isomer of P.R.4; P.O.2, Orthonitroaniline Orange, which is the positional isomer of the para toner P.R.1 are only of regional importance. Table 16 lists the commercially available β-naphthol pigments. The Colour Index numbers are listed along with the common names, since older products are frequently referred to by these names.

Tab. 16: Commercially available β-naphtol pigments.

C.I. Name	C.I. Constitution No.	R^2	R^4	Common name
P.O.2	12060	NO$_2$	H	Orthonitraniline Orange
P.O.5	12075	NO$_2$	NO$_2$	Dinitraniline Orange
P.R.1	12070	H	NO$_2$	Parachlor Red/Para Toner
P.R.3	12120	NO$_2$	CH$_3$	Toluidine Red
P.R.4	12085	Cl	NO$_2$	chlorinated Parachlor Red
P.R.6	12090	NO$_2$	Cl	Parachlor Red

It is impossible to clearly assign a distinct shade to each of these pigments because each compound produces a very different range of red, depending on the manufacture, the additives, and the particle size distribution. Variation of choice and rate of reactant addition during diazotization and coupling, pH, and concentration, choice and amount of additives for the coupling reaction are the factors determining the exact shade of each pigment. Commercially available β-naphthol pigments possess a notable feature: they all carry at least one nitro group, either in ortho or in para position relative to the azo bridge. The two most important species are obtained by coupling with o-nitroaniline derivatives as diazo components.

Individual Pigments

Pigment Orange 2

This pigment has only limited impact outside Europe; it is used especially in the USA. Compared to its positional isomer P.R.1, it performs slightly better as far as fastness to various solvents, such as aliphatic hydrocarbons, is concerned. Commercial P.O.2 types afford clean orange shades with good hiding power. Resinated and transparent types are not known. The pigment is primarily used to color aqueous flexo inks, paper, air drying coatings, and artists' colors.

Pigment Orange 5

P.O.5 is one of the most significant organic pigments. Two product lines with different particle sizes are available which differ considerably in their coloristic properties. The varieties with coarser particle sizes and specific surface areas between ca. 10 and 12 m^2/g are much more reddish and duller than the types with somewhat finer particle sizes and specific surface areas between 15 and 25 m^2/g.

P.O.5 is principally applied in air drying systems. The type with a coarser particle size provides good hiding power. In full shades, there is some darkening upon exposure to light, although its lightfastness equals step 6 on the Blue Scale. Similarly, the pigment exhibits very good weatherfastness. Its lightfastness, however, deteriorates rapidly in white reductions, although this trend is much less pronounced than with other pigments within the same family. 1:5 reductions with TiO_2 equal step 5 on the Blue Scale for lightfastness; at 1:40 TiO_2 reduction, the pigment only reaches step 4. Consequently, P.O.5 is used only in full or similarly deep shades, partially in combination with inorganic pigments such as Molybdate Orange.

The yellower, cleaner types with finer particle sizes are almost as lightfast as opaque varieties. Good lightfastness and weatherfastness make for good coloration of emulsion paints; only the deeper shades are used for exterior paints. P.O.5 is not entirely fast to alkali and lime. It shows a strong tendency to bloom in baking paints. The concentration limits (Sec. 1.6.3.1) are 0.5% at 120°C, and 1% at 140°C. At higher baking temperatures, P.O.5 always blooms, irrespective of the concentration. It is not suited to use in epoxy resin coatings, the color of which changes to a brownish shade.

P.O.5 is used to an appreciable extent in printing inks, especially for offset, flexo, and packaging gravure printing inks. Transparent types with fine particle sizes are particularly important for these purposes. They are considerably stronger, some by more than 30%, and afford much yellower, cleaner prints which are glossier. These prints are tinctorially only about half as strong as similarly colored P.O.34, but much more lightfast.

Lightfastness at common pigment area concentrations, i.e., at 1/1 to 1/25 SD, equals approximately step 6 on the Blue Scale. P.O.5 is a typical member of its class in that the prints are more or less sensitive to organic solvents and to the standard DIN 16 524 solvent blend (Sec. 1.6.2.1). While the pigment lacks butter and paraffin fastness, it is completely resistant to soap. P.O.5 is not used for metal printing or for decorative printing.

P.O.5 is in high demand in the field of the textile printing. As far as most of the relevant fastness properties are concerned, P.O.5 performs less well than the somewhat yellower and much more expensive perinone pigment P.O.43. Compared to the somewhat yellower P.O.34, however, its fastness to light is superior. 1/3 SD P.O.5 samples equal step 7 on the Blue Scale, as opposed to step 5-6 reached by P.O.34. In other respects, such as fastness to dry-cleaning with perchloroethylene or petrolether, to laundering with peroxide bleach or alkali, P.O.5 performs less well than P.O.34.

In the plastics industry, P.O.5 is limited to application in rigid PVC. Transparent colorations (0.1% pigment) afford a lightfastness that equals step 8 on the Blue Scale, although the shade darkens somewhat upon exposure. 1/1 to 1/25 SD opaque samples equal step 6 on this scale. 0.5% pigmented films (Sec. 1.6.2.3) pass the coconut test. P.O.5 is also suitable for a variety of other applications, including office articles and artists' colors. In the latter, it is used particularly where good lightfastness is required: in pigmented drawing inks, in colored pencils, wax and marking crayons, and water colors. Likewise, P.O.5 is used in the mass and surface coloration of paper.

Pigment Red 1

Known also as para red or para toner, P.R.1 has proven to be incompatible with ever increasing industrial demands and has lost most of its significance. It provides a very dull, somewhat brownish red hue. P.R.1 is less stable to organic solvents than other members within the same family. Moreover, it is also less fast to light. In air drying paints, full shades of P.R.1 equal step 5 on the Blue Scale; however, there is some darkening. Even trace amounts of TiO_2, reduce the lightfastness considerably. Poor solvent resistance is accompanied by a tendency to bloom and bleed in baking enamels.

P.R.1 is used in printing inks for low-cost articles. It used to be important for newsprint purposes, although the prints are affected by a large number of media. In contrast to other pigments of its class, P.R.1 is sensitive to acids, bases, soaps, and even water.

Pigment Red 3

Toluidine Red, like P.O.5, is by volume one of the 20 largest organic pigments in the world. It shows insufficient fastness towards solvents; in fact, it is partially inferior even to monoazo yellow pigments, which is also true for other members of this class. Its stability to alcohols, aliphatic and aromatic hydrocarbons, and dibutyl phthalate equals step 3 on the 5 step scale; P.R.3 is even less fast to esters and ketones.

P.R.3 is used primarily in air drying paints. Its hue varies considerably with the particle size, therefore pigment manufacturers usually offer a range of brands. The products with coarser particle sizes are bluer. Very bluish varieties, however, partially contain chemically modified P.R.3.

As paints films with coarse particle sizes are dried, the volume of the air drying system contracts with increasing drying time and the glossy surface turns hazy. A phenomenon known as "Toluidine Red haze" appears (Sec. 1.7.6). Coarse pigment particles, at the surface with adsorbed binder components, protrude more and more from the surface of the coating, scatter light, and reduce the gloss. Varieties with finer particle sizes, which are also more yellowish, show less of a tendency to turn hazy. Intensive dispersion, however, improves the gloss, even of bluish types, resulting in more yellowish materials. This implies that the gloss can only be improved by reducing the size of the coarser particles, which is confirmed by electron microscopic studies of ultrathin layers.

Full shades of Toluidine Red are extremely lightfast and weatherfast, but deteriorate rapidly as the pigment is reduced with white pigment. In full shades, the lightfastness equals step 7 on the Blue Scale, while 1:4 TiO_2 reductions only reach step 4. The pigment is therefore used preferably in full or similarly deep shades. Recommendations include emulsion paints for interior application or short-term advertisement and marking purposes.

P.R.3 is very likely to bloom in baking enamels. At 120°C, the concentration limit for blooming is 2.5%; beyond about 140°C, blooming occurs with certainty, irrespective of the concentration. Therefore, only full shades of P.R.3 are used to color baking enamels, and only where application temperatures are low. The pigment is used extensively in combination with Molybdate Red.

P.R.3 application in printing inks is restricted; the pigments are being displaced by their stronger Naphthol AS pigment counterparts. P.R.3 is used primarily in flexo printing. The problem with offset application is that it frequently provides poor gloss. The prints are very sensitive to organic solvents, including the standard DIN 16 524 blend; but they are soap, alkali, and acid fast.

In plastics, Toluidine Red is practically limited to rigid PVC. Its lightfastness in full shades and slight white reduction is fair. Besides, the pigment is also used to color a number of specific media, such as normal wax crayons and pastel chalks or low-cost watercolors.

Pigment Red 4

P.R.4 is also known as chlorinated Para Red. It has lost much of its commercial impact in recent years. The pigment affords a yellowish red shade, somewhere between the more yellowish P.O.5 and the more bluish P.R.3. Tinctorially, P.R.4 is the weakest of the three. In order to formulate equal depth of shade in an air drying alkyd system, three equivalents of TiO_2 are necessary per equivalent of P.R.4; while P.O.5 and P.R.3 require 5 and 6 parts of TiO_2, respectively, per equivalent of organic pigment; provided all other parameters are equal. Full shades exhibit good lightfastness (step 6), but darken somewhat upon exposure. Addition of even small amounts of TiO_2 reduces the lightfastness considerably. Likewise, only full and similarly deep shades tolerate weather.

In the print field the pigment is almost exclusively used in air drying systems. P.Y.4 is very likely to bloom in stoving enamels. At only 120 °C, the concentration limit for blooming is as low as 2.5%, and at 140 °C the limit is at 5%, which makes it necessary to carry out test experiments. Application is consequently restricted to baking enamels targeted for low temperature purposes. In epoxy resins, the pigment turns brown, as does P.O.5, and is therefore unsuitable for use in these media.

In the printing inks field, P.R.4 is employed as a clean, yellowish red colorant for purposes where no solvent fastness is required. The prints do not tolerate most organic solvents, including the standard DIN 16 524 solvent blend. They are even more sensitive to some solvents than prints made with P.O.5 or P.R.3. A series of other fastness properties, such as fastness to paraffin, butter, and grease, are poor. Nevertheless, P.R.4 is of regional importance for both offset printing and packaging gravure and flexo printing. Its prints are strong compared to other members of this class that cover the same range of shades. Under standard conditions, 1/1 SD P.R.4 letterpress proof prints are prepared with printing inks containing 13% pigment; while P.R.3 samples require between between 16 and 20%; or, alternatively, 16% P.O.5.

At typical pigment area concentrations, i.e., standard depths of shade in the range between 1/1 and 1/25 SD, the lightfastness equals step 4 or 3 on the Blue Scale, respectively. Application in metal deco printing is not feasible, which is also true for other pigments within the same class.

Application in rubber blends, which used to be an important market for P.R.4, has suffered from increasingly stringent requirements. P.R.4 bleeds considerably during curing into overlaid white rubber sheets, while it hardly bleeds into the fabric backing (Sec. 1.8.3.6).

P.R.4 is also used in specialized applications. The pigment lends color to cleaning agents and detergents, including shoe polish and floor polish, office articles and artists' colors, including colored pencils, wax crayons for schools and artists, and pastel chalks, and to low price watercolors. It is also used in decorative cosmetics, for which types are available which meet the legal requirements.

Pigment Red 6

This pigment, also known as Parachlor Red, has lost most of its commercial impact. It provides a yellowish red hue, somewhat on the yellow side of Pigment Red 3 and bluer than Pigment Red 4. P.R.6 parallels these two pigments in its fastness properties; the only exception is its stability to light. In air drying alkyd systems, full and similarly deep shades of P.R.6 are as lightfast as Toluidine Red, while in white reductions the pigment tolerates light much better. Compared to P.R.3, 1/3 SD P.R.6 samples equal step 6–7 as opposed to step 4 on the Blue Scale, while 1/25 SD samples reach step 5–6 as opposed to step 1.

References for Section 2.5

[1] C.T. Grainger and J.F. McConnell, Acta Crystallogr. Sect. B (1969) 1962–1970.
[2] A. Whitaker, Z. Kristallogr. 152 (1980) 227–238 and 156 (1981) 125–136.
[3] A. Whitaker, Z. Kristallogr. 145 (1977) 271–288.
[4] A. Whitaker, Z. Kristallogr. 147 (1978) 99–112.

2.6 Naphthol AS Pigments

For Naphthol AS pigments (Naphthol Red pigments), the same chemistry is involved as with monoazo yellow pigments, except that coupling is carried out with arylides of 2-hydroxy-3-naphthoic acid. The reference compound, Naphthol AS* is the anilide of 2-hydroxy-3-naphthoic acid:

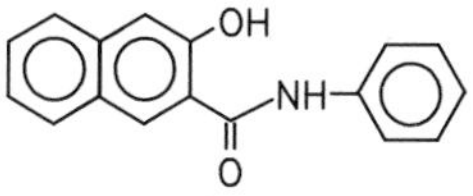

Substitution in the anilide ring affords a series of Naphthol AS derivatives, although only a limited number are commercially recognized.

In 1892, the chemist Schöpf, in an attempt to prepare 2-phenylamino-3-naphthoic acid, developed a synthetic route leading to the anilide of 2-hydroxy-3-naphthoic acid. His method continues to be used today, if only in a slightly modified form. He added phosphorus trichloride to a molten reaction mixture containing aniline and 2-hydroxy-3-naphthoic acid (*beta-oxynaphthoic acid*, also known as BONA) and received Naphthol AS in good yield. Modern processes differ from this principle only in terms of reaction control; the synthesis is now carried out in the presence of organic solvents, such as aromatic hydrocarbons.

Naphthol AS was not rediscovered until 1909, when BASF in Germany claimed a patent for a diazotization dye which could be developed by diazotizing primuline** on the fiber and then coupling withNaphthol AS in an alkaline solution.

---

* Naphthol AS is the condensation product of β-hydroxynaphthoic acid and aniline; Naphtol AS is the trade name of this product sold by Hoechst AG, Germany. Pigments that are known as Naphtol AS compounds in Germany are referred to in the USA as Naphthol Reds.
** Primuline is obtained by melting p-toluidine with sulfur and subsequently sulfonating the resulting product. It has the following chemical structure:

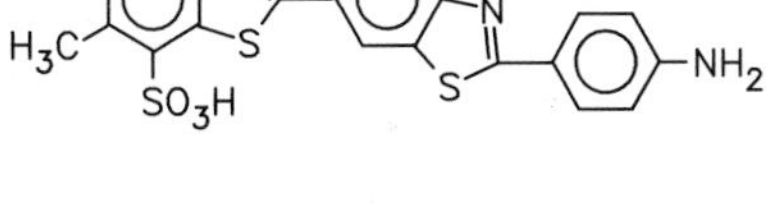

In 1911, A. Winter, H. Laska, and A. Zitscher at Griesheim-Elektron, now the Offenbach site of Hoechst AG in Germany, made a discovery that was to prove an important breakthrough in the Naphthol AS field. They synthesized azo colorants from diazotized anilines or toluidines (with Cl or NO_2 substituents) and Naphthol AS as the coupling component. The resulting pigments ("Grela Reds"), although superior to β-naphthol pigments in terms of lightfastness and solvent stability, were initially disregarded, since the pigment industry at the time focussed on the more economical and immensely commercially successful β-naphthol pigments (Toluidine Reds).

In 1912, Griesheim-Elektron first replaced β-naphthol pigments by Naphthol AS, to be used for ice dyeing. Naphthol AS has a much higher substantivity and can therefore be fixed much more evenly than β-naphthol. Intermediate drying on the fiber prior to coupling can thus be eliminated. Moreover, alkaline naphthol solutions are much more stable in air than the corresponding β-naphthol solutions. This discovery initiated the rapid development of Naphthol AS dyeing into a well-established technology within a few years, followed by continuous introduction of new Naphthol AS pigments. These events stimulated the discovery of a vast number of substituted anilides of 2-hydroxy-3-naphthoic acid.

Naphthol AS derivatives that have retained commercial importance are listed in Table 10, together with their name, constitution, and Colour Index number (Sec. 2.1.2, p. 195).

During the 1920s and 1930s, the development of Naphthol AS technology in Germany was initiated by IG Farben. New Naphthol AS pigments were synthesized, based on combinations previously used in Naphthol AS dyeing. In the USA, the development of Naphthol Red pigments commenced in the 1940s. The pigments were developed further by varying the substitution pattern of the diazo component, attempting to improve solvent and migration fastness. Sulfonamide and, to an even higher extent, carbonamide groups as substituents were most successful.

Technically significant Naphthol AS pigments are therefore divided into two groups, according to whether their diazo component carries

    I. simple substituents, such as Cl, NO_2, CH_3, or OCH_3,
    II. sulfonamide groups and/or carbonamide groups.

In the latter case, more than one carbonamide group may be introduced through the coupling component (Sec. 2.6.2).

Today, 80 years after their discovery, Naphthol AS pigments continue to play a major role among organic pigments. Although comparatively many derivatives are known, only few are produced in large volume.

2.6.1 Chemistry, Manufacture

Naphthol AS pigments have the following general chemical structure:

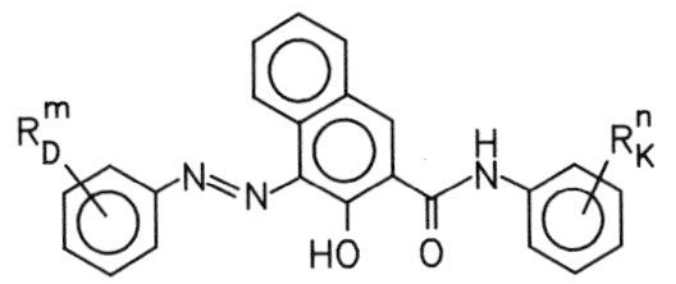

Commercial derivatives are primarily substituted as follows:
R_D = R_K, COOCH$_3$, CONHC$_6$H$_5$, SO$_2$N(C$_2$H$_5$)$_2$,
R_K = CH$_3$, OCH$_3$, OC$_2$H$_5$, Cl, NO$_2$, NHCOCH$_3$;
m and n are numbers between 0 and 3.

The reaction sequence is the usual one. The hydrochloride of the aromatic amine is diazotized with sodium nitrite/hydrochloric acid and subsequently coupled onto a Naphthol AS derivative.

In the earlier stages of development of these pigments, the coupling components often presented solubility problems. Although sodium naptholates dissolve most easily in alcohol/water mixtures, organic solvents increase the price of pigment manufacture and also present ecological problems. The coupling component is therefore heated to 60 to 90 °C in the presence of a 7 to 10% aqueous sodium or potassium hydroxide solution and thus converted to the soluble dialkali salt of the amine enolate:

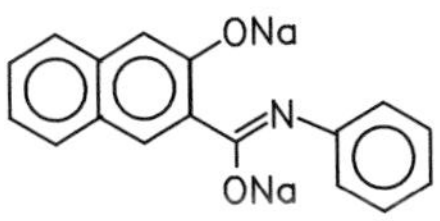

Acetic acid or hydrochloric acid, possibly together with a tenside, reprecipitates the compound. This method affords the Naphthol AS derivative as a material with a very fine particle size, ready for coupling. A slightly acidic pH can be optimized and maintained by adding a sodium acetate buffer.

Coupling is usually carried out at 10 to 25 °C, although it sometimes requires as much as 40 to 70 °C – provided the diazonium salt tolerates such temperatures.

For group I Naphthol AS pigments, further treatment is not necessary. The pigment suspension may be heated to 60 to 80 °C for a short time before it is filtered.

Manufacture of group II pigments, on the other hand, is typically followed by intensive thermal aftertreatment in water or water/organic solvents to produce a pigment which is easy to disperse. The first attempts to elucidate the three dimensional structure of Naphthol AS Pigments by X-ray diffraction analysis were carried out by Hoechst [1]. Studies by Whitaker [2] followed. Analyses of single crystals confirmed the structural features which had been found earlier in β-Naphthol pigments (Sec. 2.5.1):

- planar structure of the molecule within the unit cell,
- prevalence of the quinonehydrazone form over the hydroxyazo form,
- presence of all possible intramolecular hydrogen bridges, but absence of intermolecular ones.

This is to prove once more that these pigments adopt the hydrazo form and should be considered "hydrazone" rather than "azo pigments".

The chlorinated derivatives of Pigment Red 9 (R = H) and Pigment Brown 1 (R = OCH$_3$), respectively,

H: chloro derivative of Pigment Red 9
R: OCH$_3$: chloro derivative of Pigment Brown 1

were used to study the relation between the three dimensional structure and the hue (Sec. 1.4.1).

2.6.2   Properties

Commercially available Naphthol AS pigments afford a variety of shades ranging from yellowish to very bluish red; including all possible intermediate shades, such as bordeaux, maroon, violet, and brown. Most are tinctorially very strong compared to other pigments that cover the same range of shades.

Among diazo components which are used to obtain Naphthol AS pigments with a simple substitution pattern (Cl, CH$_3$, NO$_2$, OCH$_3$), i.e., group I Naphthol AS pigments, chloroanilines produce orange to scarlet hues, chlorotoluidines primarily make bluish reds, and nitrotoluidine and nitroanisidine frequently afford bordeaux shades. Most of the Naphthol AS pigments that have been developed in the USA are obtained from diazo components which possess nitro groups.

A variety of Naphthol AS pigments are polymorphous: they display at least two crystal modifications. The list includes P.R.9, 12, and 187. A wide variety of diazonium compounds and coupling components with different substitution patterns have been developed in order to alter the properties of the resulting Naphthol AS pigments. This is particularly true for parameters such as fastness to solvents, in which most Naphthol AS pigments perform poorly (group I). Sulfonamide and carbonamide moieties in particular confer higher solvent resistance on their parent structure (group II). As pigment powders, various members of this class show a resistance to ethylacetate which equals step 1 or 2 on a five step scale; while other pigments, such as P.R.146, reach step 4 to 5 on the same scale (Sec. 1.6.2.1). Each

Tab. 17: Structural principles of Naphtol AS pigments.

C.I. Name	C.I. Constitution No.	Structure	Number of CONH groups
P.R.3	12120		0
P.R.2	12310		1
P.R.170	12475		2
P.O.38	12367		3
P.R.187	12486		3

type is affected differently by plasticizers such as dioctyl phthalate or dibutyl phthalate that are frequently used in polymers.

The fastness of a pigment to solvents controls a variety of its resulting properties, such as the tendency to migrate, which differs considerably between types. There are media in which pigment behavior may range from strongly migrating, blooming and bleeding pigments to nonblooming and more or less bleed resistant types.

Table 17 lists a number of commercially available pigments, along with their chemical structures, in order to illustrate the different structural types of Naphthol AS pigments. Fastness to solvents and migration resistance improve from top to bottom, i.e., with increasing number of CONH groups in the molecule. The first example, a simple β-naphthol pigment, is the skeleton from which all other species are derived.

The tendency of a pigment to migrate may be reduced further by introducing heterocyclic substituents into the coupling component (Sec. 2.8). Naphthol AS pigments are fairly lightfast. Even though a few members, do not quite reach the fastness of β-naphthol pigments, such as full shades of P.O.5 or P.R.3, they are dinstinctly superior in white reduction. Full shades of P.R.9, for instance, equal step 7 for lightfastness; while 1:6 reduction with TiO_2 equals step 6, and 1:500 formulations are still lightfast enough to reach step 4–5 on the Blue Scale.

The heat sensitivity of a Naphthol AS pigment is largely a function of its chemical constitution and may, for instance in print, range from under 120 °C to 200 °C.

2.6.3   Application

Increasingly stringent application requirements have in the course of the last two decades forced a number of Naphthol AS pigments out of the market. Only a few of them continue to be produced locally or regionally. Others, such as P.R.22 or 23, have long been produced in large volume in certain areas, while they are almost nonexistent in other countries. Requirements vary considerably, and it is frequently impossible to refer to a standard. A number of particularly high grade Naphthol AS pigments, however, have enjoyed increasing commercial recognition in recent years.

Naphthol AS pigments are used primarily in paints and printing inks. The applicability of a pigment in paints is determined or restricted largely by its fastness to organic solvents. Very fast pigments are used not only in air drying paints, nitrocellulose combination lacquers, and other paints that are processed at room temperature; but they find additional application in baking enamels. A few Naphthol AS pigments are even used in high grade systems, such as those used for various automotive finishes, including original equipment manufacture (O.E.M.) finishes. Chemically simple species bloom in baking enamels, while high grade pigments are nonblooming in all concentrations. Others may be used within a certain concentration range in baking enamels (Sec. 1.6.3.1); provided the pigment concentration is sufficiently high and the baking temperature is as low as possible. Depending on their chemical constitution, however, these pigments show a more or less pronounced tendency to bleed; none of them is entirely fast to overcoating.

Many Naphthol AS pigments derive most of their commercial importance from the application in printing inks. Some pigments which show high tinctorial strength produce exceptionally brilliant shades. Corresponding to the solvent fastness, other fastness properties limit the application of Naphthol AS pigments in certain fields, such as packaging printing inks. Fastness to soap and paraffins is almost always excellent or at least good, which is also true for water, acid, and alkali resistance. However, the pigments exhibit a certain lack of stability to clear lacquer coatings and sterilization, which is very poor in some types. Application in print is therefore possible only where no overlacquering fastness is required. Naphthol AS pigments are primarily used in offset, flexo, and packaging gravure printing inks.

Almost all Naphthol AS pigments, with a few exceptions, are excluded from metal deco printing. However, a number are used in large volume by the textile printing industry. Poor migration and heat resistance make most members of this class inapplicable to plastics.

Naphthol AS pigments are used to an appreciable extent in special areas, such as in office articles, artists' colors, cleaning agents and detergents, including soaps. They are used to color paper, both mass colored paper and surface coated paper.

2.6.4   Commercially Available Naphthol AS Pigments

General

The number of Naphthol AS pigments that have gained commercial recognition is comparatively large (Table 18). There is a dual classification system according to the substitution pattern (Sec. 2.6.2).

Somewhat of an intermediate position is assumed by a compound which may be considered both a disazo and a Naphthol AS pigment. Pigment Blue 25 is an interesting example of a "cross" compound, derived from coupling bisdiazotized 3,3'-dimethoxy-4,4'-diaminodiphenyl with Naphthol AS.

As far as intermediates for the manufacture of group I Naphthol AS pigments are concerned, 2-methyl-5-nitroaniline and 2,5-dichloroaniline reign supreme among

Tab. 18: Commercially available Naphthol AS pigments.

C.I. Name	C.I. Constitution No.	R_D^2	R_D^4	R_D^5	R_K^2	R_K^4	R_K^5	Shade
Group I								
P.R.2	12310	Cl	H	Cl	H	H	H	red
P.R.7	12420	CH_3	Cl	H	CH_3	Cl	H	bluish red
P.R.8	12335	CH_3	H	NO_2	H	Cl	H	bluish red
P.R.9	12460	Cl	H	Cl	OCH_3	H	H	yellowish red
P.R.10	12440	Cl	H	Cl	H	CH_3	H	yellowish red
P.R.11	12430	CH_3	H	Cl	CH_3	H	Cl	ruby
P.R.12	12385	CH_3	NO_2	H	CH_3	H	H	bordeaux
P.R.13	12395	NO_2	CH_3	H	CH_3	H	H	bluish red
P.R.14	12380	NO_2	Cl	H	CH_3	H	H	bordeaux
P.R.15	12465	NO_2	Cl	H	OCH_3	H	H	maroon
P.R.16	12500	OCH_3	NO_2	H	*)	–	–	bordeaux
P.R.17	12390	CH_3	H	NO_2	CH_3	H	H	red
P.R.18	12350	NO_2	CH_3	H	H	H	NO_2	maroon
P.R.21	12300	Cl	H	H	H	H	H	yellowish red
P.R.22	12315	CH_3	H	NO_2	H	H	H	yellowish red
P.R.23	12355	OCH_3	H	NO_2	H	H	NO_2	bluish red
P.R.95	15897	OCH_3	H	$SO_2OC_6H_4NO_2(p)$	CH_3	H	H	carmine
P.R.112	12370	Cl	Cl	Cl	CH_3	H	H	red
P.R.114	12351	CH_3	H	NO_2	H	H	NO_2	carmine
P.R.119	12469	CH_3	H	$SO_2OC_6H_4CO_2CH_3$	OCH_3	H	H	yellowish red
P.R.136	–	–	–	–	–	–	–	bordeaux
P.R.148	12369	Cl	Cl	H	CH_3	H	H	orange
P.R.223	–	–	–	–	–	–	–	bluish red
P.O.22	12470	Cl	H	Cl	OC_2H_5	H	H	orange
P.O.24	12305	H	H	Cl	H	H	H	orange
P.Br.1	12480	Cl	H	Cl	OCH_3	H	OCH_3	brown
P.V.13	–	–	–	–	–	–	–	violet
Group II								
P.R.5	12490	OCH_3	H	$SO_2N(C_2H_5)_2$	OCH_3	OCH_3	Cl	carmine
P.R.31	12360	OCH_3	H	$CONHC_6H_5$	H	H	NO_2	bluish red
P.R.32	12320	OCH_3	H	$CONHC_6H_5$	H	H	H	red
P.R.146	12485	OCH_3	H	$CONHC_6H_5$	OCH_3	Cl	OCH_3	carmine
P.R.147	12433	OCH_3	H	$CONHC_6H_5$	CH_3	H	Cl	pink
P.R.150	12290	OCH_3	H	$CONHC_6H_5$	**)			carmine
P.R.164	–	***)						yellowish red
P.R.170	12475	H	$CONH_2$	H	OC_2H_5	H	H	red
P.R.184	12487	OCH_3	H	$CONHC_6H_5$	$\begin{cases} CH_3 & H \\ OCH_3 & Cl \end{cases}$		$\left.\begin{matrix} Cl \\ OCH_3 \end{matrix}\right\}$ruby	
P.R.187	12486	OCH_3	H	$CONHC_6H_4\text{-}(p)CONH_2$	OCH_3	OCH_3	Cl	bluish red
P.R.188	12467	$COOCH_3$	H	$CONHC_6H_3\text{-}Cl_2(2,5)$	OCH_3	H	H	yellowish red

Tab. 18: (Continued)

C.I. Name	C.I. Constitution No.	R_D^2	R_D^4	R_D^5	R_K^2	R_K^4	R_K^5	Shade
P.R.210	12474	****)	CONH$_2$	H	{ OCH$_3$	H	H }	red
	12475	H			{ OC$_2$H$_5$		}	
P.R.212	–	–	–	–	–	–	–	very bluish red
P.R.213	–	–	–	–	–	–	–	bluish red
P.R.222	–	–	–	–	–	–	–	bluish red
P.R.238	–	–	–	–	–	–	–	bluish red
P.R.245	12317	OCH$_3$	H	CONH$_2$	H	H	H	bluish red
P.R.253	12375	Cl	SO$_2$NHCH$_3$	Cl	CH$_3$	H	H	red
P.R.256	–	–	–	–	–	–	–	yellowish red
P.R.258	12318	OCH$_3$	H	SO$_2$CH$_2$C$_6$H$_5$	H	H	H	
P.R.261	12468	OCH$_3$	H	CONHC$_6$H$_5$	OCH$_3$	H	H	
P.O.38	12367	Cl	H	CONH$_2$	H	NHCOCH$_3$	H	reddish orange
P.V.25	12321	OCH$_3$	NHCOC$_6$H$_5$	OCH$_3$	H	H	H	violet
P.V.44	–	–	–	–	–	–	–	violet
P.V.50	12322	OCH$_3$	NHCOC$_6$H$_5$	CH$_3$	H	H	H	violet
P.Bl.25	21180			Formula see p. 287			–	reddish blue

   * P.R.16 contains 2-hydroxy-3-naphthoic acid-1-naphthylamide as coupling component.
  ** P.R.150 contains 2-hydroxy-3-naphthoic acid amide as coupling component.
 *** P.R.164 is a disazo pigment.
**** P.R.210 as mixed coupling product has Constitution No. 12477.

diazo components. The list of important coupling components, apart from the an-ilide of 2-hydroxy-3-naphthoic acid itself, includes primarily derivatives which are substituted in the ortho position by methyl or methoxy groups. All of the group II pigments with known structures carry either an SO$_2$N group in the 5 position on the aromatic ring of the diazo component; or, alternatively, a CONH group in 4 or 5 position.

Individual Pigments

Group I Pigments (Sec. 2.6.2)

Pigment Red 2

P.R.2 provides a medium red shade, which is somewhat yellower than the Naph-thol AS pigment P.R.112. Its main area of application is the printing inks field. In some systems, P.R.2 is even slightly stronger than P.R.112, although it does not quite achieve the same lightfastness. 1/1 SD letterpress proof prints equal step 5 on the Blue Scale, 1/25 SD prints reach step 4. The corresponding values for P.R.112 are 1/2 to 1 step higher. Most of the commercially available products have specific surface areas between ca. 20 and 30 m^2/g. They afford prints with correspondingly poor transparency. The types with comparatively coarse grains, on the other hand, which are frequently used to formulate highly pigmented printing inks, show good flow properties and satisfy the requirements for corresponding application. The prints exhibit good gloss.

P.R.2 does not show perfect performance in special applications (Sec. 1.6.2.3) in prints, which is also true for a number of other members of this class. In this respect, P.R.2 is inferior to P.R.112. This may have a particular impact on marginal areas of pigment applicability. P.R.112 letterpress proof prints, for instance, tolerate mineral spirits and soap; while P.R.2 prints only reach step 4 on the 5 step fastness scale. P.R.2 specimens are also sensitive to clear lacquers coatings and to sterilization.

P.R.2 is primarily used in offset and packaging gravure and flexo printing inks. The pigment also lends itself to textile printing. Besides, P.R.2 also functions as a pigment in connection with the spin dyeing of viscose rayon and viscose cellulose and in the mass coloration of viscose sponges and films. Leather coverings are also frequently colored with P.R.2.

The paint industry has little demand for P.R.2. The pigment finds some use in household paints, especially in air drying systems.

Pigment Red 7

P.R.7 provides bluish red shades and is used in a variety of media in the printing inks, paints, and plastics field. Only one or two decades ago, P.R.7 was considered the leading Naphthol AS pigment. The pigment is now in the process of being removed from the market altogether, because one of the starting materials for its synthesis is no longer available. There are other Naphthol AS pigments covering the same range of shades, such as P.R.170, which may replace P.R.7.

Pigment Red 8

Pigment Red 8 affords clean, bluish shades of red. It is primarily used in the printing ink industry. P.R.8 exhibits high tinctorial strength and produces brilliant prints. Commercially available types with specific surface areas between about 50 and 60 m^2/g afford transparent prints. P.R.8 is used in prints which require no particular solvent resistance. However, the pigment tolerates solvents much better than the yellower P.R.7; in this respect, P.R.8 matches the yellower, but more lightfast P.R.5. The prints are fast to soap but not entirely stable to butter and paraffin. P.R.8 is sensitive to clear lacquers coatings and to sterilization. It tolerates exposure to 140°C for 30 minutes.

The lightfastness (such as in letterpress proof prints) deteriorates drastically with decreasing pigment area concentration. 1/1 SD prints equal step 5, while 1/25 SD specimens only reach step 3 on the Blue Scale. P.R.8 is used in letterpress and offset inks, in packaging gravure printing inks, and in a variety of flexo inks. Types with finer particle sizes, dispersed in vehicles for various printing techniques, may recrystallize and drastically deteriorate in tinctorial strength and transparency. P.R.8 is also used in textile printing, although its lightfastness does not satisfy high demands.

P.R.8 is used in a variety of special media outside the paints, printing inks, and plastics field, which is also true for other members of this class of pigments. One such application is in the paper industry, where the pigment is used for mass coloration and surface coating formulations. It also lends itself to application in artists' colors and office articles.

Pigment Red 9

P.R.9 affords clean, yellowish shades of red, which show very good lightfastness. The fact that the pigment is supplied in the form of an unstable crystal modification, which is very sensitive to aromatic hydrocarbons and some other organic solvents, restricts its use to media that do not contain aromatic solvents. P.R.9 is therefore used in letterpress and offset printing and aqueous flexo inks. Although the prints are comparatively sensitive to a number of solvents, they are soap and butter fast and almost fast to paraffin, dibutyl phthalate, and mineral spirits. Compared to the slightly yellower P.R.10, P.R.9 is more lightfast if the pigment area concentration is low. In comparison with the slightly bluer 1/3 SD P.R.53:1 prints, P.R.9 is more lightfast by a few steps on the Blue Scale; it is also alkali and acid resistant. Commercial types with low specific surface areas around 20 m^2/g afford little transparency in print.

The paints field is another area in which P.R.9 is restricted to media containing aliphatic hydrocarbons only. Application in systems containing aromatic hydrocarbons, ketones, esters, or glycol ethers changes the crystal modification and causes recrystallization, shifts the color considerably, and drastically reduces the tinctorial strength. In suitable finishes, however, P.R.9 demonstrates very good stability to light, even in white reduction. This qualifies it for use in emulsion paints, where it shows equal perfection in lightfastness and weatherfastness. Even its full shades, however, are frequently not fast enough for exterior application.

Applications outside the above-mentioned areas include colored pencils.

Pigment Red 10

This pigment exhibits a clean, yellowish red shade, similar to that of P.R.9. It is supplied as a stable crystal modification and is thus not as sensitive to aggressive solvents as P.R.9. This broadens its scope to include packaging gravure and flexo printing inks containing solvents. The pigment provides good lightfastness. 1/3 or 1/25 SD letterpress proof prints, however, score 1/2 to 1 step less on the Blue Scale than corresponding P.R.9 prints. P.R.10 prints are not suitable for clear lacquer coatings and may not be sterilized; there heat stability is low.

Textile prints score poorly in dry cleaning tests involving agents such as perchloroethylene, a feature which is typical of Naphthol AS pigments. P.R.10 is affected by dry heat and is equally unsuitable for PVC coatings. The prints equal step 6–7 on the Blue Scale for lightfastness, depending on the depth of shade.

P.R.10 is used to a certain extent in office articles, artists' colors, and cleaners.

Pigment Red 11

P.R.11 has largely been replaced by other products. Its hue is a bluish red which resembles the color of ruby; P.R.11 is more bluish than its positional isomer P.R.7. Both pigments behave similarly as far as fastness properties are concerned, which includes fastness to most of the solvents that are used in paints and printing inks, fastness to overlacquering, heat stability, and others. However, P.R.11 is much less lightfast than P.R.7; incorporated in medium-oil alkyd systems at 1/3 SD, P.R.11 scores 2 steps less on the Blue Scale (step 4).

Pigment Red 12

P.R.12 is one of the more significant bordeaux colored pigments. Two crystal modifications are known, of which only the thermodynamically unstable bordeaux one is commercially used. Little activation energy is required to convert the bordeaux modification into the stable red modification. Conversion is, for instance, initiated as the pigment is dispersed in an oily binder containing zinc oxide using low shear equipment, such as a Hoover muller. Change of crystal modification may also occur if P.R.12 is milled in dry state with dry baryte. The transition occurs also very fast in the presence of chlorohydrocarbons or ketones in the dispersion medium. Exceptionally large pigment crystals are formed as the crystal modification changes, considerably reducing the viscosity of the dispersion medium. Poor lightfastness renders the stable red modification unsuitable for commercial application.

The graphic industry uses P.R. 12 in offset inks, for packaging gravure, and flexo printing inks. The pigment is comparatively lightfast; 1/1 to 1/25 SD letterpress proof prints equal step 5–6 to step 4 on the Blue Scale. P.R.12 demonstrates very high tinctorial strength; the fastness of the prints to solvents and other media is, however, moderate. The prints are affected by paraffine, butter, and other fats. P.R.12 is similarly sensitive to alkali.

P.R.12 textile prints are excellently lightfast; 1/1 to 1/6 SD prints equal step 7 and 6–7, respectively, on the Blue Scale. However, the pigment fails dry cleaning tests and migrates considerably in PVC coatings. P.R.12 tolerates dry heat up to 150 °C; at 180 °C, the pigment still largely retains its initial color value.

The paint industry uses P.R.12 primarily in air drying paints. Baking may be a problem, since a pigment concentration of less than 1% will cause blooming between 140 and 180 °C; at 200 °C, the concentration limit is 2.5%. P.R.12 is therefore only used in highly concentrated formulations, especially in full and related shades. Ketones and other agressive solvents induce a change of the crystal modification, accompanied by a drastic color change.

However, P.R.12 demonstrates a satisfactory degree of lightfastness. Air drying paints equal step 6 on the Blue Scale in full shade; while baking enamels reach step 6–7, but show some darkening. In white reductions (1:10 TiO_2), the pigmented systems equal step 4–5 and 5–6, respectively. In baking enamels, P.R.12 bleeds considerably, irrespective of the temperature. The pigment is, however, occasionally used in emulsion paints wherever its fastness properties satisfy the specifications.

P.R.12 is also employed in a series of special applications, such as automotive cleaners, floor polish, shoe polish, etc., and it is frequently used to color office articles and leather.

Pigment Red 13

Produced in the USA, this pigment is of limited commercial interest. Its color is a bluish red.

Pigment Red 14

P.R.14 provides an intense, lightfast bordeaux shade, which is much yellower than that of P.R.12. Like P.R.12, it also possesses two crystal modifications. Again, the commercially available form is thermodynamically unstable; but, in contrast to P.R.12, conversion to the stable modification, which is technically less interesting because of its poor lightfastness, requires somewhat of an effort. It is facilitated, for instance, by aromatic hydrocarbon and ketone solvents.

The main market for P.R.14 is in the paint industry, but the pigment is also used in printing inks and in some other applications.

Apart from air drying systems, P.R.14 is also used in baking enamels; however, it blooms beyond certain concentration limits. In an alkyd-melamine resin system, for instance, the pigment shows resistance to blooming up to 160 °C, independent of the concentration; the concentration limit at 180 °C is 0.2%, and at 200 °C it is 1%. Blooming is accompanied by considerable bleeding.

P.R.14 is very lightfast; its full shade lightfastness in air drying systems equals step 6–7 on the Blue Scale; in baking enamels, it reaches step 7–8. In white reductions (1:7 TiO_2), the lightfastness of the same systems equals step 5–6 and 6–7, respectively.

Very lightfast red shades can be produced by combining P.R.14 with Molybdate Red. P.R.112 is a member of the same class of pigments, behaves similarly, and is particularly suitable as a partner for P.R.14 in blends which approach the colors of Toludine Red (P.R.3) but provide much improved lightfastness and weatherfastness.

In applications where the pigment is fast enough to solvents to satisfy the requirements, it may also be used for offset, packaging gravure, and flexo printing inks. The resulting prints are soap, alkali, and acid resistant. They are not completely fast to paraffin and quite sensitive to butter and a number of other fats. P.R.14 prints are not fast to clear lacquer coatings and may not be sterilized. Heat stability is less than 140 °C. 1/1 SD letterpress proof prints equal step 6 on the Blue Scale for lightfastness, 1/3 SD prints equal step 5–6, and 1/25 SD prints reach step 4.

Pigment Red 15

P.R.15 continues to be offered in the Japanese market, but is of limited commercial value. Its hue is a brilliant medium maroon.

Pigment Red 16

The bluish bordeaux crystal modification of this polymorphous pigment used to be marketed, but has recently been discontinued. It was used in printing inks wherever tolerance to solvents was unimportant. Although the pigment exhibits average lightfastness, it fails to meet current industrial standards. The shade can be adjusted by tinting P.R.12, which is a member of the same pigment family.

Pigment Red 17

P.R.17 provides medium reddish shades. As a result of poor fastness properties, its commercial significance is somewhat limited and it is sold only in small volume. P.R.17 has the advantage of being fast to acid, alkali, and soap. It is therefore used in offset, gravure, and flexo printing inks wherever tolerance to alkali and soap is a major concern. Moreover, P.R.17 is also employed in connection with mass coloration and surface coloration of paper.

Pigment Red 18

P.R.18, a positional isomer of P.R.114, is currently unavailable in Europe. Production abroad, especially in Japan and Central America, is limited. P.R.18 produces shades of maroon, with properties that parallel those of other members of its class with somewhat poorer fastness levels. The pigment is registered in the USA as D&C Red No. 38.

Pigment Red 21

P.R.21 is produced and sold in the USA and in Japan. It affords yellowish shades of red. Since it fails to meet the increasingly stringent industrial requirements, P.R.21 is no longer extensively used.

Pigment Red 22

P.R.22 affords yellowish shades of red. Its commercial significance varies considerably with the region. Less common in Europe, the pigment maintains an important position in the USA and especially in Japan. Its main market is in textile printing, but it is also used throughout the graphics field, such as in offset and gravure printing, especially in NC-based inks. P.R.22 is advantageous where soap and alkali resistance are required. A pigment with high tinctorial strength, it is somewhat lighter than P.R.9 in high pigment area concentrations; while at lower pigment area concentrations it is yellower. Compared to P.R.2, P.R.22 is distinctly yellower and less lightfast; this tendency is even more pronounced in comparison with P.R.112. As opposed to Lake Red C pigments (P.R.53:1), P.R.22 is bluer to yellower, depending on the choice of printing ink; but its lightfastness is better by about 1 step on the Blue Scale.

The paint industry employs P.R.22 in air drying systems, in emulsion paints, and occasionally in industrial finishes; although there is some danger of blooming, and the appropriate limit has to be observed. Again, P.R.22 is much less lightfast in these media than P.R.112. Areas of application include paper mass and surface coloration, colored pencils, artists' colors, and other purposes.

Pigment Red 23

P.R.23 provides bluish, dull red shades, yellower and much duller than P.R.146. Its commercial significance, like that of P.R.22, varies considerably with the region. Fastness to solvents and application performance are mostly inferior, which is also true for lightfastness: 1/1 SD letterpress proof prints equal step 3 on the Blue Scale, while P.R.146, at equal depth, reaches step 5. Commercial P.R.23 types usually have smaller particles than P.R.146; they are stronger and more transparent, but at the same time provide higher viscosity in ink. Inadequate solvent resistance and, as a result, a tendency to recrystallize is somewhat of a disadvantage, especially in packaging gravure printing inks. Compared to P.R.146, P.R.23 will produce more opaque, weaker prints. The pigment lends itself to application in aqueous printing inks and NC-based inks. It is also used for textile printing.

P.R.23 is recommended for use in industrial paints; although this is one of the markets where it competes primarily with P.R.146. Despite being tinctorially stronger and somewhat yellower, P.R.23 is also less fast to overpainting than P.R.146 and considerably less lightfast than the latter.

Pigment Red 95

P.R.95 produces a bluish red; a carmine.

The pigment is used for printing inks and paints, where it satisfies average requirements. 1/1 SD letterpress proof prints equal step 4–5 on the Blue Scale for lightfastness; the prints are not fast to butter and various fats. They are acid and alkali fast but not entirely so to soap. They are also not completely resistant to the DIN 16 524/1 standard solvent mixture. The pigment tolerates heat up to 180 °C for 10 minutes, but it is not entirely stable to calandering and sterilization, a feature which is typical of its class.

P.R.95 is suitable for a variety of printing techniques. Poor migration resistance and insufficient fastness to plasticizers render the pigment unsuitable as a colorant in special gravure inks for plasticized PVC.

The paint industry employs P.R.95 primarily in a variety of industrial finishes and in paints. Relatively lightfast in full shade, its fastness to light rapidly deteriorates upon reduction of the shade with TiO_2. 5% full shade finishes (medium-oil alkyd resin) equal step 6–7 on the Blue Scale; 1/3 SD coatings reach step 4–5, and 1/25 SD samples afford step 3. P.R.95 is also frequently used in blends with Molybdate Red. The commercial variety shows good transparency and is therefore suited to purposes such as metallic coatings and "hammertone finish". These systems, however, do not satisfy the stringent requirements with regard to lightfastness; besides, they are not fast to overcoating. At high temperature and low pigment concentration, blooming is observed. Incorporated in an alkyd-melamine resin system, for instance, at a pigment concentration below 0.1%, blooming may be observed above 140 °C. P.R.95 is heat stable up to 140 °C for 30 minutes.

Pigment Red 112

P.R.112 produces highly brilliant medium red shades. Patented as early as 1939, it has only been marketed for a few decades. However, production has increased rapidly due to its superior application properties. Its main market is in printing inks, finishes, and paints; but it is also used in a variety of other areas.

The printing inks field employs P.R.112 primarily in letterpress and offset inks, in packaging gravure inks, and in flexo inks. It is tinctorially very strong. At standard thickness of layer, 1/1 SD letterpress proof prints are formulated at 17% pigment concentration; but in practice, the amount is often much higher. The prints show excellent lightfastness. 1/1 to 1/25 SD letterpress proof prints, which, after all, cover a wide range of pigment concentrations per surface area unit, uniformly equal step 5–6 or step 6 on the Blue Scale. The fastness to light increases with increasing pigment concentration, until it reaches step 6–7 on the Blue Scale (25% printing inks). The prints are fast to soap, although they do not completely tolerate paraffin and are only moderately fast to butter and other fats (step 3). They are also affected by clear lacquer coatings and sterilization. The pigment demonstrates poor heat stability, the prints do not withstand exposure to 140 °C.

P.R.112 is very lightfast on textiles; 1/1 to 1/3 SD (deep shade) prints equal step 7 on the Blue Scale. Exposure to dry heat at 150 and 180 °C has no effect. However, like other members of its class, P.R.112 fails the dry cleaning test with perchloroethylene. The pigment may not be brought into contact with PVC coatings, into which it bleeds.

In paints, P.R.112 produces a shade which is referred to as signal red. It may be used not only in air drying paints but also in baking enamels, provided appropriate conditions are maintained to avoid blooming. At baking temperatures between 140 and 160 °C, the concentration threshold is 0.1%; between 180 and 200 °C, the limit is 2.5%. However, the fastness to overcoating is unsatisfactory. Lightfastness and weatherfastness are excellent, even in white reductions. Air drying finishes in full shade equal step 7–8 on the Blue Scale, while baking enamels reach step 8. There is not much difference in white reduction: 1:10 reductions with TiO_2 afford step 6–7 and step 7, respectively.

The shade of P.R.112 approaches that of Toluidine Red (P.R.3), which it may consequently replace in applications with demanding requirements. P.R.112 may also be combined with the bordeaux P.R.12, which behaves similarly, to provide a broad range of colors. It is somewhat bluer than P.R.9 and considerably more lightfast and weatherfast. Types which provide optimized flow are used particularly in opaque full shade finishes. Excellent fastness properties make P.R.112 suitable for high grade paints, such as automotive finishes and general industrial coatings. Electrophoretic painting is one of its special applications; the pigment is also found in emulsion paints, despite the disadvantage of comparatively poor fastness in exterior application.

Rigid PVC is one of the polymers which are sometimes colored by P.R.112. Transparent systems (0.1% pigment) equal step 8 on the Blue Scale for lightfastness; while the lightfastness of white reductions is only step 5–6. The spin dyeing industry employs P.R.112 for viscose rayon and viscose cellulose, in which the pigment exhibits excellent lightfastness and performs, if not perfectly, then almost satisfactorily.

Besides, P.R.112 is also used in a variety of special media and applications. It is encountered in aqueous wood stains, in which it may also be combined with yellow pigments, such as P.Y.83, or with violet colors, such as P.V.23, or with carbon black to produce shades of brown. In these media, its lightfastness equals approximately step 5 on the Blue Scale. However, the pigment bleeds when oversprayed by a white nitrocellulose combination paint; but it may be coated with a nitro or acid hardening varnish or a polyester varnish. The list of special applications includes office articles and artists' colors, such as pigmented felt-tip pen inks, watercolors and colored pencils, poster paints, as well as cleaning agents and detergents. In paper mass coloration, optimum lightfastness is only realized in full shade.

Pigment Red 114

P.R.114, an isomer of P.R.18, is sold only in Japan, where it is of minor importance. It produces a bluish red shade, carmine. The pigment largely parallels P.R.22 in terms of application properties, especially in its fastness properties.

Pigment Red 119

Although it is considered a member of the Naphthol AS pigment series, the exact chemical constitution of P.R.119 has not yet been published. The pigment affords a brilliant yellowish red shade and is used throughout the paint and printing ink industry. A typical representative of group I Naphthol AS pigments as far as application performance goes, the pigment exhibits very good lightfastness in finishes. Incorporated in medium-oil alkyd varnishes, for instance, or in alkyd-melamine resin systems, the full shade lightfastness of a 5% formulation equals step 7–8 on the Blue Scale; although some darkening is observed. In white reduction, at 1/25 SD, the pigment still scores step 5 on the Blue Scale for lightfastness, in which it resembles the somewhat bluer P.R.112. Comparisons of the weatherfastness, however, are in favor of P.R.112.

P.R.119 is not resistant to overcoating; however, no blooming has been observed at low pigment concentrations and typical baking conditions. The pigment is also used in emulsion paints, although exterior application is not recommended.

Throughout the printing ink industry, P.R.119 is in direct competition with other members of the Naphthol AS pigment series. Its shade, for instance, closely resembles that of P.R.10; although it is somewhat weaker. Regarding fastness to light, P.R.119 scores 1/2 to 1 step less on the Blue Scale than P.R.112. P.R.119 is mainly used in flexo and wallpaper inks. It migrates in gravure inks printed on plasticized PVC, a typical feature of members of the Naphthol AS series. P.R.119 is resistant to clear lacquer coatings, may be sterilized, and is highly suitable for metal deco printing. The pigment is heat stable up to 180°C for 10 minutes. It is also found in crayons and artists' colors.

Pigment Red 136

Although P.R.136 is a Naphthol AS pigment, its exact chemical constitution remains to be published. Two versions are commercially available. The transparent type has a specific surface area of almost 70 m^2/g and shows a considerable degree of structural viscosity. Its shade is a cherry red. The bordeaux grade, on the other hand, has a specific surface area of about 25 m^2/g, is much more opaque, and exhibits a more favorable flow behavior. It is used, primarily in close to full shade colorations, for applications where hiding power is an issue. The shade of RAL 3004 is matched by adding appropriate amounts of TiO$_2$.

Both types, however, enjoy limited commercial success. They are found in paints, in which they are heat stable up to 160°C. The fastness of the pigment to solvents is typical of a member of the Naphthol AS pigment series, which includes inadequate fastness to overcoating. At low pigment concentrations, high baking temperatures may induce blooming. The pigment will bloom, for instance, at 160°C in an alkyd-melamine resin baking enamel formulated at less than 0.05% pigment concentration. Both P.R.136 types demonstrate good lightfastness and weatherfastness, but darken in full shade. The more opaque version is more lightfast and weatherfast and is recommended for automotive refinishes.

Pigment Red 148

P.R.148 was only recently withdrawn from the market. It used to be employed in printing inks, paints, colored pencils, and also for the spin dyeing of viscose rayon and viscose cellulose. Although P.R.148 is lightfast, its behavior towards solvents and its corresponding fastness properties, such as migration resistance, fails to meet current industrial standards. The pigment is very heat stable, but shows a strong tendency to bloom, which precludes its use even in rigid PVC. It affords a very yellowish red or reddish orange shade, which may also be produced by appropriate blends of other Naphthol AS pigments.

Pigment Red 223

Its chemical constitution is not yet published. P.R.223 used to have some commercial importance before it was recently withdrawn from the market. The pigment produces a bluish red shade, bluer even than P.R.170. Its main market was in high grade industrial finishes. Good lightfastness and durability in full shades and the fact that it meets certain specifications make it suitable for use on public transportation vehicles. However, the pigment is not fast to overpainting.

Pigment Orange 22

A reddish orange pigment, P.O.22 is used for spin dyeing viscose rayon and viscose cellulose, for which it is sold in the form of a pigment preparation. Properties which have a bearing on textile printing and coloration, such as fastness to light, as well as fastness to wet and dry rubbing, dry cleaning, and bleaching with peroxide are excellent.

Pigment Orange 24

P.O.24 is manufactured exclusively in Japan and enjoys only minor recognition. Its resistance to organic solvents is inferior to that of other Naphthol AS pigments. The pigment fails to meet most lightfastness requirements.

Pigment Brown 1

P.Br.1 is a neutral brown pigment of very good lightfastness. However, it has lost most of its importance in recent years. The pigment is not stable to organic solvents. Its main market is in printing inks wherever solvent resistance is not required. The prints are not fast to soap and butter, do not tolerate clear lacquer coatings, and cannot be calandered or sterilized without color change. 1/1 to 1/25 letterpress proof prints equal step 6–7 on the Blue Scale for lightfastness. In high grade prints, P.Br.1 is being displaced successively by newer pigments, such as P.Br.23 and 25, which have much better application properties and are calander and sterilization resistant. P.Br.1 is also falling prey to blends of orange, red, and yellow pigments with carbon black, which show more acceptable performance.

The pigment is not used in paints. Rigid PVC is one of the few polymers in which it is employed. P.Br.1 is found in bottles, in which its transparent formulations (0.1%) equal step 7 on the Blue Scale for lightfastness; while opaque versions reach step 6–7.

P.Br.1 is also used in polystyrene, in which it produces an orange shade as it dissolves. P.Br.1 used to play somewhat of a role in the spin dyeing of viscose rayon and viscose cellulose. The pigment continues to be employed in connection with textile printing, where it exhibits excellent lightfastness and durability. Its lightfastness in full shade (1/1 SD) equals step 8 on the Blue Scale, while in white reduction (1/6 SD) it reaches step 7. Although the pigment does not tolerate dry cleaning with tetrachloroethylene, other facets of its performance are excellent or almost excellent. The pigment also lends color to colored pencils.

Pigment Violet 13

Although a Naphthol AS pigment, its exact chemical constitution has not yet been published. P.V.13 produces clean shades of violet. Its significance is somewhat limited, but it is used in printing inks, primarily to modify shades. Compared to other group I Naphthol AS pigments, P.V.13 shows poor general fastness properties. 1/1 SD letterpress proof prints only equal step 3 on the Blue Scale for lightfastness, while 1/3 SD specimens reach step 2. The prints tolerate acid but not alkali or soap. Similarly, they are not fast to overlacquering and may not be sterilized, which renders the pigment unsuitable for metal deco printing and for printing on PVC or polyolefin foils. The pigment is not known to be used to a considerable extent outside the printing industry.

Group II Naphthol AS Pigments (Sec. 2.6.2)

Pigment Red 5

P.R.5 affords a bluish red shade, somewhat similar to that of P.R.8, although the former is much more lightfast. Its commercial significance and its primary use depend on the region: In Europe it is primarily employed in printing inks, while the American and Japanese markets mainly utilize the pigment in the paint field.

Compared to other members of its class, P.R.5 exhibits good fastness to solvents.

In the printing inks field, P.R.5 is in direct competition with the much more bluish P.R.146, which provides more brilliant prints and is considerably more solvent resistant. P.R.146, on the other hand, is less lightfast. P.R.5 letterpress proof prints score between step 6–7 and step 5 on the Blue Scale for lightfastness, depending on the depth of shade (1/1 to 1/25 SD). The corresponding values for P.R.146 are 5 and 4, respectively. Prints made from P.R.5 are fast to soap and resistant to clear lacquer coatings, but are not entirely stable to butter and paraffin wax.

Since the prints do not tolerate calandering, the pigment is primarily used in offset inks, as well as in packaging gravure inks and in flexo inks.

In finishes, P.R.5 does not bloom under normal processing conditions, but it bleeds in baking enamels. The pigment is lightfast in full shade and in white reductions. Full shade lightfastness equals step 7 on the Blue Scale, while some darkening is observed. At 1:6 reduction with TiO_2, the lightfastness equals step 6 in air drying systems and step 7 in baking enamels. At 1:60 TiO_2 reduction, the values are slep 4–5 and step 5 on the Blue Scale, respectively. P.R.5 is generally used in industrial and in trade sales paints. It may be blended with Molybdate Orange to produce opaque shades of red.

Types containing comparatively large particles and therefore featuring very low specific surface areas produce duller shades; these varieties are used mainly in Japan. They are employed to formulate dark red shades for automotive finishes, in which they compete with more opaque types of P.R.170, which produce a cleaner shade, tolerate solvents better, and are fast to overlacquering.

Due to the disadvantage of comparatively poor migration resistance, P.R.5 is not used in plasticized PVC, but it can be applied in rigid PVC. Its lightfastness is excellent in this medium, transparent and opaque colorations (up to 0.01% pigment + 0.5% TiO_2) equal step 7 and, respectively, step 6–7 on the Blue Scale. Dispersed pigment preparations are available for the mass coloration of viscose films as well as for spin dyeing of viscose rayon and viscose cellulose.

The pigment performs satisfactorily on textiles, it shows good lightfastness. Depending on the SD, its lightfastness equals step 5 to step 7 on the Blue Scale (1/12 and 1/1 SD).

P.R.5 also lends itself to a variety of special applications. The list includes (decorative) cosmetics such as lipstick, eye shadow, powder, nail polish, and others. Application in this area depends on special purity requirements, and the commercially available types are tested accordingly. The pigment is listed in the European Cosmetics List.

Pigment Red 31

The pigment enjoys regional significance on the North American Continent and in Japan, but has vanished from the European market. It provides bluish shades of red, bordeaux. Its principal application is in rubber, in which it shows good lightfastness. No bleeding into natural rubber or into the fabric backing is observed (Sec. 1.8.3.6). The pigmented products resist cold and boiling water, soap, detergent solutions, 5% aqueous acetic acid, and 50% aqueous ethanol.

In polystyrene, polymethacrylate, unsaturated polyester resins, or similar media, P.R.31 affords highly transparent colorations of medium red shades, which are used as automobile tail lights and for other signalling purposes. Its heat stability in PS and PMA extrusion products (5 minutes) is good up to about 280 °C. P.R.31 is stable to the usual peroxide catalysts. The pigment is lightfast (0.025% formulations in PMA, 1.5 mm thickness of layer: step 7 on the Blue Scale). P.R.31 is also used in textile printing.

Pigment Red 32

P.R.32 is made in Japan and enjoys only limited local significance. Its fastness properties and application performance largely parallel those of the chemically related Naphthol AS pigment P.R.31.

Pigment Red 146

P.R.146 is a bluish red pigment with very good solvent resistance, even compared to other group II Naphthol AS pigments. It is primarily used in printing inks, coatings, and paints.

The printing ink industry uses P.R.146 for letterpress and offset inks and also in packaging gravure and flexo printing inks. The pigment lends itself to a variety of special applications: it is used, for instance, to print bank notes and securities. A certain tendency to migrate, however, precludes its use in print on plasticized PVC films.

P.R.146 is somewhat yellower than P.R.57:1; it therefore fails to match the standard magenta for three and four color printing. However, P.R.146 prints have the advantage of good application performance (Sec. 1.6.2.3), corresponding to the above-mentioned solvent fastnesses.

The prints tolerate white spirit, dibutyl phthalate, butter, soap, alkali, and acid. Moreover, they are also stable to clear lacquer coating; their fastness to sterilization, on the other hand, exceeds that of P.R.57:1, but is not perfect. Where sterilization fastness is required, P.R.146 is superseded by P.R.185, a benzimidazolone pigment, which is its coloristically closest neighbor.

As far as heat stability goes, P.R.146 retains its color value for 10 minutes at 200 °C or for 30 minutes at 180 °C; it thus scores 20 °C higher than P.R.57:1. 1/1 and 1/3 SD letterpress proof prints equal step 5 on the Blue Scale for lightfastness, while 1/25 SD specimens match step 4. P.R.146 is thus 1/2 to 1 step on the Blue Scale more lightfast than P.R.57:1. Prints made from commercially available types are semitransparent.

P.R.146 also shows good lightfastness in textile printing. 1/1 SD prints equal step 7 on the Blue Scale, while 1/3 SD samples score step 6–7 and are therefore somewhat less lightfast than those of the yellower P.R.7, but superior to the likewise yellower P.R.170. Weatherfastness in print, however, is much inferior to that of P.R.7 and 170. P.R.146 prints are not entirely stable to dry cleaning and dry heat; they match step 4 on the 5 step evaluation scale.

P.R.146, which is also used in paints, is primarily applied in emulsion and architectural paints; it also lends color to general industrial paints in applications where durability is unimportant.

In air drying systems, its full shade lightfastness equals step 5–6 on the Blue Scale; in baking enamels, the lightfastness equals step 6 and some darkening is observed. In white reduction, at 1:7 reduction with TiO$_2$, however, fastness to light drops to step 3–4. The pigment thus satisfies the majority of requirements for emulsion paints in interior application. As far as durability is concerned, full shade paints only score step 2 to 3 (Sec. 1.6.6) on the 5 step scale. P.R.146 exhibits high tinctorial strength and does not bloom in baking enamels. Its fastness to overcoating is much superior to that of group I Naphthol AS pigments, but it is not perfect. P.R.146 may be used in combination with Molybdate Orange pigments to produce bright opaque shades of red.

In polymers, P.R.146 is only used to color rigid PVC. Transparent colorations (0.1%) afford a lightfastness which matches step 8 on the Blue Scale, while opaque specimens equal between step 6–7 and step 6, depending on the standard depth of shade and on the TiO$_2$ content. Insufficient heat resistance in polyolefins (less than 200 °C) precludes its use in such media.

P.R.146 is a suitable candidate for a variety of special applications. The list includes wood stains, in which it is frequently blended with yellow pigments, especially with P.Y.83, and also with black to afford shades of brown. The products are fast to overcoating and stable to nitro and acid catalyzed and polyester varnishes. Intense shades match step 5 on the Blue Scale for lightfastness. Other areas of application include office articles and artists' colors, cleaning agents, paper mass coloration, laundry markers, etc. In connection with cosmetics, the pigment frequently lends color to soaps.

Pigment Red 147

P.R.147 produces a very bluish, clean red, a pink. Its main field of application is in printing inks. Inks formulated at ca 15% pigment concentration afford the standard magenta for multicolor printing on the European Scale CIE 12-66. As far as fastness to solvents is concerned, the pigment may be compared to the yellower P.R.184 or to the even yellower P.R.146. The prints are correspondingly fast to soap, paraffin, dibutyl phthalate, white spirit, and toluene, but are not entirely fast to butter and other fats. They are suitable for clear lacquer coatings, but may not be sterilized. The heat stability of the pigment equals 200 °C for 10 minutes or 190 °C for 30 minutes, which is considered excellent.

Compared to P.R.146, lower pigment area concentrations of P.R.147 are less lightfast. 1/1 SD letterpress proof prints equal step 5 on the Blue Scale for lightfastness, and 1/3 SD specimens match step 4, while 1/25 SD prints are equal to step 3. P.R.147 is used in offset inks, in packaging gravure printing inks, and in various flexo inks. Like other Naphthol AS pigments, P.R.147 is also employed in a series of special applications.

Pigment Red 150

P.R.150 is of limited regional significance. Depending on the area of application and on the reduction ratio, the pigment affords shades from bluish purple to carmine. P.R.150 is used in textile printing. In PVC coloration, which used to be its main market, it has been superseded by other products.

Pigment Red 164

The chemical constitution of P.R.164 has not yet been published. The pigment provides a somewhat dull, yellowish red shade of little strength. Used in printing inks, paints, and plastics, the pigment is produced only in small volume. Lack of light-fastness generally precludes its use in exterior application. Incorporated in air drying alkyd paint, its full shade lightfastness (5%) equals step 6–7 on the Blue Scale, while addition of TiO_2 (1:5 TiO_2) reduces this value to 5–6.

P.R.164 is heat stable up to 200 °C, which makes it a suitable candidate for applications where good temperature resistance is required. At baking temperatures of 140 to 160 °C and higher, the fastness to overcoating begins to decrease.

Letterpress proof prints with 30% P.R.164 in the inks reach step 5–6 on the Blue Scale for lightfastness. The prints are inert to paraffin and butter, as well as to a variety of other fats, but they do not tolerate acid and alkali. The same is true for soap. P.R.164, however, is stable to sterilization.

Good heat stability makes P.R.164 an interesting colorant for a series of polymers. It may be worked into polystyrene, in which full shades retain their color value up to 270 °C before turning yellow; white reductions are stable up to 250 °C. The pigment shows average lightfastness. P.R.164 also lends itself to the coloration of rigid PVC. Its heat stability in PE is only 200 to 220 °C. A tendency to bloom makes it necessary to observe certain concentration limits.

P.R.164 is also found in cast resin composed of methacrylate and unsaturated polyester. The pigment does not affect the hardening process of such media, which may be carried out, for instance, by using peroxides. An important field of application is in the coloration of various polyurethanes, for which the pigment is also sold in the form of a pigment preparation.

Pigment Red 170

P.R.170 provides medium shades of red, which in tints are somewhat bluish. Developed only in the 1960s, excellence in application performance soon made this one of the most recognized pigments.

Like some other Naphthol AS pigments, P.R.170 is polymorphous; the shades of the known crystal modifications are all within the same range of colors.

Commercially available types, which are made from two crystal modifications, differ primarily in terms of opacity. The more transparent version is also somewhat more bluish. The very opaque modification is much more stable to a variety of agents than the more transparent type. The opaque type is, for instance, slightly more resistant to organic solvents than the transparent one. It should be noted, however, that even transparent varieties are very resistant to solvents, compared to other members of this class of pigments. They are not affected by white spirit and are largely inert to alcohol, esters, xylene, and other solvents. The stability of the more transparent version to ethylglycol and methylethylketone equals step 3 on the 5 step scale, while the opaque variety matches step 4.

As a result of its excellent fastness properties, P.R.170 is used in high grade industrial paints. The pigment lends color to finishes for tools, to implements, agricultural machinery, and commercial vehicles; the opaque varieties are also used for automotive finishes, such as automotive refinishes. Thorough testing is necessary before a product can be used in original automotive finishes, for which full shades of P.R.170 are sometimes employed.

P.R.170 does not bloom in stoving enamels, but it does bleed. Opaque varieties show better overcoating fastness than the more transparent ones. Very pure shades of red, which exhibit good hiding power, are produced by blending P.R.170 with Molybdate Orange. Combinations with quinacridone pigments afford bluer red shades. P.R.170 may, as a result of its excellent lightfastness and durability, be incorporated into high grade systems. The standard transparent type, for instance, equals step 6 on the Blue Scale in full shade, while the opaque version matches step 7–8 for lightfastness. Full shades show some darkening upon exposure to light and weathering. Recently, opaque types have been introduced to the market which are characterized by noticeably improved weatherfastness compared to previously available grades. P.R.170 is less costly than similarly colored perylene tetracarbonic acid pigments or diketopyrrolopyrrole pigments such as P.R.254, which are completely fast to overpainting and also more weatherfast. The opaque versions exhibit good flow behavior, which makes it possible to enhance the hiding power of a product by increasing the pigment concentration without affecting the gloss.

The transparent type is preferred for printing inks. It is tinctorially very strong and lends itself to use in a variety of high grade formulations. Inks formulated at only 15% pigment concentration will under standardized conditions produce 1/1 SD letterpress proof prints; the resulting prints are very lightfast. 1/1 to 1/3 SD letterpress samples equal step 6 on the Blue Scale. The opaque, yellower and somewhat weaker version is more lightfast than the transparent one; the difference is about 1/2 step on the Blue Scale.

P.R.170 and 5 are closely related as far as hue is concerned: 1/1 SD prints made from the bluish crystal modification resemble each other coloristically. The fact that P.R.5, depending on the depth of shade, is more lightfast by 1/2 to 1 step on the Blue Scale is somewhat compromised by its considerably inferior fastness in application and by its appreciably lower tinctorial strength. The slightly bluer P.R.8 is much less lightfast and solvent resistant than P.R.170; prints made from P.R.170 also have the advantage of being fast to butter, soap, alkali, and acid. Fastness to clear lacquer coatings and stability to sterilization are almost perfect (step 4-5). P.R.170 is also very heat stable; it retains its color strength for 10 minutes at 200°C or for 30 minutes at 180°C, which makes it a valuable pigment for metal deco printing.

P.R.170 is also found in decorative printing inks for polyester films. It is almost completely inert to styrene monomer; in this respect, it matches step 4–5 on a 5 step scale. Gravure prints made from 8% pigment formulations almost equal step 7 on the Blue Scale, which indicates excellent lightfastness. There is no danger of plate-out during the production of laminated films (Sec. 1.6.4.1). Another area of application is in textile printing, in which the pigment shows very good fastness properties to a number of agents. It is almost completely fast to dry cleaning with tetrachloroethylene or white spirit, to washing with peroxide at 95 °C, and to alkali.

The plastics industry uses P.R.170 almost exclusively to color rigid PVC. Transparent variations (0.1%) equal step 8 on the Blue Scale for lightfastness, while opaque types (1/1 to 1/25 SD) equal step 8 to step 6–7, depending on depth of shade and reduction with TiO_2. In plasticized PVC, the pigment is also very lightfast, although it does tend to bleed. At low pigment concentrations, less than 0.01% pigment, P.R.170 may not be used in plasticized PVC. However, it is suitable for printing on films made of plasticized PVC or on PVC or PUR simulated leather, such as in automobiles.

P.R.170 is not always heat stable enough to allow application in polyolefins. In HDPE systems formulated at 1/3 SD, the pigment tolerates exposure to 220 to 240 °C for one minute. Its tinctorial strength, on the other hand, is excellent. P.R.170 is also occasionally used in polypropylene and polyacrylonitrile spin dyeing; in the latter medium, it satisfies the specifications of the clothing and home textiles industries. Besides, P.R.170 lends color to viscose rayon and viscose cellulose; it is used for the mass coloration of semisynthetic fibers made of cellulose; last but not least, it colors yarns, fibers, and films made of secondary acetate.

P.R.170 is broad in scope. It is found in wood stains, including solvent-based stains; it is blended with carbon black and yellows to produce a variety of interesting shades of brown. The colorations are fast to overcoating in these media and resist nitro and acid hardening varnishes and polyester coatings. Its lightfastness in these media equals step 7 on the Blue Scale.

Pigment Red 184

P.R.184 affords a red which is somewhat on the bluish side of P.R.146, to which it is closely related in terms of chemical constitution. Both products also behave very similarly in application. Their prints are fast to soap, butter, paraffin, dibutyl phthalate, white spirit, and toluene. P.R.184 produces a shade which matches that of the standard magenta for multicolor printing on the European Color Scale CIE 12-66. This shade results from formulating an ink at 15% pigment concentration and printing the ink in a standard layer (1 µm).

P.R.184 is used especially in applications where P.R.57:1 fails the requirements for alkali, acid, or soap fastness, or where the pigment lake is not lightfast enough to satisfy the demand. Depending on the standard depth of shade, the lightfastness of P.R.184 exceeds that of P.R.57:1 by approximately one step on the Blue Scale. Excellent fastness to solvents makes P.R.184 a suitable candidate for letterpress and offset application, as well as for packaging gravure printing inks and flexo inks. Its metal deco prints tolerate clear lacquer coatings, but not sterilization. The pigment is heat stable up to 170 °C for 10 minutes or up to 160 °C for 30 minutes. P.R.184 is also employed in rubber.

Pigment Red 187

Two crystal modifications are known, which differ considerably in terms of shade and fastness properties. Only the bluish red form continues to be commercially available. Its high specific surface area (about 75 m^2/g) makes it a transparent pigment.

The main market for P.R.187 is in the plastics field. The pigment does not migrate in plasticized PVC and shows very good lightfastness: depending on the depth of shade and on the amount of TiO$_2$, it equals step 6 to step 8 on the Blue Scale. However, in PVC its white reductions are frequently replaced by the yellower and brighter benzimidazolone pigment P.R.208. P.R.187 exhibits high heat stability in polyolefins; in HDPE, 1/3 SD samples, reduced with 1% TiO$_2$, retain their color upon exposure to 250 °C for 5 minutes; in LDPE, the temperature may be raised to 270 °C, and the heat stability of 1/25 SD colorations is 10 °C higher. The shrinkage of the polymer is only slightly affected by the pigment. As a result, P.R.187 may be used in polyolefin extrusion products which do not show rotational symmetry, such as bottle crates. The pigment exhibits equally satisfactory lightfastness in polyolefins: at 1/3 SD, transparent and opaque versions equal step 7 on the Blue Scale.

Good thermostability is an asset in media like polystyrene. P.R.187 is a very interesting product for polypropylene spin dyeing, for which it is available as a special preparation. 1/3 SD colorations tolerate temperatures up to 290 °C for 5 minutes. The pigment is also very lightfast in polyacrylonitrile spin dyeing: 1/3 SD specimens equal step 6–7 on the Blue Scale. Dry and wet crocking fastness is satisfactory and makes it a suitable pigment for home textiles, such as upholstery and carpeting.

The printing industry uses P.R.187 inks for all printing techniques. Although its shade is somewhat more bluish, its lightfastness in print equals that of P.R.208, which is stronger and also much less transparent. The thermostability of P.R.187 is about 20 °C higher: it is stable at 220 °C for 10 minutes and at 200 °C for 30 minutes. The pigment is very fast to chemicals and to clear lacquer coatings and may be sterilized, which makes it a useful transparent product for metal deco printing. Likewise, P.R.187 offers some advantage in paper lamination, i.e., in decorative prints. Fastness to light and other fastness properties in compression molded melamine sheets are comparable. The pigment is also used to color polyester films.

In paints, P.R.187 is fast to overcoating, even at high baking temperatures. This is another application in which high thermostability is an asset, which is especially true for coil coating (Sec. 1.8.2.2). The pigment lends color to industrial paints in general; its high transparency facilitates application in transparent paints, such as in films or bicycle paints, and in metallic finishes. P.R.187 tolerates exposure to light very well; incorporated in an alkyd-melamine system, full shades equal step 7–8 on the Blue Scale, while 1:600 TiO_2 reductions still reach step 6–7. Pigment weatherfastness is equally excellent, which makes P.R.187 a suitable candidate for automotive refinishes. The fact that the pigment is chemically inert makes it an interesting product for application in a variety of powder coatings, such as amine accelerated epoxy powders. The list of applications also includes artists' colors, especially wax colors.

Pigment Red 188

P.R.188 is an intense yellowish red pigment with very good fastness properties.

Its main fields of application are in printing inks and paints; the pigment is suited to all printing techniques. 1/1 to 1/25 SD letterpress proof prints equal step 5–6 to step 5 on the Blue Scale for lightfastness. The pigment shows relatively poor tinctorial strength. However, the prints are very fast to organic solvents, fats, paraffin, soap, alkali, and acid. They are also fast to clear lacquer coatings and may be sterilized. Fastness to the DIN 16 524/1 solvent mixture, however, is not perfect. Regarding heat stability, P.R.188 prints tolerate exposure to up to 220 °C for 10 minutes or to 180 °C for 30 minutes. Similarly shaded but somewhat more dull Toluidine Red pigments are less lightfast by one step on the Blue Scale, and they are considerably inferior in terms of solvent and heat stability, as well as other special application requirements.

The paints industry uses P.R.188 for its excellent fastness properties in high grade industrial finishes. Although some darkening is observed upon exposure to light, its full shades equal step 7 on the Blue Scale for lightfastness. TiO_2 reductions up to 1:5 exhibit the same fastness to light. Pigment weatherfastness is equally excellent in full shade and related deep shades. The pigment does not bleed at baking temperatures of 160 °C or less, and it retains its color in a coating up to 200 °C. P.R.188 is frequently found in decorative paints, an area in which it is used to a great extent in the USA.

A highly opaque P.R.188 grade has recently been introduced to the market. It provides brilliant, still yellower shades of red than the previously known types. It also shows improved lightfastness and weatherfastness, as well as very good rheological properties and high gloss, compared to these types. The pigment is mainly used to produce lead-free, clean shades of red in automotive and industrial paints.

In plasticized PVC systems, P.R.188 blooms at low concentrations, which is also true for rigid PVC. Its thermostability in polyolefins is only about 220 °C, which makes it an unsuitable pigment for most polymers. However, the pigment is used to some extent in PVC and PUR plastisols.

P.R.188 is also employed in paper mass coloration, paper surface coloration, paper pulp, and paper spread-coating formulations, as well as in wallpaper and wax crayons.

Pigment Red 210

P.R.210, a chemical modification of P.R.170, is used primarily in printing inks. Its hue is considerably bluer than that of P.R.170.

P.R.210 is inferior to P.R.170 in various aspects of pigment performance. 1/3 SD and 1/25 SD letterpress proof prints, for instance, score 1/2 to 1 step less on the Blue Scale for lightfastness. In contrast to P.R.170, which is almost entirely fast to clear lacquer coatings and sterilization, P.R.210 is very sensitive in this respect. The two pigments behave similarly in print; P.R.210 prints are equally fast to a number of organic solvents, paraffin, soap, alkali, and acid. They are also heat stable up to 200 °C.

Besides printing inks, P.R.210 is primarily found in aquarelle colors.

Pigment Red 212

P.R.212, whose chemical constitution has not yet been published, is rarely sold outside Japan. It produces considerably bluish shades of red, which might be considered pink. Compared to the similarly colored quinacridone pigment P.R.122, P.R.212 is duller, weaker, and less fast. 1/3 SD letterpress proof prints, for instance, equal step 2–3 on the Blue Scale for lightfastness; while prints containing P.R.122 equal step 6–7. P.R.212 is also less lightfast than the somewhat more bluish and duller benzimidazolone pigment P.V.32, a coloristically related maroon; the difference is 2 to 3 steps on the Blue Scale.

P.R.212 is used throughout the graphics industry and for textile printing, as well as in specialized media, such as colored pencils.

Pigment Red 213

P.R.213 is a recent product and is rarely encountered outside Japan. Its exact chemical constitution has not yet been published.

The pigment affords very bluish shades of red, which are too much on the blue side to match the standard magenta for three and four color printing. The commercially available type is considerably opaque but not very lightfast compared to other Naphthol AS pigments. 1/3 SD letterpress proof prints, for instance, equal only step 3–4 on the Blue Scale for lightfastness; while 1/25 SD specimens match step 3.

Pigment Red 222

Although the pigment is known to be a member of the Naphthol AS pigment series, its chemical constitution remains to be published. It affords a bluish red shade, which may be considered magenta or ruby. The commercially available type is very transparent.

Its main market is in printing inks, where it is used in three and four color printing. P.R.222 is heat stable up to 180 °C, which makes it a suitable candidate for metal deco printing. It also tolerates organic solvents well. Prints containing P.R.222 are fast to overcoating but may not be sterilized. Gravure prints on plasticized PVC films show a slight tendency to migrate.

The plastics industry uses P.R.222 primarily in polyurethane. The pigment exhibits average tinctorial strength. 1/3 SD colorations in HDPE, for instance (1% TiO_2), are formulated at 0.23% pigment concentration.

P.R.222 is heat stable up to 240 °C; above this temperature, the color shifts appreciably towards the bluish side of the spectrum. P.R.222 bleeds in plasticized PVC.

Pigment Red 238

This pigment, which was introduced to the market a few years ago, enjoys only limited regional importance; its exact chemical constitution has not yet been published. It is recommended for printing inks and for industrial paints. Its shade in print is much bluer than the CIE 12-66 standard magenta for three color printing on the European Color Scale (Sec. 1.8.1.1). A pigment of low tinctorial strength, P.R.238 is neither fast to clear lacquer coatings, nor does it tolerate sterilization. In paints, P.R.238 is of average tinctorial strength and it bleeds.

The pigment is recommended for textile printing.

Pigment Red 245

P.R.245 is of little commercial significance and is rarely encountered in Europe. It provides bluish shades of red, ruby, and carmine. P.R.245 is used in printing inks, particularly in packaging inks. Suitable inks contain nitrocellulose polyamide-based polymers or VC/VAc copolymers; in these media, the pigment exhibits high tinctorial strength, but only average to poor lightfastness.

Pigment Red 253

P.R.253, which was introduced to the market a few years ago, provides medium red shades, which are considerably yellower than those of P.R.170. The commercial grade of P.R.253 exhibits good transparency. The pigment is recommended especially for use in paints.

Its fastness to organic solvents and chemicals corresponds to that of other representatives of group II Naphthol AS pigments (Sec. 2.6.2). Consequently, P.R.253 is almost completely fast to overpainting.

In full shade and in similarly deep shades, P.R.253 shows good lightfastness and weatherfastness with some darkening. White reductions with TiO_2, however, exhibit noticeably less weatherfastness.

P.R.253 is used in packaging printing inks. For a pigment of its class, it shows medium tinctorial strength. The prints demonstrate good fastness to organic solvents, but they are not completely fast to the DIN 16524 solvent mixture (Sec. 1.6.2.1) (step 4–5). 1/1 SD letterpress proof prints show a lightfastness corresponding to step 5 on the Blue Scale.

Pigment Red 256

The exact chemical constitution of P.R.256, which was introduced to the market a few years ago, has not yet been published. P.R.256 is used especially for industrial and architectural paints in applications where fastness to overcoating is not required. The pigment provides a yellowish red shade, a scarlet. Compared to other pigments covering the same range of shades and fastness properties, P.R.256 shows medium tinctorial strength. It exhibits good lightfastness; its full shade and similarly deep shades reach step 7 on the Blue Scale. Similar values are found for reduced shades with TiO_2, up to a ratio of approximately 1:4.

Pigment Red 258

This pigment was reported to the Colour Index by a Japanese producer. The chemical constitution has been disclosed. P.R.258 has not yet been introduced into the market.

Pigment Red 261

P.R.261, together with its chemical constitution, was recently reported to the Colour Index by an American manufacturer. It has not yet been introduced into the market.

Pigment Orange 38

P.O.38, a clear yellowish red, is mainly used in printing inks and in plastics.

In printing inks, P.O.38 shows good fastness properties. 1/1 SD to 1/25 SD letterpress proof prints, for example, equal step 6 to step 5 on the Blue Scale for lightfastness, depending on the depth of shade. The prints are stable to the standard DIN 16 524 solvent mixture, to paraffin, fats, oils, soap, alkali, and acid; they also tolerate clear lacquer coatings and sterilization. They are heat stable up to 220 °C for 10 minutes or to 200 °C for 30 minutes. P.O.38 is therefore suitable for offset printing inks as well as for metal deco printing; it is used in special printing inks for substrates such as plasticized PVC and in various flexo inks.

Besides, P.O.38 lends color to decorative inks for laminated sheets. Inks formulated at a pigment concentration of 8% and printed in 20 μm cell melamine sheets equal step 6–7 on the Blue Scale for lightfastness; at 40 μm, they match step 7. Laminated sheets do not exhibit plate-out.

P.O.38 is stable to styrene monomer, but not entirely fast to acetone, which occasionally leads to bleeding if the pigment is processed according to the diallyl-phthalate method (Sec. 1.8.1.2). This is also true if the pigment is treated with a melamine resin solution.

The plastics industry uses P.O.38 primarily to color PVC, polyolefins, and polystyrene. In plasticized PVC, the pigment does not bleed at concentrations up to about 0.3%. Its lightfastness is good. Transparent colorations in rigid PVC (0.1%) equal step 7–8 on the Blue Scale; while opaque versions (0.1% pigment + 0.5% TiO_2) equal step 7.

P.O.38 is frequently employed in combination with P.Y.83 to provide brilliant and very lightfast shades of orange. P.O.38 is particularly useful to produce shades of brown in PVC and PUR imitation leather. 1/3 SD HDPE formulations containing 1% TiO_2 withstand temperatures up to 240°C for 5 minutes. At this temperature, the pigment does not noticeably affect the polymer shrinkage. Transparent specimens equal step 6 on the Blue Scale in terms of lightfastness; while opaque versions equal step 5–6.

P.O.38 is also used in spin dyeing, where it colors filament, fibers, and films made of secondary acetate. The pigment possesses excellent fastness properties in this medium.

P.O.38 is especially lightfast in unsaturated polyester resins. Transparent samples equal step 8 on the Blue Scale, opaque versions (0.1% pigment + 0.5% TiO_2) equal step 7; however, it should be noted that the hardening of the resin is slowed down considerably.

As far as the paints field is concerned, P.O.38 is solvent resistant enough to qualify for application in high grade industrial paints wherever lightfastness and durability requirements are not too stringent. The lightfastness of close to full strength shades in an alkyd-melamine resin system, for instance, equals step 6–7 on the Blue Scale (1:1 TiO_2); further addition of white pigment reduces the lightfastness to step 5–6. The pigment does not bloom; however, it ceases to be completely fast to overcoating as the baking temperatures rise. It is heat stable up to 180°C.

P.O.38 is broad in scope. The list of applications includes special media, such as wax crayons, artists' colors, and wood stains, including those that are solvent based. The products are very lightfast (step 7 on the Blue Scale) and fast to overcoating. Blends of P.O.38 with yellow pigments, such as P.Y.83 or P.Y.120, or with carbon black produce useful shades of brown.

Pigment Violet 25

At present, P.V.25 is only produced in Japan and enjoys limited regional significance. It is used in printing inks.

Pigment Violet 44

A Japanese product like P.V.50, the pigment is registered in the Colour Index as P.V.44. The two pigments are extremely similar, both coloristically and in terms of performance in application, even in detail. It is very likely that the products are chemically identical and listed twice in the Colour Index.

Pigment Violet 50

P.V.50, produced in Japan, is only of limited regional importance, since it fails to satisfy the increasingly stringent application requirements for organic pigments. Its fastness to light, for instance, is particularly poor: 1/3 SD and 1/1 SD letterpress proof prints equal only step 2 on the Blue Scale. Compared to P.V.23, Dioxazine Violet, P.V.50 is redder at greater depth of shade, while samples which are further reduced with TiO_2 appear bluer and considerably duller. P.V.50 is tinctorially weaker than P.V.23, which in some media is as much as twice as strong as the former.

P.V.50 is used in printing inks and in office articles. Poor lightfastness and a strong tendency to migrate makes it an inadequate product for most plastics materials. Lack of lightfastness also precludes its use in paints (see also P.V.44).

Pigment Blue 25

P.Bl.25 is only produced in small volume in Europe, Japan, and the USA. It is used for the spin dyeing of secondary acetate, lends color to rubber, and is found in inks for packaging purposes.

Its hue is a somewhat reddish navy blue, which varies considerably if the pigment is chemically modified.

P.Bl.25 is very fast in application; it is fast to fats, oils, soap, and paraffin, which makes it a suitable candidate for packaging inks. Its lightfastness, however, is not excellent. In natural rubber, P.Bl.25 tolerates curing very well, and it bleeds neither into the rubber nor into the fabric backing (Sec. 1.8.3.6). In rubber, the pigment is fast to cold and hot water, to soap, soda, and alkali solutions, and to acetic acid.

In spin dyed secondary acetate threads, fibers,and films, P.Bl.25 exhibits good textile fastness properties; the only problem is a certain lack of fastness to bleaching with sodium hypochlorite (Sec. 1.6.2.4). Its fastness to light in 0.1% spin dyed specimens equals step 3–4 on the Blue Scale, while 1% samples equal step 5.

References for Section 2.6

[1] D. Kobelt, E.F. Paulus and W. Kunstmann, Acta Crystallogr. Sect. B 28 (1972) 1319–1324 and Z. Kristallogr. 139 (1974) 15–32.
[2] A. Whitaker, Z. Kristallogr. 146 (1977) 173–184.

2.7   Red Azo Pigment Lakes

The term refers to azo colorants bearing sulfonic and/or carboxylic acid functions, which are used as pigments after being rendered insoluble by conversion into insoluble alkali earth or manganese salts.*

Red azo pigment lakes may be classified according to the coupling component. There are four industrially important groups. Pigments are based on the following coupling components:

- β-naphthol,
- 2-hydroxy-3-naphthoic acid (BONA),
- Naphthol AS derivatives,
- naphthalene sulfonic acid derivatives.

If β-naphthol is used as a coupling component, the diazo component must be substituted with sulfonic or carboxylic acid groups. Commercially important pigments are frequently produced by diazotization with sulfonic acid derivatives.

In contrast to red azo pigment lakes, which have stimulated considerable technical interest, the correponding yellow pigments are much less important (Sec. 2.3.1.2; 2.3.4.1; Table 13).

2.7.1   β-Naphthol Pigment Lakes

The development of azo pigment lakes was initiated by the discovery of "Lithol Red" by Julius (BASF) in 1899. Lithol Red, which is synthesized by using 2-naphthylamine-1-sulfonic acid as a diazonium compound, was initially employed in the form of its calcium and barium salts, which were precipitated onto inorganic carrier materials. The pigment was used in its pure form after it became apparent that the carriers contribute very little to the application properties of the product. Lithol Red is one of the earliest colorant developed specifically for application as pigment.

Lithol Red was followed by Lake Red C pigments, which were discovered by Meister Lucius & Brüning, now Hoechst AG, in 1902.

β-Naphthol pigment lakes have lost much of their importance, although there are still some types that maintain a very important position in the market.

---

* Pigment Yellow 104 is listed in this chapter because of its chemical constitution.

2.7.1.1 Chemistry, Manufacture

β-Naphthol-based red azo pigment lakes are characterized by the following general structure:

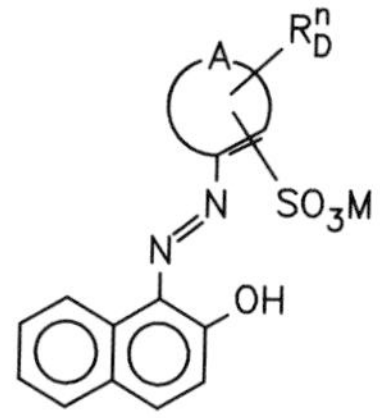

In commercially important pigments, A typically stands for a benzene or a naphthalene ring, $R_D = Cl, CH_3, C_2H_5$, COOM, n = 0 through 2, and M is usually an alkaline earth metal, sometimes a manganese, aluminum, or sodium atom.

The pigments are manufactured by the conventional method of first diazotizing an aromatic amine and subsequently coupling the product onto a suspension of β-naphthol. The latter is prepared by precipitating the β-naphthol sodium salt with dilute acid, such as acetic acid. The coupling initially affords the usually water soluble sodium salt of the respective azo dye. Addition of a suitable alkaline earth salt, such as calcium, barium, or strontium chloride, or manganese (II) sulfate, exchanges the metal ion, i.e., precipitates the insoluble alkaline earth or manganese salt as a pigment. The laking conditions must be precisely specified in order to afford the desired crystal modification and particle size distribution. Additives such as colophony-based rosin are frequently used to advantage.

In the past it used to be quite common for customers of the pigment industry, especially for printing ink manufacturers throughout western Europe, to be supplied with the sodium salts of the sulfonated β-naphthol compounds. Pigment formation, e.g., precipitation was then typically carried out by the printing ink manufacturer.

2.7.1.2 Properties

β-Naphthol pigment lakes vary in shade from yellowish to bluish red; it should be noted, however, that these products tend to be yellower than corresponding BONA pigment lakes based on the same diazonium component (Sec. 2.7.2; p. 323). β-Naphthol pigment lakes produce pure shades. Most afford varying shades, depending on the synthetic method and also on the metal. The occurrence of a variety of red shades in these pigments is also attributed to the influence of different crystal modifications. The calcium salt of Pigment Red 49, for instance, exists in several modifications.

As a result of their salt character, β-naphthol pigment lakes are faster to solvents and more resistant to migration than β-naphthol pigments, but also less lightfast. They are only moderately fast to alkaline agents. The polar character of these pigments is responsible for their good heat stability.

2.7.1.3   Application

Good general fastness properties make β-naphthol pigment lakes suitable candidates primarily for the printing inks and plastics industry. Their primary area of application varies according to the type. β-Naphthol pigment lakes are also employed in paints and in emulsion paints, but to a lesser degree.

2.7.1.4   Commercially Available Pigments

General

Pigment Red 53 has for some time dominated the β-naphthol pigment lake market in Europe and Japan. Pigment Red 49, employed especially as the barium and less frequently as the calcium salt, is mainly used in the USA. Other pigments of this type are less important.

A certain amount of emphasis is placed on Pigment Red 68, with its two acidic groups available for salt formation.

The diazo components for all of the commercial pigments of this type, with the exception of P.O.17, 17:1, and P.R.51:1, typically carry a sulfonic acid group in ortho position to the NH$_2$ function on the aromatic ring.

In Pigment Red 50:1, C.I. No. 15500, the sulfonic group in the diazo component is replaced by the barium salt of a carboxy group:

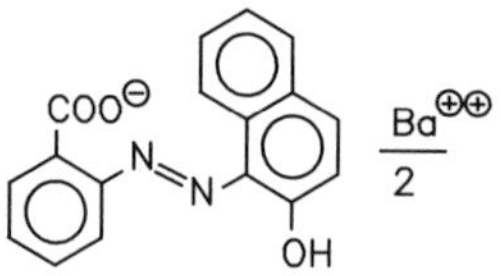

Individual Pigments

Pigment Red 49 Types

Products of the Pigment Red 49 range differ considerably in their commercial importance, depending on the region in which they are marketed. Although of little importance in Europe and Japan, P.R.49 types play a major role in the USA. The pigments are supplied as sodium (P.R.49) or as calcium salts (P.R.49:2); but the largest fraction is sold as barium salts (P.R.49:1). The product line has recently been extended to include the strontium salt (P.R.49:3). The shade of the barium lake is yellower than that of P.R.57:1 and bluer than that of P.R.53:1. Calcium types are somewhat bluer than P.R.49:1, while the sodium salt is more yellowish.

Pigment Red 49 types are often referred to as Lithol Red pigments, especially in the USA. They are used in all types of printing inks. Throughout the USA, Pig-

Tab. 19: Commercially available β-naphthol pigment lakes.

General chemical structure:

$\widehat{DC}$ = diazo component group

C.I. Name	C.I. Constitution No.	DC	$\frac{M}{2}$	Shade
P.R.49	15630	SO_3M (naphthalene)	2 Na	yellowish red
P.R.49:1	15630:1		Ba	yellowish red
P.R.49:2	15630:2		Ca	bluish red or maroon
P.R.49:3	15630:3		Sr	red
P.R.50:1	15500:1	Formula see p. 316	Ba	scarlet
P.R.51	15580	H_3C–, SO_3M (benzene)	Ba	scarlet
P.R.53	15585	SO_3M, Cl, CH_3 (benzene)	2 Na	scarlet
P.R.53:1	15585:1		Ba	scarlet
P.R.53:–	15585:–		Sr	yellowish red
P.R.68	15525	SO_3M, Cl, $COOM$ (benzene)	2 Ca	yellowish red
P.O.17	15510:1	SO_3M (benzene)	Ba	orange
P.O.17:1	15510:2		$\frac{2}{3}$ Al	reddish orange
P.O.46	15602	SO_3M, Cl, C_2H_5 (benzene)	Ba	yellowish red

ment Red 49 grades are frequently employed to lend red shades to publication gravure printing inks targeted for three or four color printing. P.R.49:1 is slightly too yellow to match the CIE 12-66 standard for process printing inks on the European Color Scale for offset printing.

Pigment Red 49

P.R.49, sodium salt, is used to a limited extent. A pigment with high tinctorial strength, it is employed particularly in economical, solvent-based flexographic printing inks. Both as a pigment and in print, it is much less fast to various solvents than other metal salts. The same is true for its fastness to water, which precludes its use in aqueous printing inks. Moreover, P.R.49 is also less lightfast than other types.

Pigment Red 49:1 and 49:2

Both the barium lake P.R.49:1 and the calcium lake P.R.49:2 resemble Lake Red C pigments (P.R.53:1) in terms of fastness to organic solvents, to alkali and acid, and to fastness in application. However, P.R.49 grades are much less heat stable, which narrows their applicability in plastics. Areas of application, especially in the USA, include elastomers, to a lesser extent also inexpensive industrial paints, air drying and nitro paints. Their main market, however, is in printing inks.

According to production figures compiled by the US International Trade Commission, about 2/3 out of the 2106 t of P.R.49 type produced in the USA in 1988 were used for publication gravure printing inks, while the major share of the remaining amount was targeted for packaging gravure and newsprint inks. Resinated grades are also available. These are typically more transparent, more brilliant, and show less of a tendency to bronze in print than resin-free varieties. Special-purpose types, on the strength of their good rheological properties, are supplied for use in water-based printing inks.

Pigment Red 50:1

P.R.50:1, patented as early as 1905 by Meister Lucius & Brüning, is only of regional importance today; its sale has been discontinued in Europe. Marketed first as a sodium lake (P.R.50) to be transformed into the barium salt by the consumer, the barium lake (P.R.50:1) is now exclusively used. This salt affords clean shades of scarlet.

P.R.50:1 is used primarily in packaging printing inks, which mostly afford bronzing prints with good fastness to organic solvents. Weak alkaline agents induce a color shift to blue shades, while acids have a yellowing effect. In the presence of drying agents based on cobalt salts, the pigment undergoes a reformulation of its salt form to produce shades of brown. P.R.50:1 has accordingly lost its importance in offset inks, which used to be its major market. The pigment fails to meet modern requirements.

Pigment Red 51

P.R.51, a barium pigment lake, affords shades of scarlet similar to those of Pigment Red 53:1. P.R.51 performs like P.R.49 in terms of fastness, but it is particularly sensitive to alkaline agents. P.R.51, which used to be quite important, especially in printing inks and rubber, has largely been replaced by other products. It continues to be used in typewriter and printer ribbons for its shade.

Pigment Red 53 Types

Barium and sodium lakes are commercially available (P.R.53:1 and 53). Calcium salts (P.R.53:2) have only enjoyed commercial success for a limited time; they have largely vanished from the market. The production of strontium lakes has recently been discontinued.

Pigment Red 53

P.R.53, the sodium salt which used to be transformed into the barium salt by the customer, continues to be used occasionally for its very yellowish shade. Its use is restricted by poor solvent and application fastness properties.

Pigment Red 53:1

P.R.53:1, a barium lake, is one of the most important red pigments for use in printing inks. Its scarlet shade is much yellower than that of P.R.57:1, which is the DIN 16 539 standard magenta for three and four color printing. This precludes its use for this purpose. P.R.53:1 has stimulated special interest as a so-called warm red. It has also become increasingly important as a replacement for P.R.49, although its importance varies considerably with the region. P.R.53:1 is used especially in disposable printed products, especially in sheet and web offset, gravure, and flexographic printing inks. 1/1 SD letterpress proof prints equal step 3 on the Blue Scale for lightfastness, while 1/25 SD specimens reach only step 1. The Naphthol AS pigments P.R.9 and 10, which are somewhat yellower in equally deep shades, are more lightfast than the less expensive Lake Red C types. The difference comprises several steps on the Blue Scale; 1/25 SD samples, for instance, equal step 5 (P.R.9) and, respectively, step 4 (P.R.10). Both pigments are close in shade to P.R.53:1.

P.R.53:1 is a comparatively strong and brilliant pigment within its range of shades. Inks formulated at about 19 and 11% pigment concentration, respectively, suffice to print standard letterpress layers of 1/1 and 1/3 SD samples. For comparison, it should be noted that the corresponding values for P.R.9 and 10 are approximately 20.5% and 13.5% pigment, respectively. Yellowish types of P.R.53:1 are usually less strong than bluish ones.

In print, P.R.53:1 shows good fastness to a series of organic solvents: it is almost completely stable to the DIN 16 524 solvent mixture. The pigment is entirely fast to clear lacquer coatings. As a result of the chemical constitution of the pigment, the prints are sensitive to alkali and acids. Moreover, there are some drawbacks to pigment performance in a number of special applications: the prints are, for instance, affected by agents such as soap or butter. Their heat stability, on the other hand, is excellent. P.R.53:1 prints tolerate exposure to more than 200 °C for 10 minutes. In contrast to the similarly colored P.R.68, which is a member of the same class of pigments, P.R.53:1 is not fast to sterilization. The commercially available P.R.53:1 types differ considerably in their degree of transparency. Specific surface areas are around 16 to 50 m^2/g. Today, resinated grades have somewhat less of a technical value.

P.R.53:1, which is used in large volume in aqueous flexographic printing inks, may present problems in terms of storage stability. Such basic printing inks tend to show a more or less pronounced viscosity increase, almost invariably combined with a shift to yellowish shades. The latter, however, is not very noticeable in deep shades. The lack of storage stability is attributed to interaction between pigment or, more precisely, between the alkaline-earth metal of the pigment on the one hand and the alkaline agent or the base of the printing ink on the other hand, which ultimately exchanges the metal.

P.R.53:1 is used for its excellent heat stability as a colorant for plastics. 1/3 SD samples in HDPE withstand exposure to about 260 °C for 5 minutes.

P.R.53:1 specimens exhibit medium tinctorial strength. P.R.68, a member of the same group of pigments, is somewhat yellower and even more heat stable than P.R.53:1. The disazopyrazolone pigments P.R.37 and 38 are, respectively, slightly and considerably bluer, but much less heat stable in PE. Under common processing conditions, up to 260 °C, P.R.53:1 has practically no effect on the shrinkage of injection-molded polyolefins. Its lightfastness in PE is approximately between step 3 and step 1 on the Blue Scale, depending on the depth of shade and on the pigment concentration.

P.R.53:1 is also very strong in PVC. It is not very lightfast, but a number of disposable articles are colored to advantage with P.R.53:1. However, it has a poor fastness to bleeding. For economical reasons, P.R.53:1 is also employed in polystyrene, in which it is heat stable up to 280 °C and moderately lightfast (step 1–2 on the Blue Scale).

P.R.53:1 may also be used in PUR foam products wherever the fastness requirements, especially regarding water, soap, and solvents, are not too stringent.

The pigment also lends itself to application in elastomers, such as natural rubber blends. It is migration fast enough to satisfy most specifications. P.R.53:1 is also completely bleed resistant in natural rubber, although some color is transferred into the wet cotton cloth liner, the wrapper (Sec. 1.8.3.6). P.R.53:1 is lightfast enough for most applications. Products containing P.R.53:1 are not always entirely fast to hot water or alcohol.

Where it satisfies the fastness specifications, especially regarding lightfastness, P.R.53:1 may also be employed in paints. It is also found in applications outside the large pigment consuming industries. Less stable types lend color to products such as cleaning agents and office articles, where P.R.53:1 is found in inexpensive color pencils and water colors. However, in several countries there are legal limits concerning the amount of soluble barium traces which are permissible in the pigment (Sec. 5.3.5).

A new crystal modification of P.R.53:1 has recently been introduced to the market, which is slightly weaker but considerably yellower than conventional barium or strontium lakes. Both modifications perform similarly in application, and they are used in similar areas. Pigment mixtures afford a comparatively large variety of intermediate shades.

Pigment Red 53, Strontium Lake

Production of the strontium lake has recently been discontinued. The strontium lake performs like barium lakes of P.R.53:1 and is equally fast to organic solvents, to alkali and acid. Both types are equally heat stable and lightfast, for instance in print. Coloristically, the strontium salt of P.R.53 is yellower, but it does not quite match the shade of the above-mentioned novel crystal modification of the barium lake. This pigment, while on the market, was used for the same purposes as P.R.53:1.

Pigment Red 68

P.R.68 is supplied as a calcium salt. It produces a yellowish red shade, referred to as scarlet, which at equal depth of shade is slightly bluer than P.R.53:1, a member of the same class of pigments. P.R.68 has gained considerably less commercial recognition than the widely used Lake Red C pigments.

The main area of application for P.R.68 is in the coloration of plastics. The pigment, whose tinctorial strength is average, is fast to blooming but not quite bleed resistant in plasticized PVC. Its lightfastness, especially in white reduction, fails to satisfy more stringent requirements. 1/3 SD samples which contain 5% TiO_2 equal step 4 on the Blue Scale for fastness to daylight, while 1/25 SD specimens are equal to step 3. P.R.68 is heat stable enough to be applied safely in polyolefins. Incorporated in HDPE, for instance, it is stable up to 300°C. The pigment exhibits average tinctorial strength. P.R.68 does not affect the shrinkage of injection-molded polymers. The lightfastness of 1/3 to 1/25 SD specimens equals between step 2–3 and step 3 on the Blue Scale. P.R.68 is also found in polystyrene.

The pigment is also used as a colorant for printing inks, but its importance in this area has diminished considerably. P.R.68 is somewhat weaker than P.R.53:1, but more lightfast. 1/3 SD letterpress proof prints equal step 4 (2) on the Blue Scale for lightfastness, while 1/25 SD specimens equal step 3 (1). The values in brackets represent the lightfastness of P.R.53:1. In print, P.R.68 shows excellent fastness to a number of organic solvents, including the standard DIN 16 524 solvent blend. Its resistance to soap, alkali, and acid, however, is unsatisfactory. P.R.68, in contrast to P.R.53:1, is completely fast to sterilization. Since P.R.68 also provides excellent heat stability (1/3 SD samples withstand exposure to 220 °C for 10 minutes), it lends itself to metal deco printing, although it is not fast to clear lacquer coatings. Like in other application media, insufficient fastness of P.R.68 inks to water may affect the results of offset printing.

P.R.68 is used in paints for applications where it performs well, especially in terms of lightfastness. Incorporated in an air drying alkyd resin vehicle, full shades equal step 5 on the Blue Scale for lightfastness. 1:5 TiO_2 reductions, however, only score as high as step 2. P.R.68 exhibits excellent resistance to a large number of organic solvents, while it is quite sensitive to alkali and acid, as well as to water.

P.R.68 is also recommended for use in decorative cosmetics, such as nail polish, lipstick, powders, and shading creams. For these purposes, types are supplied which fulfil legal purity standards and are tested accordingly.

Pigment Orange 17

Besides being offered as a barium salt, P.O.17 is also available as a precipitate onto aluminum oxide hydrate. These types, especially in the USA, are referred to as Persian Orange. The products have been discontinued in Europe, but play a certain role in the Japanese market. P.O.17 affords brilliant, orange-red shades. Commercial types are tinctorially strong and highly transparent. They are used in inexpensive packaging prints, including metal deco printing. Varieties which contain a large amount of aluminum oxide hydrate often adversely affect the drying properties of oily media, such as offset printing inks. In a highly acidic binder, the pigment may lose its metal ion as it reverts to the free acid and consequently undergo considerable color shift. In the past, P.O.17 has also been used in large volume to color natural rubber.

Pigment Orange 17:1

Aluminum lakes of P.O.17:1 are available in the USA. Properties which are pertinent to the coloristics and application of the pigment parallel those of the barium lakes. P.O.17:1, an equally high strength pigment, also affords brilliant shades and is used in packaging prints, especially for paraffin-based wrapping paper for bread.

Pigment Orange 46

P.O.46, which has only recently gained minor commercial recognition in Europe, plays somewhat of a role in the US market. According to the US International

Trade Commission, 386 t were produced nationwide in 1988. The application properties of the pigment are quite similar to those of the Lake Red C types, P.R.53:1. P.O.46 produces a shade which is appreciably yellower than that of conventional P.R.53:1 types. Its shade is much more comparable to the color of the new, only recently marketed crystal modification of the barium salt of Lake Red C.

Commercially available P.O.46 types are usually quite transparent. They are employed primarily in packaging printing inks, also in offset and metal deco printing. Publication gravure inks, plastics, especially PVC, LDPE, and elastomers, as well as general industrial paints are suitable media for pigment application. P.O.46 is less solvent resistant than P.R.53:1, but it is faster to alkali and acid. In terms of lightfastness, P.O.46 performs poorly: 1/3 and 1/25 SD prints equal only step 1 on the Blue Scale.

2.7.2 BONA Pigment Lakes

These pigments derive their name from 2-hydroxy-3-naphthoic acid, which is used as a general coupling component for the entire group. The compound is also known as **beta-oxynaphthoic acid (BONA)**.

The history of BONA pigment lakes parallels that of β-naphthol pigment lakes. Literary evidence of the use of 2-hydroxy-3-naphthoic acid, which was first synthesized in 1887 by Schmitt and Burkard as a coupling component, dates from 1893 (Kostanecki). However, it was not until 1902 that the compound began to be employed in dye synthesis by AGFA (aniline → BONA).

Pigment Red 57, discovered in 1903 by R. Gley and O. Siebert at AGFA, developed into one of the most important organic pigments in the market. It was first supplied in the form of its yellowish red sodium salt, to be converted to the calcium or barium salt by the consumer.

Initially, these "lakes" were precipitated onto inorganic carrier materials. It is interesting to note that this group of colorants was originally used for pigments in paints. Application in the textiles market followed later. Today, it is mainly the calcium and barium salts, but also manganese and, less frequently, strontium salts that have the largest sales volume.

Technically important pigments in this group – similar to β-naphthol pigment lakes – contain aminosulfonic acids as diazonium components. Therefore, these pigments always contain two acidic groups for salt formation.

2.7.2.1 Chemistry, Manufacture

BONA pigment lakes are monoazo pigments, characterized primarily by the following general chemical structure:

In commercially important pigments, R_D usually stands for hydrogen, chlorine, or a methyl group. M typically represents a divalent metal atom from the alkaline-earth series, including calcium, barium, strontium, or manganese.

As a diazonium component, aniline sulfonic acid may occasionally be replaced by 2-aminonaphthalene-1-sulfonic acid or by aniline.

Aromatic aminosulfonic acids are synthesized by a sequence of important industrial processes, including sulfonation of benzene. This is followed, wherever necessary, by chlorination, nitration, and reduction, or by aniline sulfonation, possibly involving subsequent "baking" [1,2].

BONA pigment lakes are basically synthesized like β-naphthol pigment lakes.

Diazotization of the aminosulfonic acid and subsequent coupling onto the sodium salt of 2-hydroxy-3-naphthoic acid initially affords the monoazo compound in the form of its soluble sodium salt. Subsequent reaction with chlorides or sulfates of alkaline earth metals or with a manganese salt, frequently in the presence of a dispersion agent, rosin or its derivatives, at elevated temperature yields the insoluble BONA pigment lake.

2.7.2.2   Properties

In terms of performance, BONA pigment lakes are most closely related to β-naphthol pigment lakes. BONA pigment lakes afford more bluish red shades ("ruby", "maroon") than their β-naphthol counterparts and are also more lightfast. This is especially true for the manganese salts. Other aspects of pigment properties, however, such as fastness to alkali, soap, and acid, solvent and migration fastness, and heat stability, are very similar. BONA pigment lakes are products with high tinctorial strength.

2.7.2.3   Application

The majority of pigments within this group find extensive use in printing inks. There are also some products which are used primarily in paints. BONA pigment lakes play a role as colorants for plastics as well as for a variety of other media.

2.7.2.4   Commercially Available BONA Pigment Lakes

General

As with β-naphthol pigment lakes, there is only a limited number of BONA pigment lakes that are marketed in large volume. Two of these pigments, however, maintain an important position within the pigments industry. Table 20 lists the currently available BONA pigment lakes.

Tab. 20: Commercially available BONA pigment lakes.

C.I. Name	C.I. Constitution No.	R_D^2	R_D^4	R_D^5	M	Shade
P.R.48:1	15865:1	$SO_3^\ominus$	CH_3	Cl	Ba	red
P.R.48:2	15865:2	$SO_3^\ominus$	CH_3	Cl	Ca	bluish red
P.R.48:3	15865:3	$SO_3^\ominus$	CH_3	Cl	Sr	red
P.R.48:4	15865:4	$SO_3^\ominus$	CH_3	Cl	Mn	bluish red
P.R.48:5	15865:5	$SO_3^\ominus$	CH_3	Cl	Mg	red
P.R.52:1	15860:1	$SO_3^\ominus$	Cl	CH_3	Ca	ruby
P.R.52:2	15860:2	$SO_3^\ominus$	Cl	CH_3	Mn	maroon
P.R.57:1	15850:1	$SO_3^\ominus$	CH_3	H	Ca	bluish red (magenta, ruby)
P.R.58:2	15825:2	H	Cl	$SO_3^\ominus$	Ca	bluish red
P.R.58:4	15825:4	H	Cl	$SO_3^\ominus$	Mn	medium red
P.R.63:1	15880:1	2-aminonaphthalene-1-sulfonic			Ca	bordeaux
P.R.63:2	15880:2	acid as diazo component			Mn	bordeaux
P.R.64	15800	H	H	H	$\frac{Ba}{2}$	red
P.R.64:1	15800:1	H	H	H	$\frac{Ca}{2}$	yellowish red
P.R.200	15867	$SO_3^\ominus$	Cl	C_2H_5	Ca	bluish red
P.Br.5	15800:2	H	H	H	$\frac{Cu}{2}$	brown

All pigments within this series, with the exception of P.R.64 and P.Br.5, contain two acidic functions: a sulfonic acid group and a carboxylic acid moiety. Precipitation is accomplished primarily with calcium or manganese, more rarely with barium, strontium, or magnesium. The sulfonic acid function is found almost invariably in ortho position relative to the azo group. Commercially important pigments, including Pigment Red 48 and 57, also bear a methyl group in para position relative to the azo bridge. Manganese lakes are always bluer than the strontium or barium lakes. They often appear even more bluish than the calcium salts.

Individual Pigments

Pigment Red 48 Types

Among the commercially available types are "lakes" with various metals, including barium (P.R.48:1), calcium (P.R.48:2), strontium (P.R.48:3), manganese (P.R.48:4), and magnesium salts (P.R. 48:5). There are also types, such as barium/calcium salts, which are obtained by "mixed" laking. The commercial interest in each type depends on the area of application, but the pigments enjoy overall importance. Pigments of this type are known under a variety of common names, especially in the English language. The best known designation, "2B toner", is clearly derived from the name under which the pigment was first marketed, namely as Permanent Red 2B, a sodium salt.

Pigment Red 48:1

P.R.48:1, the barium salt, is a coloristically versatile product. It affords light yellowish to medium red shades, depending on the specific surface area of the product. Fastness to a number of common organic solvents, such as esters, ketones, and aliphatic and aromatic hydrocarbons, is good. However, P.R.48:1 shows only poor resistance to soap, alkali, and acid.

The main field of application for P.R.48:1 is in printing inks and plastics. In order to enhance the transparency in print and to reduce the strong tendency of the prints to bronze, pigments that are targeted for use in printing inks are frequently supplied in resinated form.

P.R.48:1 types show high tinctorial strength, but are less strong than P.R.53:1 grades. P.R.48:1 is considerably yellower than P.R.57:1 and noticeably bluer than P.R.53:1. Although the pigment has been recommended for use in all types of printing inks, the prints of many types lack resistance to the DIN 16 524 standard solvent mixture and to clear lacquer coatings and also lack fastness to sterilization. Paints of P.R.48:1 withstand exposure to 180 °C for 10 minutes or to 160 °C for 30 minutes. Barium lakes are much less lightfast than salts of other metals. Besides, a certain lack of storage stability may create problems in aqueous media: the printing inks tend to thicken.

Incorporated in plastics, P.R.48:1 exhibits medium tinctorial strength. Its fastness to migration in plasticized PVC is good. The pigment is stable to blooming and almost completely bleed resistant. Its lightfastness in full shades equals step 3 on the Blue Scale, while 1/3 SD specimens only reach step 1–2. In terms of heat stability, 1/3 SD formulations in PE withstand 200 to 240 °C for 5 minutes, depending on the type. Higher temperatures quickly shift the color towards bluer, duller red shades. Transparent PE systems are less heat stable.

P.R.48:1 is recommended especially for use in LDPE. Besides, it also affords colorations of medium tinctorial strength in PE. 0.34% pigment is required to formulate 1/3 SD colorations with 1% TiO_2. Comparative values are listed for a number of other pigments. Problems may occur during vulcanization with open steam in the presence of P.R.48:1. The pigment bleeds into the wrapper (Sec. 1.8.3.6).

Paint manufacturers frequently use P.R.48:1 in inexpensive industrial paints, in which the pigment exhibits good fastness to overpainting. The lightfastness in full shade equals step 5–6 on the Blue Scale, but the pigment loses its fastness to light rapidly as the TiO_2 content increases. P.R.48:1 is not recommended for exterior application.

Pigment Red 48:2

In 1986, the Gravure Technical Association in the USA adopted P.R.48:2 as its standard process red for gravure packaging inks. According to the US International Trade Commission, 760 t were produced nationwide in 1988. P.R.48:2, the calcium salt, affords bluish red shades, referred to as ruby. P.R.48:2 is substantially bluer than P.R.48:1, distinctly bluer than P.R.48:4, and still noticeably yellower than P.R.57:1. In its fastness properties, for instance in terms of fastness to organic solvents, the pigment comes close to the corresponding barium salts. It is inferior to P.R.48:1 in terms of sterilization fastness in print, and the same is true for its fastness to soaps. On the other hand, P.R.48:2 is distinctly more lightfast than P.R.48:1. Exposed 1/1 SD P.R.48:2 letterpress proof prints, for instance, equal step 4–5 on the Blue Scale, while corresponding P.R.48:1 specimens only reach step 3–4.

The two types are used in almost exactly the same areas of application. P.R.48:2 is also found in printing inks, especially for packaging inks based on NC. It is employed, especially in the USA, in three and four color printing for areas such as magazine covers, where P.R.49 varieties are not lightfast enough to satisfy the specifications. Resinated grades are more transparent and bronze less in print. The viscosity of aqueous printing inks often rapidly increases during storage, a problem which may lead to thickening.

P.R.48:2 is also used in plastics, where it affords good tinctorial strength. 0.21% pigment is, for instance, needed to produce 1/3 SD HDPE colorations containing 1% TiO_2. Such colorations are heat stable up to 230°C for 5 minutes, but higher temperatures shift the color towards bluer, duller shades. At higher temperature, on the other hand, more reduced colorations (1/9 SD) shift the shade to yellower and cleaner hues. P.R.48:2 is used to an appreciable extent in LDPE, in which it is more lightfast than corresponding P.R.48:1 samples: the difference is about 2 1/2 steps on the Blue Scale. Incorporated in plasticized PVC, P.R.48:2 is not entirely bleed resistant. The pigment may be used in PP spin dyeing only if polymers with a low melting point are used. In order to ensure sufficient lightfastness, the pigment is used to advantage in deep shades, i.e., at high concentration.

P.R.48:2 is less commonly found in paints. In paints, like in other areas of application, the calcium salt performs like the barium lake. Both are, for instance, equally fast to overcoating. The list of suitable application areas for both pigments is the same. P.R.48:2 is also used in oven drying paints, nitro paints, and in similar systems. Besides, it is also found in emulsion paints. While barium and calcium salts exhibit equal lightfastness in full shades, there is a considerable difference in white reductions. Increasing amounts of TiO_2 render P.R.48:2 much more sensitive to light than P.R.48:1.

Pigment Red 48:3

P.R.48:3, the strontium salt, is distinctly bluer than P.R.48:1, noticeably yellower than P.R.48:2, and slightly yellower than P.R.48:4. The main field of application for P.R.48:3 is in plastics. Incorporated in plasticized PVC, P.R.48:3 is the most bleed resistant of all metal salts, although it does show a certain tendency to migrate. P.R.48:3 is quite often more lightfast than other P.R.48 types. Transparent samples, for instance, formulated at a 0.2% pigment concentration, equal step 6 on the Blue Scale. This is 3 steps higher than the lightfastness of corresponding dispersions of P.R.48:1 and 1/2 to 1 step higher than that of P.R.48:2 and 48:4 samples. 1/3 SD colorations containing 5% TiO_2 equal step 4 on the Blue Scale. Only white reductions of P.R.48:4 score one step higher.

P.R.48:3 is of average tinctorial strength. 1/3 SD HDPE colorations containing 1% TiO_2, for instance, are produced by using about 0.25% pigment. In terms of heat stability, P.R.48:3 withstands up to 240°C. At temperatures in excess of this value P.R.48:3 rapidly undergoes a color shift towards bluer and duller shades. The pigment is used primarily in LDPE, but also in PP spin dyeing if polymers with a low melting point are processed. The list of application media also includes polystyrene, polyurethane, and elastomers.

P.R.48:3 is also recommended for the coloration of printing inks, especially for packaging purposes, and also for the paint field. It is recommended for use in architectural paints and general industrial finishes. The pigment demonstrates overall fastness properties which parallel those of P.R.48:1 and 48:2.

Pigment Red 48:4

P.R.48:4, the manganese salt, affords red shades, which are noticeably on the bluish side of P.R.48:3 and yellower than P.R.48:2. The pigment is used in a variety of applications, especially in paints. In order to produce opaque shades of red, P.R.48:4 is frequently combined with Molybdate Orange. The pigment is considerably more lightfast and durable than other P.R.48 types, especially in full shades. In oven dyring and air drying systems, for instance, P.R.48:4 equals step 7 on the Blue Scale, while its weatherfastness after one year of exposure matches step 4–5 on the Gray Scale. Its shade darkens upon weathering. In such deep shades the pigment satisfies more stringent requirements and may even be used in applications such as automobile refinishes.

In terms of solvent fastness, however, P.R.48:4 performs less well than other P.R.48 types. It is noticeably less acid and alkali resistant and less fast to lime. Its fastness to overcoating, on the other hand, is satisfactory. Since the pigment contains manganese metal, it may accelerate the drying process in oxidatively drying paint systems. The plastics industry uses P.R.48:4 primarily in PVC and in polyo-lefins. For a pigment with good tinctorial strength in PE, only 0.18% P.R.48:4 is needed in order to formulate a 1/3 SD sample containing 1% TiO_2. Such colorations are heat stable up to 200 to 290°C for 5 minutes, depending on the type. Higher temperatures shift the color of less heat stable varieties partly reversibly towards yellower shades and also make for dull colors. However, economical considerations frequently make it necessary to compromise between coloristic perfection and price.

Incorporated in PP, P.R.48:4 often causes aging, which makes the plastic brittle and precludes the use of P.R.48:4 in this medium. The effect on PE is less noticeable. The shrinkage of PE extrusion products at higher processing temperatures is practically unaffected by P.R.48:4.

P.R.48:4 does not bloom in plasticized PVC and is almost completely fast to bleeding. Its tinctorial strength in this medium is equally good. Desirable dielectric properties make P.R.48:4 a suitable candidate for use in PVC cable insulations. The pigment is also used for mass colored secondary acetate threads, fibers, and films wherever it meets the requirements for application.

In printing inks, P.R.48:4, due to its manganese content, may accelerate the drying process in oxidative drying systems, for instance in offset printing inks. This effect must be taken into consideration in calculating the siccative content of such inks. In print, the pigment is insufficiently fast to alkali, acid, and soap, and it also fails to tolerate clear lacquer coatings and sterilization. P.R.48:4 does, however, afford good tinctorial strength in print. 18% pigment suffice to produce 1/1 SD letterpress proof prints, printed in films of standardized thickness. 1/3 SD prints, on the other hand, are produced using about 9% pigment. The prints equal step 4 on the Blue Scale for lightfastness. Exposure of the sample to light causes a certain yellow shift, the extent of which depends on the depth of shade and on the printed substrate. This coloristic change is frequently neglected in the evaluation of lightfastness: samples consequently appear several steps higher on the Blue Scale than their actual lightfastness warrants.

P.R.48:4 is also found in packaging gravure printing inks and in flexographic inks. It should be noted, however, that the manganese content in the pigment may oxidatively destroy the resin in polyamide vehicle systems, rapidly causing the printed sheets to stick together and producing an unpleasant odor. Problems may also occur as P.R.48:4 is used in different binder systems to be printed on PE films. In PE mass coloration, these materials may become brittle, i.e., age prematurely.

P.R.48:4 is also supplied in the form of resinated grades.

Pigment Red 48:5

P.R.48:5, a magnesium salt, is a comparatively recent product. It is considerably yellower and at the same time more brilliant than P.R.48:4. The pigment is also much weaker in white reductions. Compared to the manganese version, P.R.48:5 does not perform as well if exposed to light and/or weather, and it darkens in full shades. Its primary field of application is the pigmentation of polyolefins, but it is also employed to color PVC and polystyrene. Its application properties and fastness properties in print and in paints largely parallel those of P.R.48:2.

Pigment Red 52 Types

Both calcium and manganese salts are commercially available. The US market also offers strontium salts and mixed barium/calcium lakes. Although they are important pigments in the USA, P.R.52 types generally have little commercial importance in Europe. US production figures for 1983 were 300 t of P.R.52:1 and 145 t of P.R.52:2 types. These types are referred to in the USA as IBB toners.

Pigment Red 52:1

P.R.52:1, the calcium salt, covers the same range of shades as P.R.57:1. The pigment is somewhat bluer than P.R.48:2 and also duller. Both grades are very similar in terms of fastness to organic solvents. P.R.52:1 is used for printing inks, especially for solvent-based packaging gravure and flexographic printing inks. The pigment is tinctorially strong in these media. P.R.52:1 prints are only poorly to moderately fast to alkali and acid and are particularly sensitive to soap and sterilization. However, some grades which have recently been introduced in Europe are fast to alkaline agents and are almost completely acid resistant.

Resinated types are more transparent and, as a result of smaller particle sizes, are considerably bluer. In the USA they are often used instead of P.R.57:1 to produce standardized process inks for offset printing. In terms of lightfastness, P.R.52:1 does not perform as well as P.R.48:2 and 57:1. The difference is 1 to 2 steps on the Blue Scale in the first case and roughly 1 step in the latter, depending on the standard depth of shade. Various P.R.52:1 types, in contrast to P.R.57:1, have the advantage of being nonthickening in aqueous flexographic inks.

P.R.52:1 exhibits equally poor lightfastness in paints. Full shades, like those of P.R.57:1, match step 4–5 on the Blue Scale, a value which decreases considerably in white reduction. P.R.52:1 is of little interest for the coloration of plastics.

Pigment Red 52:2

P.R.52:2, the manganese lake, affords shades of maroon. Its main area of application is in oven drying systems, in which full and related shades exhibit good lightfastness and weatherfastness. In this respect, the pigment is much superior to P.R.52:1 and P.R.57:1, but does not quite reach the qualities of the yellower manganese lake P.R.48:4. Like the latter, P.R.52:2 may accelerate the drying process in oxidatively drying systems. Despite the fact that the pigment is only moderately fast to common solvents, a large number of paint systems containing P.R.52:2 exhibit excellent fastness to overpainting; however, its fastness to acid and alkali is very poor. P.R.52:2 is therefore unsuitable for use in acid- or amine-cured systems.

The pigment is recommended particularly for use in blends with Molybdate Red pigments.

Pigment Red 57:1

P.R.57:1 is known under a variety of names, the most common among which is 4B toner. P.R.57:1, the calcium salt, ranks high among organic pigments in production volume and use: according to the US International Trade Commission, 1989 production figures in the US alone were 6689 t of P.R.57:1 types. The pigment is used primarily in printing inks. Its shade matches the standard magenta of a number of color scales for three and four color printing, including the CIE 12-66 European Color Scale (Sec. 1.8.1.1).

There are enough commercially available grades of P.R.57:1 to cover a comparatively wide range of shades. Adding particular mixed coupling components, for instance to the diazonium component, will produce any desired hue within this range. A similar effect is achieved by varying the particle size distribution. Components such as Tobias acid, which are chemically related, are incorporated in the crystal lattice of the pigment. The resulting blue shift is attributed to an expansion of the crystal lattice through the comparatively large substituents of these agents. The rate of crystal growth is also inhibited and a chemical color shift is observed. These are also factors contributing to the blue shift.

P.R.57:1 is a pigment with high tinctorial strength. Approximately 19% pigment is needed to formulate offset and letterpress inks which produce 1/1 SD proof prints at standard film thickness. 1/3 SD prints, on the other hand, require 7.8% pigment, and 1/25 SD specimens are made with 2.1% pigment. Comparative values are listed under the respective pigments. Inks for prints which satisfy the specifications of the so-called European Standard are prepared under standardized conditions at pigment concentrations of about 14%.

The lightfastness of 1/1 SD P.R.57:1 letterpress proof prints equals step 4–5 on the Blue Scale, while 1/3 and 1/25 SD specimens match step 4 and step 3, respectively. In print, P.R.57:1 is fast to many of the most common organic solvents, including the DIN 16 524 standard solvent mixture, but it performs poorly in terms of soap, alkali, and acid resistance. The pigment is equally sensitive to clear lacquer coatings. Wherever prints are expected to be fast to such media, P.R.57:1 is frequently replaced by the coloristically closely related Naphthol AS pigment P.R.184, which has the added advantage of being more lightfast. However, prints made from P.R.184 are as sensitive to sterilization as those obtained from P.R.57:1. P.R.57:1 is heat stable enough to withstand exposure to 180 °C for 10 minutes or to 160 °C for 30 minutes.

A number of resinated grades are produced in order to provide higher transparency and to optimize other aspects of pigment properties in application. For reasons connected with process engineering, the resin is typically incorporated as a metal (calcium) resinate. In the past, types of P.R.57:1 additionally contained certain amounts of barium sulfate.

P.R.57:1 is found in all types of printing inks. Like with other pigment lakes, however, there is a certain possibility that storage problems may arise, a fault which is independent of the resination of a pigment. The viscosity of an alkaline printing ink often increases more or less upon storage, i.e., the ink thickens. This is a problem which is largely attributed to the interaction between the pigment lake and the cation of the base and leads to an exchange of the metal ion. Reprecipitation also plays a certain role. This phenomenon is based on the fact that the pigment lake, being a salt, is somewhat soluble in water. In an aqueous alcoholic medium, the binder converts dissociated metal ions (calcium ions) into similarly sparingly soluble metal/binder salts, forming the partially soluble sodium pigment lake. The process is repeated as more of the calcium lake dissolves, until the solubility product is reached. In the face of these problems, P.R.184 is also a suitable alternative to P.R.57:1.

Temperatures of up to 100 °C and higher may be observed as P.R.57:1 is dispersed in offset printing inks, especially in web offset inks by means of modern dispersion equipment. This is especially true for agitating pearl mills. Printing inks and prints may undergo color shifts if they are exposed to these conditions. The process, which is reversible, is caused by release of the water of crystallization from the pigment.

The different grades of P.R.57:1 vary considerably, not only in their coloristics but also in their rheological behavior. Individual grades are frequently optimized for special printing inks, for instance for publication gravure inks. In contrast to corresponding diarylide yellow pigments, amine treated types of P.R.57:1 are not available for publication gravure printing inks.

The plastics industry uses P.R.57:1 primarily wherever performance in application is a minor concern, especially with respect to lightfastness. In plasticized PVC, for instance, transparent coloration (0.1% pigment) of P.R.57:1 equal only step 2 on the Blue Scale, while samples containing TiO_2 never score higher than step 3. The exact value depends on the standard depth of shade and on the pigment concentration. P.R.57:1 is not suitable for use in concentrations below approximately 0.03%.

P.R.57:1 is employed in cable insulations because of its good dielectrical properties. It is much more lightfast in rigid PVC: transparent colorations (0.1% pigment) equal step 6 on the Blue Scale, while white reductions with TiO_2 match between step 4 and step 2, depending on the standard depth of shade and on the TiO_2 content.

P.R.57:1 is heat stable up to about 250 °C, which makes it a suitable candidate for use in polyethylene, particularly in LDPE types. Incorporated in LDPE, the pigment only slightly affects the shrinkage of extruded articles (Sec. 1.6.4.3). However, the lightfastness of such products is below expectation, which considerably restricts the versatility of the pigment. The same is true for PP spin dyeing. Apart from being used in these areas, P.R.57:1 also lends color to polystyrene.

P.R.57:1 is also used for PUR foam products. Problems arise with steam vulcanization, because some color is transferred to the wrapper as the pigment is worked into rubber (Sec. 1.8.3.6). The pigmented rubber articles are not completely fast to a number of organic solvents, to soap and sodium carbonate solutions, and to acids and SO_2 (Sec. 1.6.2.2).

Poor lightfastness considerably restricts the use of P.R.57:1 in coatings and emulsion paints. The pigment is, however, occasionally found in industrial paints.

P.R.57:1 is a useful colorant for a variety of special purpose media, including colored pencils and crayons. A number of countries have legally defined purity regulations concerning the use in decorative cosmetics articles, such as face powder and lipstick. The same is true for cheese casings [3]. Suitable grades are commercially available.

Pigment Red 58 Types

The following salts have been produced: the sodium salt (P.R.58), the barium salt (P.R.58:1), the calcium salt (P.R.58:2), the strontium salt (P.R.58:3), and the manganese lake (P.R.58:4). The commercial success of each type varies considerably by the region but has generally decreased considerably as the pigments were replaced by other lakes. P.R.58:4 is unique in that it is the only representative which has remained in the European market.

Pigment Red 58:2

East Asia is the only part of the world where P.R.58:2, a calcium lake, continues to be used. The pigment affords bluish red shades, somewhere between those of P.R.53:1 and 57:1. Its main field of application is in printing inks, where its lightfastness equals that of P.R.57:1; the two pigments also exhibit similar tinctorial strength.

Pigment Red 58:4

P.R.58:4 is primarily used in general industrial paints, where its full shade and related shades are of interest. This is also true for other manganese pigment lakes. In full shade, P.R.58:4 affords deep carmine shades, but white reductions provide only very dull, bluish shades of red. The pigment is fast to organic solvents and shows good overspraying fastness, but its resistance, especially to alkali, soap, and lime, is poor. Even in full shades, it remains inferior to P.R.48:4 and 52:2 in terms of lightfastness and weatherfastness. The difference is even more noticeable in white reductions. P.R.58:4 also exhibits moderate heat stability.

Like other manganese lakes, P.R.58:4 accelerates the drying process of oxidatively drying systems, which rapidly thickens the paint. This clearly precludes its use in such systems.

Pigment Red 63 Types

The following cations have been used for salt formation: sodium, calcium, barium, and manganese, as well as blends of calcium and manganese. European manufacturers continue to supply the calcium and manganese lakes, while the Japanese market also provides a barium lake. The latter, however, is of limited regional importance.

Pigment Red 63:1

The use of P.R.63:1, the calcium lake, is steadily declining, even in the USA, where the pigment used to be very important until the mid-1950s. P.R.63:1 produces a deep, bluish bordeaux shade and shows good fastness to organic solvents, including aliphatic and chlorinated hydrocarbons and plasticizers. It bleeds slightly into alcohols, ketones, and aromatic hydrocarbons (Sec. 1.6.2.1). Its primary area of application is in low-cost industrial and trade sales paints, and it also lends color to leather finishes. The pigment is not acid or alkali resistant and it is sensitive to lime. Full shades approximately equal step 4 on the Blue Scale for lightfastness, while 1/3 SD samples in white reductions match approximately step 2. The pigment is therefore not recommended for exterior use.

Pigment Red 63:2

P.R.63:2, the manganese lake, affords a dark maroon shade. It is a pigment with high tinctorial strength. P.R.63:2 is much less fast to organic solvents than the corresponding calcium lake, P.R.63:1. The pigment equals step 3 on the 5 step stability scale (Sec. 1.6.2.1), due to its bleeding into alcohols, ketones, and aliphatic and chlorinated hydrocarbons. The pigment is even less stable to aromatic hydrocarbons and esters; it is consequently not fast to overspraying. Its acid and alkali fastness are even inferior to that of P.R.63:1.

Like other manganese lakes, P.R.63:2 may accelerate the drying process of oxidatively drying resin systems. Its full shade lightfastness, which matches step 6–7 on the Blue Scale, is good. The pigment reaches step 4–5 in white reductions (1/3 SD), which is considerably better than that of the corresponding calcium lake. P.R.63:2 is also considerably more durable than P.R.63:1. The pigment is used regionally to afford low-cost shades of red in industrial finishes, and it is also recommended for use in printing inks.

Pigment Red 64 Types

Several metal lakes have been prepared from this parent structure of all β-oxynaphthoic acid pigments. The list includes the barium salt (P.R.64), the calcium salt (P.R.64:1), and the copper lake, which is registered as Pigment Brown 5. The pigments are rarely used in Europe, and their impact on the market in Japan and the USA has also decreased considerably.

Pigment Red 64:1

P.R.64:1, a low-cost calcium lake, exhibits good lightfastness. It provides brilliant, yellowish shades of red, which are referred to as scarlet. P.R.64:1 is registered in the USA as D&C Red No. 31. It is used in cosmetics.

Pigment Red 200

P.R.200, which was first introduced in the USA, is as yet only supplied in the form of its calcium salt. It produces a clean bluish shade of red, which is close to that of P.R.57:1. The two pigments are also similar in their resistance to organic solvents. P.R.200 shows average fastness to overcoating. It is recommended for use in air drying and heat drying systems, in which it only exhibits moderate lightfastness. P.R.200 is also found in various different types of printing inks, especially those for offset and gravure printing. The pigment is used to lend color to plastics wherever lightfastness is a minor issue. The pigment is not fast to acid, alkali, or soap.

Pigment Brown 5

P.Br.5 only enjoys limited regional importance. It is the poorest performer of all pigments within its class. P.Br.5 is used in special applications, such as flexographic printing inks, textile inks, and wood stains.

2.7.3   Naphthol AS Pigment Lakes

2.7.3.1   Chemistry, Manufacture, and Properties

Naphthol AS pigment lakes form a comparatively small group of pigments, all of which feature one or two sulfonic acid functions. These groups may be introduced into the pigment molecule either through the diazo component or through the arylide moiety of the coupling component. The sulfo groups, apart from providing a site for salt formation in order to form the insoluble pigment, also contribute to the performance of their host by noticeably improving the solvent and migration fastness of Naphthol AS pigments.

Pigment synthesis follows the typical route to azo pigment lakes: the aniline derivative or the aniline sulfonic acid is diazotized with sodium nitrite in an acidic medium (hydrochloric acid), followed by coupling on the Naphthol AS derivative, which is initially dissolved in an alkaline solution and then precipitated by adding inorganic or acetic acid. If a Naphthol AS sulfonic acid is used as a coupling component, it must be neutralized with alkali for dissolution and coupled directly.

The resulting alkali salts (usually sodium salts) of the sulfonated Naphthol AS pigments are treated as described in 2.7.1.1 and reacted with calcium or barium salts.

2.7.3.2   Commercially Available Naphthol AS Pigment Lakes

General

There is no common structural principle to these pigments beyond the basic Naphthol AS pigment skeleton. Table 21 lists the commercially available Naphthol AS pigment lakes.

Tab. 21: Commercially available Naphthol AS pigment lakes.

C.I. Name	C.I. Constitution No.	R_D^2	R_D^4	R_D^5	R_K^2	R_K^4	M	Shade
P.R.151	15892	$SO_3^\ominus$	H	H	H	$SO_3^\ominus$	Ba	red
P.R.237	–	–	–	–	–	–	–	yellowish red
P.R.239	–	–	–	–	–	–	–	bluish red
P.R.240	–	–	–	–	–	–	–	maroon
P.R.243	15910	$SO_3^\ominus$	CH_3	Cl	OCH_3	H	$\frac{Ba}{2}$	red
P.R.247	15915	CH_3	H	$CONHC_6H_4SO_3^\ominus$(p)	H	OCH_3	Ca	bluish red
P.R.247:1	15915:1	CH_3	H	$CONHC_6H_4SO_3^\ominus$(p)	H	OCH_3	Ca	red

Individual Pigments

Pigment Red 151

P.R.151, a barium lake, affords a bluish red color, which may be referred to as carmine. Of medium tinctorial strength, the pigment is used primarily in plastics.

P.R.151 is completely migration resistant in plasticized PVC. Transparent colorations (0.1% pigment) equal step 6 on the Blue Scale for lightfastness, while the pigment only matches step 3–4 in white reduction (0.01% pigment + 0.5% TiO_2). Pigmented rigid PVC systems are appreciably more lightfast, the corresponding values are 7 and 6, respectively. P.R.151 is frequently used for the coloration of synthetic leather made of PVC. However, the pigment is not fast to acid or alkali. Good dielectrical properties make it a suitable candidate for PVC cable insulations.

Incorporated in PE, P.R.151 shows only average tinctorial strength. 0.24% pigment is required to produce 1/3 SD colorations in PE containing 1% TiO_2. In HDPE, the pigment is heat stable up to 290 °C, but this advantage is somewhat compromised by the fact that it affects the shrinkage to a considerable extent. Caution should therefore be exercised if P.R.151 is to be used in larger, nonsymmetrical injection-molded objects. The lightfastness of P.R.151 in PE equals that in rigid PVC.

One of the primary fields of application for P.R.151 is in polystyrene, although there is a slight color change at temperatures above 260 °C, at which the pigment partially dissolves. It is also used to a considerable extent in ABS. Cast resins based on methylmethacrylate and unsaturated polyesters are also frequently colored with P.R.151, which is resistant to the peroxide catalysts that are used to harden the plastic. The lightfastness in these media is good; it equals step 6–7 on the Blue Scale.

Pigment Red 237

P.R.237, introduced some years ago, enjoys only limited regional importance. Its chemical constitution has not yet been published. The pigment affords a yellowish red shade, referred to as scarlet. P.R.237 is particularly recommended as a colorant for PVC, in which it shows good bleed resistance. Its coloristic properties in this medium parallel those of the β-naphthol pigment lake P.R.68. P.R.237 is only of average tinctorial strength: 0.36% pigment is needed in order to formulate a 1/3 SD sample (1% TiO_2). The pigment is heat stable up to 260 °C. Good fastness to overlacquering makes it a suitable candidate for industrial finishes.

Pigment Red 239

P.R.239 is a comparatively recent product. It enjoys limited regional impact and is used very little in Europe. Its exact chemical constitution remains to be published. P.R.239 affords dull, bluish shades of red. Its main application media are plastics. The pigment is bleed resistant in plasticized PVC. Its shade in this medium equals that of P.R.247, a member of the same class of pigments, which, however, has the advantage of providing a much cleaner shade. Incorporated in HDPE, P.R.239 is of average tinctorial strength and is heat stable up to 270 °C.

Pigment Red 240

The exact chemical constitution of this pigment, which is a recent product of the Japanese market, has not yet been published. P.R.240 enjoys only limited regional importance. Its dull bluish shades of red are referred to as maroon. P.R.240 is recommended for use in industrial finishes for its good lightfastness and weather-fastness as well as for its bleed resistance. The commercial type is quite transparent and lacks tinctorial strength. P.R.240 is equally weak in plastics. 0.42% pigment is needed, for instance, to afford 1/3 SD HDPE samples (1% TiO_2). In this medium the pigment withstands temperatures up to 300°C.

Pigment Red 243

P.R.243, the barium salt of a Naphthol AS pigment, is a comparatively recent product. It affords a dull, yellowish to medium red. The pigment is somewhat sensitive to solvents. P.R.243 is recommended for use in plastics. It is almost completely bleed resistant in plasticized PVC, but fails to satisfy more stringent lightfastness requirements. Transparent samples (0.1% pigment) equal merely step 4 on the Blue Scale, while opaque formulations (1/3 SD) match only step 3. Incorporated in HDPE, the pigment affects the shrinkage of the polymer. It is recommended specifically for use in LDPE.

Pigment Red 247

The primary area of application for P.R.247 is in plastics, in which it produces medium to bluish, brilliant and opaque shades of red. Its tinctorial strength is moderate. 0.28% pigment is needed, for instance, to formulate 1/3 SD HDPE samples containing 1% TiO_2. Such samples equal step 6–7 on the Blue Scale for lightfastness.

P.R.247 exhibits excellent heat stability, it withstands temperatures up to 300°C. The shrinkage of such partially crystalline plastics is not noticeably affected. P.R.247 is not completely bleed resistant in plasticized PVC, but it shows good lightfastness. Transparent (0.1% pigment) and opaque colorations (0.01% pigment + 0.5% TiO_2) equal step 6–7 on the Blue Scale for lightfastness. The same values apply if the pigment is incorporated in rigid PVC. P.R.247 is also used to color PS, ABS, and polyacetal. Excellent heat stability makes it a suitable candidate for PP spin dyeing.

Pigment Red 247:1

P.R.247:1, a recent commercial product, has the same chemical structure as Pigment Red 247, but a different crystal modification. Its main area of application is also in plastics. P.R.247:1 affords a much yellower shade than P.R.247 and shows similar tinctorial strength. 1/3 SD HDPE samples (1% TiO_2), for instance, are formulated at 0.32% pigment. Such samples equal step 5–6 on the Blue Scale for lightfastness; in this respect they score approximately one step less than the other modification.

The heat stability of P.R.247:1 is equally excellent, the pigment withstands temperatures up to 300 °C in HDPE. Such partially crystalline plastics show no noticeable shrinkage through P.R.247:1 incorporation. P.R.247:1, which is almost completely bleed resistant, is superior in this respect to P.R.247. P.R.247:1 exhibits good lightfastness, but is not quite as lightfast as P.R.247.

P.R.247:1 is also recommended for use in rigid PVC, PS, and ABS, as well as for PP spin dyeing.

2.7.4   Naphthalene Sulfonic Acid Pigment Lakes

2.7.4.1   Chemistry, Manufacture, and Properties

Naphthalene sulfonic acid pigment lakes are monoazo pigment lakes, obtained by using a naphthalene derivative bearing one or two sulfo groups as a coupling component. Typical diazonium compounds are monosubstituted anilines carrying either another SO_3H function or a COOH group.

Metal cations are Ba, Na, or Al. Some grades are precipitated onto aluminum oxide hydrate.

The pigments are synthesized by diazotizing the aniline derivative or the aniline sulfonic acid with sodium nitrite in hydrochloric acid and subsequently coupling onto the naphthalene sulfonic acid derivative, which is previously dissolved by neutralizing with a sodium hydroxide solution, producing the corresponding azo dye solution.

The pigment is then laked according to the procedure described for β-naphthol pigments (Sec. 2.7.1.1). Aluminum lakes are an exception. A soluble aluminum salt is first converted to aluminum oxide hydrate, which is washed to remove salt. The moist product is then combined with the dye solution, while a more soluble aluminum salt is added simultaneously. The insoluble pigment is finally washed salt-free and dried.

2.7.4.2   Commercially Available Pigments

General

These pigments have the following general structures:

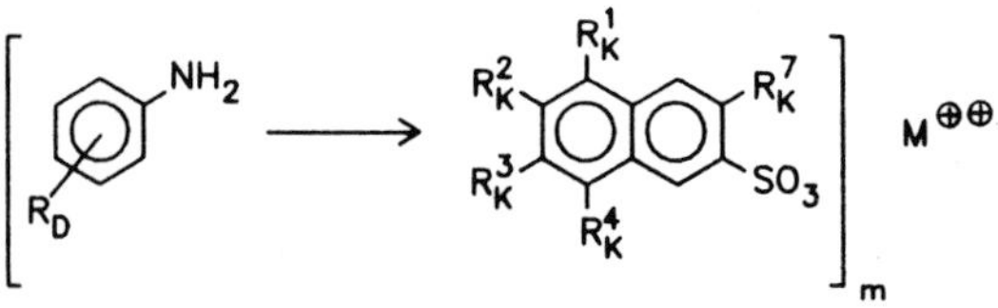

R_D: Cl, CH_3, CH_3O, $COO^{\ominus}$ or $SO_3^{\ominus}$

M: Ba, Al, Na

m = 1 to 3

R_K^1, R_K^4, R_K^7: H; R_K^2: OH and R_K^3: H or $SO_3^{\ominus}$

if coupling occurs in 1-position

R_K^1: NHCO (S: other simple substituents)

R_K^2, R_K^3, R_K^7: H; R_K^4: $SO_3^{\ominus}$ if coupling occurs in 7-position

Table 22 shows the monoazo pigment lakes which are based on naphthalene sulfonic acid (CP = coupling position).

Tab. 22: Commercially available naphthalene sulfonic acid pigment lakes.

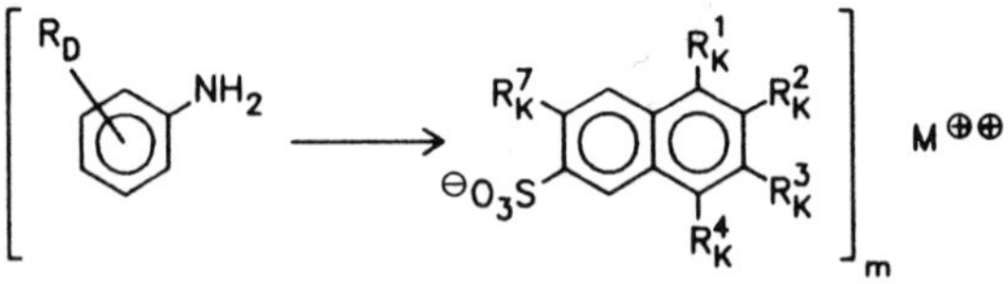

C.I. Name	C.I. Constitution No.	R_D	R_K^1	R_K^2	R_K^3	R_K^4	R_K^7	m	M	Shade
P.Y.104	15985:1	4-$SO_3^{\ominus}$	CP	OH	H	H	H	3	2 Al	reddish yellow
P.O.19	15990	2-Cl	CP	OH	H	H	H	1	$\dfrac{Ba}{2}$	orange
P.R.60:1	16105:1	2-$COO^{\ominus}$	CP	OH	$SO_3^{\ominus}$	H	H	2	3 Ba	bluish red
P.R.66	18000:1	3-CH_3	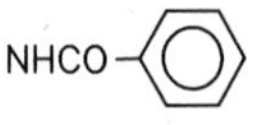 NHCO–	H	H	$SO_3^{\ominus}$	CP	2	Ba, 2 Na	red
P.R.67	18025:1	2-OCH_3	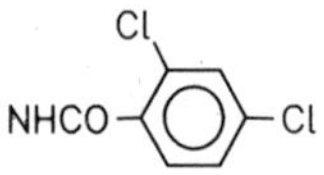 NHCO––Cl	H	H	$SO_3^{\ominus}$	CP	2	Ba, 2 Na	bluish red

Individual Pigments

Pigment Yellow 104

P.Y.104, an aluminum salt, has been approved for use in food, pharmaceuticals, and cosmetics, and appropriate purity standards have been developed. The pigment is known under the designation E 110 throughout the European Community and as FD&C Yellow No. 6 in the USA. It produces reddish shades of yellow. Like a number of other aluminum lakes of simple dyes, P.Y.104 exhibits poor fastness to organic solvents and limited bleeding fastness. Similarly, it is not fast to soap and alkali. P.Y.104 exhibits very little tinctorial strength in a variety of media and is not lightfast.

Pigment Orange 19

P.O.19 has stimulated only limited industrial interest and is rarely used in Europe. It provides a shade which in white reductions resembles that of P.O.34, but which is much duller. Its full shade is noticeably yellower than that of P.O.34. P.O.19, incorporated in medium-oil alkyd resin systems, is less lightfast, both in full shade and in white reductions, than opaque P.O.34 types. The commercially available type exhibits poor rheological behavior.

Pigment Red 60:1

P.R.60:1, which was patented as early as 1902 by Meister Lucius & Brüning, has lost much of its importance in recent years. Today, it is mostly used throughout the USA and Japan. According to the US International Trade Commission, 144 t of this type of pigment were produced nationwide in 1983. P.R.60:1 affords bright bluish red shades, which are referred to as shades of scarlet. Its tinctorial strength is poor compared to that of other pigment lakes with similar shades. P.R.60:1 shows limited fastness to acid, alkali, and soap, and exhibits poor lightfastness. A number of commercial types, which feature the pigment as a barium salt laked onto aluminum oxide hydrate, provide improved coloristic and fastness properties. Zinc oxide, added as the pigment precipitates, affects only the coloristic properties of the pigment. The exact chemical composition of such types is therefore very complex.

P.R.60:1 is used throughout the printing ink industry, especially for inexpensive packaging and metal deco printing inks. It does not bronze. Since the pigment is sensitive to water, problems may occur in offset printing. P.R.60:1 is used increasingly as a colorant for plastics, such as PVC, polyethylene, and especially for LDPE and polystyrene. Its lightfastness is satisfactory in these media.

P.R.60:1 is also used for emulsion paints and in paper mass coloration.

Pigment Red 66

P.R.66, a barium salt, is sold only in the USA. The pigment is also available as an aluminum oxide hydrate precipitate. Its shade is considered a brilliant medium red, which is somewhat yellower than that of the chemically related P.R.67. Commercial types of P.R.66 are very transparent. The pigment is highly sensitive to acid, alkali, and soap. Its fastness to organic solvents is poor, its fastness to overcoating as well. P.R.66 exhibits limited lightfastness. Its main application is in metal deco printing.

Pigment Red 67

P.R.67, a barium salt, is also available in the form of an aluminum oxide hydrate precipitate. Its shade is bluer compared to the chemically related P.R.66, it is referred to as a bright bluish red. Commercial types are transparent and tinctorially strong. P.R.68 is used especially in metal deco printing. The prints do not tolerate acid, alkali, or soap. They show only limited fastness to organic solvents and to clear lacquer coatings. P.R.68 prints are not fast to sterilization and only poorly lightfast. The pigment is also used for the coloration of rubber. It exhibits good resistance to common oxidants and does not tend to migrate.

References for Section 2.7

[1] U. Blank et al. in Winnacker-Küchler: Chemische Technologie, 4. Ed., Vol. 6, 143–310, Carl Hanser Verlag, München, 1982.
[2] N.N. Woroshzow: Grundlage der Synthese von Zwischenprodukten und Farbstoffen, 4. Aufl. Akademie-Verlag, Berlin, 1966.
[3] Lebensmittelzusatz-Zulassungsverordnung, 30. December 81, EG-Nr. E 180i
DFG-Farbstoffkommission (Dyestuff Commission) / Ringbuch – Farbstoffe für Lebensmittel, Colours for Foods (1988), LB-Rot 2
Kosmetikverordnung vom 21. Dezember 77 / Anl. 3, Teil A
DFG-Farbstoffkommission (Dyestuff Commission) / Ringbuch – Kosmetische Färbemittel, Colours for Cosmetics (1991), C-Rot 12.

2.8   Benzimidazolone Pigments

Benzimidazolone pigments [1,2] derive their name from the 5-aminocarbonyl benzimidazolone group (**22**), which is common to all pigments within this group:

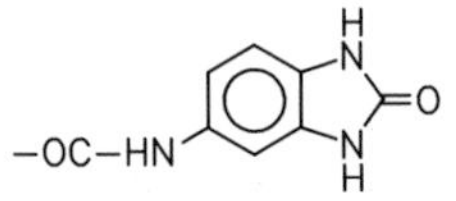

22

Although, strictly speaking, it would be more precise to refer to such compounds as benzimidazolone azo pigments, the convention of listing them as benzimidazolone pigments will be followed.

Basically, there are two ways of improving the solvent and migration fastness of an organic pigment:

1. Enlarging the pigment molecule, as with disazo condensation pigments (Sec. 2.9).
2. Introducing substituents into the pigment molecule which reduce the solubility of their host structure.

For benzimidazolone pigments the second option was chosen.

Initially, pigments had been found which could be rendered insoluble in organic media if they were converted into polar structures: molecules containing sulfo or carboxylic acid groups form insoluble salts with alkaline earth metals or manganese ("lakes", Sec. 2.7).

The next step consisted of introducing groups to improve the hydrophilicity of the parent structure to a certain extent, but not enough to render it soluble in water. Best results were achieved by the carbonamide function. Additional introduction of several such groups, for instance into Naphthol AS pigments, resulted in very solvent fast and migration resistant pigments (Sec. 2.6.2).

Later, 5- and 6-membered heterocyclic rings were introduced into the pigment molecule. The most effective among these are tetrahydroquinazoline-2,4-dione (23) and tetrahydroquinoxaline-dione (24)

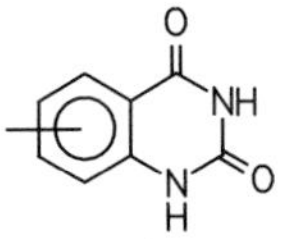

23 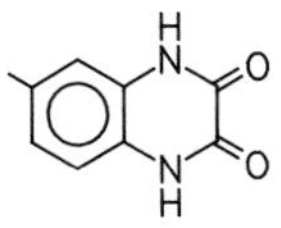   24

and especially the benzimidazolone group. It is most advantageous to introduce the heterocyclic ring as part of the coupling component.

The most important coupling component for pigments in the yellow range is 5-acetoacetylamino-benzimidazolone (25), while red pigments are derived primarily from 5-(2'-hydroxy-3'-naphthoyl)-aminobenzimidazolone (26).

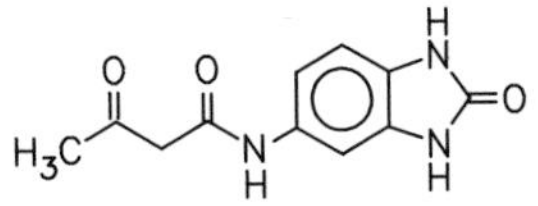

25 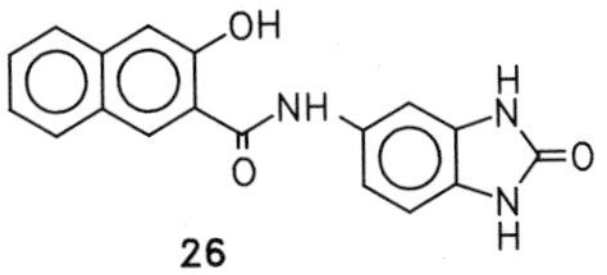   26

Compound (25) parallels the acetoacetarylides of the monoazo yellow pigment series. Red and brown pigments, on the other hand, which are obtained with (26), find their counterpart in the corresponding Naphthol AS pigments.

Introducing a benzimidazolone moiety into a monoazo yellow or Naphthol AS pigment structure is exceptionally useful in improving the properties of the parent pigments. This is especially true not only for the solvent and migration fastness but also for the lightfastness and weatherfastness of these "host" pigments. It is on the basis of their improved fastness properties that benzimidazolone pigments are suitable for applications with more stringent fastness requirements. Benzimidazolone pigments are among the products with the highest fastness standards in the azo range.

2.8.1   Chemistry, Manufacture

Various synthetic pathways have been proposed to prepare the benzimidazolone skeleton, from which all pigments in this series are derived. The 5-amino derivative should be mentioned in particular. It is prepared advantageously from 4-nitro-1,2-diaminobenzene as a starting material, which reacts with phosgene or with urea in the melt to form 5-nitrobenzimidazolone. Subsequent reduction yields 5-aminobenzimidazolone:

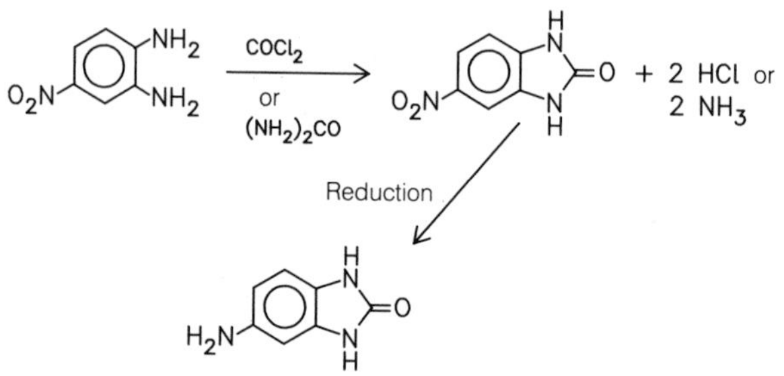

The chemistry and manufacture of the coupling components for yellow/orange and red benzimidazolone pigments are treated separately.

2.8.1.1   Yellow and Orange Benzimidazolone Pigments—
      Coupling Component

These pigments are derived from the following general structure:

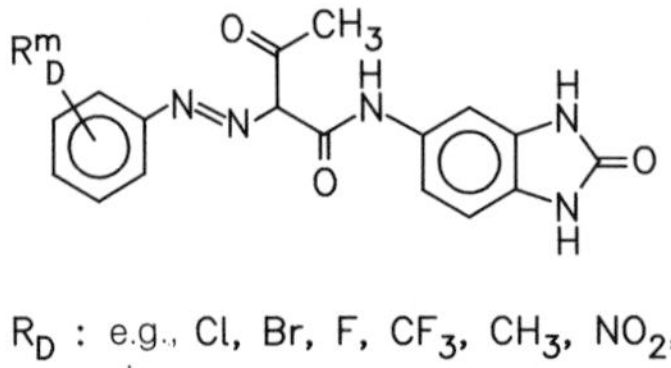

R_D : e.g., Cl, Br, F, CF_3, CH_3, NO_2, OCH_3, OC_2H_5,
COOH, COOAlkyl, $CONH_2$, $CONHC_6H_5$, $SO_2NHAlkyl$, $SO_2NHC_6H_5$

m = 1 to 3

The synthesis of the coupling component 5-acetoacetylaminobenzimidazolone (**25**) corresponds to the preparation of acetoacetarylides (Sec. 2.1.2) from 5-aminobenz-imidazolone by reaction with diketene or acetoacetic ester:

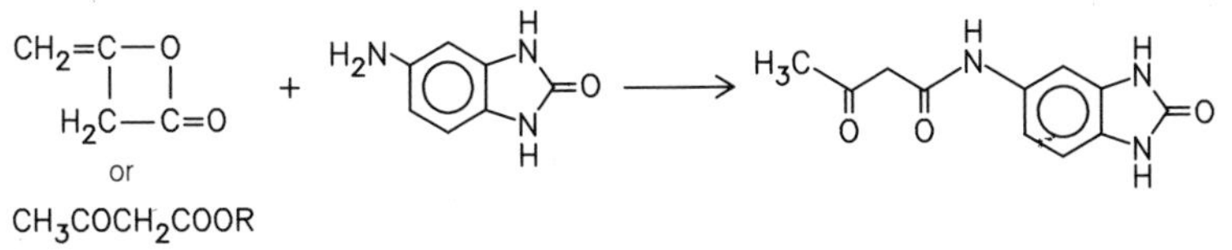

2.8.1.2   Red Benzimidazolone Pigments—Coupling Component

Red benzimidazolone pigments, which cover the entire range of shades in the red and brown part of the spectrum, are based on the following general structure:

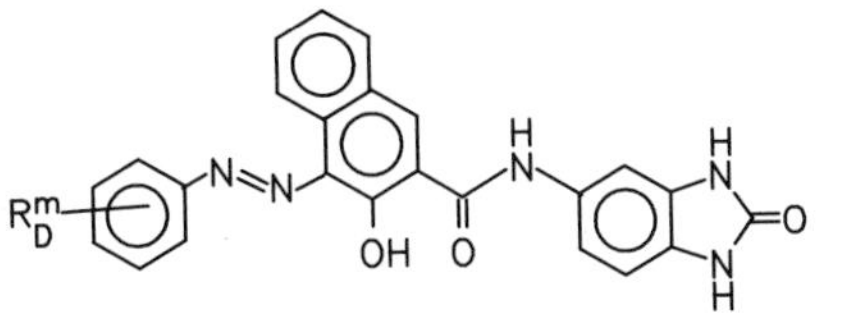

R_D and m represent the same atoms or groups as in the corresponding yellow pig-ments (Sec. 2.8.1.1).

The coupling component **26** is prepared like a Naphthol AS derivative. 5-Ami-nobenzimidazolone is treated with 2-hydroxy-3-naphthoic acid chloride or with 2-hydroxy-3-naphthoic acid and phosphorus trichloride in an organic solvent (Sec. 2.1.2).

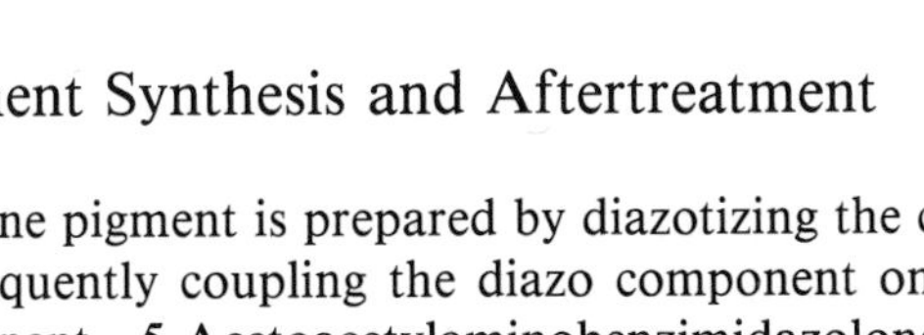

2.8.1.3   Pigment Synthesis and Aftertreatment

A benzimidazolone pigment is prepared by diazotizing the corresponding aromatic amine and subsequently coupling the diazo component onto a suspension of the coupling component. 5-Acetoacetylaminobenzimidazolone or 5-(2'-hydroxy-3'-naphthoyl)aminobenzimidazolone is dissolved in an alkaline solution. It is neces-sary to reprecipitate the product with acid (usually with acetic or hydrochloric acid) in the presence of a surfactant, since coupling reactions in alkaline media do not afford uniform products. The suspended coupling component now consists of particles small enough to make it react in a coupling reaction.

Coupling reactions, which are commonly carried out in aqueous media, afford benzimidazolone pigments, usually in the form of hard particles. Thermal after-treatment is necessary to adapt these crude products to the demands of technical application (Sec. 2.2.3).

Subsequent finishing of the crude product typically involves heating the aqueous pigment suspension, frequently to temperatures of 100 to 150°C. The crude pigment slurry is thus heated under pressure. This technique may be varied to a certain extent. It is possible, for instance, to add or to exclusively use either water soluble or water immiscible organic solvents or to add surface-active nonionic, anionic, or cationic agents.

2.8.1.4   Results of Crystal Structure Analyses

In recent years, single crystals [3,4,5] of a yellow (27) and a red (28) benzimidazolone pigment have been studied by X-ray diffraction analysis.

27   28

At a first glance, these studies appear to afford results which are comparable to the known data on the crystal structures of monoazo yellow, β-naphthol, and Naphthol AS pigments:

- presence of the oxohydrazone form,
- largest possible number of intramolecular hydrogen bonds,
- hydrogen bonds connecting three atoms (forked H bonds),
- almost entirely planar molecules.

X-ray diffraction studies, however, have shown that all of the investigated yellow and red azo pigments feature exclusively **intra**molecular hydrogen bonds and that the distances between the molecules within a crystal are so large that they can only be held together by van der Waals forces. The two studied benzimidazolone types, on the other hand, were the first azo pigments found to feature additional **inter**molecular hydrogen bonds. Both pigments exhibit these bonds in one dimension only, so that the molecules are connected to each other in one direction to form bands. In the yellow benzimidazolone pigment, pairs of molecules are attached to each other via identical hydrogen bonds, resulting in "bands" of consecutive pigment pairs (29). In the red pigment, these "bands" consist of a sequence of individual

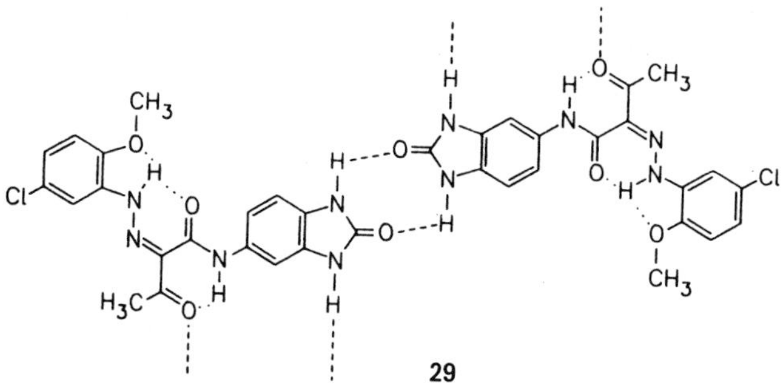

29

molecules which are interlocked with each other (**30**). In terms of molecular structure, benzimidazolone pigments thus represent an entirely new type of azo pigment.

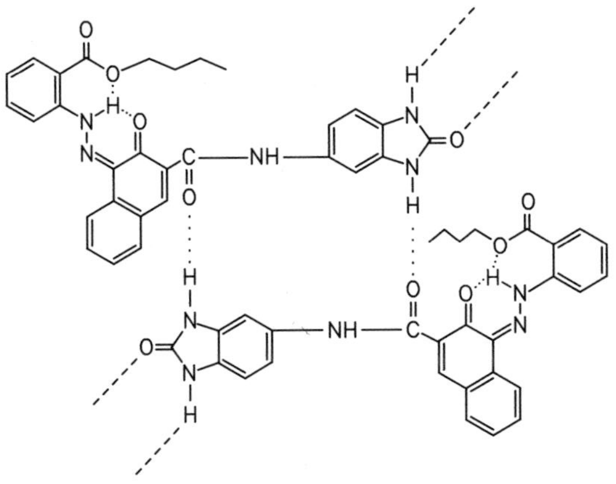

30

The excellent solvent and migration fastness which is typical of all benzimidazolone pigments probably results from this singular structural principle, which has
not yet been discovered in any other azo pigment class.

2.8.2 Properties

Pigments which are obtained by using 5-acetoacetylaminobenzimidazolone as a
coupling component cover the range from very greenish yellow to orange shades.
Products derived from 5-(2′-hydroxy-3′-naphthoylamino)-benzimidazolone, on the
other hand, extend this range towards the yellowish red region to include all major
shades of red, including bordeaux, maroon, and carmine. Technically important

brown pigments are also obtained by using 5-(2'-hydroxy-3'-naphthoylamino)-benzimidazolone as a coupling component.

Benzimidazolone pigments vary considerably in their tinctorial strength. The different physical characteristics of the various types, especially the wide spectrum of particle size distributions, contribute appreciably to the differences in tinctorial strength. A number of pigments produce relatively strong colorations, while others are comparatively weak. Formulating 1/3 SD samples with 5% TiO_2 in plasticized PVC, for instance, requires between 3.6% (of the weakest type) and 0.4% pigment (of the strongest representative). Benzimidazolone pigments in the yellow and orange series are somewhat comparable to monoazo yellow pigments in terms of tinctorial strength, while the red types parallel Naphthol AS pigments in strength.

The benzimidazolone moiety, a fundamental feature of all benzimidazolone pigments, is responsible for the high fastness of these pigments to the solvents which are typically found in application media. In this respect, benzimidazolone pigments perform much better than their counterparts in the monoazo yellow and Naphthol AS pigment series. Excellent fastness to solvents and chemicals is accompanied by good migration fastness. Benzimidazolone pigments do not bloom, and most of them show good and some even excellent bleed fastness and fastness to overcoating. All benzimidazolone pigments, with one exception (P.Y.151), are inert to alkali and acid. Most of them disperse easily in the common application media.

A considerable number of benzimidazolone pigments meet the major heat stability standards for practical application, while some are even among the most heat stable of organic pigments known. Moreover, benzimidazolone pigments as a group are characterized by excellent lightfastness, especially those covering the yellow and orange range, and a high degree of weatherfastness.

Different types are available as standard products featuring optimized physical parameters to satisfy the demands of particular applications. A custom-made product may be adapted to the customer's demand regarding transparency or opacity, flow behavior, or certain fastness properties.

Polymorphism is a common phenomenon in benzimidazolone pigments. However, each pigment is marketed in only one of its crystal modifications.

2.8.3   Application

Representatives of the benzimidazolone pigment series are used on a broad scale in almost all areas of pigment application, primarily for their excellent fastness properties. Benzimidazolone pigments satisfy higher or even very high specifications in application, especially regarding lightfastness and weatherfastness, heat stability, chemical inertness, and migration fastness.

Benzimidazolone pigments are used throughout the paint industry to lend color to all types of industrial finishes.

A large number of representatives fulfil the standards for use in automotive refinishes. Some even satisfy highest durability specifications and are used in original automotive finishes, sometimes even as shading pigments, or in metallic finishes. Areas of application include finishes for commercial vehicles, agricultural machinery and implements, as well as other high grade industrial finishes.

Certain commercial varieties with particularly high transparency provide transparent or metallic effects in finishes. Other grades are optimized in terms of high opacity and may, for instance, offer a technical alternative to inorganic Chrome Yellow or Molybdate Red pigments. Excellent rheological properties make it possible to increase the pigment concentration in a paint without affecting the gloss. The hiding power of the paint is thus open to further improvement. Such high opacity products are frequently used in combination with inorganic pigments, such as Chrome or Nickel Titanium Yellow, or iron oxide pigments. Opaque grades show excellent lightfastness and weatherfastness over a range of shades. However, in a number of pigments these fastness properties deteriorate rapidly as the concentration of white pigment increases. Pigment application frequently requires fastness to overcoating, a specification which at common baking temperatures is satisfied by a number of benzimidazolone pigments in many systems.

Several benzimidazolone pigments are suitable candidates for use in powder coatings based on polyester, acrylic, or polyurethane resin. These pigments satisfy the heat requirements of this application and do not plate out in these media (Sec. 1.6.4.1). Various benzimidazolone pigments even meet the particularly high thermal standards of coil coating. These pigments are also suitable for use in architectural and emulsion paints.

Benzimidazolone pigments, especially those covering the red range of the spectrum, were originally developed and used mostly for plastics. None of them were found to adversely affect the physical characteristics of their host medium. Benzimidazolone pigments do not bloom in plasticized PVC and other polymers. They are usually bleed resistant under typical application conditions.

In PVC, most benzimidazolone pigments are heat stable up to 220 °C. They have excellent to outstanding lightfastness. Some representatives are very weatherfast in impact resistant PVC types and in rigid PVC and even withstand long-term weathering. Various benzimidazolone pigments are used, for instance by the automobile industry, in PVC plastisols to lend color to synthetic leather.

Incorporated in polyolefins, benzimidazolone pigments vary considerably in terms of heat stability. These pigments may tolerate temperatures from less than 200 °C to 300 °C. Specific grades are available for use in various types of polyolefins, depending on the respective heat requirements. Pigments are thus custommade for use in HDPE, LDPE, or PP. There are many varieties that do not affect the shrinkage of polyolefins in injection molding. Benzimidazolone pigments are therefore used to advantage in thick-walled, large, nonsymmetrical injection-molded articles, such as bottle crates.

Benzimidazolone pigments are also used to color polystyrene, ABS, and other polymers which are processed at high temperature. A number of grades show excellent lightfastness and also satisfy the heat requirements for use in unsaturated polyester without affecting the hardening of the polymer.

Various benzimidazolone pigments are heat stable enough to be used in polypropylene spin dyeing. Several types find extensive use in the spin dyeing of other fibers, such as polyacrylonitrile, viscose reyon, and vicose cellulose, or secondary acetate.

The printing ink industry employs benzimidazolone pigments to color high grade printing inks. The transparent grades have stimulated particular interest in connection with this area. Benzimidazolone pigments usually perform well in print, most of them are fast to clear lacquer coatings and may safely be sterilized. More heat stable grades withstand exposure to up to 220 °C for 30 minutes and satisfy the stringent requirements of metal deco printing. In print, like in other applications, most grades are very lightfast. It is their lightfastness which makes them suitable colorants for long-term products, such as posters or other advertising items. Benzimidazolone pigments are frequently selected for their good solvent fastness, fastness to plasticizers, and migration fastness in printing inks for PVC films. Some products are particularly interesting as colorants in decorative printing inks for plastic laminates.

Commercial types of benzimidazolone pigments are also sold for use in solventbased wood stains and other special-purpose media which have not been mentioned previously.

2.8.4   Commercially Available Benzimidazolone Pigments

General

Table 23 lists commercially available benzimidazolone pigments. It should be noted that pigments in the red range are often derived from a diazonium component carrying a methoxy group in ortho position to the azo group.

Individual Pigments

The Yellow and Orange Series

Pigment Yellow 120

P.Y.120 affords a medium yellow shade. The pigment powder shows good fastness to solvents, in which it is similar to other yellow pigments within this class.

Tab. 23: Commercially available benzimidazolone pigments.

Yellow and orange series

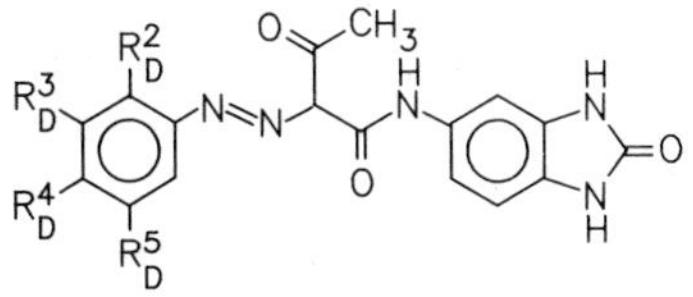

C.I. Name	C.I. Constitution No.	R_D^2	R_D^3	R_D^4	R_D^5	Shade
P.Y.120	11783	H	COOCH$_3$	H	COOCH$_3$	yellow
P.Y.151	13980	COOH	H	H	H	greenish yellow
P.Y.154	11781	CF$_3$	H	H	H	greenish yellow
P.Y.175	11784	COOCH$_3$	H	H	COOCH$_3$	very greenish yellow
P.Y.180	21290	*)	H	H	H	greenish yellow
P.Y.181	11777	H	H	**)	H	reddish yellow
P.Y.194	11785	OCH$_3$	H	H	H	yellow
P.O.36	11780	NO$_2$	H	Cl	H	orange
P.O.60	11782	Cl	H	H	CF$_3$	reddish yellow
P.O.62	11775	H	H	NO$_2$	H	yellowish orange

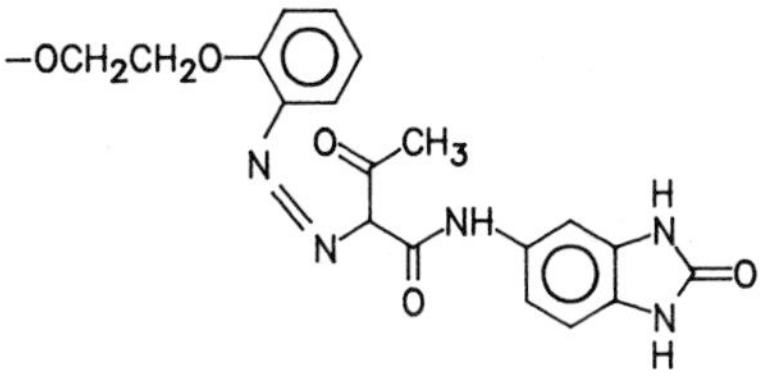

*) $-OCH_2CH_2O-$

**) $-OCNH-$⟨⟩$-CONH_2$

Red and brown series

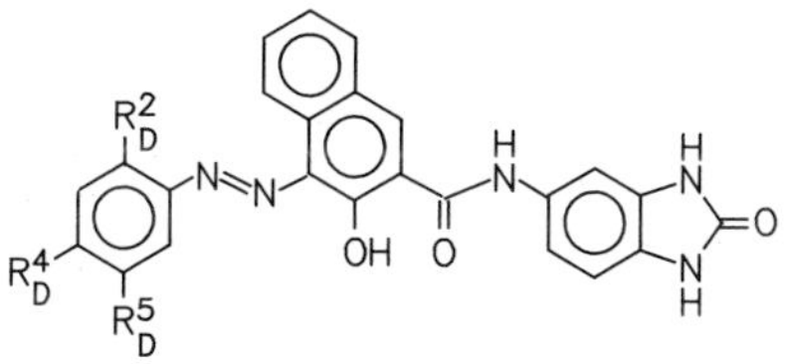

C.I. Name	C.I. Constitution No.	R_D^2	R_D^4	R_D^5	Shade
P.R.171	12512	OCH$_3$	NO$_2$	H	maroon
P.R.175	12513	COOCH$_3$	H	H	bluish red
P.R.176	12515	OCH$_3$	CONHC$_6$H$_5$	H	carmine
P.R.185	12516	OCH$_3$	SO$_2$NHCH$_3$	CH$_3$	carmine
P.R.208	12514	COOC$_4$H$_9$(n)	H	H	red
P.V.32	12517	OCH$_3$	SO$_2$NHCH$_3$	OCH$_3$	bordeaux
P.Br.25	12510	Cl	H	Cl	brown

P.Y.120 is primarily applied in plastics, especially in PVC. A pigment preparation is available for this purpose. P.Y.120 is very bleed resistant in plasticized PVC and has excellent lightfastness in both plasticized and rigid PVC. 1/1 to 1/25 SD samples (5% TiO_2) as well as transparent formulations equal step 8 on the Blue Scale for fastness to daylight. The pigment, however, is much less weatherfast in lead/cadmium stabilized rigid PVC than other representatives of its class, which precludes its use in exterior application. P.Y.120 is heat stable enough in HDPE for 1/3 SD samples (1% TiO_2) to tolerate exposure to 270 °C for 5 minutes. The shrinkage of the polymer is not affected. Incorporated in LDPE, P.Y.120 withstands approximately 220 °C, in polystyrene 240 °C. Transparent HDPE colorations (0.1%) equal step 8 on the Blue Scale for lightfastness, while opaque samples (0.01% + 0.5% TiO_2) match step 6–7.

The printing ink industry uses P.Y.120 primarily for decorative printing inks on laminated melamine and polyester resin sheets. In terms of lightfastness, 8% gravure prints in 20 and 40 μm cells on such plates correspond to step 8 on the Blue Scale. Plate-out is not observed. P.Y.120 is insoluble in monostyrene and acetone (Sec. 1.8.1.2).

P.Y.120 lends itself to all printing methods. Its lightfastness in letterpress and offset prints is also very good. Depending on the depth of shade, prints equal step 7 to step 6 on the Blue Scale. P.Y.120 thus scores approximately 1 step higher than the similarly colored P.Y.97.

P.Y.120 is used primarily to color high grade gravure inks for posters, metal sheets, and packaging purposes. The pigment has gained particular recognition in gravure inks for plasticized PVC. The prints are completely resistant to the DIN 16524 standard solvent mixture, as well as to a variety of organic solvents and media, including toluene, mineral spirits, methylethylketone, ethylacetate, paraffin, butter, soap, alkali, and acid. Moreover, the prints are fast to clear lacquer coatings and may safely be sterilized. They are heat stable up to 200 °C for 30 minutes.

P.Y.120 has less of an impact on the paint industry. In contrast to the similarly colored P.Y.97, P.Y.120 is fast to overcoating. Moreover, it exhibits noticeably higher durability. White reductions, reduced 1:1 to about 1:5, are approximately as durable as P.Y.151 systems. P.Y.120 is recommended for general industrial finishes, including automotive refinishes, and it is also suitable for use in architectural paints. P.Y.120 is completely fast to alkali.

Pigment Yellow 151

P.Y.151, available since 1971, affords a clean greenish yellow shade. Its hue is somewhat greener than that of P.Y.154 and distinctly redder than that of P.Y.175. The type which features a specific surface area of less than 20 m²/g provides good hiding power.

P.Y.151 maintains an important position throughout the pigment industry. Its main area of application is in the paint industry, which uses P.Y.151 particularly for high grade industrial finishes. Good rheological properties make it possible to incorporate up to approximately 30% pigment in a paint without affecting the gloss of the coating. The percentage value is given relative to the amount of solid binder. For comparison, it should be mentioned that full shade paints containing other pigments are usually restricted to a pigment concentration of 10 to 15%. P.Y.151 is frequently used in combination with TiO_2 and/or inorganic yellow pigments. Very lightfast and weatherfast P.Y.151/phthalocyanine green pigment combinations are equally important; they are sometimes used in conjunction with the above-mentioned inorganic pigments. As a rule, P.Y.151 is employed to produce deeper shades, which makes it an interesting product for manufacturers of automobile (O.E.M) finishes and automotive refinishes and paints for commercial vehicles.

Coatings containing P.Y.151 are very lightfast and durable. Systems based on acrylic melamine resin, for instance, were exposed to the Florida climate for one year and then evaluated. 1:1 reductions with TiO_2 equalled step 5 on the Gray Scale for weatherfastness, while 1:3 TiO_2 reductions matched step 4, and 1:35 reduced samples coincided with step 3–4. Comparative values for P.Y.154 and 175 are listed for the respective pigments. P.Y.151 is fast to overcoating up to 160 °C.

P.Y.151 is heat stable up to a maximum of 200 °C and withstands acid, but is affected by alkali under certain test conditions. Although the pigment tolerates weak bases, it will undergo a distinct color change towards reddish yellow shades if it is exposed to strong alkali. P.Y.151 is also affected by lime (Sec. 1.6.2.2), which almost entirely precludes its use in emulsion paints.

The plastics industry uses P.Y.151 to color PVC, polyolefins, and other polymers. The pigment shows excellent migration fastness in plasticized PVC. Plasticized PVC samples up to 1/25 SD equal step 8 on the Blue Scale for lightfastness. The pigment exhibits good weatherability in rigid PVC, but performs distinctly poorer in this respect than P.Y.154 and 175, and it is therefore usually unacceptable for long-term exposure. 1/3 SD P.Y.151 samples in HDPE (with 1% TiO_2) show a heat stability of 260 °C for 5 minutes. Temperatures in excess of this value shift the color towards the reddish side of the spectrum and decrease its brightness. The shrinkage of the polymer is only slightly affected at processing temperatures between 220 and 280 °C. P.Y.151 is equally heat stable in polystyrene, as long as the processing temperature does not exceed 260 to 280 °C. The pigment is very lightfast in this medium (step 8).

P.Y.151, like 154 and 175, lends itself to polypropylene spin dyeing. This is particularly true for polymer types whose flow behavior makes it possible to work at temperatures between 210 and 230 °C. P.Y.151 is very lightfast in these media.

P.Y.151 is employed for printing inks wherever lightfastness is a prime consideration. 1/1 SD letterpress proof prints equal step 7 on the Blue Scale for lightfastness, while 1/3 SD prints match step 6–7. The prints are fast to soap but not sufficiently alkali resistant. They are fast to clear lacquer coatings, but not fast to sterilization. P.Y.151 finds extensive use in offset and letterpress applications, as well as in packaging gravure inks for PVC substrates. It is equally suitable for decorative printing inks for laminated plastic sheets based on polyester resin. P.Y.151 is insoluble in monostyrene and acetone. In terms of lightfastness, 8% gravure prints (etching depth 20 and 40 µm) equal step 8 on the Blue Scale. P.Y.151 should not be used for laminated melamine resin sheets. It is not compatible with aqueous melamine resin solutions, in which it dissolves to a certain extent.

Pigment Yellow 154

P.Y.154, which was introduced in the mid-seventies, affords a somewhat greenish yellow shade of very high lightfastness and weatherfastness. Its shade is distinctly redder than that of P.Y.175 and noticeably redder than that of P.Y.151, both of which are also members of the benzimidazolone series. P.Y.154 is completely or at least almost completely resistant to the major organic solvents. The list includes alcohols, esters, such as butylacetate, aliphatic and aromatic hydrocarbons, such as mineral spirits or xylene, and dibutyl phthalate.

P.Y.154 is primarily applied in paints, in which it is one of the most weatherfast organic yellow pigments. Incorporated in the same system and tested and evaluated under the same conditions as P.Y.151, for instance, after one year of outdoor exposure in Florida, TiO_2 reductions up to 1:3 were found to equal step 5 on the Gray Scale, while 1:30 TiO_2 reductions matched step 4–5.

P.Y.154 is thus also a suitable shading component for other hues and may be used to tone reduced clean yellow and green shades.

Irrespective of the concentration, P.Y.154 is recommended for all high grade industrial paints, including automobile (O.E.M) finishes. Incorporated in baking enamels, it may safely be overcoated up to 130 °C. Temperatures in excess of this value will increase its tendency to bleed, which is only slightly noticeable at 140 °C. The pigment is heat stable up to 160 °C. Its good rheology makes it possible to increase the pigment concentration in a paint without affecting the gloss of the coating. P.Y.154 disperses easily. Areas of application also include other common media throughout the coatings and paints industry, such as architectural paints and emulsion paints. P.Y.154 is employed wherever high lightfastness and weatherfastness are required.

Its principal application within the plastics area is in PVC. P.Y.154 reigns supreme in terms of lightfastness and durability in rigid PVC and in impact resistant PVC types, which makes it a suitable product for exterior use. P.Y.154 also has excellent lightfastness and weatherfastness in PVC plastisols which are coil coated onto steel. Systems of this kind are frequently applied to the exterior of houses. P.Y.154, however, exhibits inferior tinctorial strength. 2.5% pigment is required to formulate 1/3 SD colorations with 5% TiO_2 in plasticized PVC. Only 1.4% are needed of the slightly redder P.Y.120, which is also a member of the benzimidazolone pigment series. P.Y.154 has excellent bleeding fastness in plasticized PVC. The heat stability of 1/3 SD samples in HDPE (with 1% TiO_2) is only up to 210 °C for 5 minutes, which practically precludes its use in this polymer. The pigment is equally unsuitable for polypropylene spin dyeing.

P.Y.154 is a useful pigment for the printing ink industry wherever high lightfastness is required. Letterpress proof prints up to 1/25 SD equal step 6–7 on the Blue Scale for lightfastness, which is at least 1 1/2 to 2 steps above that of similarly colored diarylide yellow pigments or representatives of the monoazo yellow pigment series. In print, however, like in plastics, P.Y.154 lacks tinctorial strength. 1/3 SD standardized letterpress proof prints are prepared from inks which contain 25% pigment. For comparison, equally deep samples are prepared from inks which contain 3.5% P.Y.13 or 8.8% P.Y.1, respectively. P.Y.154 prints are not fast to clear lacquer coatings or to sterilization.

The list of applications also includes other media, such as oil colors for artists, in which P.Y.154 is used for its high lightfastness.

Pigment Yellow 175

P.Y.175, a clean, very greenish yellow pigment, is a comparatively recent product. It has only gained limited commercial recognition. The shade of the commercially available modification is somewhat greener than that of other benzimidazolone pigments and greener than corresponding pigments of other types, such as P.Y.109, 128, or 138.

P.Y.175 is primarily used in the paint field, where it is used especially to color high grade systems, such as automobile (O.E.M) and automotive refinishes. The products are used for brilliant yellow or green shades. Shades of green are accessible, for instance, by combining P.Y.175 with phthalocyanine pigments. Despite its comparatively low specific surface area (approx. 20 m^2/g), the commercially available type exhibits good transparency. However, it is not weatherfast enough to be used in metallic finishes. On the other hand, both lightfastness and durability are excellent in solid shades. In medium white reductions with TiO_2, P.Y.175 shows excellent weatherfastness, but this fastness deteriorates rapidly as more white pigment is added. In this respect, P.Y.175 is inferior to the somewhat redder P.Y.154, which is also a member of the benzimidazolone pigment series. Incorporated in an acrylic melamine resin system and exposed to the Florida weather for one year, for instance, P.Y.175 samples which are reduced 1:1 with TiO_2 equal step 5 on the Gray Scale, while 1:2.5 samples coincide with step 4, and 1:30 reductions match step 3–4.

P.Y.175 does not bloom. It may safely be overcoated up to 140 °C. At higher temperatures, bleeding is observed to a small extent in various systems. The pigment is heat stable up to 180 °C. The commercially available type exhibits good rheological properties in paints and may therefore be used at higher concentrations.

P.Y.175 is of interest to the plastics industry wherever high lightfastness and weatherfastness are a prime concern. It is, for instance, frequently used for long-term exposure in rigid PVC in white reductions containing little TiO_2. The pigment is, however, less durable than P.Y.154. 1/3 SD HDPE samples (with 1% TiO_2) are heat stable up to 270 °C. The shrinkage of the polymer is not affected at processing temperatures between 220 and 270 °C (Sec. 1.6.4.3). However, P.Y.175 is comparatively weak. 0.55% pigment is needed to produce a 1/3 SD sample with 1% TiO_2, while only 0.13% is required of the less lightfast, still somewhat greener diarylide yellow pigment P.Y.17. 0.27% of the somewhat redder P.Y.16, on the other hand, is sufficient to achieve the same purpose, although the two latter pigments are considerably less heat stable and lightfast than P.Y.175.

The printing ink industry uses P.Y.175 only for high grade products. Up to 1/25 SD, its lightfastness equals step 6–7 on the Blue Scale. Its limited tinctorial strength, however, is somewhat of a disadvantage. The prints are fast to paraffin, butter, soap, dibutyl phthalate, toluene, mineral spirit, and other media. They are also fast to clear lacquer coatings but may not be sterilized. Prints containing P.Y.175 are heat stable up to 200°C for 10 minutes or up to 180°C for 30 minutes.

Pigment Yellow 180

P.Y.180, a greenish to medium yellow shade pigment, was introduced a few years ago. It is a disazo yellow pigment and is of particular interest to the plastics industry.

In HDPE, P.Y.180 is heat stable up to 290°C. P.Y.180, like P.Y.181, does not affect the shrinkage of the plastics, so that there is no limitation to its use in injection-molded articles. It is, however, less fast to light than the redder P.Y.181. 1/3 SD colorations (1% TiO$_2$), for instance, equal step 6 on the Blue Scale. 1/3 SD specimens are formulated at a pigment concentration of 0.15%, which indicates good tinctorial strength. P.Y.180 is also used for PP spin dyeing, since it has excellent heat stability and affords clean hues. A special pigment preparation is available for this purpose. Redder shades are accessible through combination with the very reddish P.Y.181. Deep colors have excellent lightfastness: 1/1 to 1/3 SD samples equal step 7 on the Blue Scale. The fastness to light decreases rapidly, however, as more white pigment is added: 1/25 SD specimens match only step 5 on the Blue Scale. P.Y.180 performs excellently in terms of textile fastness properties.

P.Y.180 is becoming increasingly important and is utilized in printing inks to suit particular applications where diarylide yellow pigments cannot be used. Diarylide yellow pigments decompose at temperatures in excess of 200°C (Sec. 2.4.1.3), which precludes their use in certain inks for metal deco which are baked at temperatures above 200°C.

P.Y.180 is an equally important pigment for PVC. It does not migrate in plasticized PVC. Depending on the standard depth of shade and on the TiO$_2$ content, its lightfastness in rigid and in plasticized PVC equals step 5 (0.01% pigment + 0.5% TiO$_2$) to step 6–7 on the Blue Scale (0.1% pigment + 0.5% TiO$_2$). P.Y.180 is also used to color other plastics which are processed at high temperature. Technical polymers, such as polycarbonate, PS, ABS, and polyester, are particularly common media for P.Y.180. The pigment is also used to color injection-molded and extrusion-made polyamide—in contrast to P.Y.181, which is not suitable for this purpose. These plastics melts are slightly basic and have a reducing effect, in contrast to other polymer melts which are neutral.

Pigment Yellow 181

P.Y.181, a reddish yellow pigment, was introduced to the market a few years ago. Its main area of application is in plastics, especially in polyolefins. In these media,

P.Y.181 is heat stable up to 300°C and very lightfast. 1/3 SD HDPE samples (1% TiO$_2$), for instance, equal step 8 on the Blue Scale. The pigment does not affect the shrinkage of the partially crystalline polymer. Its tinctorial strength, however, is poor.

P.Y.181 does not migrate in plasticized PVC. Besides, it shows excellent lightfastness in this medium. Both transparent samples (0.1% pigment) and white reduced samples (0.1% pigment + 0.5% TiO$_2$) equal step 8 on the Blue Scale. Even plasticized and rigid PVC samples which contain a higher concentration of white pigment (0.01% pigment + 0.5% TiO$_2$) still match step 7–8 on the Blue Scale.

P.Y.181 is extremely heat stable, which makes it a suitable pigment for other polymers which are processed at high or very high temperature. The list includes PS, ABS, polyester, polyacetal, and various other technical plastics. The pigment is equally interesting in connection with PP spin dyeing. 0.3% and 3% PP samples equal step 7 and step 7–8 on the Blue Scale, respectively. A special pigment preparation is available for this purpose. P.Y.181 also frequently colors spin dyed viscose rayon and viscose cellulose. In these media, the pigment satisfies the particularly stringent specifications regarding lightfastness and weatherfastness for use in automobile interiors. This is a purpose to which only very few organic yellow pigments are suited.

P.Y.181 is also recommended for use in paints, but in this area it is in direct competition with a large number of similarly shaded pigments within the same class as well as from different pigment classes. Inferior tinctorial strength makes P.Y.181 a less important product in this field.

Pigment Yellow 194

The pigment is currently encountered primarily in the US market. In terms of shade and fastness properties, it is in direct competition with the monoazo yellow pigment P.Y.97 and the bisacetoacetarylide pigment P.Y.16. P.Y.194 is used in industrial paints. The pigment is also suitable for use in the coloration of plastics. It provides good tinctorial strength. At similarly medium shade of yellow, 1/3 SD HDPE colorations (1% TiO$_2$) require 0.18% pigment. Such colorations are heat stable up to 230°C.

In plasticized PVC, P.Y.194 shows much better fastness to bleeding than the similarly shaded P.Y.97, but it is not completely fast.

Pigment Orange 36

P.O.36 is a reddish, somewhat dull orange pigment which provides very good lightfastness and weatherfastness. It is a highly important pigment. The commercially available types differ considerably in their hiding power/transparency. Different types vary even in terms of shade—the opaque version is redder and noticeably cleaner—and other properties, as well as fastness properties.

P.O.36 is broad in scope, its main area of application is in paints. Masstone and similar deep shades have excellent lightfastness. The opaque variety, which was introduced in the mid-seventies, performs much better in this respect than the standard grade. In deep shades, P.O.36 meets the requirements for use in automobile (O.E.M) finishes. The opaque version is the current standard grade in this range of shades for applications which require lead free pigmentation. To suit this purpose, P.O.36 is frequently used in combination with quinacridone pigments to produce paints which feature, for instance, lightfast and durable RAL shades 3000 (fire engine red), 3002 (carmine), 3003 (ruby), and 3013 (tomato red). Small amounts of TiO_2 may be added. Very good rheological behavior in paints makes it possible to increase the concentration of opaque pigment without affecting the gloss. The opaque version of P.O.36 is also used in combination with chromate pigments. Coatings which are pigmented in full and similarly deep shades do not darken as they are exposed to light and weather. The list of applications also includes automotive repair finishes, paints for commercial vehicles and agricultural machines, as well as for general industrial finishes. P.O.36 is used extensively in these areas.

P.O.36 is completely fast to overcoating. Bleeding into a white overcoat is only observed at baking temperatures above 160 °C. P.O.36 is heat stable up to 160 °C. In dispersing the opaque type, it is important to avoid excessive shearing forces and to closely monitor the temperature, especially if agitated ball mills are used. Faulty temperature control may lead to a dull color.

P.O.36 is used throughout the printing ink industry for letterpress and offset inks, all types of packaging gravure and flexographic inks, and for metal deco printing. The prints are very lightfast. Depending on the grade and on the standard depth of shade, letterpress proof prints up to 1/25 SD equal step 6–7 to step 5 on the Blue Scale. The prints are excellently fast to chemicals and resistant to solvents such as toluene, mineral spirit, methylethylketone, and to the DIN 16 524 standard solvent blend. They withstand clear lacquer coatings and may safely be sterilized. Prints made from P.O.36 tolerate exposure to 220 °C for 30 minutes. At equal standard depth, their shade is noticeably duller than that of the somewhat redder P.O.34 and also duller than that of other less fast pigments covering the same range of shades.

The plastics industry uses P.O.36 to color PVC. 1/3 to 1/25 SD plasticized PVC samples containing 5% TiO_2 equal step 8 and step 7, respectively, for lightfastness, while transparent specimens (0.1% pigment) equal step 8 on the Blue Scale. The pigment does not migrate, e.g., it is practically resistant to bleeding and blooming. Combinations with carbon black are frequently used to create shades of brown for furniture films. Suitable media include PVC plastisols.

Incorporated in rigid PVC, P.O.36 lacks lightfastness at low concentrations. Barium/cadmium stabilized 1/3 SD colorations in rigid PVC show good durability but are frequently not weatherfast enough to satisfy the requirements for plastics materials in terms of long-term exposure. Heat stability, tested in HDPE at 1/3 SD, is 220 °C. The pigment does not affect the shrinkage of the plastic. P.O.36 is normally, however, not heat stable enough to meet the standards of practical pigment application in polyolefins.

P.O.36 is also used to color unsaturated polyester resins. Both transparent and opaque samples exhibit a lightfastness in these media that equals step 7 on the Blue Scale. The pigment does not affect the shrinkage of the plastic.

Pigment Orange 60

Introduced some years ago, P.O.60 is a yellowish orange pigment with excellent lightfastness and durability.

Its main area of application is in paints, in which it is used primarily to color high grade industrial finishes, such as automotive finishes, especially automobile (O.E.M) finishes. P.O.60 is used to advantage in white reductions and as a shading pigment. Its response to accelerated weathering under a xenon arc lamp is much inferior to its actual weatherfastness in outdoor exposure, for instance in Florida. The commercially available grade is transparent and exhibits a low specific surface area (less than 20 m^2/g). This is what makes P.O.60 a suitable candidate to produce orange shades in metallic finishes. There is a certain disadvantage to its lack of tinctorial strength, which is attributed to the low specific surface area.

P.O.60 is one of the organic orange pigments with the highest fastness standards. Its durability—even in metallic finishes—equals that of P.Y.154. After one year of outdoor exposure in Florida, acrylic melamine resin systems containing up to 1:25 TiO$_2$ equal step 5 on the Gray Scale for durability. Metallic finishes containing 60% pigment and 40% aluminum paste (65%) and also 40:60 blends equal step 4–5 on the same scale. Comparative values are listed with some other representatives of this class.

P.O.60 is heat stable up to 160 °C. Baking temperatures in excess of this value will in some paint systems cause a decrease in fastness to overcoating.

P.O.60 can basically also be used in printing inks and plastics, provided the requirements concerning lightfastness and durability are high. Adding small amounts of TiO$_2$ confers long-term weatherfastness on the pigment, although this advantage is compromised by its poor tinctorial strength. Wherever the requirements are appropriate, P.O.60 may also be employed in other application media, such as artists' colors, poster colors, etc.

Pigment Orange 62

P.O.62 affords a clean, very yellowish shade of orange. The commercially available grade exhibits a low specific surface area of approximately 12 m^2/g and has, accordingly, coarse particles which make for good hiding power.

P.O.62 is utilized mostly to color paints, especially to produce very clean yellowish shades of red or reddish shades of yellow in full and similarly deep shades. Good rheological properties make it possible to employ the pigment at higher concentrations without affecting the gloss. The pigment concentration may reach up to 30% relative to the content of solid binder. Deep shades are very lightfast and weatherfast. Full and related shades of P.O.62 are also recommended for use in automotive finishes. The list of applications includes automotive refinishes, commercial vehicles, agricultural machinery, and other high grade and general industrial paints. This is especially true for areas where intense shades of red are required and Molybdate Red is to be avoided or its use restricted. At higher baking temperatures (160 °C), P.O.62 is not completely fast to overcoating. The pigment is heat stable up to 180 °C.

P.O.62 is frequently used in the printing inks field to produce lightfast offset and aqueous flexographic inks. Its lightfastness in these media equals step 5 to step 6 on the Blue Scale, depending on the standard depth of shade. The prints are not completely fast to alkali and do not tolerate clear lacquer coatings and sterilization.

P.O.62 is interesting to the plastics field because it is very lightfast and durable in rigid PVC. However, it fails to meet higher standards regarding long-term weathering.

P.O.62 also lends itself to polypropylene spin dyeing, especially at temperatures up to approximately 230 °C. This makes it a suitable colorant for PP types which exhibit good flow behavior.

The Red and Brown Series

Pigment Red 171

P.R.171 affords a dull, very bluish red, referred to as maroon. The pigment generally provides good fastness properties. The commercially available types are highly transparent.

P.R.171 is used in plastics and in paints. Its lightfastness in PVC equals step 7 to step 8 on the Blue Scale, depending on the exact composition of the tested system, the pigment concentration, and the TiO_2 content. Incorporated in plasticized PVC, P.O.171 is migration resistant and heat stable up to 180 °C. It is used in conjunction with organic yellow pigments, frequently also with iron oxides, to produce shades of brown. Shades of bordeaux are accessible in deep transparent colorations.

P.R.171 exhibits excellent tinctorial strength. Only 0.46% pigment is required to formulate 1/3 SD samples containing 5% TiO_2. 1/3 SD samples (with 1% TiO_2) in PE are heat stable up to approximately 240 °C. P.R.171 application is accordingly restricted to LDPE, which is processed at low temperature. Transparent samples (0.1% pigment) equal step 6–7 on the Blue Scale for lightfastness, while reduced colorations (0.01% pigment + 0.5% TiO_2) match step 6. Used in polyacrylonitrile spin dyeing, P.R.171 affords very lightfast products. In the concentration range between 0.1 and 3%, its lightfastness equals step 7–8 on the Blue Scale. The products are completely fast to dry rubbing but do not entirely withstand wet rubbing. Incorporated in unsaturated polyester resins, P.R.171 accelerates the hardening of the plastic. It shows very good lightfastness in this medium, too.

Paints containing P.R.171 may be safely oversprayed and exhibit excellent lightfastness and weatherfastness. Its good fastness properties make P.R.171 a suitable candidate for high grade industrial paints, including automotive repair finishes. Its high transparency in paint films is used to advantage in a host of applications, such as in transparent foil coatings and metallic finishes.

P.R.171 lends maroon shades to printing inks wherever it suits the required fastness standards.

Pigment Red 175

P.R.175, a somewhat dull red pigment, exhibits very good fastness properties. It is completely or almost completely insoluble in common organic solvents. The commercially available types exhibit high specific surface areas and are therefore very transparent.

The paint industry uses P.R.175 primarily to color industrial paints and also automobile repair finishes. High transparency makes the pigment an important product for transparent and metallic effect finishes. The pigment is suited to two-coat metallic automobile (O.E.M) finishes, also referred to as base coat/clear coat finishes, especially if the clear coat contains UV absorbants. Its lightfastness and weatherfastness are excellent. P.R.175 does not bloom, is completely fast to over-painting, and is heat stable up to 200 °C.

In plastics, P.R.175 is also very lightfast and weatherfast. Its lightfastness in plasticized and rigid PVC, for instance, equals step 7 to step 8 on the Blue Scale, depending on the composition of the polymer, the type of stabilization, the depth of shade, and possibly on the TiO_2 content. In these media, P.R.175 is frequently applied in conjunction with carbon black to afford shades of brown for furniture laminates. The pigment is migration resistant in plasticized PVC, bleeding or blooming is practically never observed. P.R.175 is also applied in PVC and PUR plastisols for synthetic leather, which is used in markets such as automobile interiors. The pigment is very durable. P.R.175, combined with carbon black but containing no TiO_2, is used in window frames made of impact resistant PVC.

In white reductions, P.R.175 does not quite satisfy the requirements of long-term outdoor exposure. PVC plastisols, coil coated onto steel plates, also provide good fastness to light and weather. 1/3 SD HDPE samples (with 1% TiO_2) tolerate exposure to up to 270 °C for 5 minutes. The pigment only affects the shrinkage of HDPE to a negligible extent. Transparent (0.1% pigment) and opaque colorations (0.01% pigment + 0.5% TiO_2) equal step 6–7 on the Blue Scale for lightfastness.

P.R.175 is also used in polypropylene spin dyeing, where it satisfies the light-fastness requirements. The pigment is an interesting colorant for polystyrene and for polyester (PETP). These systems are utilized to make bottles and other products.

P.R.175 has stimulated interest throughout the printing ink industry wherever high lightfastness, excellent solvent fastness, fastness to sterilization, and very high heat stability (up to 220 °C) are a requirement.

Pigment Red 176

P.R.176 provides a bluish red shade, which is somewhat bluer than those of P.R.187 and 208 and somewhat yellower than that of P.R.185.

P.R.176 is primarily applied in plastics and in laminated papers. The pigment exhibits very good migration fastness in plasticized PVC. 1/3 SD samples containing 5% TiO_2 equal step 6–7 on the Blue Scale for lightfastness, 1/25 SD specimens match step 6, and transparent colorations (0.1% pigment) equal step 7. Transparent rigid PVC colorations match step 7–8. P.R.176 is heat stable up to 200 °C, and it is a suitable colorant for PVC cable insulations and synthetic leather. The pigment is also used in polyolefins and in polystyrene, where its lightfastness is equally good. Transparent polystyrene samples (0.1%) are heat stable up to 280 °C, addition of large amounts of TiO_2 reduces this stability to 220 °C.

The list of applications includes decorative prints for laminated plastic sheets. P.R.176, like P.R.187 and 208, is insoluble in styrene monomer and acetone, exhibits no plate-out onto the plastic sheets, and does not bleed if it is soaked with a melamine resin solution. This is what makes P.R.176 an interesting pigment for both melamine and polyester resin sheets. Its lightfastness is inferior to that of these other pigments covering the same color range; it scores one step less on the Blue Scale. P.R.176 affords a shade which closely approaches the standard magenta for three and four color printing.

P.R.176 provides very lightfast polyacrylonitrile spin dyeing products. The samples equal step 6–7 on the Blue Scale. Dry and wet crocking may affect the objects to a certain extent. P.R.176 is also used in polypropylene spin dyeing, especially for coarse textiles, such as carpet fibers, split fibers, filaments, bristles, or tape, but also for finer denier yarns. A special pigment preparation for this purpose is commercially available. 1/3 SD samples tolerate exposure to up to 300 °C for one minute or up to 290 °C for 5 minutes. In terms of lightfastness, 0.1% colorations equal step 5–6 on the Blue Scale, while 2% samples match step 7.

Pigment Red 185

The commercially available types of this polymorphous pigment afford very clean, bluish shades of red. P.R.185 is completely or almost completely insoluble in common solvents.

Its main area of application is in graphics printing and in the mass coloration of plastics.

The printing ink industry uses P.R.185 for all printing techniques. The prints show very good solvent fastness, for instance towards the standard DIN 16524 solvent mixture and towards clear lacquer coatings. They may safely be sterilized. P.R.185 prints are heat stable up to 220 °C for 10 minutes or to 200 °C for 30 minutes, which makes the pigment a suitable product for metal deco printing. In terms of lightfastness, 1/1 SD prints equal step 6–7 on the Blue Scale, while 1/3 to 1/25 SD specimens match step 5. P.R.185 affords a color which closely approaches the DIN 16524 standard magenta for three and four color printing (Sec.1.8.1.1). Consequently, the pigment is used in process colors wherever P.R.57:1 and 184 are not fast enough to meet the standards. This is true for applications such as metal deco printing. Moreover, P.R.185 also lends itself to use in laminated polyester products.

P.R.185 is employed throughout the plastics industry to color PVC, poly(vinylidene) chloride, and polyolefins. In plasticized PVC, the pigment is migration resistant at concentrations down to 0.005%. Under standardized conditions, the extent of bleeding into a white plasticized PVC sheet which is in contact with a 1/3 SD colored plasticized PVC sheet (Sec. 1.6.3.2) is only 0.2 CIELAB units. 1/3 SD samples (5% TiO_2) equal step 6–7 on the Blue Scale for lightfastness, while 1/25 SD samples match step 6. Similar values are measured for rigid PVC. In deeper shades, P.R.185 is also found in several types of synthetic leather, including materials based on PVC and PUR for use in automobiles. In terms of heat stability, 1/3 SD P.R.185 samples in PE (1% TiO_2) withstand less than 200 °C, which restricts its use to low temperature LDPE. P.R.185 is also well suited to polypropylene spin dyeing, provided the polymer can be processed at low temperature. P.R.185 is used by the paint industry to color general industrial finishes.

Pigment Red 208

Incorporated in its application medium, P.R.208 affords medium shades of red. The pigment exhibits good fastness to chemicals and solvents. Its main area of application is in the mass coloration of plastics and in packaging gravure printing inks.

P.R.208 worked into PVC provides medium shades of red. Pigment blends with compounds such as P.Y.83 or with carbon black are frequently employed to produce shades of brown. P.Y.208 is an interesting product for synthetic leather based on PVC, which is targeted for use in automobiles. The pigment does not bloom in plasticized PVC, provided it is used at concentrations above 0.005%. It shows excellent bleed resistance. Samples in the range between 1/3 and 1/25 SD (with 5% TiO_2) equal step 6–7 on the Blue Scale for lightfastness. The pigment is heat stable up to approximately 200 °C and exhibits high tinctorial strength. 0.6% pigment is required to formulate 1/3 SD PVC samples with 5% TiO_2. Good dielectric properties make the pigment a suitable candidate for use in PVC cable insulations.

In white reductions, P.R.208/polyolefin systems only withstand temperatures below 200 °C, while transparent specimens (0.1%) are stable up to approximately 240 °C. Thus the pigment is a suitable and economical candidate for polypropylene spin dyeing, provided the temperature is kept below 200 °C. It is also possible to apply higher temperatures if a color shift towards more yellowish shades is acceptable. In terms of lightfastness, P.R.208 meets the common standards for interior application.

P.R.208 is also used in polyacrylonitrile spin dyeing. It exhibits excellent textile fastness properties and shows good lightfastness. Full shades (3% pigment concentration) equal step 7 on the Blue Scale, while very light (0.1% pigment) red specimens match step 5. The list of applications includes secondary acetate spin dyeing and mass coloration of polyurethane foam and elastomers. P.R.208 is inert to peroxides.

P.R.208 is recommended for all printing methods and is very lightfast in print. The prints exhibit very good resistance to solvents and clear lacquer coatings and may safely be sterilized. Moreover, they are heat stable up to 200 °C for 10 minutes or 180 °C for 30 minutes. For comparison, it should be noted that the cleaner, somewhat yellower P.R.112 is less lightfast, less fast to solvents and chemicals, and less heat stable. Shades of brown are accessible by blending P.R.208 with yellow pigments and with carbon black, frequently in printing inks based on vinyl chloride/acetate mixed polymer. Such materials are used especially to produce wood imitations. P.R.208 is particularly suited to use in decorative printing inks. The lightfastness of 8% gravure prints at 20 and 40 μm cell depth in laminated melamine sheets equals step 8 on the Blue Scale. Plate-out, i.e., coloration of the printing plates, is not observed. The pigment does not bleed if the print is soaked with a melamine resin solution. It is also completely inert to styrene monomer and acetone, which makes it an equally interesting candidate for polyester sheets.

The paint industry recommends P.R.208 primarily for general industrial finishes, in which it is used to a limited extent. Its lightfastness and weatherfastness do not satisfy the standards of a number of applications.

P.R.208 is also applied in a variety of specialty media besides the three main groups mentioned, such as in crayons and laundry inks, as well as in solvent-based wood stains. Applied onto a surface, these products may safely be overcoated and are resistant towards nitro varnish, acid hardening varnish, and polyester varnish. The lightfastness in these media equals step 7 on the Blue Scale, which is excellent.

Pigment Violet 32

P.V.32 is a somewhat dull, very bluish red pigment, referred to as a bordeaux. It is completely fast to a large number of organic solvents and is used in paints, plastics, and printing inks, as well as in spin dyeing.

Paints containing P.V.32 are totally resistant to overpainting. However, the pigment exhibits only moderate lightfastness. Full shade samples in an alkyd-melamine resin equal step 5–6 on the Blue Scale and darken, while 1:10 TiO_2 reductions match step 5–6 and 1:100 reductions coincide with step 3–4. The pigment is utilized in general industrial finishes wherever its lightfastness and weatherfastness satisfy the requirements. The similarly colored P.R.12, on the other hand, a Naphthol AS pigment, bleeds and blooms in baking enamels.

High transparency makes P.V.32 an interesting colorant for transparent foil coatings and similar purposes. Like P.R.171 and 175, P.V.32 shows excellent heat stability, even in long-term exposure. Incorporated in a silicone resin coating, even more reduced colorations (1:50 TiO_2) withstand up to 200 °C for 120 hours and 180 °C for 1000 hours. No color change is observed. However, the lightfastness deteriorates during exposure by approximately 1/2 step on the Blue Scale.

Transparent P.V.32 colorations in PVC equal step 7–8 on the Blue Scale for lightfastness, while opaque types (1/3 SD; 5% TiO_2) match step 6–7. The pigment is completely bleed resistant in plasticized PVC. P.V.32 exhibits high tinctorial strength. Only 0.44% pigment is required to produce 1/3 SD coloration in PVC (5% TiO_2). The similarly colored but not bleed resistant tetrachlorothioindigo pigment P.R.88 is only about half as strong (Sec. 3.5.3). P.V.32 is frequently used in combination with Molybdate Red. Its heat stability in polyolefins is somewhat below 200 °C, although the color change, as in many other cases, is not so noticeable as to absolutely preclude pigment application at a higher temperature. Therefore P.V.32 is generally recommended only for use in LDPE at low processing temperatures. It is also suitable for spin dyeing, for instance for secondary acetate, viscose rayon, or viscose cellulose. Its lightfastness in these media is good: 1/1 SD samples equal step 7 on the Blue Scale and 1/12 SD specimens match step 6. The textile fastnesses are excellent, although the pigment is not entirely fast to wet crocking and only poorly vat resistant.

Incorporated in printing inks, P.V.32 affords clean, highly transparent bluish shades of bordeaux. The pigment is used to produce high grade printing inks wherever fastness to calendering and other fastness properties are required. Moreover, P.V.32 is employed in laminated plastic sheets. Prints made from 8% gravure printing inks in 20 µm cells correspond to step 7 on the Blue Scale for lightfastness. The pigment is suitable both for polyester and for melamine sheets. Bleeding, however, may occur to a very slight extent if prints are soaked with a melamine resin solution. It should also be noted that the pigment is not entirely fast to acetone (Sec. 1.8.1.2). P.V.32 is frequently used in printing inks based on vinyl chloride/vinyl acetate copolymer.

Apart from these areas of application, P.V.32 is also used in solvent-based wood stains. The lightfastness of the products equals step 6 on the Blue Scale, which is very good. These systems may safely be overcoated. Pigment blends with yellow pigments, such as with P.Y.83, and blends with black provide interesting shades of brown.

Pigment Brown 25

The pigment affords a reddish shade of brown. The commercially available types feature a high specific surface area of approximately 80 m^2/g and are therefore highly transparent. P.Br.25 is somewhat less fast to some solvents than other pigments of the same class. It is used in paints, plastics, and printing inks and is in these areas in direct competition with the coloristically closely related but somewhat yellower and more opaque P.Br.23.

The paint industry uses P.Br.25 primarily as a colorant for high grade industrial paints, such as automobile (O.E.M) and repair finishes. High transparency makes it a suitable product wherever transparent or metallic effects are to be created, especially in two-coat (base coat/clear coat) metallic automotive finishes. For these purposes, P.Br.25 is frequently used in pigment blends with transparent iron oxides or with carbon black to afford shades of brown or maroon which are not accessible by using iron oxide pigments alone. P.Br.25 is completely fast to overcoating and is heat stable up to 200 °C. Its lightfastness and weatherfastness are equally excellent. Even in white reductions, alkyd-melamine resins systems reduced 1:300 with TiO_2 equal step 7–8 on the Blue Scale. White reductions of acrylic/melamine resin, reduced 1:1 with TiO_2 and exposed to Florida weather for one year, match step 5 on the Gray Scale for durability, while 1:6 reductions equal step 4–5 and 1:60 specimens correspond to step 4 on the Gray Scale.

P.Br.25 is a frequently used pigment in plastics. It exhibits very good but not perfect migration fastness in plasticized PVC. At concentrations down to 0.005%, P.Br.25 is heat stable up to 200 °C. It exhibits high tinctorial strength: only 0.77% pigment is required to produce 1/3 SD colorations in PVC with 5% TiO_2. Its lightfastness in PVC equals step 8 on the Blue Scale, and step 7–8 if considerable amounts of white pigment are added. P.Br.25 shows excellent weatherfastness in rigid PVC and in impact resistant PVC, which is also true for reductions with TiO_2. The pigment even satisfies the requirements of long-term exposure. It is used, for instance, in window frames made of impact resistant PVC. PVC plastisols which are coil coated onto steel plates exhibit excellent lightfastness and durability. P.Br.25 is also found in PVC and PUR leather targeted for use in automobiles.

1/3 SD HDPE colorations containing 1% TiO_2 are heat stable up to 290 °C. The shrinkage of injection-molded articles is noticeably affected in the direction of flow if the processing temperature is low (220 °C). This influence, however, is already very slight at 260 °C.

P.Br.25 also shows excellent lightfastness in polyolefins. Transparent polystyrene samples are heat stable up to 280 °C, while specimens reduced considerably with TiO_2 withstand up to 240 °C. P.Br.25 is also used in polyester, in which it provides an interesting raw material for the manufacture of bottles.

Excellent fastness properties make P.Br.25 a suitable product for polypropylene spin dyeing. It is also very lightfast in polyacrylonitrile spin dyeing: 1/3 SD samples equal step 7 on the Blue Scale for lightfastness. Pigment weatherfastness is equally excellent, which qualifies P.Br.25 for use in awnings, etc. The pigment is completely fast to dry rubbing and almost completely fast to wet crocking. Moreover, P.Br.25 is also interesting as a pigment for polyurethane foam.

The printing ink industry uses P.Br.25 for all printing methods. The prints show excellent lightfastness. 1/1 to 1/25 SD letterpress proof prints, for instance, equal step 7 to step 6–7 on the Blue Scale. Prints made from P.Br.25 are fast to the DIN 16 524 standard solvent mixture, to paraffin, butter, soap, and acid, but they are not entirely fast to alkali. The products are fast to clear lacquer coatings and may safely be sterilized. The temperature stability is up to 240 °C for 10 minutes or 220 °C for up to 30 minutes, which makes P.Br.25 a suitable candidate for metal deco printing inks. It is also frequently applied in printing inks for PVC.

Moreover, P.Br.25 is used in a variety of specialty media, for instance in oil colors for artists and in water colors. It lends itself to solvent-based wood stains. The products are very lightfast (step 7) and fast to overcoating.

References for Section 2.8

[1] E. Dietz and O. Fuchs, Farbe + Lack 79 (1973) 1058–1063.
[2] R.P. Schunck and K. Hunger, in: Pigment Handbook, ed. by P.A. Lewis, Vol. I, 523–533, Wiley, New York, 1988.
[3] E.F. Paulus and K. Hunger, Farbe + Lack 86 (1980) 116–120.
[4] K. Hunger, E.F. Paulus, and D. Weber, Farbe + Lack 88 (1982) 453–458.
[5] E.F. Paulus, Z. f. Kristallogr. 160 (1982) 235–243.

2.9   Disazo Condensation Pigments

There are various techniques which make it possible to confer higher solvent and migration resistance on a Naphthol AS pigment:

1. Salt formation, i.e., formation of pigment lakes (Sec. 2.7).
2. Introducing additional carbonamide moieties into the pigment molecule (Sec. 2.8).
3. Enlarging the pigment molecule.

In the early 1950s, Ciba research succeeded in synthesizing red disazo compounds of relatively high molecular weight. These products were known as disazo condensation pigments [1,2]. Such compounds may be structurally visualized as disazo pigments, formally composed of two monoazo units, which are attached to each other by an aromatic diamino carbonamide bridge:

$$D-N \diagdown_{N-C-CONH-Ar-HNOC-C-N} \diagup^{N-D}$$

D: diazo component
C: coupling component
Ar: bifunctional aromatic group

The number of carbonamide groups is thus doubled in relation to the monoazo pigment.

Yellow as well as red disazo pigments are derivatives of this parent structure. Both types are prepared by essentially the same but slightly modified route.

A monoazo yellow pigment obtained from acetoacetarylide (Sec. 2.3) constitutes the monoazo portion in a typical yellow product. Red types, on the other hand, are derived from a β-naphthol derivative (Sec. 2.5) or a BONA pigment using the carboxylic function. Bisacetoacetarylide pigments, although featuring a more simple structure, are also technically considered constituents of the yellow series (Sec. 2.4.2). It should be noted, however, that the term disazo condensation pigment, as it is used in this context, refers to somewhat more complex compounds. The presence of additional carbonamide functions in the diazo component produces a marked influence on the fastness of these pigments to solvents and migration. It is not possible to synthesize disazo condensation pigments by traditional methods. M. Schmid at Ciba discovered a suitable pathway and thus laid the foundation for the industrial manufacture of disazo condensation pigments.

2.9.1   Chemistry, Manufacture

The formal structure is exemplified by the red disazo condensation pigments. The structure may be visualized as resulting from the dimerization of two monoazo pigments of the Naphthol AS type:

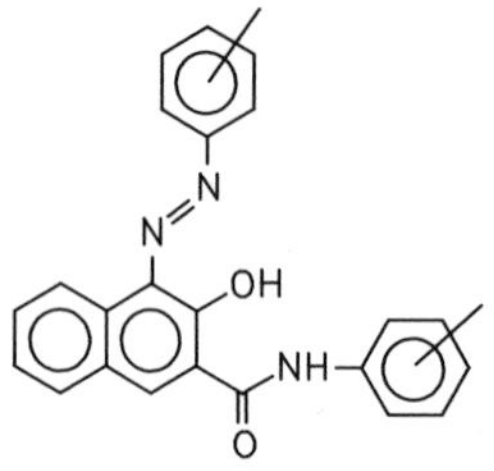

Naphthol AS type 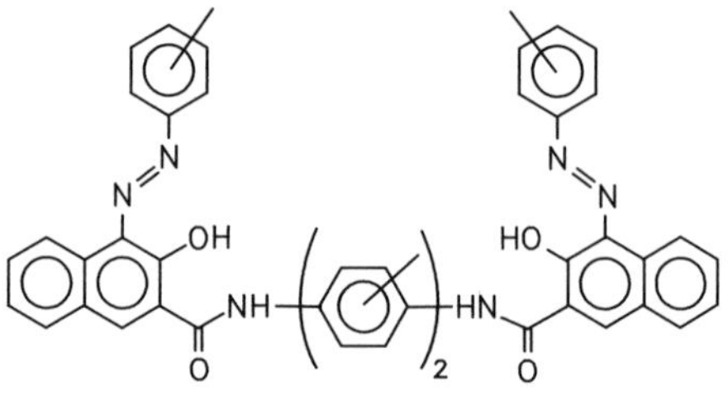   Type of red disazo condensation pigments

Red disazo condensation pigments therefore have the following general structure:

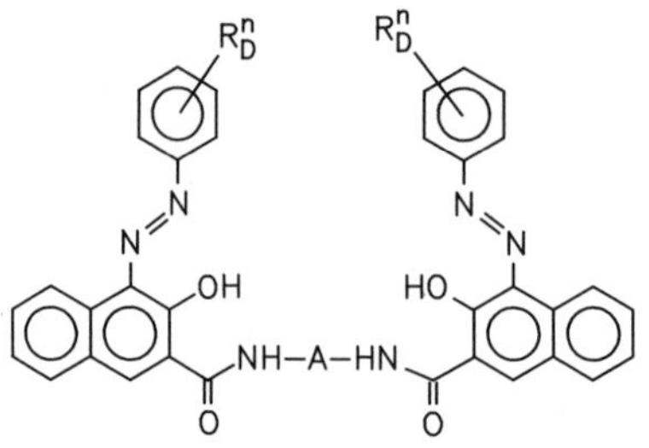

R_D^n stands for the usual substituents for diazo components (see, for instance, Sec. 2.6.1). However, it may also comprise additional carbonamide groups, connected via phenyl moieties. In commercially important pigments, A commonly represents a phenylene or a diphenylene group, and n is any number between 1 and 3. This general structure also covers orange and brown pigments.

Two possible chemical structures have been found for yellow disazo condensation pigments. The two monoazo units may be connected either via the coupling component (type 1) or via the diazo moiety (type 2).

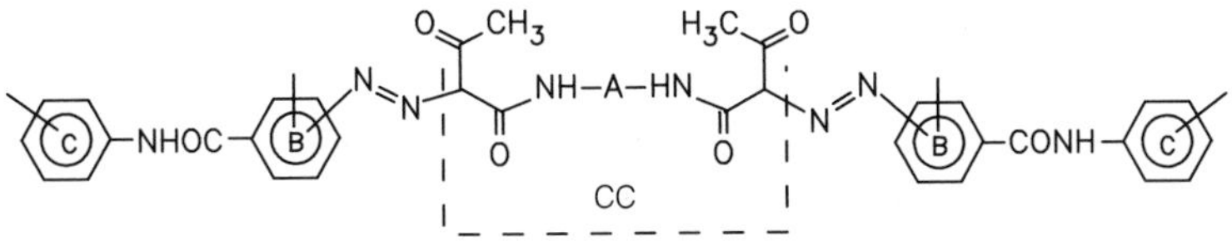

Type 1

CC: coupling component

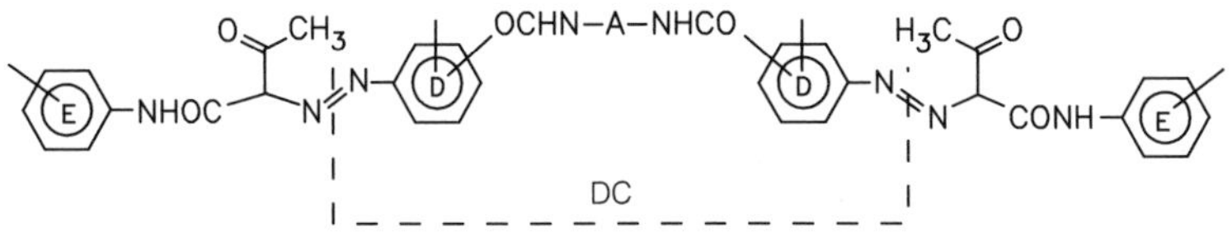

Type 2

DC: diazo component

Derivatives in this series have been prepared in which rings B through E in structures 1 and 2 carry substituents such as CH_3, OCH_3, OC_2H_5, Cl, NO_2, $COOCH_3$, CF_3, or OC_6H_5.

There are basically two routes leading to disazo condensation pigments. The two schemes will be exemplified by the synthesis of red pigments.

Using amine, 2-hydroxynaphthoic acid, and aromatic diamine as starting materials, there are two pathways I and II that lead to the desired products:

Pathway I is the method that first comes to mind but presents somewhat of a problem. It is not possible to synthesize the product in a straightforward two-step sequence by simply coupling two equivalents of diazotized amine with the bifunctional coupling component. This reaction does not afford a definitive product, since the monoazo compound is frequently insoluble enough to precipitate and is thus eliminated from further reaction.

Although difficulties of this kind may be obviated, the necessary procedures are comparatively demanding. Applying physical action, such as thoroughly milling the starting materials, vigorously stirring the reaction mixture, or using a mixing jet and working with an excess of dinaphthol are among the techniques used to facilitate this route. An alternative scheme involves diazotizing a mixture of amine and dinaphthol in an organic solvent with alkylnitrite. Another suitable starting material is the diazoamino compound of the amine to be coupled. It may be treated with dinaphthol in an organic solvent in the presence of glacial acetic acid. All these techniques present serious disadvantages, since they are not generally applicable and do not always afford a uniform product.

The decisive step towards a practical synthesis was taken by performing the reverse reaction II: coupling preceding condensation.

This method proceeds via the carboxylic acid of the monoazo dye, which is obtained from diazotized amine and 2-hydroxy-3-napthoic acid in an alkaline medium. The prepared intermediate is then dried (azeotropically) in an organic solvent, such as mono or dichlorobenzene, with a phosphorus halogenide or with thionyl chloride to form the acid chloride. This step only proceeds uniformly and under

DC: diazo component

very mild conditions in the presence of catalytic amounts of N,N'-dimethylform-amide to afford the desired product. Subsequent condensation of two equivalents of the monoazo dye acid chloride with one equivalent of diamine $H_2N-A-NH_2$ is achieved in an organic solvent. This is the step which has given these pigments their name. Some publications also recommend adding an acid scavenger, such as sodium acetate or a tertiary organic base. Since the reaction is performed in an organic solvent system, the pigment is most efficiently separated in a closed filtering apparatus, preferably by using a suction filter.

The technique of having the coupling step precede the condensation is similarly applied to the two types of yellow pigments. Type 1 (see p. 369) is synthesized by coupling two equivalents of diazotized aminobenzoic acid onto bisacetoacetylated aromatic diamine, especially diaminobenzene. Conversion to the disazo acid chloride, followed by condensation with two equivalents of usually carbonamide substituted amine, finally affords the desired pigment.

Type 2 (see p. 369) is obtained by coupling diazotized aminobenzoic acid onto acetoacetarylide. The resulting acid is converted into the acid chloride and condensed with an aromatic diamine (see structure in the appendix).

The condensation principle makes it possible to synthesize a wide range of disazo condensation pigments.

2.9.2   Properties

Disazo condensation pigments provide a variety of shades ranging from very greenish yellow to orange to bluish red and violet. Brown pigments are also available. The commercial types exhibit good to average tinctorial strength. It is especially the yellow representatives which, although members of the disazo series, are tinctorially weak.

The doubling of the molecule and, in some grades, larger pigment particles enhance the fastness properties of these products, especially their resistance to a variety of organic solvents. Pigments in the yellow range, for instance, are fast to alcohols and aliphatic and aromatic hydrocarbons, but show some bleeding in ketones and esters. The extent of bleeding is a function of the substitution pattern of the molecule. Some of the Naphthol AS type disazo condensation pigments are somewhat less fast to solvents. Tested relative to conventional standards (Sec. 1.6.2.1), these pigments have shown to be not entirely resistant to alcohols and aromatic hydrocarbons. As a rule, however, good overall solvent fastness makes this variety of disazo condensation pigments largely migration resistant. The pigments do not bloom, and many exhibit good to excellent bleed resistance. The same is true for fastness to overcoating. Moreover, disazo condensation pigments are also fast to acid and alkali. Easy dispersion in a variety of media, especially in plastics, is attributed to the fact that the pigments are prepared in organic media. A large number of representatives provide excellent lightfastness, some exhibit good to very good weatherfastness. Besides, disazo condensation pigments often possess excellent heat stability.

2.9.3   Application

Disazo condensation pigments, whose manufacture is somewhat more demanding than that of diarylide yellow and Naphthol AS pigments, are correspondingly expensive.

As a result of their higher price, due to the demanding synthesis, these pigments are used primarily in high grade media and in quality products. Disazo condensation pigments are broad in scope, they are found in various types of plastics, in spin dyeing products, printing inks, industrial finishes and paints, as well as in special media.

The pigments are primarily applied by the plastics industry, which uses many of these pigments as colorants for PVC and PO. Large-sized molecules make these pigments very migration resistant in plasticized PVC. The yellow types are typically more bleed resistant than the red types. Disazo condensation pigments are heat stable enough to satisfy the requirements for use in plasticized and rigid PVC. Some of the yellow types are tinctorially weak. Approximately 0.7 to 2% of a yellow disazo condensation pigment is needed to produce 1/3 SD plasticized PVC systems with 5% TiO$_2$ on a two-roll mill. Similarly colored diarylide yellow pigments, on the other hand, afford the same result at a concentration of approximately 0.3 to 1.0% pigment. This advantage, however, is compromised by the fact that diarylide yellow pigments are generally less fast. The corresponding concentration range for red disazo condensation pigments is approximately between 0.5 and 1.4% pigment. In plasticized PVC, disazo condensation pigments satisfy almost any requirement regarding lightfastness, but this is not entirely true for their weatherfastness. Incorporated in rigid PVC, for instance, disazo condensation pigments (with the exception of Pigment Brown 23) do not meet the demands of long-term exposure.

Some of these pigments are also used to an appreciable extent in polyolefins. The tinctorial strength may vary considerably between pigments. Between 0.18 and 0.44% pigment is required to produce a 1/3 SD HDPE sample containing 1% TiO$_2$, depending on the choice of pigment. A tinctorially weak representative may thus differ from a pigment with high tinctorial strength by a factor of 2.5. 1/3 SD formulations are heat stable enough to withstand temperatures between 250 and 300°C, the exact value is dependent on the type of pigment. Several disazo condensation pigments are considerably affecting the shrinkage of polyethylene (Sec. 1.8.3.2). Coloring larger symmetrical articles made from polyolefins or other partially crystalline polymers with disazo condensation pigments may therefore result in distortion. Other application media, such as elastomers, polystyrene, or PUR, are subject to similar distortion.

As a result of their good heat fastness, various disazo condensation pigments are used to advantage in polypropylene spin dyeing. Like other classes of pigments, disazo condensation pigments are therefore also supplied in the form of special-purpose preparations which contain predispersed pigment. A number of grades are also used to an appreciable extent in polyacrylonitrile spin dyeing. The pigments demonstrate good textile fastness in this medium.

The paint industry is interested in individual representatives of this class, such as P.Y.128.These are used to color high grade paints, such as automobile (O.E.M) finishes and automotive refinishes. Other types are used in general industrial paints. As a rule, the fastness of disazo condensation pigments to overcoating is good to excellent in some media. The pigments are also used in architectural paints, sometimes also in emulsion paints.

Disazo condensation pigments have stimulated interest throughout the printing ink industry, they are suited to all types of printing techniques. Their main market in this area, however, is in high grade packaging inks. Good migration and solvent fastnesses make these pigments suitable candidates for special-purpose ketone/ester-based gravure inks intended for use on PVC films. Disazo condensation pigments are also used in printing inks targeted for other substrates. Special pigment preparations are available for this purpose.

In print, most pigments in this series are completely fast to many of the packaged goods: butter, cheese, and soap. Prints made from disazo condensation pigments are also acid and alkali resistant and fast to overcoating, calendering, and sterilization. The prints are completely heat stable up to 160°C for 60 minutes. Exposure to 200°C for 15 minutes results in minor color changes. It is their high heat stability that makes these pigments suitable candidates for metal deco printing. As a rule, the prints provide excellent lightfastness.

Various disazo condensation pigments are utilized in decorative printing on laminated plastic sheets. They are also applied in a number of special-purpose media outside the above-mentioned groups, such as oil colors for artists and crayons.

2.9.4   Commercially Available Pigments

General

Among yellow disazo condensation pigments, it is especially the ones that are based on the type I structure which enjoy widespread use. These pigments cover the spectral range from very greenish to reddish shades of yellow. Members of the Naphthol series produce colors from orange to red and violet as well as shades of brown.

The product lines offered by pigment manufacturers include a considerable number of commercially interesting disazo condensation pigments. The most important representatives in the yellow range are Pigment Yellow 93, 94, 95, and 128. Orange, red, and brown shades are covered particularly by Pigment Orange 31, Pigment Red 144, 166, 214, 220, 221, 242, 248, and Pigment Brown 23. Table 24 lists the commercially available disazo condensation pigments.

Individual Pigments

Yellow Series

Pigment Yellow 93

P.Y.93 affords a light greenish to medium yellow shade which is similar to that of P.Y.16.

The pigment is primarily used to color plastics, including PP spin dyeing. Incorporated in plasticized PVC, P.Y.93 shows good to average tinctorial strength and good bleed resistance. Approximately 0.85% pigment is required to formulate a 1/3 SD samples with 5% TiO_2. P.Y.93 demonstrates very good lightfastness. In PVC, it is as lightfast as P.Y.94 and in this respect even slightly superior to P.Y.95, but it is less stable to light than the greener P.Y.128.

1/3 SD pigment formulations in rigid PVC and in PVC plastisols which are intended for coil coating lose weatherfastness much more rapidly as more TiO_2 is added than do P.Y.94 and 128. There is no difference in the response of the respective pigmented systems to weather, as long as they contain only 1% TiO_2, while a

TiO_2 content of 5% renders the P.Y.93 system considerably less durable than its counterparts.

1/3 SD P.Y.93 formulations in HDPE (1% TiO_2) withstand exposure to up to 290 °C for one minute or up to 270 °C for 5 minutes. The pigment causes distortion at processing temperatures between 220 and 280 °C. This effect, however, is so minor as to be tolerable for most purposes. Distortion phenomena therefore rarely restrict the extent of pigment application. In terms of lightfastness, 1/3 to 1/25 SD samples (1% TiO_2) equal step 7 on the Blue Scale. Good heat fastness makes P.Y.93 a suitable candidate for polypropylene spin dyeing. The pigment is also

Tab. 24: Commercially available disazo condensation pigments.

Yellow series

Structure		Shade	C.I. Name	C.I. Constitution No.	Reference
A	B				
H₃C / Cl	Cl CH₃	yellow	P.Y.93	20710	[3, 4]
Cl / Cl	Cl / CH₃	greenish yellow	P.Y.94	20038	[5]
H₃C / CH₃	Cl / CH₃	reddish yellow	P.Y.95	20034	[3, 6]
H₃C / Cl	CF₃ ... Cl	greenish yellow	P.Y.128	20037	
Cl / Cl	Cl CH₃	yellow	P.Y.166	20035	

Tab. 24: (Continued)

Orange, red, and brown series

Structure		Shade	C.I. Name	C.I. Constitution No.	Reference
R_D	B				
Cl (structure)	[Cl structure]$_2$	orange	P.O.31	20050	[3]
Cl, Cl (structure)	Cl (structure)	bluish red	P.R.144	20735	[3, 7]
Cl, Cl (structure)	(structure)	yellowish red	P.R.166	20730	[3]
Cl, Cl (structure)	Cl, Cl (structure)	bluish red	P.R.214		[8]
CH$_3$, COOCH$_2$CH$_2$Cl (structure)	H$_3$C, CH$_3$ (structure)	bluish red	P.R.220	20055	
Cl, COOCH(CH$_3$)$_2$ (structure)	Cl, Cl (structure)	bluish red	p.R.221	20065	
Cl, CF$_3$ (structure)	Cl, Cl (structure)	reddish orange (scarlet)	P.R.242	20067	[8, 9]
NO$_2$, Cl (structure)	Cl (structure)	reddish brown	P.Br.23	20060	

very lightfast in this medium. P.Y.93 shows good to excellent textile fastnesses in textile printing.

P.Y.93 has no importance in the paint industry. The graphic ink field uses P.Y.93 only where fastness standards are high, such as in high grade packaging and metal deco inks.

Pigment Yellow 94

P.Y.94 is the greenest one of all available yellow disazo condensation pigments. With approximately equal cleanness, it is somewhat greener than P.Y.128, which is a member of the same class of pigments. However, P.Y.94 is only about half as strong tinctorially. 0.44% P.Y.94 is required to produce a 1/3 SD HDPE sample containing 1% TiO_2, while the same end result is achieved by using only either 0.22% P.Y.128 or 0.2% of the somewhat redder P.Y.93.

P.Y.94 shows excellent heat stability. 1/3 SD to 1/25 SD samples in HDPE (1% TiO_2) withstand up to 300 °C. The shrinkage of the plastic, however, is considerably affected. P.Y.94 is equally weak in PVC. 2.1% pigment is needed to formulate 1/3 SD samples containing 5% TiO_2. P.Y.94 exhibits excellent lightfastness and weatherfastness, which is also true for PVC plastisols targeted for coil coating. The pigment is completely fast to migration, even in highly plasticized PVC. Its excellent heat stability makes it a suitable candidate for PP spin dyeing. Besides, P.Y.94 is also used in polyacrylonitrile spin dyeing.

The printing ink industry uses P.Y.94, like other disazo condensation pigments, only for high grade printed products, especially in metal deco printing. Economical pigment formulation is compromised by somewhat poor tinctorial strength. The commercial type, which features a comparatively low specific surface area of approximately 35 m^2/g, is consequently less transparent than other products in this class of pigments. P.Y.94 is noticeably less lightfast than P.Y.128, which has a finer particle size; the difference is approximately one step on the Blue Scale. In this respect, P.Y.94 resembles the diarylide yellow pigment P.Y.113, which is somewhat greener and distinctly stronger tinctorially. P.Y.94 exhibits plate-out on the metal plates if it is used in decorative printing, i.e., on laminated papers for laminated plastic sheets (Sec. 1.6.4.1).

Pigment Yellow 95

P.Y.95 affords reddish shades of yellow. At 1/3 SD, its shade may be located somewhere between those of the diarylide yellow pigments P.Y.13 and 83.

Its main field of application is in plastics and in spin dyeing, which is also true for other members of this class of pigments. P.Y.95 is a pigment with high tinctorial strength, it is the strongest yellow disazo condensation pigment. 0.7% pigment is required to formulate a 1/3 SD sample in plasticized PVC containing 5% TiO_2. For comparison, 0.8% are needed of the greener P.Y.93 and 2.1% of P.Y.94, which is even greener at 1/3 SD.

0.35% must be added of the similarly shaded, but generally less fast P.Y.13 in order to achieve the same result. P.Y.95 is very lightfast. However, incorporated in plasticized or in rigid PVC, P.Y.95 is less lightfast than P.Y.93.

The pigment shows excellent heat stability in polyolefins. 1/3 SD samples may safely be exposed to 290 °C, while 1/25 SD specimens (1% TiO_2) withstand 270 °C. Incorporated in polyolefins, P.Y.95 demonstrates much higher tinctorial strength than most other yellow pigments within its class. Tested in HDPE, the pigment was found to influence the shrinkage of the polymer in injection molding only to a very negligible extent. In terms of lightfastness, 1/3 SD colorations (1% TiO_2) of this medium equal step 6 on the Blue Scale.

P.Y.95 is also used for PUR as well as for polypropylene spin dyeing, where it is employed for its good heat stability and its high tinctorial strength. However, the pigment is not lightfast and durable enough to meet the standards for application in polyacrylonitrile spin dyeing products, such as canvas awnings (Sec. 1.8.3.8). Besides, P.Y.95 is also utilized in textile printing.

Like other representatives of its class, P.Y.95 has only limited impact on the printing ink industry. It is only used in special-purpose printing inks for high grade prints. Its lightfastness in offset and in metal deco printing corresponds to that of P.Y.93. P.Y.95 exhibits good heat stability. The prints are fast to overcoating and sterilization and in most cases show excellent fastness in special applications (Sec. 1.6.2). P.Y.95 therefore lends itself to metal deco printing and packaging gravure printing inks, i.e., printing inks based on nitrocellulose, polyamide, and vinylchloride copolymers.

Pigment Yellow 128

At medium tinctorial strength, P.Y.128, after P.Y.94, is the second greenest pigment in the disazo condensation range, and it exhibits very good general fastness properties. The pigment is completely or at least almost completely fast to almost all organic solvents which are used in coloristically important media. The pigment possesses excellent lightfastness and weatherfastness, and it is completely fast to overcoating. This makes it a suitable product for automotive paints, even for original automotive (O.E.M) finishes. The commercial type features a relatively high specific surface area of approximately 70 m^2/g and is correspondingly transparent. It is mainly used in high grade industrial paints, on which it confers good tinctorial strength and high gloss, and also in architectural paints, for instance in green shades. P.Y.128 is, however, also used in emulsion paints.

The plastics industry uses P.Y.128 mainly in PVC, in which the pigment exhibits average tinctorial strength. 1/3 SD colorations containing 5% TiO_2, for instance, are formulated at 1.35% pigment. The lightfastness is very good and equals step 7–8 on the Blue Scale for 1/3 SD samples. Pigment durability is equally very good in this medium and also in PVC plastisols for coil coating. P.Y.128 also lends color to elastomers. 1/3 to 1/25 SD colorations in HDPE (1% TiO_2) withstand exposure to 250 °C for 5 minutes. The shrinkage of polyethylene is only slightly affected. P.Y.128 can also be used for polyacrylonitrile spin dyeing, an area in which it exhibits very good lightfastness and performs excellently in terms of the more important textile fastnesses. Its weatherfastness, however, does not satisfy the requirements for outdoor application. P.Y.128 is therefore used particularly in home textiles such as upholstery.

The printing ink industry uses P.Y.128 particularly in high quality products. The resulting prints, like those made from other disazo condensation pigments, are fast to calendering and sterilization. P.Y.128 lends itself particularly to metal deco printing. It shows good lightfastness. In this respect, it scores approximately one step higher under comparable conditions than the greener P.Y.94 and the somewhat redder P.Y.93, which are members of the same class of pigments.

Pigment Yellow 166

P.Y.166 is available in Japan, but has failed so far to gain commercial importance.

Orange, Red, and Brown Pigments

Pigment Orange 31

P.O.31 affords a reddish, somewhat dull orange shade. It is recommended particularly for brown shades.

P.O.31 is a useful colorant for plastics and is employed especially in polypropylene spin dyeing. Compared to other members of its class, the pigment exhibits good to average tinctorial strength.

P.O.31 is not completely fast to bleeding in plasticized PVC. Its lightfastness in this medium is only satisfactory at higher pigment concentrations, but the pigment possesses excellent heat stability. 1/3 SD HDPE samples (1% TiO_2) are fast up to 300 °C. P.O.31 considerably affects the shrinkage of polyolefins in injection molding (Sec. 1.8.3.2). 1/3 SD specimens equal step 6 on the Blue Scale for lightfastness. Good heat fastness makes the pigment a suitable candidate especially for polypropylene spin dyeing, mainly to produce shades of brown and beige. P.O.31 is also used in textile printing, most of its textile fastnesses are very good.

Pigment Red 144

P.R.144 is a medium to slightly bluish red pigment, which probably reigns supreme within its class. It is broad in scope and is mainly used to color plastics, including spin dyeing products. The commercially available types of the acicular pigment differ considerably in terms of particle size, consequently demonstrating a range of coloristic properties. This is especially true for the tinctorial strength. P.R.144 is less stable to a variety of organic solvents than other disazo condensation pigments of its class, such as P.R.166.

P.R.144 is almost completely fast to migration in plasticized PVC. One of the most tinctorially strong disazo condensation pigments, only about 0.7% P.R.144 is needed to produce a 1/3 SD PVC sample containing 5% TiO_2. Comparative values are listed for other pigments within the same class and also in other classes. P.R.144 is very fast to light. Its weatherfastness in rigid PVC is less satisfactory, it fails to meet the standards of long-term exposure.

Within this group of pigments, P.R.144, incorporated in PE, ranks second only to P.R.221 regarding tinctorial strength. 0.13% pigment is required to produce 1/3 SD specimens containing 1% TiO_2. In HDPE, P.R.144 withstands temperatures up to 300°C at 1/3 to 1/25 SD, both with and without TiO_2. It considerably affects the shrinkage of the plastic at processing temperatures between 220 and 280°C, which should be taken into consideration if the pigment is to be used in injection molding. P.R.144 has excellent lightfastness: 1/3 SD colorations in HDPE containing 1% TiO_2 equal step 7–8 on the Blue Scale, while transparent samples even reach step 8.

P.R.144 is also suited to use in other plastics, such as polystyrene, polyurethane, elastomers, or cast resins, including those made from unsaturated polyester.

Good heat stability and lightfastness make P.R.144 a suitable candidate especially for polypropylene spin dyeing. It is equally important for polyacrylonitrile spin dyeing, but does not satisfy the demands for use in canvasses. The more important textile fastnesses are very good, but the pigment is not completely stable to dry and wet crocking. In this respect, it scores as high as step 4 on the 5 step Gray Scale.

The printing ink industry uses P.R.144, like other pigments of its class, primarily in high quality products. The commercial types exhibit specific surface areas between approximately 50 and 90 m^2/g, which makes for more or less transparent prints. A large number of fastnesses which are important for packaging purposes (Sec. 1.6.2), such as resistance to soap, butter, cheese, paraffin, acid, and alkali, are perfect. P.R.144 is fast to clear lacquer coatings and sterilization, which makes it suitable for metal deco printing. The pigment has very good lightfastness; letterpress proof prints equal step 6 to step 7 on the Blue Scale, depending on the pigment area concentration. In this respect, P.R.144 (at low pigment area concentrations such as 1/25 SD) is somewhat inferior to P.R.166.

Throughout the paint industry, P.R.144 functions as a pigment in general industrial coatings, automotive finishes, and architectural paints.

Pigment Red 166

P.R.166 affords clean yellowish shades of red. It is broad in scope and in this respect resembles the somewhat bluer disazo condensation pigment P.R.144. Its main area of application, however, is in plastics and in spin dyeing.

In the plastics sector, P.R.166 is used primarily to color PVC and polyolefins. The pigment is almost completely fast to bleeding in plasticized PVC. Similarly colored pigments of other classes perform poorer in terms of migration and lightfastness and also regarding heat stability. These pigments are considered suitable alternatives to P.R.166 only where the application requirements are less stringent.

P.R.166 exhibits medium to good tinctorial strength compared to other pigments covering the same range of shades. Between 0.9 and 1.2% pigment is needed, for instance, to produce 1/3 SD PVC samples containing 5% TiO_2. The exact amount depends on the grade. Commercial types of this pigment provide medium to high hiding power. The lightfastness and weatherfastness of P.R.166 in PVC are very good. However, its weatherfastness is often not sufficient to meet the standards of long-term outdoor exposure.

P.R.166 shows excellent heat stability in polyolefins, for instance in HDPE. 1/3 SD samples containing 1% TiO_2, for instance, are fast up to 300 °C. This high heat stability in HDPE is compromised by the fact that P.R.166 causes considerable distortion in this medium, irrespective of the processing temperature. Transparent HDPE samples equal step 8 on the Blue Scale for lightfastness, while corresponding polypropylene colorations match step 7. 1/3 SD colorations containing 1% TiO_2 reach step 6–7 and step 7, respectively. High lightfastness makes P.R.166 a suitable candidate for polypropylene spin dyeing. Its application in polyacrylonitrile spin dyeing is restricted by the fact that it fails to satisfy the high lightfastness and durability standards of materials such as canvasses. P.R.166 exhibits perfect or almost perfect textile fastnesses. It is also used in polystyrene, rubber, and other elastomers. P.R.166 is encountered in polyurethane which is intended for use in coatings, and applied in cast resins made from unsaturated polyester or aminoplast resins.

P.R.166 is recommended in the paint field for use in high grade industrial paints, for original automotive finishes, and for automobile refinishes, as well as for architectural paints and emulsion paints. Incorporated in baking enamels, deep shades (1:1 TiO_2) and white reductions (1:20 TiO_2) exhibit good weatherability. However, P.R.166 is not quite as weatherfast in these media as the somewhat yellower and considerably cleaner Anthanthrone pigment P.R.168 (Sec. 3.6.4.2). Dispersed in an alkyd-melamine resin which is baked at 120 °C, the pigment is almost completely fast to overcoating.

Like other members of its class, P.R.166 is used throughout the printing ink industry for high grade prints, especially for packaging purposes. It basically possesses all-round suitability for various printing techniques. P.R.166 is fast to clear lacquer coatings and to sterilization, which qualifies it particularly for use in metal deco printing. It provides very lightfast prints: 1/1 to 1/3 SD letterpress proof prints equal step 7 on the Blue Scale. P.R.166 is thus faster to light than the somewhat bluer disazo condensation pigment P.R.144. P.R.166 shows excellent fastness to agents such as alkali, acids, soap, fats, paraffin, and others, which are frequently encountered in packaging printing inks. The pigment is also used in textile printing.

Pigment Red 214

P.R.214 affords a medium to bluish shade of red. It exhibits average stability to organic solvents, an aspect in which it resembles other members of its class.

P.R.214 is very lightfast and fast to overcoating. Its weatherfastness, however, does not quite reach the standards of other types of pigments within its class: it is somewhat inferior, for instance, to that of P.R.242.

The paint industry uses P.R.214 primarily for various types of industrial finishes. The pigment demonstrates very good tinctorial strength in plastics. 0.13% pigment is required, for instance, to produce 1/3 SD HDPE coloration containing 1% TiO_2. P.R.214 produces a shade which, although cleaner, is very similar to that of P.R.144, a member of the same class of pigments. The two pigments also show similar tinctorial strength. The fact that P.R.214 considerably affects the shrinkage of the plastic makes it less important in injection molding. It should be emphasized that P.R.214 possesses excellent heat stability. 1/3 to 1/25 SD HDPE samples containing 1% TiO_2, as well as transparent specimens of the same depth of shade, are fast to temperatures up to 300 °C. P.R.214 is also used in PP spin dyeing. It shows good textile fastness. A pigment with high tinctorial strength in plasticized PVC, P.R.214 is not completely fast to bleeding in this medium. The pigment is also recommended for use in polystyrene and in a number of engineering plastics.

Like other representatives of its class, P.R.214 lends itself particularly to use in high grade printing inks which are targeted for purposes such as posters, packaging gravure printing inks for PVC films, and metal deco printing. The prints exhibit good fastness properties, they are stable to soap, alkali, and acids and withstand temperatures up to 200 °C. They are also fast to clear lacquer coatings and to calendering.

Pigment Red 220

P.R.220 affords a yellowish red shade, which is somewhat bluer than that of P.R.166 and slightly yellower than that of P.R.144. P.R.220 is considerably weaker, however, than the latter: 0.31% pigment is required to produce 1/3 SD HDPE samples containing 1% TiO_2. Although P.R.220 affects the shrinkage of injection-molded HDPE at processing temperatures of 220 °C, it does this to an extent which is acceptable for most purposes. The effect is negligible at 260 °C. 1/3 SD specimens are fast to temperatures up to 300 °C and equal step 7 on the Blue Scale for lightfastness. P.R.220 is thus a suitable candidate for polypropylene spin dyeing and is also used in polyacrylonitrile spin dyeing.

Pigment Red 221

P.R.221 is a slightly bluish red pigment. Its main field of application is also in plastics, especially in PVC and in PUR. P.R.221 is sufficiently fast to migration in plasticized PVC to satisfy the requirements for most applications. The pigment demonstrates very high tinctorial strength in plasticized PVC; its full shade and related shades are the strongest in its class. Only 0.58% pigment is needed to produce a 1/3 SD sample containing 5% TiO_2. The tinctorial strength of lighter shades corresponds to that of other disazo condensation pigments in the red range. On the other hand, P.R.221 is less lightfast than many other types. The pigment does not exhibit plate-out in PVC.

P.R.221 is not recommended for use in polyolefins, since it also causes considerable distortion. It has, however, gained recognition as a pigment for high grade printing inks, especially for metal deco printing.

Pigment Red 242

P.R.242 affords a yellowish red shade, referred to as scarlet. It exhibits good to excellent resistance to organic solvents, such as alcohols, esters, ketones, and aliphatic hydrocarbons. The pigment is more soluble in aromatic hydrocarbons. P.R.242, like other members of its class, is fast to alkali and acid.

P.R.242 is used throughout the plastics industry as a colorant for PVC as well as for polyolefins, polystyrene, and other engineering plastics. It exhibits average tinctorial strength. 0.2% pigment is required to formulate 1/3 SD HDPE samples containing 1% TiO_2. P.R.242 is heat stable up to 300 °C. 1/25 SD specimens containing 1% TiO_2, as well as transparent samples at 1/3 to 1/25 SD are stable to temperatures up to 280 °C. The pigment has a considerable effect on the shrinkage of the plastic. It is of interest to PP spin dyeing, for which a special-purpose pigment preparation is available.

In plasticized PVC, P.R.242 shows good fastness to bleeding and provides average tinctorial strength. 1/3 SD colorations containing 5% TiO_2 require 1.1% pigment. Good heat stability makes P.R.242 a suitable candidate for use in polystyrene, ABS, PMMA, and polyester.

P.R.242 is an equally valuable product for paints, especially for various types of industrial paints. It is also recommended for use in automotive finishes. Both lightfastness and weatherfastness are excellent, but do not quite reach the levels of the appreciably yellower anthanthrone P.R.168. P.R.242 is fast to overcoating and heat stable above 180 °C. It is also employed in emulsion paints based on synthetic resin.

In the printing ink field P.R.242 is utilized to color high grade systems, such as printing inks for PVC films or metal deco printing. P.R.242 exhibits good fastness properties in application. It shows some bleeding in styrene monomer, which makes the pigment unsuitable for decorative printing for laminated plastic sheets based on styrene/polyester.

Pigment Red 248

P.R.248 is a comparatively recent product which is used to color plastics, especially polyolefins and polystyrene. It provides bluish shades of red.

1/3 SD HDPE samples containing 1% TiO_2 are heat stable up to 290 °C. The distortion problem in injection-molded HDPE gains significance as the processing temperature increases. While at 220 °C the material is distorted to a still acceptable degree, considerable shrinkage is observed at 250 °C and higher temperatures.

P.R.248 is one of the types in its class which provide high tinctorial strength. Its strength is comparable to that of P.R.166: 1/3 SD systems containing 1% TiO_2 are formulated at 0.16% pigment. 1/3 SD HDPE samples equal step 7 on the Blue Scale for lightfastness. P.R.248 is not recommended for polypropylene spin dyeing.

P.R.248 exhibits excellent bleed resistance in plasticized PVC. Transparent specimens equal step 8 on the Blue Scale for lightfastness, while 1/3 SD samples match step 7. The pigment does not perform as well in terms of weatherfastness, for instance in PVC plastisols for coil coating. P.R.248 is also recommended for use in elastomers, polyurethane, and unsaturated polyester.

Pigment Red 262

P.R.262 is currently in the process of being introduced to the market. It produces bluish shades of red. P.R.262 is recommended especially for polypropylene spin dyeing, but possesses all-round suitability for other applications, as long as the requirements do not exceed the fastness properties of this class of pigments.

Pigment Brown 23

P.Br. 23 confers reddish shades of brown on its application media. In a large number of areas, this pigment is in direct competition with the coloristically similar, somewhat redder and more transparent P.Br.25.

In terms of stability to organic solvents, P.Br.23 performs like other red pigments within its class. It is thus somewhat inferior to the yellow products. Regarding fastness to various ketones, esters, and alcohols, as well as to dioctyl phthalate and dibutyl phthalate, P.Br.23 equals step 3–4 and step 4, respectively, on the 5 step scale. P.Br.23 is broad in scope, but its main field of application is in plastics.

Incorporated in plasticized PVC, P.Br.23 exhibits average to good tinctorial strength. 0.75% pigment is required to produce 1/3 SD samples containing 5% TiO_2. The pigment is not entirely bleed resistant (step 4), but exhibits excellent lightfastness and weatherfastness. In rigid PVC, for instance, P.Br.23 is fast to long-term exposure. It is primarily used to color window frames and similar articles.

P.Br.23 shows excellent heat stability in polyolefins. 1/3 SD samples containing 1% TiO_2, as well as transparent colorations at 1/3 SD in HDPE are stable to exposure to 300°C for 5 minutes. In injection molding, P.Br.23 considerably affects the shrinkage of the plastic at 220°C, an effect which diminishes with increasing temperature (Sec. 1.8.3.2).

P.Br.23, like other members of this series of pigments, is a suitable colorant for elastomers. However, due to the low fastness of rubber itself, P.Br.23 is only used in high grade elastomers. The pigment is also recommended for use in polystyrene. Transparent polystyrene samples are fast to exposure to 280°C for 5 minutes, while light shades containing TiO_2 (0.01% pigment + 0.5% TiO_2) are only stable up to 240 °C. In terms of lightfastness, these samples equal step 7 and step 5, respectively, on the Blue Scale. Good fastness properties make P.Br.23 a suitable candidate for polyacrylonitrile spin dyeing, although its lightfastness and weatherfastness do not satisfy the stringent requirements for use in applications such as canvasses.

The paint industry utilizes P.Br.23 particularly in industrial paints. As a result of its lack of solvent fastness, P.Br.23 is not completely fast to overcoating if it is used in baking enamels at a curing temperature of 120 °C. It is, however, very fast to light and weather. In deep shades (1:1 TiO_2), P.Br.23 equals step 8 on the Blue Scale for lightfastness, while light shades (1:20 TiO_2) match step 7–8. In order to assess their weatherfastness, baking enamels containing P.Br.23 were subjected to two years of outdoor exposure. The colorations were found to equal step 4–5 and step 3–4 on the Gray Scale, respectively (Sec. 1.6.6). In the automotive industry, P.Br.23 is recommended for use in original automotive and repair finishes. The list of applications also includes emulsion paints and high grade architectural paints. Moreover, the pigment is also found in various types of wood stains and other special-purpose media.

P.Br.23 possesses all-round suitability for all printing applications. Its shade and its fastness properties make it particularly suitable for printing imitation wood, for instance on PVC. The prints show good fastness to clear lacquer coatings. Sterilization causes somewhat of a yellow shift. P.Br.23 prints demonstrate good lightfastness: under standardized conditions, 1/1 to 1/25 SD letterpress proof prints equal step 6–7 and step 6 on the Blue Scale, respectively, depending on the substrate.

Pigment Brown 41

This disazo condensation pigment, is a very recent product. It affords very yellowish shades of brown, covering the range of the displaced benzimidazolone pigment P.Br.32. No information is yet available about its preferred areas of application. In terms of fastness and application properties, P.Br.41 largely resembles other representatives of its class.

Incorporated in PVC, P.Br.41 provides medium tinctorial strength and good hiding power. The pigment is not completely fast to bleeding in plasticized PVC. In polyolefins, P.Br.41 also exhibits medium tinctorial strength. 0.21% pigment is required to formulate 1/3 SD HDPE samples containing 1% TiO_2. Such specimens are stable up to 300 °C.

Incorporated in paints, P.Br.41 exhibits equally high hiding power, which makes it a less attractive product for use in metallic finishes. Its tinctorial strength in these media is comparatively poor. However, P.Br.41 is completely fast to overcoating. The commercial grade shows considerable flocculation in various paint systems.

Pigment Brown 42

P.Br.42 is an equally recent product. It provides yellowish shades of brown, which are on the yellowish side of P.Br.41. The pigment is particularly recommended for use in rigid PVC, in which it exhibits medium tinctorial strength. 0.85% pigment is required to produce 1/3 SD samples containing 5% TiO_2. P.Br.42 is very lightfast and weatherfast, a quality which makes it suitable for long-term exposure. Incorporated in plasticized PVC, P.Br.42, like other brown pigments within its class, is not completely bleed resistant (step 3–4).

1/3 SD P.Br.42 samples in HDPE containing 1% TiO_2 are stable up to 280°C for 5 minutes, which indicates very good heat stability. 0.22% pigment is needed to formulate such 1/3 SD samples.

References for Section 2.9

[1] Max Schmid, DEFAZET Dtsch. Farben Z. 9 (1955) 252–255.
[2] H. Gaertner, J. Oil and Colour Chem. Assoc. 46 (1963) 13–46.
[3] NPIRI (National Printing Ink Research Institute) Raw Materials Data Handbook, Vol. 4 Pigments; National Print. Ink. Res. Inst., Bethlehem, Penn. USA, 1983.
[4] DE-PS 1 150 165 (Ciba) 1957; DE-PS 1 544 453 (Ciba) 1964.
[5] DE-OS 2 312 421 (Dainichi Seika) 1972.
[6] DE-PS 1 150 165 (Ciba) 1957.
[7] DE-OS 1 644 117 (Ciba) 1966.
[8] Pigments Handbook, ed. by P.A. Lewis, Vol. 1, p. 724, J. Wiley, New York 1988.
[9] B.L. Kaul, J. Col. Chem. 12 (1987) 349–354; B.L. Kaul Soc. Plastics Eng. 1989, 213–226.

2.10 Metal Complex Pigments

Most metal complex pigments, with the exception of the large group of copper phthalocyanine pigments (3.1), have only recently stimulated interest [1,2] for their commercially interesting properties. Water-soluble complexes, especially those of the azo series, have for a long time maintained an important position as colorants for textiles. Besides a few complex metal salts of azo pigments, it is primarily azomethine metal complexes which have gained commercial significance.

Apart from alizarin "lake", which is now being formulated as an aluminum/calcium complex [3] (Sec. 3.6.2), the oldest known metal complex pigment is an iron complex. In 1885, O.Hoffmann reported on the iron complex of 1-nitroso-2-naphthol, which under the name of Pigmentgrün B (Pigment Green 8, 10006) was first industrially exploited in 1921 by BASF.

Largely displaced by copper phthalocyanine green pigments, this product has lost most of its commercial importance.

In 1946, Du Pont described a nickel/azo complex pigment, which was introduced in 1947 as Green Gold (Pigment Green 10, 12775). For a long time, this pigment was the most lightfast and weatherfast product within the greenish yellow range of the azo series. The promotion of metal complex pigments in the azo and azomethine series with improved fastness properties stalled until the early 1970s, as chemically novel structures were developed.

This group of metal complex pigments does not only exhibit good fastness properties, but has proven economically advantageous enough to maintain limited but solid ground within the ranges of organic pigments.

2.10.1   Chemistry, Synthesis

The commercially most interesting metal complex pigments within the azo series are those obtained from aromatic o,o′-dihydroxyazo compounds, while important products within the azomethine series are nickel or copper complexes of aromatic o,o′-dihydroxyazomethine compounds.

The aromatic moieties are possibly substituted benzene or napthaline rings. In azomethine pigments, only one form of metal complex is possible. This is in contrast to azo metal complexes, which may assume either structure **31** or **32**:

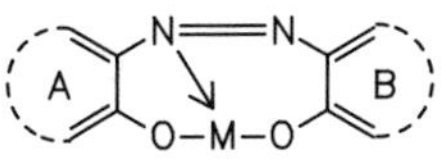

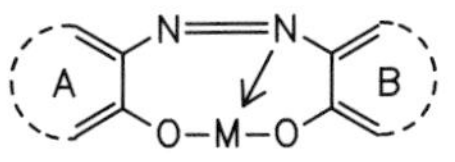

The nitrogen atom which is connected to the least nucleophilic aromatic moiety is always the one to serve as a ligand. As is the case with yellow and red monoazo pigments, which have been studied by three dimensional X-ray diffraction analysis (Sec. 2.3.1.1 and 2.5.1), the chelate-6-rings of azo metal complexes prefer the quinonehydrazone structure over the hydroxyazo form [4].

The commercially interesting metal complex pigments usually contain the coordinative tetravalent Cu^{++} or Ni^{++} ions, less commonly Co^{++} ions. The fourth coordination site is typically occupied by a solvent molecule with a free electron pair. It may also be engaged by the second nitrogen atom of a different pigment molecule, a phenomenon which is observed in azo complexes and similar materials. In the latter case, sandwich structures are obtained [5]. The copper and nickel complexes are mostly planar molecules.

It is important for metal complex compounds to be free from solubilizing groups in order to provide the necessary pigment characteristics.

2.10.1.1   Azo Metal Complexes

The above-mentioned Pigment Green 10 is the 1:2 nickel complex of the azo pigment obtained from p-chloroaniline and 2,4-dihydroxy-quinoline (**33**):

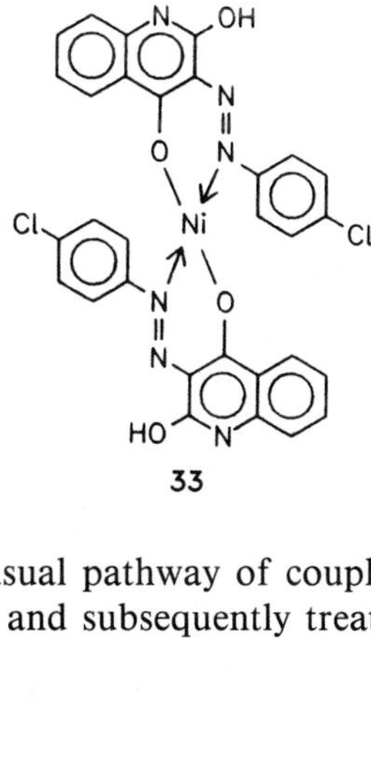

Its synthesis follows the usual pathway of coupling diazotized 4-chloroaniline onto 2,4-dihydroxy-quinoline and subsequently treating the product with a nickel (II) salt.

The fact that protons are released through complexation makes it sometimes possible to enhance the reaction and to improve the yield by adding a base (such as sodium acetate).

An interesting product in the range of azo metal complexes is the nickel complex of azo barbituric acid (34) [6]:

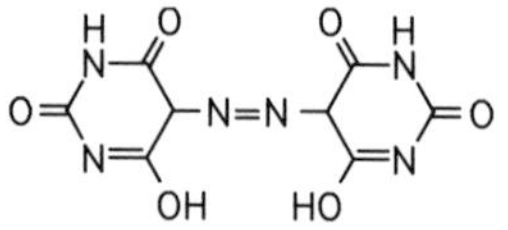

34

This molecule probably assumes the structure of a 1:1 sandwich complex [5].

A reaction known as diazo group transfer produces diazo barbituric acid from barbituric acid and p-toluene sulfonyl azide. Additional barbituric acid affords azo barbituric acid [7]. Subsequent complexation with a nickel (II) salt yields a greenish yellow pigment.

Other Complexation Methods

Apart from the reaction of o,o′-dihydroxyazo compound with a metal salt, there are two more complexation techniques which are worth mentioning, although these are primarily used to synthesize azo metal complex dyes.

Demethylating while copperizing is achieved only under severe conditions, i.e., at temperatures in excess of 100 °C, frequently under pressure. Suitable starting materials are o-hydroxy-o′-methoxyazo compounds.

During the oxidative copperizing process, an o-hydroxyazo compound reacts with a copper(II) salt in the presence of hydrogen peroxide. Both methods broaden the scope of metal complexation reactions by extending the selection of suitable diazo components [8,9].

2.10.1.2   Azomethine Metal Complexes

Copper, nickel, and cobalt complexes reign supreme among industrially interesting products within this class. The complexes are manufactured by condensing aromatic o-hydroxyaldehydes with aromatic o-hydroxyamines, frequently o-aminophenols, in an aqueous medium or in organic solvents at 60 to 120 °C. The intermediate products, usually orange to red azomethines, are either separated or immediately reacted further by adding a Cu, Ni, or Co(II) salt. The conversion is carried out at similarly elevated temperature, although the prime concern is the presence of a solvent to afford the largely insoluble azomethine metal complexes. As above, the presence of a base may sometimes be useful.

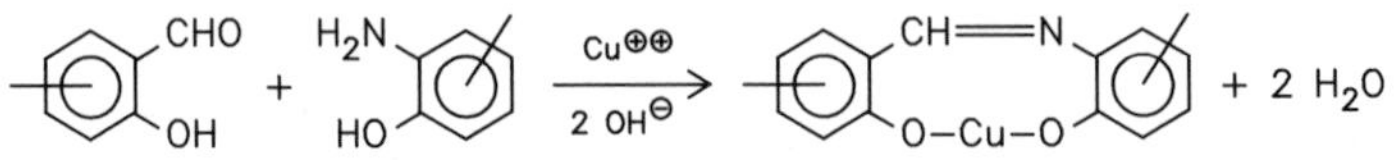

There is another structure which, although it does not represent a typical azomethine compound, is synthesized from nickel complexes of the anilide of diimino
butyric acid (**35**).

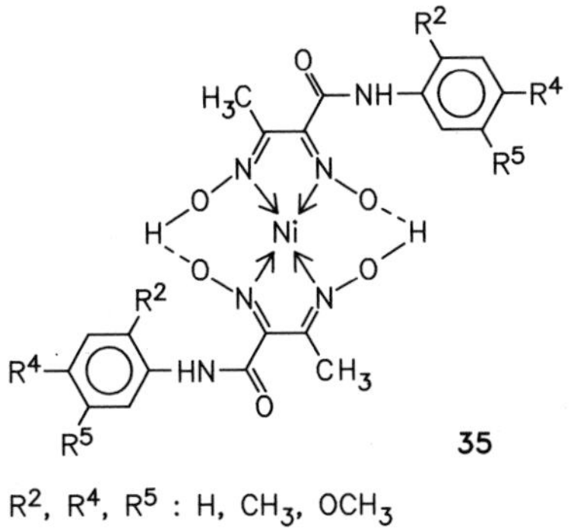

R², R⁴, R⁵ : H, CH₃, OCH₃

The synthesis involves nitrosating the corresponding substituted acetoacetic anilide with sodium nitrite in acetic acid; and subsequently, by adding hydroxylamine
to the same reaction vessel, converting the compound to the oxime. Finally, complexation is achieved by means of a Ni(II) salt.

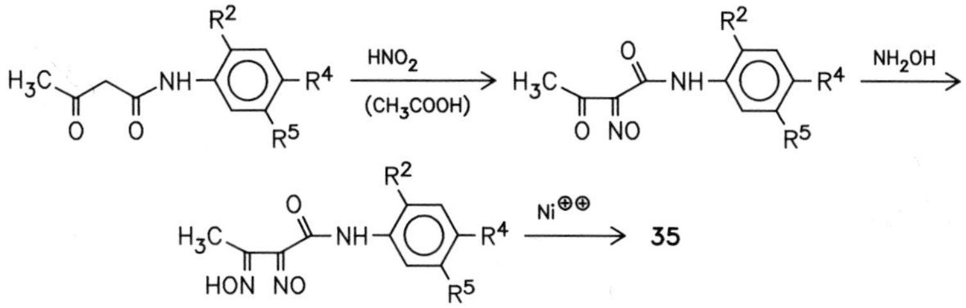

Yet another structural principle is represented by metal complex pigments based
on isoindolinones. Condensation of amino-iminoisoindolinones (iminophthalimide) with 2-aminobenzimidazole in a high boiling solvent affords an azomethine
(**36**). This compound reacts with salts of divalent metals, such as Co, Cu, Ni, to
yield yellow azomethine metal complex pigments [10]:

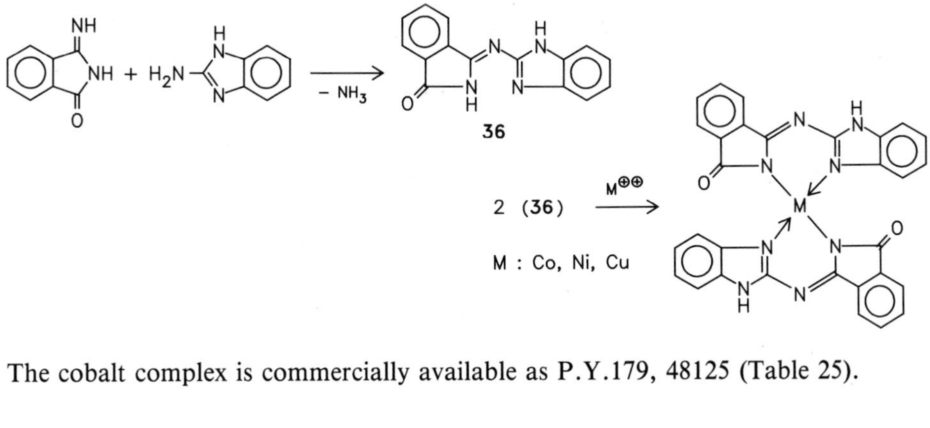

M : Co, Ni, Cu

The cobalt complex is commercially available as P.Y.179, 48125 (Table 25).

Tab. 25: Examples of azo and azomethine metal complex pigments.

Structure	Shade	C.I. Name	C.I. Constitution No.	Reference
Azo series				
	green	P.Gr.8	10006	
	very greenish yellow	P.Gr.10	12775	
Ni-complex	greenish yellow	P.Y.150	12764	[6, 12]
Azomethine series				
	greenish yellow	P.Y.117	48043	[12, 13]
	very greenish yellow	P.Y.129	48042	
R^2, R^4, R^5 : H R^2 : OCH$_3$, R^4, R^5 : H R^2, R^4 : CH$_3$, R^5 : H	greenish yellow, orange, red	P.Y.153	48545	[12, 14, 15]

Tab. 25: (Continued)

Structure	Shade	C.I. Name	C.I. Constitution No.	Reference
	dull yellow	P.Y.177	48120	
	reddish yellow	P.Y.179	48125	
	dull reddish orange	P.O.65	48053	
	orange	P.O.68		[16]
	red violet	P.R.257		[16]

Pigment Yellow 177, 48120 is synthesized by using 2-cyanomethylene benzimidazole instead of 2-amino benzimidazole to react with iminophthalimide. For more information about cobalt complexes, refer to [11].

2.10.2   Properties

Commercially available azo and azomethine metal complex pigments cover the spectral range from considerably greenish to reddish yellow and yellowish orange. Compared to their parent structures (the corresponding azo and azomethine compounds), azomethine metal complexes frequently exhibit a distinctly duller shade. Formation of the metal complex often shifts the color of an originally yellow material in the greenish yellow direction.

There are a number of advantages to complexation. It imparts better weatherfastness and lightfastness, as well as enhanced solvent resistance and migration fastness compared with the metal free compound.

A number of the commercially available members of this class are characterized by high transparency. Their tinctorial strength, however, is not always satisfactory. Both lightfastness and weatherfastness are generally very good, sometimes excellent. While some grades are fast to overcoating under application conditions, which include temperatures up to 160°C, other products bleed at less than 120°C. Likewise, the pigments vary considerably in terms of heat stability.

2.10.3   Application

Metal complex pigments are mainly used in paints. The products are fast enough to be applied especially in industrial finishes. Some representatives, particularly azomethine copper complex pigments, are very weatherfast, which makes them suitable candidates for automotive finishes. High transparency in combination with good weatherfastness is an asset for use in metallic finishes. It is not uncommon for metal complexes to lose much of their brilliance in white reductions. Some are also recommended for use in architectural paints, especially for emulsion paints.

Besides, metal complex pigments are also used in printing inks as well as in other areas of application.

2.10.4   Commercially Available Pigments

General

Recent developments in this class prefer azomethine complexes as chemical structures rather than azo metal complexes. The list of commercially available types includes Pigment Green 8 and 10, Pigment Yellow 117, 129, 150, 153, 177, 179, and Pigment Orange 59, 65, and 68, as well as P.R.257.

Table 25 shows examples of the above type of metal complex pigments.

Individual Pigments

Pigment Green 8

P.Gr.8, also referred to as Pigment Green B, provides dull yellowish shades of green or, respectively, greenish shades of yellow up to the olive hues. Over the last few decades, P.Gr.8 has largely given way to the cleaner phthalocyanine green pigments.

P.Gr.8 exhibits good fastness to organic solvents, but performs poorly towards esters, such as ethyl acetate, ketones, such as methylethylketone, and ethyl glycol (ethylene glycol monoethylether). In contrast to possessing excellent fastness to alkali and lime, P.Gr.8 is not stable to acids. The pigment exhibits good lightfastness, but it is not as lightfast as its copper phthalocyanine counterparts. 1/3 SD P.Gr.8 emulsion paint samples, for instance, equal step 7 on the Blue Scale for lightfastness, while more reduced specimens up to 1:200 equal step 6–7. Corresponding phthalocyanine green pigments match step 8 on the Blue Scale.

With its moderate price, P.Gr.8 is used in emulsion paints and also in concrete mass coloration. Like P.Gr.7, it is one of the few organic pigments which are suitable for a large variety of concrete formulations. P.Gr.8 in general is not lightfast and weatherfast enough to be used in outdoor application. Its dull shade is of no consequence to indoors application, the shade is cleaner than that of Chrome Oxide Green. As always in this type of application, preliminary testing is recommended.

An important application of P.Gr.8 is in the coloration of rubber. The pigment, however, is not suitable for use in blends which contain large amounts of basic fillers. It is somewhat sensitive to cold vulcanization. The colored articles usually perform well in general application but are not entirely fast to aromatic hydrocarbons and to some fats, and they are sensitive to acid and sulfur dioxide. P.Gr.8 also colors some plastics, especially LDPE and polystyrene. Heat stable up to 220 °C, P.Gr.8 grades equal step 2–3 on the Blue Scale for lightfastness. Other areas of application include wallpaper and artists' colors.

Pigment Green 10

P.Gr.10, a nickel azo complex which has been known since 1947, affords a dull, very greenish yellow shade which, as in the Colour Index, may also be considered a very yellowish green (in white reduction and in full shade, respectively). It is a comparatively weak product and is used primarily in high grade industrial paints, especially as a shading pigment for blue and green colors. The commercial types are highly transparent, which makes them suitable candidates for metallic, transparent, and hammer tone finishes. P.Gr.10, however, fails to satisfy the requirements for use in automotive finishes. The colorations remain very lightfast even in highly reduced white reductions: reductions up to 1:200 SD equal step 8 on the Blue Scale for lightfastness. The weatherfastness level is equally very good. In this respect, P.Gr.10 performs similarly to P.Y.117, an azomethine copper complex pigment.

Incorporated in finishes, P.Gr.8 is heat stable up to about 180°C. It is not entirely fast to overcoating. Most systems containing P.Gr.8 will fail to maintain their color value if they are exposed to acids: there is a shift towards yellower hues, a phenomenon which is attributed to demetallation of the pigment.

Pigment Yellow 117

P.Y.117, an azomethine copper complex, provides very greenish shades of yellow which resemble those of P.Y.129. Combination with TiO_2, for instance at a ratio of 1:50, considerably reduces the cleanness of the system, resulting in very dull olive green shades. P.Y.117 exhibits high transparency, therefore it is recommended for use in metallic finishes, especially throughout the automotive industry. It is also recommended for olive shades in automotive finishes. Blends with copper phthalocyanine blue and green pigments are equally recommended for automotive finishes.

Compared to other pigments covering the greenish yellow range of the spectrum, P.Y.117 exhibits good tinctorial strength. It shows good stability to organic solvents and under standardized conditions exhibits complete fastness to mineral spirits. Its fastness to toluene, alcohol, or esters, such as ethyl acetate, equals step 4 on the 5 step scale; and the pigment equals step 3 in terms of stability to ketones, such as methylethylketone or ethyl glycol.

P.Y.117 demonstrates good fastness to alkali and acids. It is completely fast to overcoating up to 160°C and heat stable up to 180°C. For a system consisting of P.Y.117 incorporated in a silicone resin paint, the manufacturer lists a temperature stability of up to 300°C for one hour exposure.

P.Y.117 is very lightfast and weatherfast. 1/3 to 1/25 SD formulations in an alkyd-melamine resin equal step 8 on the Blue Scale for lightfastness, while 1:200 SD samples still reach step 7. The weatherfastness of the pigment in this range of standard depths and in this binder system equals approximately that of P.Y.153, but is not quite as high as that of P.Y.129. P.Y.117 is also recommended for use in emulsion paints.

P.Y.117, like P.Y.129, undergoes recomplexation if it is incorporated in plasticized PVC in conjunction with tin stabilizers, such as dibutyltin thioglycolate. The color shifts towards reddish shades, and the pigment begins to exhibit an appreciable tendency to migrate in the polymer. If suitable stabilizers are added, P.Y.117 is very weatherfast in PVC and shows high tinctorial strength. 0.65% pigment is required to produce 1/3 SD samples containing 5% TiO_2.

Pigment Yellow 129

P.Y.129 exhibits a clean, very greenish yellow shade. Its only field of application is in paints, especially in automotive and industrial finishes. High transparency makes it a useful pigment to produce interesting metallic finishes. Much of the cleanness is lost as white pigment, such as TiO_2, is added, resulting in olive shades. The coatings are very lightfast and weatherfast. Up to 140 °C, P.Y.129 is fast to overcoating. Reduced 1:50 with TiO_2, the systems are heat stable up to 150 °C. P.Y.129 is also recommended for use in architectural paints.

P.Y.129 is not employed in plastics. The type of metal in the stabilizer defines how much of a color change is observed as the chelated metal in the pigment molecule is exchanged. Tin stabilizers produce red complexes, which, like corresponding lead or zinc chelates, respond very poorly to light and weather (see also Sec. 1.6.7).

Pigment Yellow 150

This pigment, an azo/nickel complex, affords dull, medium shades of yellow. The pigment is recommended for use in paints and printing inks. P.Y.150 produces very lightfast paints: samples up to 1/25 SD equal step 8 on the Blue Scale. The weatherfastness of these specimens, however, deteriorates as more white pigment, such as TiO_2, is added. Systems up to 1/3 SD are heat stable up to 180 °C. P.Y.150 is used as a colorant for general industrial and architectural paints wherever the durability requirements are not too high. It exhibits good resistance to organic solvents but is not entirely fast to overcoating.

P.Y.150 is also very lightfast in printing inks. The prints, however, are neither acid nor alkali resistant. Moreover, they are somewhat sensitive to a number of organic solvents, including the DIN 16 524/1 standard solvent mixture, and soap. The systems are heat stable up to 140 °C. Prints made from P.Y.150 are almost completely fast to sterilization. A pigment with high tinctorial strength, P.Y.150 is particularly suitable for use in decoration printing inks for laminates.

A new special-purpose type of P.Y.150 has recently been introduced to the market. It is recommended for use in spin dyeing polypropylene and polyamide fibers. In this type of application, the pigment exhibits good heat stability and also good lightfastness and weatherfastness. Under common processing conditions in injection molding, P.Y.150 is likely to react with the zinc sulfide which is often found in polyamide, a tendency which precludes its use in polyamide injection molding.

Pigment Yellow 153

P.Y.153 is a nickel complex which was introduced to the market in the late 1960s. It produces slightly dull reddish shades of yellow. Although not fast to acids, the pigment may safely be exposed to alkali. It is fast to mineral spirits and alcohols, but only moderately so to aromatic solvents, such as xylene, and to esters, such as ethyl acetate.

Medium to light shades of this relatively weak pigment are used to color high grade industrial paints, especially automotive solid shades, as well as metallic finishes. P.Y.153 withstands exposure to up to 160°C for 30 minutes. Incorporated in an alkyd-melamine resin, full shades (10%) and 1/3 SD samples equal step 8 on the Blue Scale for lightfastness, while 1/25 SD specimens match step 7–8. In terms of lightfastness, P.Y.153 performs approximately like the quinophthalone pigment P.Y.138. P.Y.153 is fast to overcoating up to 120°C, but bleeds at higher temperatures.

High durability makes P.Y.153 a preferred product in emulsion paints, especially for exterior house paints. It is a relatively important pigment for these purposes.

Pigment Yellow 177

P.Y.177 has only recently been introduced to the market. It is a special-purpose pigment for polypropylene and polyamide spin dyeing. As a colorant for these media, P.Y.177 has the added advantage of enhancing the stability of the fibers.

Baking enamels containing P.Y.177 are very sensitive to overcoating. Their shade depends largely on the depth of shade. Full shades are typically reddish brown, similar to iron oxide systems, while white reductions are greenish yellow.

Pigment Yellow 179

P.Y.179, an isoindolinone/cobalt complex pigment, was introduced to the market a few years ago. It is recommended for use in paints, especially in automotive finishes. The pigment produces a reddish yellow shade. High lightfastness and excellent weatherfastness are an asset in pastel colors. Besides, good transparency makes P.Y.179 a suitable product for metallic finishes. Yet, it is not quite as weatherfast as the equally reddish yellow P.Y.24, a flavanthrone pigment.

Pigment Orange 59

P.O.59 is a nickel complex which affords very dull yellowish shades of orange. The pigment is used to match deep to medium shades in the yellow and orange part of the spectrum for high grade industrial finishes, as well as for architectural paints.

P.O.59 exhibits good lightfastness and weatherfastness. Incorporated in a baking enamel based on alkyd-melamine resin, for instance, 10% full shades equal step 8 on the Blue Scale for lightfastness, while 1/3 to 1/25 SD samples correspond to step 7, and 1:200 SD specimens (with TiO_2) match step 6 on the Blue Scale. The systems rapidly decrease in weatherfastness as more TiO_2 is added. Up to 180°C, the pigment is heat stable and fast to overcoating, as long as the baking temperature does not exceed 140°C. P.O.59 is not alkali and acid resistant; pigment demetallation fades or lightens the shade.

Pigment Orange 65

P.O.65, which was released to the market a few years ago, is an azomethine/nickel complex pigment. It produces dull, reddish shades of orange, but very brilliant shades of copper in metallic finishes. P.O.65 is therefore considered a specialty product for such purposes. The pigment is not entirely fast to overcoating. Its weatherfastness in metallic finishes is very good and satisfies the requirements for use in original automotive finishes. This is despite the fact that it fails to reach the durability standards of the reddish P.R.168, a dibromanthanthrone pigment.

Pigment Orange 68

P.O.68 lends itself for use in industrial paints, including automotive finishes, as well as for plastics. It provides dull, reddish shades of orange. Two varieties are commercially available: a version with a coarse particle size and a type with a fine particle size.

The pigment with a fine particle size provides good transparency at high tinctorial strength. It is recommended primarily for use in metallic finishes to produce shades of copper, gold, and brown. Transparent P.O.68 is not completely fast to overcoating. This type exhibits good lightfastness and durability, but tends to darken. This is especially true for deep solid shades and color intense metallic finishes.

The variety with a coarse particle size is a pigment with equally high tinctorial strength and provides good hiding power. It is bluer and somewhat duller than the version with a fine particle size. The opaque type, like the transparent variety, is not completely fast to overcoating. Lightfastness and durability equal those of the pigment with a fine particle size. Opaque P.O.68 is used in combination with quinacridone pigments, magenta, or violet pigments to produce dull shades of red and maroon.

P.O.68 is also used in plastics for its high heat stability. 1/3 SD colorations in HDPE containing 1% TiO_2 withstand temperatures up to 200 °C. The shade is a dull orange. 0.15% pigment is required to formulate 1/3 SD samples containing 1% TiO_2. Such specimens match step 7 on the Blue Scale for lightfastness. In some plastics (polycarbonate), which are processed at high temperature, the pigment withstands as much as 320 °C. This makes P.O.68 one of the most heat stable organic pigments known. The list of recommendations also includes polyamide coloration.

Pigment Red 257

P.R.257, a heterocyclic nickel complex, was first introduced to the market a few years ago. P.R.257 covers the range of reddish violet shades, its hue is distinctly yellower than that of the P.R.88 type thioindigo pigments and of the β-modification of P.V.19, an unsubstituted quinacridone pigment. In paints, P.R.257 is tinctorially weaker than these pigments. In combination with Molybdate Orange or opaque organic red pigments, it is also less intense. The commercial grade demonstrates good hiding power, good rheology, and good resistance to flocculation. Both lightfastness and weatherfastness are equally good and, both in full shades and in white reductions, comparable to the performance of P.R.88. In baking enamels, P.R.257 is fast to overcoating. Its full shade and similarly deep shades are used in original automotive and industrial finishes.

References for Section 2.10

[1] H. Baumann and R. Hensel, Fortschr. Chem. Forsch. 7 (1967) 4.
[2] O. Stallmann, J. Chem. Educ. 37 (1960) 220–230.
[3] E.G. Kiel and P.M. Heertjes, J. Soc. Dyers Colour. 79 (1963) 21.
[4] G. Schetty, Helv. Chim. Acta 53 (1970) 1437.
[5] G. Stephan, 6. Internat. Color Symposium, Freudenstadt 1976.
[6] DE-OS 2 064 093 (Bayer) 1970.
[7] M. Regitz, Angew. Chem. 79 (1967) 786–801.
[8] K.H. Schindehütte in Houben-Weyl, Methoden der org. Chemie, Bd. X/3, 213, 4. Ed., Stuttgart 1965.
[9] H. Pfitzner and H. Baumann, Angew. Chem. 70 (1958) 232.
[10] DE-OS 2 546 038 (Ciba-Geigy) 1974; F.A. L'Eplattenier, C. Frey and G. Rihs, Helv. Chim. Acta 60 (1977) 697–709.
[11] C. Frey and P. Lienhard, XVII. Congr. FATIPEC Lugano 1984, Congress book 283–301.
[12] NPIRI Raw Materials Data Handbook, Vol. 4, Pigments, National Printing Ink Research Institute, Bethlehem, Penn. USA, 1983.
[13] DE-OS 1 544 404 (BASF) 1966.
[14] H. Sakai et al. J. Jpn. Soc. Colour Mater. 55 (1982) 685.
[15] DE-OS 1 252 341 (BASF) 1966; DE-OS 1 569 666 (BASF) 1967.
[16] Pigment Handbook, ed. by P.A. Lewis, Vol. 1, p. 723, 725, J. Wiley, New York 1988.

2.11   Isoindolinone and Isoindoline Pigments

It is only for reasons of simplified classification that the pigments which are described in this section, like azomethine metal complex pigments (Sec. 2.10.1.2), are listed in Chapter 2. Actually, rather than being azo pigments, these are azo methine and methine pigments, which in the classification system adopted in this book are formally located between azo pigments and polycyclic pigments.

The common structural feature underlying these pigments is the isoindoline ring $(X^1, X^3 = H_2)$:

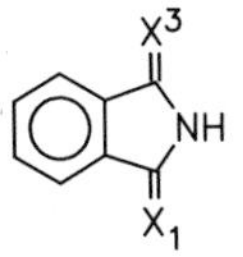

Positions 1 and 3 are substituted further.

Depending on the substitution pattern, either azomethine derivatives (X^1 = O or N-, X^3 = N-) or methine derivatives (X^1 = X^3 = C<) must be defined.

The **azomethine** type represents the important group of tetrachloroisoindolinone pigments, which were introduced in the mid-1960s.

Formally, these pigments are disazomethine types, obtained by condensation of primary aromatic diamines with two equivalents of tetrachloroisoindoline-1-one:

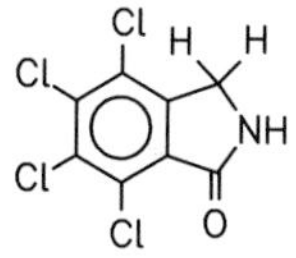

In fact, two tetrachlorophthalimide molecules are linked via a disazomethine bridge (structure see Sec. 2.11.1).

Patents concerning isoindolinone type dyes and pigments were first published by ICI in 1946. Routes to other such pigments followed in 1952 and 1953 (see ref. [1]).

All of these routes started either from the unsubstituted phthalimide or from a phthalimide backbone carrying only one or two substituents (**37**):

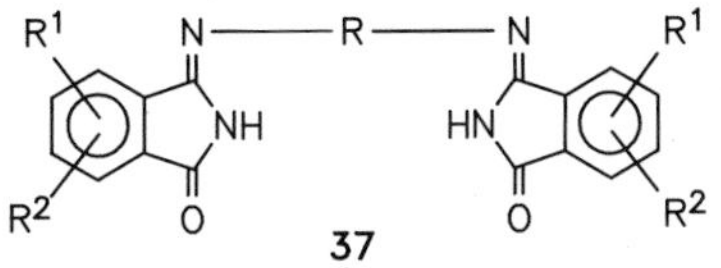

R : aromatic or heterocyclic diamine group
R^1, R^2 : H or common substituents, such as CH_3, OCH_3, NO_2, Cl

Only very few of these yellow and orange pigments have stimulated interest in terms of technical application. The patent literature, for instance, describes a pigment with the following structure (**38**) [2]:

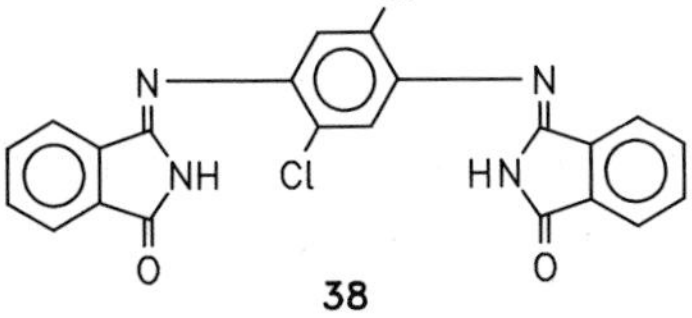

This is a yellow pigment, obtained by condensing 2,5-dichloro-1,4-diaminobenzene-dihydrochloride with 3-iminoisoindolinone in chlorobenzene at 140 °C:

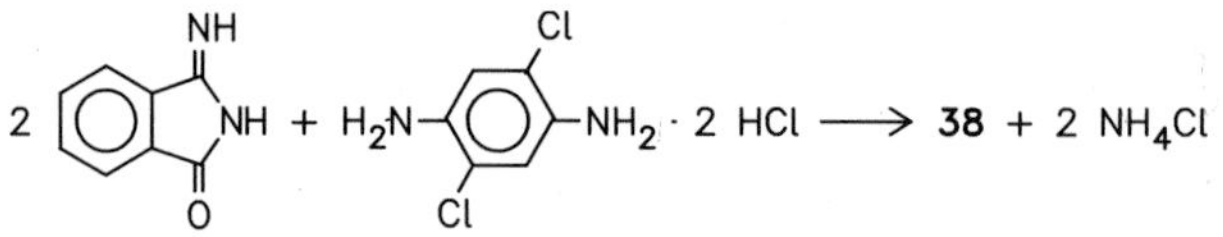

A decisive step towards the development of isoindolinone pigments is credited to Geigy. Patents published in the late 1950s describe pigments with the basic structure **37**, using tetrachlorophthalimide as starting material.

The technique of "perchlorinating" the two aromatic rings surprisingly enhances both the tinctorial strength and the fastness properties of the product considerably. This quality increase is notably attributed to factors which are described in detail in Sec. 2.11.2.

The **methine** type comprises isoindoline derivatives which, like tetrachloroisoindolinone pigments, have only recently been described. One or usually two equivalents of a compound containing an activated methylene moiety are attached to one equivalent of isoindoline. The list of compounds featuring activated methylene groups includes cyanacetamide or heterocycles such as barbituric acid or tetrahydroquinolinedione.

It should be noted that none of these pigments feature chlorine substitution on the aromatic system.

2.11.1   Chemistry, Synthesis, Starting Materials

2.11.1.1   Azomethine Type: Tetrachloroisoindolinone Pigments

Commercially used tetrachloroisoindolinone pigments have the general chemical structure **39** [1]:

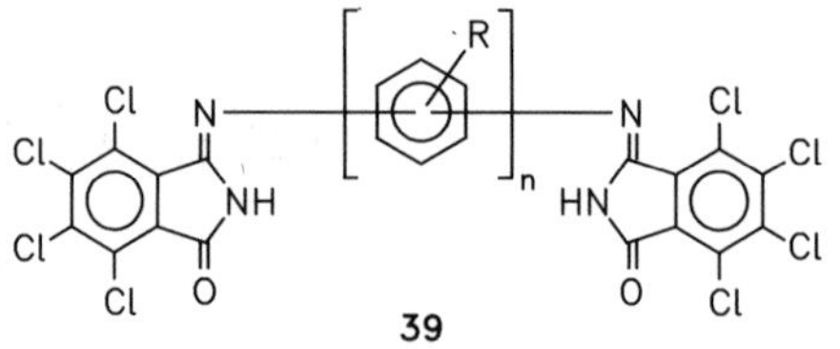

R : H, CH$_3$, OCH$_3$, Cl

n = 1,2

The synthetic route generally involves condensing two equivalents of 4,5,6,7-tetrachloroisoindoline-1-one derivatives with one equivalent of an aromatic diamine in an organic solvent. Suitable tetrachloroisoindoline-1-one derivatives are substituted in 3-position, which is occupied either by two monovalent groups (A) or one divalent group (B). A may represent a chloro atom or a CH$_3$O group, while B usually represents a NH moiety.

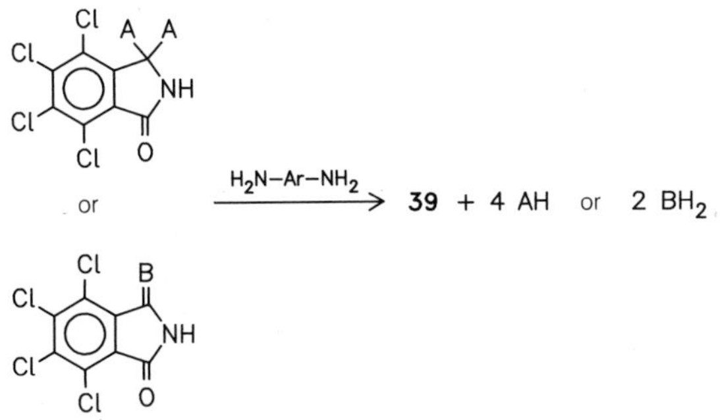

$$\xrightarrow{\text{H}_2\text{N-Ar-NH}_2} \quad \textbf{39} + 4 \text{ AH} \quad \text{or} \quad 2 \text{ BH}_2$$

Primary starting materials for this condensation reaction are 3,3,4,5,6,7-hexachloroisoindoline-1-one (**40**) and 2-cyano-3,4,5,6-tetrachlorobenzoic acid methyl ester (**41**):

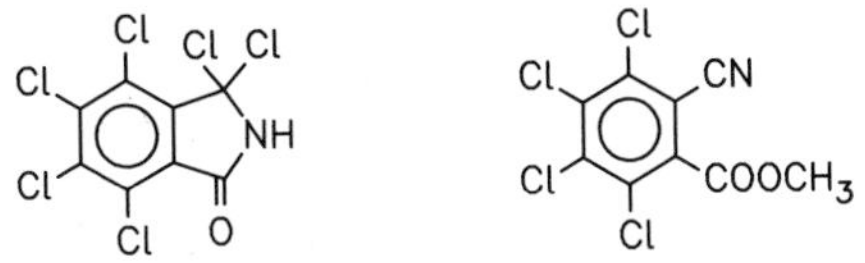

40 directly reacts with a diamine to afford the desired pigment. Two pathways have been found effective for the synthesis of **40**. Tetrachlorophthalimide may be chlorinated either

– with one equivalent of phosphorus pentachloride in phosphorus oxychloride, or
– with two equivalents of phosphorus pentachloride, affording 1,3,3,4,5,6,7-heptachloroisoindolenine (**42**) as an intermediate, which is to be converted into **40** with one equivalent of water or alcohol:

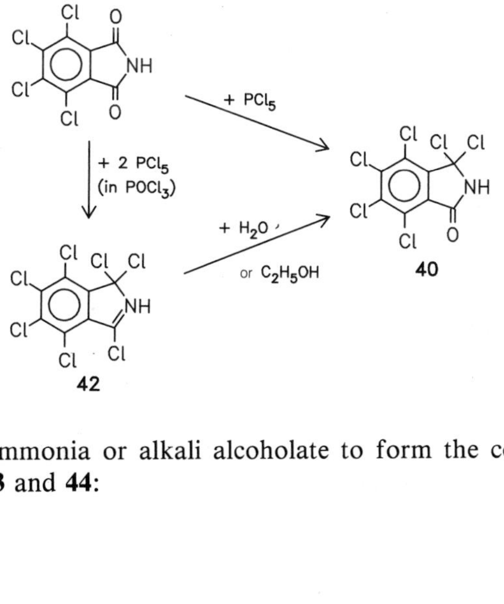

41 reacts with ammonia or alkali alcoholate to form the corresponding activated compounds **43** and **44**:

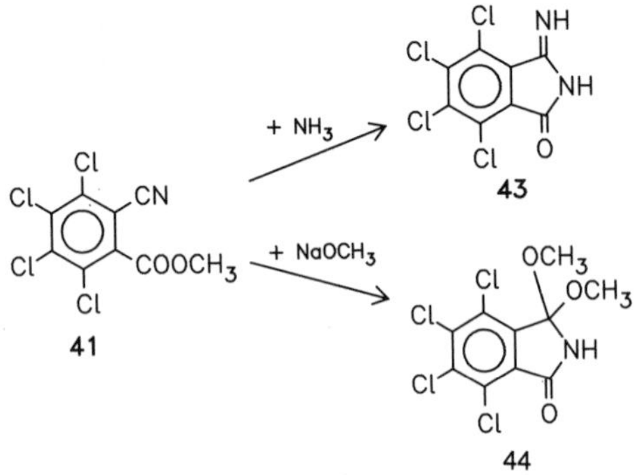

41 is synthesized from 3,3,4,5,6,7-hexachlorophthalic anhydride, which is treated with ammonia to yield the ammonium salt of 2-cyanotetrachlorobenzoic acid. After being converted to the sodium salt, the thus prepared intermediate is reacted with methylhalogenides or dimethylsulfate to form the methylester 41.

The synthetic route may be exemplified by the preparation of a tetrachloro-isoindolinone pigment. A mixture of one mole of 1,4-diaminobenzene in o-di-chlorobenzene with a solution of two moles of 3,3,4,5,6,7-hexachloroisoindoline-1-one in o-dichlorobenzene is heated to 160 to 170 °C for 3 hours. Closed filtration equipment is used to filter the hot product, which is then washed with o-dichloro-benzene and alcohol, dried, and milled. The resulting product is a reddish yellow pigment with the structure 45:

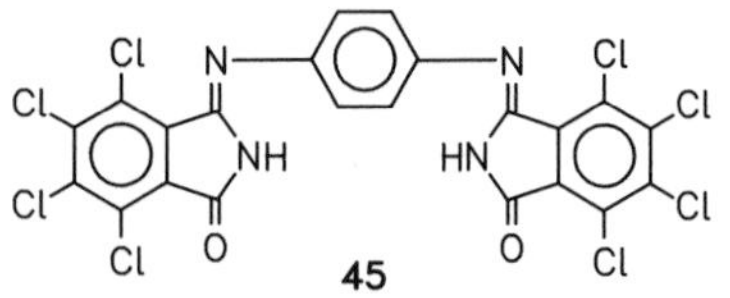

The following published examples illustrate some of the more recent develop-ments in the field of tetrachloroisoindolinone pigments.

In 1976, a patent describing a new synthetic method was issued to Dainippon Ink [3]. According to this publication, the bisacylated compound 46 is prepared from easily accessible tetrachlorophthalic anhydride reacting with an aromatic diamine at a 2:1 molecular ratio in the presence of ammonia and phosphorus pen-tachloride. Additional phosphorus pentachloride in a high boiling solvent, such as trichlorobenzene, effects ring closure of 46 to form the tetrachloroisoindolinone pigment:

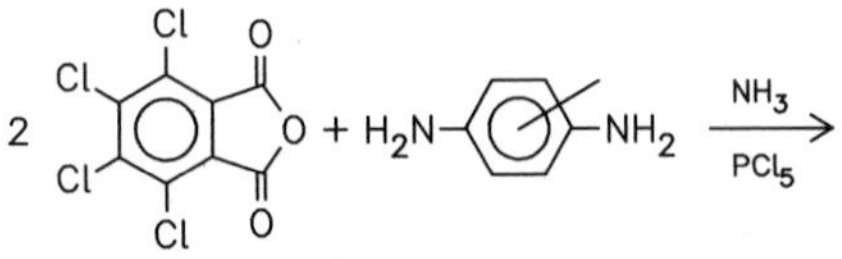

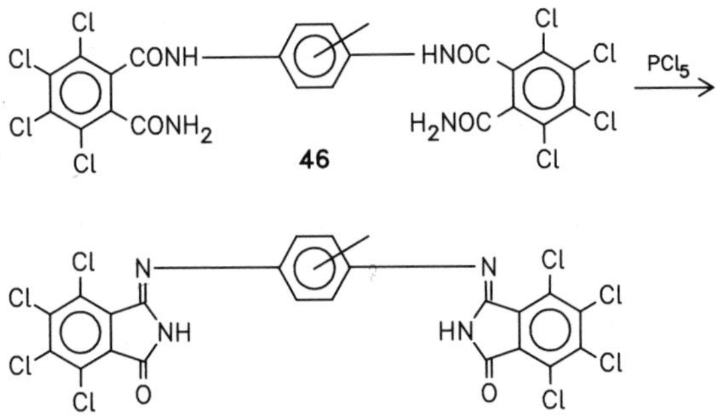

In the field of improved finishing techniques, a patent was issued to Toyo Soda [4]. A new γ-modification of the following pigment

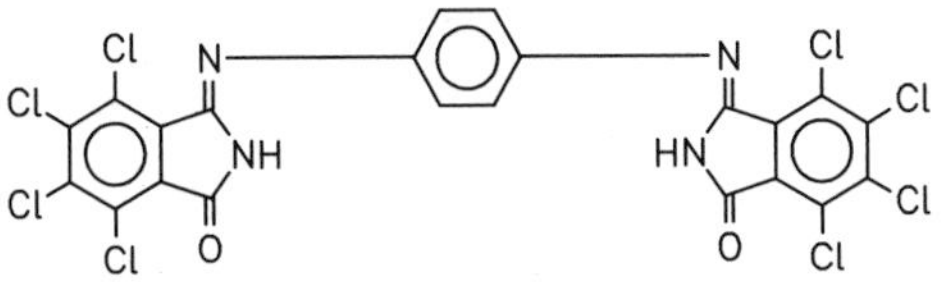

is apparently accessible by slowly hydrolyzing the corresponding potassium salt, which is obtained by usual methods, with water.

A few years earlier, Dainichi Seika [5] had published the synthesis of a β-modification of the same pigment, obtained by heating the known α-modification to temperatures between 200 to 350 °C in one of a variety of high boiling solvents. Comparative studies of the properties of the three modifications remain to be published.

Other publications describe mono and disazomethine pigments featuring heterocyclic ring systems [6]. There are also patents concerning bisiminoisoindolenine pigments [7].

There is a large number of studies concerning the synthesis of new chemical species based on the isoindolinone system. The most notable publications are those which refer to modifications of the bridging diamine between the two isoindolinone systems. The following diamine derivatives have been mentioned:

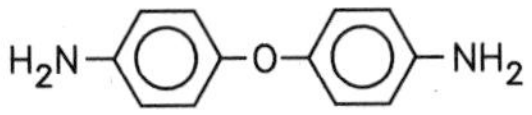

[8]

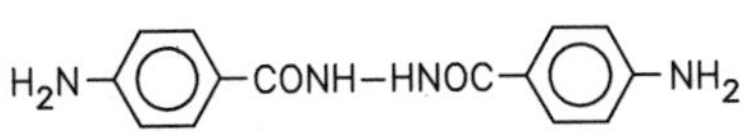

[9]

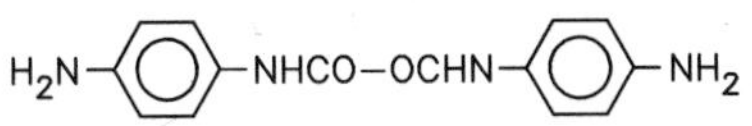

[10]

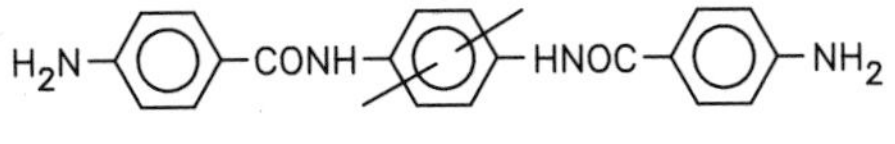

[11]

Moreover, there are patents claiming monoazo and disazo pigments, starting from compounds such as acting as diazo components [12], or, after appropriate acetoacetylation with dike-tene, as coupling components [13].

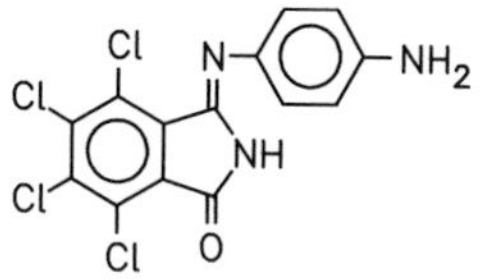

2.11.1.2   Methine Type: Isoindoline Pigments

Recent publications describe a variety of isoindoline pigments with different chem-ical structures. Such species comprise an isoindoline ring attached to two methine bridges.

These pigments have the following general chemical structure:

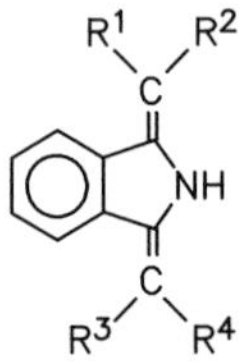

R^1 through R^4 represent CN, CONH-alkyl, or CONH-aryl. R^1 and R^2 on the one hand and R^3 and R^4 on the other hand can be also members of a heterocyclic ring system.

Examples, such as **47** type pigments, have been described by BASF and Ciba-Geigy [14, 15]:

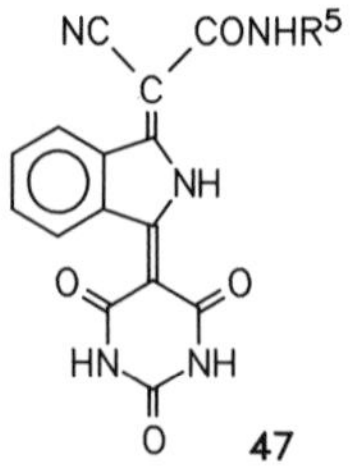

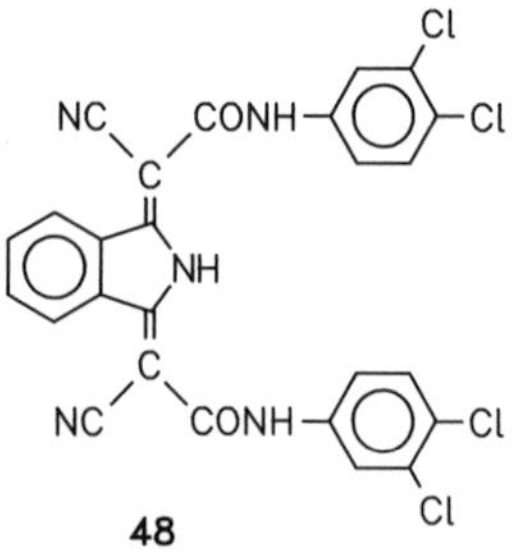

in which R^5 stands especially for a methyl group or possibly for a substituted phe-nyl moiety, or for the symmetrically substituted compound **48** [16]. Pigments with this structure afford yellow shades.

Likewise, suitable starting materials are iminoisoindolines, especially diimino-isoindoline (1-amino-3-iminoisoindolenine), which reacts with a cyanoacetamide $NCCH_2CONHR^5$ to afford a mono-condensation product **49**. Further reaction with a compound containing an activated methylene group (such as cyanoacetam-ide derivatives, barbituric acid) yields the desired pigment **47**:

NH   NH
|| ⇌ ||
NH   N + NC–CH₂CONHR⁵ ⟶
NH   NH₂

NC   CONHR⁵
\ /
C
||
NH   ——— Active methylene compound ———→ e.g., **47** or **48**
NH   H⊕
49

It is also possible to perform a simple one-step synthesis by adjusting the pH (initially pH 8 to 11, then pH 1 to 3).

The manufacture of P.Y.139 may serve as an example for the synthesis of a methine pigment. P.Y.139 is the condensation product of diiminoisoindoline with 2 moles of barbituric acid [17].

Gaseous ammonia is introduced into o-phthalodinitrile suspended in ethylene glycol, forming diiminoisoindoline. This compound is transferred into a mixture of an aqueous solution of barbituric acid, formic acid, and a surfactant. After being refluxed for 4 hours, the hot reaction mixture is filtered, washed to neutrality and free of auxiliaries, and dried. The resulting pigment is a greenish yellow compound. It is also possible to synthesize the same compound from one equivalent of 1-amino-3-cyaniminoisoindolenine

N–CN
||
N
NH₂

and two equivalents of an aqueous solution of barbituric acid [18].

A one-step synthetic route to methine pigments, starting from o-phthalonitrile, has been described by BASF [19].

Ciba-Geigy [20] has investigated other methine pigments, including bismethine compounds with one or two nonhalogenated isoindolinone ring systems.

2.11.2 Properties

Tetrachloroisoindolinone pigments are available in shades from yellow to orange to red and brown. Commercially important pigments cover the spectral range from greenish to reddish yellow. It is interesting to note that the corresponding nonchlorinated compounds produce yellow shades, which undergo a bathochromic shift as chlorine atoms are introduced. In terms of explanation, two arguments should be considered. Sterical problems on the one hand prevent the pigment molecules from assuming a coplanar arrangement, but, on the other hand, absorption in the visible portion of the spectrum is possible only in the presence of electron conjugation. It is therefore reasonable to assume that the typical pigment molecule represents a donor-acceptor-complex, with the amine functioning as a donor and the tetrachlorinated nucleus acting as an acceptor. The central part, i.e., the diamine moiety, contributes to the color insofar as higher conjugation and a large number of π-electrons cause a noticeable bathochromic shift in diamines.

The pigments are only sparingly soluble in most solvents, which makes for good migration fastness. Stability to acids, bases, and oxidizers, as well as to reducing agents is very good. The pigments exhibit good heat stability and melt around 400 °C. They are also excellently fast to light and weather, especially in white reductions.

Isoindoline pigments afford yellow, orange, red, and brown shades. They match the above-mentioned tetrachloroisoindolinone pigments in terms of solvent and migration fastness, heat stability, and chemical inertness. Likewise, isoindoline pigments demonstrate good lightfastness and weatherfastness. Several azomethine pigments have also been found to be polymorphous.

2.11.3   Application

Isoindolinone and isoindoline pigments are high quality products. They are used in general and high grade industrial paints including original automobile and automotive refinishes, in plastics and for spin dyeing, and in high grade printing inks, especially for metal deco, laminated plastic sheets, and in printing inks for bank notes and securities.

2.11.4   Commercially Available Isoindolinone and Isoindoline Pigments

General

Heading the list of commercially available pigments are yellow representatives, such as Pigment Yellow 109, 110, 139, 173, and 185, as well as Pigment Orange 61, 66, and 69, and Pigment Red 260. Another available type is Pigment Brown 38. Pigments providing other shades have failed to gain much commercial impact.

Table 26 lists examples of tetrachloroisoindolinone and isoindoline pigments (for general literature, see also [21]).

Individual Pigments

Pigment Yellow 109

P.Y.109 affords clean, greenish shades of yellow. It is recommended for use in paints, plastics, and printing inks. The paint industry uses P.Y.109 primarily to color a variety of high grade industrial finishes. The list includes, for instance, original automotive and automotive refinishes. P.Y.109 imparts very deep colors on automobile (O.E.M) finishes, particularly in combination with inorganic pigments. Green shades are produced in conjunction with blue compounds, such as

Tab. 26: Tetrachloroisoindolinone and isoindoline pigments.

Structure	Shade	C.I. Name	C.I. Constitution No.	Reference
a) Examples of the azo methine type				
	greenish yellow	P.Y.109	56284	[22, 23]
	reddish yellow	P.Y.110	56280	[22, 23]
	greenish yellow	P.Y.173		[21]
	orange	P.O.61	11265	
	red			[21]

Tab. 26: (Continued)

Structure	Shade	C.I. Name	C.I. Constitution No.	Reference

b) Examples of the methine type

	reddish yellow	P.Y.139	56298	[23, 24, 25]
	greenish yellow	P.Y.185	56280	
	yellowish orange	P.O.66	48210	[16]
	yellowish orange	P.O.69	56292	[26]
	yellowish red	P.R.260	56295	

phthalocyanine pigments. Full shades and related shades, reduced up to 1:1 with TiO$_2$, darken upon exposure to light. Depending on the type, the lightfastness in full shades (5%) equals step 6–7 to step 7 on the Blue Scale. Samples that are reduced with TiO$_2$ at a ratio of 1:3 to 1:25 match step 7–8. Both lightfastness and weatherfastness deteriorate rapidly with further reduction. P.Y.109 is fast to overcoating, its heat stability satisfies almost all requirements. Incorporated in an alkyd-melamine system, for instance, 1:50 reduction ratios with TiO$_2$ still afford products which are stable up to almost 200°C. P.Y.109 is used not only in industrial finishes but also in architectural and emulsion paints.

Its main field of application within the plastics industry is in polyolefin coloration. 1/3 SD samples containing 1% TiO$_2$ withstand up to 300°C, while 1/25 SD specimens are heat stable up to 250°C. The shade, however, is less clean at this temperature, it becomes duller. In injection molding, P.Y.109 considerably affects the shrinkage of the plastic. It is a tinctorially weak pigment. 0.4% pigment is required to produce 1/3 SD HDPE samples containing 1% TiO$_2$.

P.Y.109 is fast to migration in plasticized PVC and shows good heat stability. In white reductions (1/25 SD), it is fast up to 160°C. A noticeable color change is observed after a 10 minute exposure to 180°C, while full and related shades change color only at 200°C. In terms of lightfastness, P.Y.109 equals step 6–7 on the Blue Scale in white reductions (1/3 to 1/25 SD with 5% TiO$_2$). This is slightly inferior to the lightfastness of the redder disazo condensation pigments P.Y.128, 94, and 93. P.Y.109 exhibits good weatherfastness in rigid PVC, but does not satisfy the requirements of long-term exposure. This is especially true for white reductions with a higher TiO$_2$ content. Moreover, P.Y.109 is used to color polystyrene and other plastics, to rubber, polyurethane, aminoplastic resins, and unsaturated polyesters. It is also recommended for use in polypropylene spin dyeing.

P.Y.109 colors high grade products throughout the printing ink industry, but it is in direct competition with a large number of similarly colored products of various classes. Prints containing P.Y.109 exhibit good fastness to a number of organic solvents. They are not, however, entirely fast to the DIN 16 524/1 standard solvent mixture. The prints show good heat stability, are fast to overcoating, and may safely be sterilized. This makes P.Y.109 a suitable candidate also for metal deco printing. Its lightfastness is good.

Pigment Yellow 110

P.Y.110 affords very reddish shades of yellow. Good fastness properties make it a widely used pigment.

The paint industry uses the relatively weak P.Y.110 frequently as a colorant for industrial finishes, especially for high grade finishes. The pigment is very lightfast and weatherfast, which also makes it a suitable product for automotive finishes, for instance original automotive finishes. High transparency is an asset in metallic shades. Deep colors, for instance at 1:1 TiO$_2$ reduction, initially brighten as they are exposed to light and weather, but further exposure only has a minor effect on the intensity of the color. P.Y.110 turns redder as it is combined with TiO$_2$. It is completely fast to overcoating and heat stable up to 200 °C for 30 minutes. The pigment is also applied in emulsion paints and in architectural paints.

Incorporated in plastics, P.Y.110 is highly heat stable and possesses excellent lightfastness and weatherfastness. Full shades and related shades up to 1/3 SD in plasticized PVC withstand exposure to 200 °C for 30 minutes. Colorations up to 1/25 SD containing 5% TiO_2 equal step 7–8 on the Blue Scale for lightfastness. P.Y.110 also shows excellent durability in rigid PVC and impact resistant PVC types, as well as in plastisols for coil coating. It satisfies the requirements for long-term exposure. P.Y.110 is one of the most lightfast and weatherfast organic yellow pigments known. It shows average tinctorial strength. Between 1.4 and 1.9% pigment is needed to formulate 1/3 SD colorations with 5% TiO_2, depending on the type. Comparative values are listed for a number of other pigments covering the same range of shades.

P.Y.110 is fast to bleeding. Its high heat stability is used to advantage in poly-olefins. 1/3 SD HDPE samples maintain their color up to 290 °C, while 1/25 SD specimens tolerate up to 270 °C for 5 minutes. The color turns duller, indicating pigment decomposition. In HDPE melt extrusion, P.Y.110 has considerable effect on the shrinkage of this partially crystalline polymer at processing temperatures between 220 and 280 °C. The pigment is also very lightfast in polyolefins.

P.Y.110 lends color to polystyrene and styrene containing plastics. It is a suitable candidate for unsaturated polyester and other cast resins, as well as for polyurethane. P.Y.110 is used to an appreciable extent in polypropylene spin dyeing, it is very lightfast in this medium. It is utilized in polyacrylonitrile spin dyeing and sometimes also in polyamide. Its fastness properties, however, especially its lightfastness, do not meet special application conditions (Sec. 1.8.3.8).

The printing ink industry applies P.Y.110 in all types of printing, provided the pigment satisfies the demands. The prints are resistant to many organic solvents, including the DIN 165224/1 standardized solvent mixture. Prints made from P.Y.110 are fast to clear lacquer coatings, sterilization, and are very heat stable. 1/1 to 1/25 SD letterpress proof prints equal step 7 on the Blue Scale for lightfastness.

Likewise, P.Y.110 is found in a variety of other media, such as solvent-based wood stains.

Pigment Yellow 139

P.Y.139 is a reddish yellow pigment, used in plastics, paints, and printing inks. The commercial types exhibit a wide variety of particle size distributions and accordingly demonstrate very different coloristic properties, which is especially true for the hiding power. The opaque version is considerably redder. Incorporated in a paint, it is less viscous, which makes it possible to increase the pigment concentration without affecting the gloss of the product.

P.Y.139 is sometimes used in conjunction with inorganic pigments for paints, especially to replace Chrome Yellow pigments. The systems are fast to overpainting (up to 160 °C for 30 minutes), but they are not entirely fast to acids. The pigment performs very poorly in contact with alkali, therefore it is not suitable for use in amine hardening systems or in emulsion paints which are to be applied on alkaline substrates.

The paint industry uses P.Y.139 to color general and high grade industrial finishes, including automotive finishes, up to a 1:5 TiO_2 reduction ratio. The pigment is very lightfast and weatherfast, but its full shade and related shades darken noticeably upon exposure to light and weather. The opaque version performs better than the transparent one. The difference between the two, depending on the depth of shade, is 1/2 to 1 step on the Blue Scale. Incorporated in an alkyd-melamine resin, for instance, 1/3 SD samples of the opaque type equal step 7–8 on the Blue Scale, while corresponding transparent specimens match step 6–7.

P.Y.139 shows average tinctorial strength in plastics. Approximately 1% pigment is required to produce 1/3 SD samples in plasticized PVC containing 5% TiO_2. 1/3 SD HDPE specimens containing 1% TiO_2 are made from 0.2% pigment. Comparative values are listed for various other pigments. P.Y.139 is bleed resistant in plasticized PVC. It is often used in combination with inorganic and organic pigments, both in plasticized PVC and in PE. HDPE systems containing P.Y.139 in its transparent form (0.1%) and in white reductions (0.1% + 0.5% TiO_2) equal step 7–8 on the Blue Scale for lightfastness. Full shades darken considerably. P.Y.139 is not durable enough to satisfy more stringent requirements. 1/3 SD HDPE samples withstand exposure to 250°C, while 1/25 SD specimens are heat stable up to 260°C. The color becomes duller at higher temperatures, which is a result of pigment decomposition. P.Y.139 is an equally suitable candidate for PP spin dyeing. It is also found in PUR and in unsaturated polyester.

The printing ink industry utilizes P.Y.139 to color high grade printing products. 1/3 SD letterpress proof prints equal step 7 on the Blue Scale for lightfastness.

Pigment Yellow 173

P.Y.173, an isoindolinone pigment, affords somewhat dull, greenish yellow shades. It shows average fastness to organic solvents, especially to alcohols (ethanol), esters (ethyl acetate), and ketones (including methylethylketone and cyclohexanone). Its solvent resistance equals step 3 on the 5 step scale. P.Y.173 is almost completely to completely fast to mineral spirits and xylene.

P.Y.173 is weatherfast enough to be recommended for high grade industrial finishes, including automotive finishes, and for architectural paints. It should be noted that P.Y.173 is used advantageously to produce metallic effects in coatings. It is specifically recommended as a shading pigment, sometimes in conjunction with transparent iron oxide, to adjust flops of hue (Sec. 3.1.5, P.Bl.15:1). It is a colorant with high tinctorial strength compared to other pigments covering the same range of shades.

Alkyd-melamine resin systems may safely be overcoated, they withstand exposure up to 160°C for 30 minutes. 1/3 SD specimens are heat stable up to a maximum of 180°C. P.Y.173 shows excellent lightfastness and weatherfastness in white reductions, even as larger amounts of TiO_2 are added.

The plastics industry uses P.Y.173 to color various polymers. Incorporated in plasticized PVC, P.Y.173 is one of the tinctorially weaker pigments of its class. 1.8% pigment is required to produce 1/3 SD specimens containing 5% TiO_2. The systems are almost completely fast to overcoating. P.Y.173 demonstrates excellent lightfastness and durability in rigid PVC and in PVC plastisols intended for coil coating.

Tinctorially, P.Y.173 is equally weak in PE. It exhibits good heat stability. 1/25 SD samples reduced with TiO_2 retain their color up to 300°C. 1/3 SD specimens containing 1% TiO_2 increase in strength as the temperature exceeds approximately 260°C. At 300°C, for instance, as more of the pigment dissolves, such colorations not only exhibit higher cleanness but are also approximately 35% stronger than at 260°C.

Pigment Yellow 185

P.Y.185, a comparatively recent product, is an isoindoline pigment and provides clean greenish shades of yellow. Its main field of application is in packaging printing inks. P.Y.185 confers good tinctorial strength and high gloss on prints made from NC-based inks. Its tinctorial strength in print exceeds that of P.Y.17, a pigment of the diarylide yellow pigment series, which provides the same hue. In terms of tinctorial strength, P.Y.185 approaches the equally strong but redder P.Y.74. Likewise, the pigment exhibits high tinctorial strength in offset application. Prints made from P.Y.185 perform like those made from monoazo yellow pigments in terms of resistance to organic solvents, including fastness to the standardized DIN 16 524 solvent mixture. The same is true for the fastness to packaged goods, such as butter. However, P.Y.185 performs unsatisfactorily towards alkali (step 2). The prints are heat stable and show good lightfastness. 1/3 SD letterpress proof prints, for instance, equal step 5–6 on the Blue Scale for lightfastness. The commercial type provides good transparency.

Pigment Orange 61

P.O.61 affords yellowish shades of orange. It is noticeably less fast to organic solvents than comparative yellow types within the same class of pigments.

The paint industry uses P.O.61, like other isoindolinone pigments, to color high grade industrial finishes, including automotive (O.E.M) finishes, especially metallic shades. Other areas of application include general industrial and architectural paints and emulsion paints. The pigment is highly lightfast and weatherfast, but does not perform as well as the yellow types within the same class. Incorporated in an alkyd-melamine resin lacquer, the pigment may safely be overcoated up to 140°C.

P.O.61 is migration resistant in plasticized PVC. 1.4% pigment is needed to formulate 1/3 SD colorations containing 5% TiO_2. P.O.61 is thus one of the weaker pigments within its range of shades. Its lightfastness is very good. At depths of shade up to 1/25 SD in plasticized PVC, the pigment equals step 7–8 on the Blue

Scale. In terms of durability, P.O.61 performs equally well if it is incorporated in PVC plastisols intended for coil coating. This is especially true for transparent samples and those that contain small amounts of TiO_2. P.O.61 PVC plastisol systems are heat stable up to 200 °C. Particularly high heat stability is observed in polyolefins. 1/3 SD HDPE samples containing 1% TiO_2, for instance, withstand exposure up to 300 °C. P.O.61 produces a shade which is very similar to that of P.O.13. P.O.13 is, on the whole, less fast, but twice as strong tinctorially as P.O.61. P.O.61 causes an uncommon degree of shrinkage in HDPE, which is of great interest to injection molding. The pigment is recommended as a colorant for polystyrene, polyurethane, unsaturated polyesters, and elastomers. High heat stability and good lightfastness make it a suitable candidate for spin dyeing polypropylene and polyacrylonitrile.

The printing ink industry employs P.O.61 mainly in products which are to be printed on PVC films.

Pigment Orange 66

P.O.66, a member of the methine series, has been introduced to the market a few years ago and is recommended especially for paints. Its color is a yellowish orange. P.O.66 is a pigment with high tinctorial strength. Excellent lightfastness and weatherfastness qualify the pigment for use in high grade industrial finishes, especially in original automotive and refinishes. The currently available grade is used in metallic finishes for its high transparency, but may also be utilized as a shading pigment. At baking temperatures of 140 °C and higher, P.O.66 is not entirely fast to overcoating. It shows excellent lightfastness in plasticized PVC. P.O.66 dissolves in polyolefins if the temperature exceeds 240 °C.

Pigment Orange 69

P.O.69 has only been introduced to the market in recent years. P.O.69 is primarily applied in various types of industrial paints, and the pigment is also recommended for original automotive finishes. The commercial grade, which features a specific surface area of 30 m^2/g, has a comparatively coarse particle size. It provides good hiding power. The shade is a yellowish orange. The type is used especially in conjunction with other opaque organic pigments to produce lead-free colorations in the red and brown range. P.O.69 is fast to overcoating up to 140 °C. Both lightfastness and weatherfastness are good, but fail to meet more stringent use requirements. Full shades bleach noticeably and lose much of their gloss.

Pigment Red 260

The commercial grade of this very recent isoindolinone pigment has a coarse particle size and thus provides good hiding power. The pigment is recommended for use in various types of industrial paints, including automotive finishes.

P.R.260 may safely be overpainted up to 140°C. It provides somewhat dull shades in the yellowish red region of the spectrum. P.R.260 is comparatively strong. Full shades and related shades exhibit a slight haze, which is probably due to pigment flocculation. This problem may be met by using suitable additives. Both lightfastness and weatherfastness are good, although full shades darken slightly upon exposure and lose some of their gloss. P.R.260 bleaches considerably in white reductions.

Pigment Brown 38

P.Br.38, a comparatively recent product, is an isoindoline pigment. Its chemical constitution remains to be published. The commercial type is recommended as a colorant for plastics, especially for PVC and LDPE. The pigment provides yellowish shades of brown and is comparatively strong. 0.64% pigment is needed, for instance, to produce 1/3 SD colorations in plasticized PVC. P.Br.38 shows good migration fastness in plasticized PVC, but is not entirely fast to migration. Its lightfastness in rigid PVC is excellent. Full shade and related shades up to a 1:10 TiO_2 reduction ratio equal step 8 on the Blue Scale. P.Br.38 is heat stable up to 240°C, which makes it a suitable pigment, especially for LDPE.

References for Section 2.11

[1] A. Pugin and J.v.d. Crone, Farbe + Lack 72 (1966) 206–217.
[2] DE-OS 2 322 777 (Sandoz) 1972.
[3] Off. Gazette of the Jap. Pat Off., DE-OS 2 321 511 (Dainippon Ink) 1973.
[4] JA-PS J 5 5012106 (Toyo Soda) 1978.
[5] JA-PS J 5 1088 516 (Dainichi Seika) 1975; JA-PS J 5 20005840; J 5 2005 841 (Dainichi Seika) 1975.
[6] JA-PS J 5 5034 268 (Dainichi Seika) 1978; DE-OS 2 438 867 (Ciba-Geigy) 1973; JA-PS 7 330-659/658 (Sumitomo) 1970; CH-PS 613 465 (Ciba-Geigy) 1976; DE-OS 2 909 645 (BASF) 1979.
[7] DE-OS 2 548 026 (Ciba-Geigy) 1975.
[8] DE-OS 2 606 311 (Ciba-Geigy) 1975.
[9] EP-PS 29413 (Ciba-Geigy) 1979.
[10] DE-OS 2 518 892 (Ciba-Geigy) 1974.
[11] DE-OS 2 733 506 (Ciba-Geigy) 1976.
[12] DE-OS 2 901 121 (Dainichi Seika) 1978.
[13] EP-PS 22076 (Ciba-Geigy) 1979.
[14] DE-AS 2 914 086 (BASF) 1979; EP-PS 35672 (BASF) 1980; EP 29007 (Ciba-Geigy) 1979.
[15] J.v.d. Crone, J. Coat. Techn. 57, 725 (1985) 67–72.
[16] DE-OS 2 814 526 (Ciba-Geigy) 1977.
[17] DE-OS 2 628 409 (BASF) 1976; DE-OS 2 041 999 (BASF) 1970.

[18] DE-OS 3 022 839 (Bayer) 1980.

[19] DE-OS 2 757 982 (BASF) 1977.

[20] DE-OS 2 924 142 (Ciba-Geigy) 1978; EP-PS 19588 (Ciba-Geigy) 1979.

[21] Pigment Handbook, ed. by P.A. Lewis, Vol. 1, pp. 713, 722, J. Wiley, New York, 1988.

[22] DE-PS 1 098 126 (Geigy) 1956; DE-OS 2 804 062 (Ciba-Geigy) 1977.

[23] H. Sakai et al., J. Jpn. Soc. Col. Mat. 55 (1982) 683.

[24] DE-OS 3 022 839 (Bayer) 1980.

[25] W. Kurtz, Symposium Druckfarbenindustrie, BASF (1982) 31.

[26] H. Würth, FATIPEC XIX. Congress book, Aachen, Vol. 4 (1988) 49–65.

3 Polycyclic Pigments

Pigments are assigned to this class not on the basis of a uniform chemical constitution but on the basis of being nonazo pigments. It is thus that polycyclic pigments are distinguished from the even larger group of azo pigments.

All polycyclic pigments, with the exception of triphenylmethyl derivatives, comprise anellated aromatic and/or heteroaromatic moieties. In commercial pigments, these may range from systems such as thioindigo derivatives, which feature a benzene ring and a five-membered heteroaromatic fused ring (thionaphthenone) to such eight-membered ring systems as flavanthrone or pyranthrone. The phthalocyanine skeleton with its polycyclic metal complex is somewhat unique in this respect.

Triphenylmethane pigments, which have developed into an established technology since they were first discovered in 1914, are the oldest commercially used organic pigments. These compounds evolved from early syntheses in the form of sparingly soluble salts of sulfonic acids. The nitroso-β-naphthole iron complex ("Pigment Green"), one of the oldest "pure" nonazo pigments, was first prepared in 1921 and used as an inexpensive green pigment. It should be noted that for the purposes of this book Pigment Green is considered a metal complex pigment (Sec. 2.10) rather than a polycyclic pigment. Phthalocyanine derivatives were the first colorants which were launched directly as pigments without having previously been used as dyes. This is one aspect in which the history of the phthalocyanine series differs considerably from that of many vat pigments. The latter became commercially accessible only when synthetic methods were found to improve the quality of vat dyes in such a way that they exhibited adequate pigment properties. The coupling reaction affords azo pigments in the form of very fine powders. Polycyclic pigments, on the other hand, which are frequently prepared in organic solvents, evolve as relatively large crystals (up to 100 µm). Appropriate milling and fine dispersion processes are required (finishing) to obtain a form which is suitable as a pigment.

Most pigments derived from vat dyes are structurally based on anthraquinone derivatives such as indanthrone, flavanthrone, pyranthrone, or dibromoanthanthrone. There are other polycyclic pigments which may be used directly in the form in which they are manufactured. This includes derivatives of naphthalene and perylenetetracarboxylic acid, dioxazine (Carbazole Violet), and tetrachlorothioindigo. Recently, dioxazine pigments, quinacridone pigments, which were first introduced in 1958, and DPP pigments have been added to the series.

3.1   Phthalocyanine Pigments

The phthalocyanine [1,2,3,4] system is structurally derived from the aza-[18]-annulene series, a macrocyclic hetero system comprising 18 conjugated π-electrons. Two well known derivatives of this parent structure, which is commonly referred to as porphine, are the iron(III)complex of hemoglobin and the magnesium complex of chlorophyll. Both satisfy the Hückel and Sondheimer $(4n + 2)\pi$ electron rule and thus form planar aromatic systems.

Copper Phthalocyanine Blue is the copper(II) complex of tetraazatetrabenzoporphine. As shown below, the mesomeric structures indicate that all of the pyrrole rings simultaneously contribute to the aromatic system:

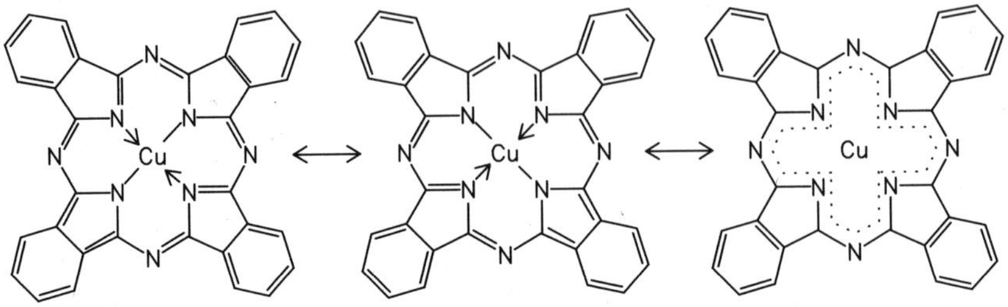

The molecule adopts a planar and completely conjugated structure which exhibits exceptional stability. The unit cell contains two centrosymmetric molecules.

Phthalocyanines were intially produced only as pigments. It was not until later that techniques such as sulfonation, chlorosulfonation, and chloromethylation made it possible to produce and commercially apply these compounds in dye form.

At present, synthetic routes to more than 40 metal complexes other than the copper complex are known. None of the resulting products, however, has stimulated commercial interest as a pigment. Nickel complexes, however, are found in reactive dyes, while cobalt complexes of this basic structure are employed as developing dyes.

At present, Phthalocyanine Blue and Phthalocyanine Green are among the most important organic pigments in the market and are sold in large volume.

As early as 1907, A.V.Braun and J.Tscherniak first obtained phthalocyanine from phthalimide and acetic anhydride. The prepared blue substance, however, was not investigated further. In 1927, de Diesbach and von der Weid, in an attempt to synthesize phthalonitrile from o-dibromobenzene and copper cyanide in pyridine at 200 °C, obtained a blue copper complex. The substance was found to be exceptionally fast to acid, alkali, and high temperature. Approximately one year later, in trying to manufacture phthalimide from phthalic anhydride and ammonia, researchers at Scottish Dyes Ltd detected a greenish blue impurity coated onto an apparently defective enamel reactor vessel. Analysis showed the product to be the phthalocyanine iron complex.

Coordination complexes of zinc, cobalt, and platinum show a stability similar to complexes involving copper and iron. They are stable to concentrated, nonoxidizing acids and bases. The fact that such complexes sublimate at 550 to 600 °C without decomposing points to their extreme heat stability.

In contrast to the ionic complexes of sodium, potassium, calcium, magnesium, barium, and cadmium, the ease with which transition metal complexes are formed (high constant of complex formation) can partly be attributed to the suitably sized atomic radii of the corresponding metals. Incorporated into the space provided by the comparatively rigid phthalocyanine ring, these metals fit best. An unfavorable volume ratio between the space within the phthalocyanine ring and the inserted metal, as is the case with the manganese complex, results in a low complex stability.

In 1929, Linsted obtained samples of this complex from ICI chemists (Scottish Dyes Ltd was now owned by ICI). ICI had developed two routes leading to the phthalocyanine iron complex. One method started from phthalic anhydride, iron, and ammonia, while the second pathway proceeded from phthalimide, iron sulfide, and ammonia. In 1933/34, elucidation of the phthalocyanine structure was credited to Linstead. The corresponding copper and nickel phthalocyanines had been prepared in the meantime. ICI introduced the first Copper Phthalocyanine Blue to the market as early as 1935, and the Ludwigshafen subsidiary of the IG Farbenindustrie followed suit with a corresponding product.

In Germany, Phthalocyanine Green was first prepared commercially in 1938, followed by US companies in 1940.

Most phthalocyanine green pigments are derived from copper polychlorophthalocyanines. Chlorobromo derivatives provide a yellowish green shade.

Copper Phthalocyanine Blue exhibits more than one crystal modification. This is also true for the metal-free ligand whose greenish blue crystal phase was used on a large industrial scale for a certain period of time (Sec. 3.1.2.6). Free-base Phthalocyanine Blue was largely displaced by β-Copper Phthalocyanine Blue as it became possible to produce the latter more economically (Sec. 3.1.2.3).

3.1.1   Starting Materials

Heading the list of suitable organic starting materials [5] for phthalocyanine pigments are phthalic anhydride and phthalonitrile.

Phthalic Anhydride

Phthalic anhydride is prepared by oxidizing o-xylene. The oxidation may be performed either in the gas phase with vanadium pentoxide as a catalyst or in the liquid phase with dissolved manganese, molybdenum, or cobalt salts as catalysts:

At the present time, gas-phase oxidation is the favored technique.

A third option involves oxidizing naphthalene, possibly with vanadium pentoxide as a catalyst.

Phthalic anhydride is a significant commercial product. Its main area of application is in synthetic resins and plasticizers. Approximately 2.7 million t were produced worldwide in 1985.

Phthalonitrile

Phthalonitrile is obtained by oxidizing o-xylene with ammonia. To this end, o-xylene is treated with ammonia, either in the presence of an oxidation catalyst or with the aid of oxygen and a catalyst, at 330 to 340 °C:

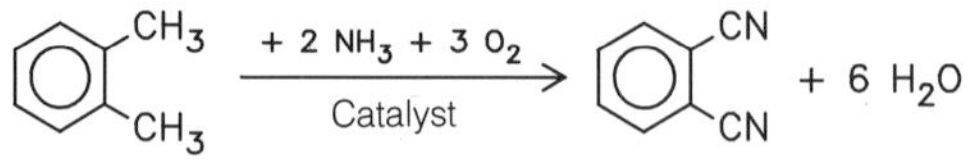

3.1.2   Manufacture

In 1937, Du Pont started producing Copper Phthalocyanine Blue in the USA after it had previously been launched in Great Britain and Germany. Other companies followed.

Copper Phthalocyanine Blue was no longer open for patent application after it had been described by de Diesbach and von der Weid in 1927. Yet a great many synthetic methods for this important compound were patented.

Metal phthalocyanine complexes may be obtained by one of the following methods:

They can be prepared from
-   phthalonitrile or substituted phthalonitriles and metals and/or metallic salts,
-   phthalic anhydride, phthalic acid, phthalic esters, diammonium phthalate, phthalamide or phthalimide and urea, metal salt, and a catalyst,
-   o-cyanobenzamide and a metal or a metallic salt.

Only two techniques have gained commercial importance. The **phthalonitrile process**, developed in England and Germany, is particularly important in Germany, while the **phthalic anhydride/urea process** has stimulated more interest in Great Britain and the USA.

The technical importance of the phthalonitrile process is only second to the phthalic anhydride/urea technique, of which two varieties are commercially used (Sec. 3.1.2.2). BASF is the primary European user of the phthalonitrile method.

Knowledge of the most important types of copper phthalocyanine pigments is useful for the understanding of the processes concepts underlying pigment manufacture. Heading the list are the α- and β-modifications of unsubstituted Copper

Phthalocyanine Blue (Sec. 3.1.2.3). The α-modification exhibits an unstabilized and a stabilized form as to change of crystal modification.

Likewise, halogenated copper phthalocyanine green pigments also have a major impact on the market (Sec. 3.1.2.5).

3.1.2.1 Phthalonitrile Process

The first technical process involved heating phthalonitrile with copper bronze or copper(I)chloride at 200 to 240 °C in copper pans. Several variations of this technique were developed in Germany prior to the Second World War. The reaction was performed either without or in the presence of a solvent. A basic distinction is commonly made between the baking process and the solvent process; both may be carried out either by continuous or by batch technique.

Baking Process

Baking the starting materials in the absence of solvents involves heating phthalonitrile with copper(I)chloride partially in a nitrogen stream to 140 to 200 °C. Copper(I)chloride is sometimes replaced by copper powder, copper(II)chloride, or pyridine/$CuCl_2$ complex. Baking was originally performed as a batch process on plates which were heated indirectly with high pressure steam. Continuous baking, on the other hand, involved using an electrically heated moving copper belt, a tunnel kiln, or a tunnel drying oven.

A considerably exothermic reaction is initiated at a starting temperature of 140 to 160 °C, and more than 300 °C may be reached during conversion. Partial cleavage under such conditions affects not only the yield but also the quality of the product. Improved temperature control is achieved by either working by continuous operation (see below), or by adding inert inorganic salts, such as sodium sulfate or sodium chloride. It is also possible to cool the reaction mixture during the appropriate phase of the process.

Baking is currently performed by continuous operation. Modern variations involve using heated crushing or milling equipment, such as kneader dispersers or oscillating mills at approximately 200 °C [6]. This technique significantly improves the reaction control over a batch process. If baking is performed by continuous process, phthalonitrile only remains within the reaction vessel for a very short period of time (between 3 and 20 minutes). It is important to remember that the temperature may not exceed 250 °C. The product which evolves from this process is usually purified by acid treatment.

The yield may range from 70 to 80%, although recent continuous techniques have claimed as much as 85%. In view of the fact that reaction control is relatively difficult during baking and that the process only affords a limited yield, it is the solvent method which has stimulated particular interest, especially as improved industrial processes were developed. Most recent trends, however, show a reverse tendency: the baking process is regaining importance. This is mainly on the grounds of economical, ecological, and physiological concerns.

Solvent Process

The solvent process involves treating phthalonitrile with any one of a number of copper salts in the presence of a solvent at 120 to 220 °C [7]. Copper(I)chloride is most important. The list of suitable solvents is headed by those with a boiling point above 180 °C, such as trichlorobenzene, nitrobenzene, naphthalene, and kerosene. A metallic catalyst such as molybdenum oxide or ammonium molybdate may be added to enhance the yield, to shorten the reaction time, and to reduce the necessary temperature. Other suitable catalysts are carbonyl compounds of molybdenum, titanium, or iron. The process may be accelerated by adding ammonia, urea, or tertiary organic bases such as pyridine or quinoline. As a result of improved temperature maintenance and better reaction control, the solvent method affords yields of 95% and more, even on a commercial scale. There is a certain disadvantage to the fact that the solvent reaction requires considerably more time than dry methods.

After completion of the reaction, the solvent is removed by filtration. Most often, however, the solvent is separated from the crude Copper Phthalocyanine Blue by distillation.

The solvent method may also be performed either by continuous (in cascades) or by batch operation. Continuous techniques in particular have gained considerable technical importance. A phthalonitrile/copper chloride solution is typically treated at 120 to 140 °C in a flow tube furnace and the temperature subsequently increased to 180 to 250 °C. The entire process requires approximately 1.5 to 2 hours and affords the pigment in practically quantitative yield. The excellent purity of the product eliminates the need for additional purification with dilute acid or base prior to finishing, a procedure which plays a major role in the baking process. These are the advantages that make the baking process economically favorable, despite the problems connected with the separation and regeneration of the solvents.

The phthalonitrile process has the particular advantage over the phthalic anhydride process of forming ring-substituted chloro-copper phthalocyanines. Using copper(I)chloride produces so-called "semi-chloro" Copper Phalocyanine Blue, a pigment which possesses a statistical average of 0.5 chlorine atoms per copper phthalocyanine molecule. Copper(II)chloride, on the other hand, affords a product which comprises an average of one chlorine atom per copper phthalocyanine molecule. A prerequisite for the formation of the chloro substituted compound, however, is the absence of ammonia or urea in the reaction mixture.

This simple one-step route leads to the starting material for the solvent-stabilized α-modification of Copper Phthalocyanine Blue.

The synthesis of the chlorine-free crude pigment by either the continuous or the batch version of the phthalonitrile process involves adding up to 20% urea or ammonia to the reaction mixture. The latter will provide an effective chlorine trap.

Moreover, the phthalonitrile process has the added advantage of being the more elegant of the two syntheses. This technique makes it possible to produce comparatively pure copper phthalocyanine without obtaining substantial amounts of side products, a phenomenon which is understandable in view of the fact that the phthalonitrile molecule provides the parent structure of the phthalocyanine ring.

Formally, rearrangement of the bonds necessitates donation of two electrons to the system:

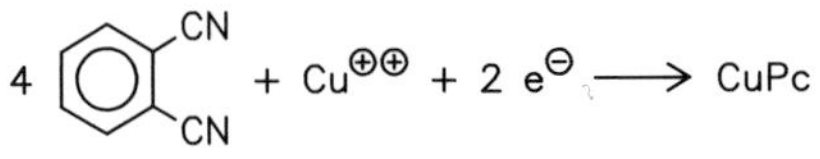

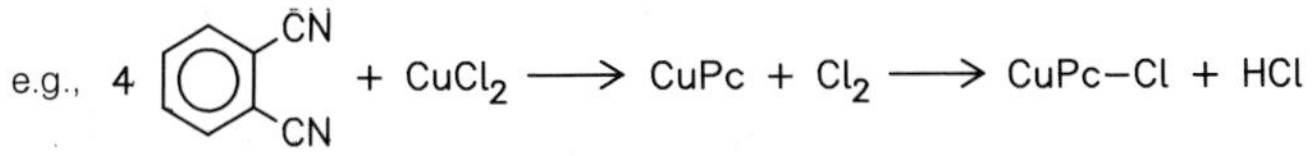

$$4 \text{ (phthalonitrile)} + Cu^{\oplus\oplus} + 2\,e^{\ominus} \longrightarrow CuPc$$

$$\text{e.g., } 4 \text{ (phthalonitrile)} + CuCl_2 \longrightarrow CuPc + Cl_2 \longrightarrow CuPc\!-\!Cl + HCl$$

The advantages of the phthalonitrile process are compromised by the fact that phthalonitrile is not only much more costly than phthalic anhydride but also less easily available. In view of the intermediates which have been found so far, the reaction mechanism may be visualized as follows:

Phthalonitrile reacts with ammonia to produce mono and diiminophthalimide. The thus prepared isoindolenines are then joined through intermolecular condensation to afford polyisoindolenines or the corresponding copper complexes. At about 180 to 200 °C, copper phthalocyanine is formed:

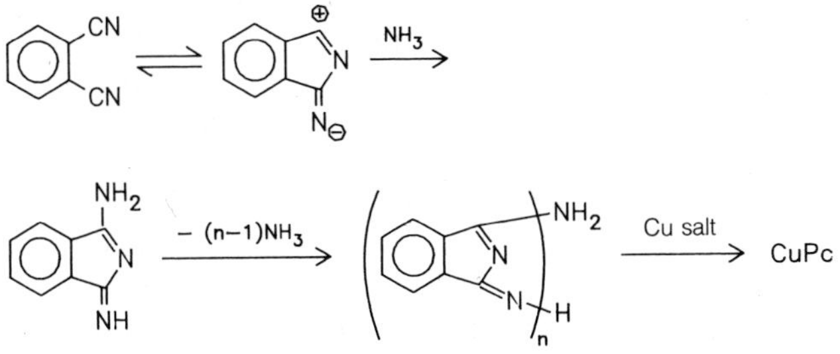

$$\text{(phthalonitrile)} \rightleftharpoons \text{(isoindolenine}^{\oplus}, N^{\ominus}) \xrightarrow{NH_3}$$

$$\text{(diiminoisoindoline)} \xrightarrow{-(n-1)NH_3} \text{(polyisoindolenine)}_n \xrightarrow{Cu\ salt} CuPc$$

3.1.2.2 Phthalic Anhydride/Urea Process

Likewise, this process [8] may also be carried out either as a solvent-free (baking) method or in the presence of solvents. Although initially performed as a solvent-free technique, it is the solvent version that currently dominates the field of copper phthalocyanine production from phthalic anhydride and urea. It should be mentioned, however, that this trend has been reversed in the very recent past and that solvent-free methods are gaining interest, especially for ecological reasons.

Baking Process

The first commercial copper phthalocyanine synthesis, a baking process, involved melting phthalic anhydride with urea at 150 °C in the presence of boric acid. Copper(II)chloride was then added and the temperature increased to approximately 200 °C until the copper phthalocyanine production was completed. The reaction mixture was cooled and the crude product milled. After being washed, first with dilute sodium hydroxide solution and then with dilute sulfuric acid, the material was filtered off and dried. The crude copper phthalocyanine obtained was then dissolved in sulfuric acid and precipitated in ice water ("acid pasting") in order to afford a form which could be used as a pigment. In subsequent years, boric acid in its role as a catalyst largely gave way to various metallic salts. Best results were achieved with molybdenum trioxide and primarily with ammonium molybdate, both of which are currently used. It thus became possible to increase the yield from approximately 50% to more than 90% of the theory. Copper(II)chloride may also be replaced by copper(I)chloride, copper carbonate, or, most commonly, copper sulfate.

The baking process has remained much the same until the present day; at a stoichiometric ratio of 1:4, phthalic anhydride or phthalic acid reacts with an ammonia releasing compound. The reaction may also start from other suitable materials, such as phthalic acid derivatives, including phthalic acid esters, phthalic acid diamide, or phthalimide. Appropriate ammonia releasing agents include urea and its derivatives, such as biuret, guanidine, and dicyanodiamide. The fact that a certain amount of urea decomposes to form side products makes it necessary to use excess urea. Approximately 0.2 to 0.5, preferably 0.25 equivalents of copper salt should be added for each mole of phthalic anhydride. 0.1 to 0.4 moles of molybdenum salt per mole of phthalic anhydride is sufficient. The reaction temperature is between 200 and 300 °C.

The baking process, particularly the batch variety, presents a number of serious disadvantages. Not only does the reaction produce solid urea decomposition products, but it also releases large amounts of ammonia and ammonium salts which escape by sublimation. The foam which is thus formed makes for a porous reaction mixture, which in turn even prevents heat conduction. Moreover, the reaction mixture tends to adhere to the surface of the reaction vessel and the stirring unit; a phenomenon which adds to the complexity of the problem.

At least some of these disadvantages may be overcome by vigorous stirring or milling during the reaction. Baking by batch process has recently been described which involves heating the reaction components in layers on metal sheets. However, such processes require precise control. If a copper phthalocyanine complex is prepared by batch process in a ball or a pin-type mill, the reaction mixture must be allowed to cool before it is discharged. Reaction times of originally 5 to 45 minutes thus create cycles lasting up to 3 hours, a duration which is highly uneconomical.

Continuous baking processes may be conducted, for instance, by treating the reaction mixture in a heated cylinder with a screw drive. The desired thin film, however, can only be produced in large elaborate reaction units, a requirement which makes its manufacture expensive. The same is true for performing the baking process in heated rotating drums. A certain amount of product (namely, the amount that is produced in two hours) must already be present as the starting materials are added in order to prevent the materials from sticking to the unit. Long reaction times and unprofitable filling levels produce a space/time yield which is unattractive for industrial purposes.

On the other hand, economically advantageous routes by continuous baking process have also been described in recent years. The processes are carried out in self-cleaning mixing apparatus such as double screw extruders. More recent patent literature claims yields as high as 80%.

Invariably, the primary product is a crude Copper Phthalocyanine Blue with insufficient properties. It is boiled with dilute hydrochloric acid or aqueous alkali and then rinsed with hot water to remove acid or base before it can be finished.

Solvent Method

There is somewhat less of a demand on the reaction equipment if phthalic anhydride and urea react in an organic solvent. In principle, the presence of a solvent makes almost no difference to the synthesis by baking process. Using suitable inert high-boiling solvents such as kerosene, trichlorobenzene, or nitrobenzene obviates many of the difficulties connected with the reaction mixtures obtained in the baking process. Solvents make it much easier to adequately mix the components and to ensure even heat transfer. Long reaction times, however, remain somewhat of a problem, although suitable temperature control during condensation makes it possible to produce the complex in almost quantitative yield.

The solvent is removed from the solid product by filtration or centrifugation to afford a crude Copper Phthalocyanine Blue of a quality that makes intermediate purification unnecessary. In contrast to the product obtained by the baking process, this material is pure enough to be used directly for further pigment manufacture. Crude Copper Phthalocyanine Blue, on the other hand, which evolves as the solvent is removed by distillation, contains so many impurities that it must be boiled before being utilized further.

Phthalic anhydride and urea, together with copper(I)chloride and ammonium molybdate, are heated to 200 °C in trichlorobenzene. The ratios between the components are the same as in the baking process. Carbon dioxide and ammonia are released to yield Copper Phthalocyanine Blue. The reaction is complete after 2 to 3 hours, producing a yield between 85% and more than 95%.

The main disadvantage of working in the presence of a solvent is the problem of regenerating this solvent.

Apart from nitrobenzene, trichlorobenzene in particular was the preferred solvent up until a few years ago. It has now been replaced by other solvents such as high-boiling hydrocarbons (kerosene, naphthalene) and also alcohols and glycols, because traces of polychlorinated biphenyls may be formed. These are not easily degradable. With hydrocarbons, however, the possibility of fire and explosion must be considered in designing suitable production units.

In the presence of a solvent, the crude pigment generally evolves in much purer form than if it is prepared by baking. Higher degrees of purity, i.e., up to 98%, are achieved by additional alkaline and/or acidic treatment. Commercially available types of crude Copper Phthalocyanine Blue typically contain more than 90% pure pigment. This preliminary product is also commonly supplied to and finished by companies who do not manufacture crude pigment themselves.

Reaction between phthalic anhydride and urea always affords chlorine-free Copper Phthalocyanine Blue. Chlorinated derivatives are obtained only in the absence of bases (ammonia) or urea. The phase stabilized α-modification is prepared by essentially the same but slightly modified route: it is derived from mixed condensation with chlorinated phthalic anhydrides (such as 4-chlorophthalic acid) (Sec. 3.1.2.4).

Despite the described disadvantages, but due to the high yield, economical production methods, and low-cost starting materials, the phthalic anhydride/urea method is presently the most significant industrial route to Copper Phthalocyanine Blue manufacture.

The gross reaction equation for the preparation of Copper Phthalocyanine Blue from phthalic anhydride and urea may be written as follows:

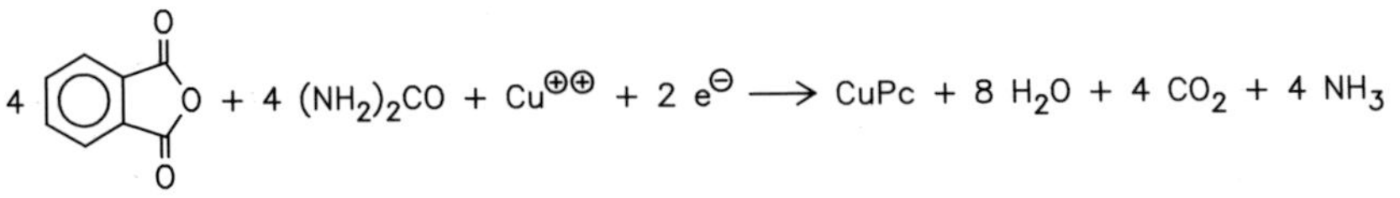

This overall reaction equation points to the fact that copper phthalocyanine formation formally requires two reduction equivalents.

Formally, copper phthalocyanine was formed by the following reaction pathway:

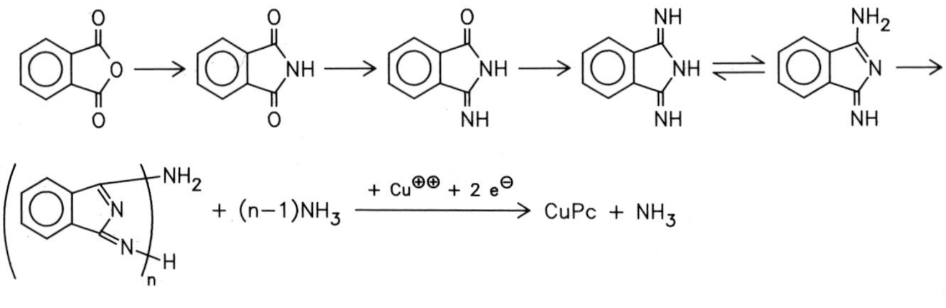

Polyisoindolenine

Phthalic anhydride initially reacts with ammonia, which in turn is liberated, for instance, by decomposition of urea. Diiminophthalimide is then produced via phthalimide and monoiminophthalimide. Subsequent self-condensation (as in the phthalonitrile process) under cleavage of ammonia affords polyisoindolenines, which form complexes with copper ions. Ring closure is achieved through further release of ammonia, and copper phthalocyanine is finally obtained by reduction.

The above-mentioned reaction mechanism is validated by proven intermediates.

Urea acts not only as an ammonia source but also forms decomposition products, such as biuret and higher condensation products. [14]C labeling has indicated that the carbon atom of the urea molecule is not incorporated into the phthalocyanine structure. Employing a phthalic anhydride molecule bearing one radioactively labeled carbonyl function affords labeled copper phthalocyanine and phthalimide (as a side product), while the liberated carbon dioxide was found not to show any radioactivity. Labeled carbon dioxide, on the other hand, has been obtained in corresponding experiments using ^{14}C labeled urea.

3.1.2.3   Manufacturing the Different Crystal Modifications

Unsubstituted Copper Phthalocyanine Blue is polymorphous. X-ray diffraction diagrams point to five different crystal modifications (α, β, γ, δ, ε) (Fig. 93). The relative thermodynamic stability of the individual crystal phases decreases in the following order: $\beta > \varepsilon > \delta > \alpha \approx \gamma$ [9, 10, 11, 12].

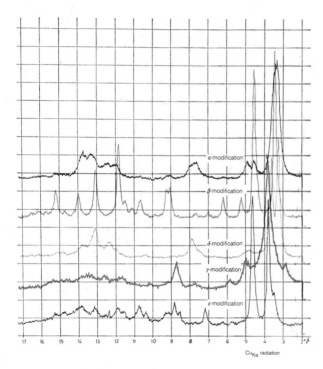

Fig. 93: Pigment Blue 15, X-ray diffraction diagrams of different crystal modifications.

Among these modifications, it is particularly the phase-stabilized α- and the β-form that have stimulated commercial interest.

Crude Copper Phthalocyanine Blue which is prepared by the phthalonitrile or urea process typically evolves as the β-modification with a coarse particle size.

The synthesis of the crystal modification is controlled primarily by the finishing technique of the crude pigment. There are basically two different methods to produce a finely dispersed pigment: treatment with acid to form copper phthalocyanine salts, followed by precipitation in water on the one hand, and mechanical treatment (milling, kneading) on the other hand. The following methods are used:

–   dissolving the crude pigment in concentrated sulfuric acid,
–   swelling in 50 to 90% sulfuric acid,

- dry milling in the presence or absence of water soluble salts, as well as with or without small amounts of organic solvents,
- milling in organic solvents,
- kneading with salt in the presence of solvent.

α-Modification

Dissolving or swelling of crude Copper Phthalocyanine Blue in sulfuric acid, followed by precipitation in water (hydrolysis) affords the α-modification with a fine particle size. Emulsifiers may be present if desired. Dry milling of the crude β-crystal phase, for instance in the presence of sodium chloride or sodium sulfate, also yields the α-phase.

β-Modification

The β-modification as a rule evolves as a more coarse-grained material than the α-phase. It is prepared by milling the crude Copper Phthalocyanine Blue with salt in the presence of a "crystallization stimulating" solvent. Aromatic hydrocarbons, esters, or ketones are normally used.

A primarily platelet-shaped β-modification may also be obtained by the phthalonitrile process if particularly pure phthalonitrile is employed [13].

γ-, δ-, ε-Modifications

Crude Copper Phthalocyanine Blue is stirred in 60% sulfuric acid and the thus obtained sulfate hydrolyzed with water. Subsequent filtration affords the γ-phase. It is also possible to knead crude Copper Phthalocyanine Blue with salt (sodium sulfate), concentrated sulfuric acid, and a third agent, which may be an alcohol, a polyalcohol, or one of the corresponding organic esters [14]. A third option is to stir α-Copper Phthalocyanine Blue with 30% sulfuric acid and glycol monobutylester or the corresponding ethyl ester or tetrahydrofuran [15].

The δ-form can be obtained by treating Copper Phthalocyanine Blue in benzene or toluene with aqueous sulfuric acid in the presence of a surfactant [16].

The ε-phase is produced by comminution of the α-, γ-, or δ-modification, for instance in a planetary ball mill. The mill base is then aftertreated in an organic solvent at elevated temperature. It is important to realize that the temperature, depending on the solvent, must be kept below the transition temperature at which the ε-phase converts to the β-modification (30 to 160 °C). The ε-modification is made best from the γ-phase, and the most preferred solvents are alcohols [17]. For the industrially hitherto insignificant π, X, and R-forms of Copper Phthalocyanine Blue, please see [1], Vol. II, 34–35.

3.1.2.4   Phase- and Flocculation-Stabilized Copper Phthalocyanine Blue Pigments

There are various methods of "stabilizing" a pigment in order to prevent conversion to a different phase, i.e., change of modification, and flocculation during pigment application. The two following techniques have been found to be most effective:

– minor chemical modification of the Copper Phthalocyanine Blue molecule, for instance by "partial chlorination",
– admixing other agents to the Copper Phthalocyanine molecule. Surface active additives may thus help to stabilize the surface.

Only a minor amount of chlorinated copper phthalocyanine, for instance, especially in the 4-position of the copper phthalocyanine molecule, prevents a change of modification from α to β. Approximately 3 to 4% chlorine is commonly used, which corresponds to the formula $CuPc\text{-}Cl_{0.5}$, also referred to as "semi-chloro-CuPc". The phthalic anhydride/urea synthesis, for instance, affords a partially chlorinated product if 4-chlorophthalic anhydride is added to the reaction mixture. Copper chlorides in the phthalonitrile process have the same effect.

Partial sulfonation or sulfamide formation may stabilize a pigment toward flocculation. Combining partial sulfonation and chlorination will further improve the effect of each individual type of modification. Similar results are observed if carboxylic acid groups are introduced into the pigment molecule.

There is an interesting technique which makes it possible to introduce carboxylic acid groups into a copper phthalocyanine structure by an economical route. Carrying out the phthalic anhydride/urea process in the presence of a small amount of trimellitic acid or another benzene polycarboxylic acid will afford a carboxylated pigment.

Flocculation may likewise be prevented through partial introduction of dialkylaminomethylene groups into the aromatic ring system of the copper phthalocyanine molecule. The copper phthalocyanine structure tightly attaches the basic groups to the surface of the pigment particles: $CuPc\text{-}(CH_2N{<}^{R^1}_{R^2})_{1-4}$ a feature which imparts partial solubility on the parent molecule. Acidic constituents within the binder may interact with the amino groups.

There are other metal complexes, such as tin, aluminum, magnesium, iron, cobalt, titanium, and vanadium complexes, which are similarly useful in stabilizing a particular phthalocyanine modification. Moreover, carboxy, carbonamido, sulfo, or phosphono-copper phthalocyanine may be admixed during fine dispersion of the pigment.

The discussed additives, which stabilize a pigment with regard to change of modification and flocculation, are usually applied at a concentration of 3 to 10%.

A flocculation-resistant β-form is obtained by milling crude Copper Phthalocyanine Blue with salt in the presence of xylene or a similar hydrocarbon. In this case, as in general, impurities in the crude Copper Phthalocyanine Blue prevent flocculation.

Polyhalogenated green copper phthalocyanine pigments are not polymorphous and thus exempt from change of modification.

3.1.2.5   Manufacture of Green Types

The early days of Copper Phthalocyanine Green synthesis were dominated by two competitive routes. One method was the synthesis of tetraphenyl copper phthalocyanine (Bayer), while the second method involved chlorination of copper phthalocyanine in carbon tetrachloride to form copper tetradeca to hexadecachloro phthalocyanine (BASF). It was on the grounds of economical considerations that manufacturers began to prefer the chlorination technique in industrial scale production.

The oldest and still most prevalent large scale synthesis for green copper phthalocyanine pigments proceeds via direct chlorination of copper phthalocyanine (1935). Chlorination is effected at 180 to 200 °C in a sodium chloride/aluminum chloride melt (eutectic blend) in the presence of a catalyst (metal chlorides such as iron(III)chloride). The temperature limits for chlorination are generally reported to be between 60 and 230 °C. There is also a battery of other media, such as chlorosulfonic acid, thionyl chloride, sulfuryl chloride, and chlorinated hydrocarbons, possibly under pressure, which are equally suitable for chlorination. All these techniques afford mixtures of multiply chlorinated copper phthalocyanines. The structure is typically chlorinated in at least 8 positions, but more often 14 to 15.5 times. The perchlorinated copper phthalocyanines produced either in $AlCl_3$/NaCl or in acid are then precipitated in water, filtered, and washed, especially to remove aluminum salts. Since the pigment already evolves from the synthesis as an extremely fine powder, it may then be subjected directly to thermal aftertreatment in an aqueous/organic medium. This is attributed to the fact that perchlorinated copper phthalocyanine interacts with aluminum chloride as well as with strong acids, leading to the formation of salts, which are cleaved through hydrolysis.

Less widely used methods include acid pasting, i.e., dissolving the material in a medium such as chlorosulfonic acid and precipitating it in water, or milling the crude pigment. These methods gain importance whenever phthalocyanine is to be chlorinated by a route other than with aluminum chloride or acids.

Commercially available Copper Phthalocyanine Green types typically contain approximately 15 chloro atoms per molecule, while yellowish green types bear not only chloro but also bromo substituents. Such derivatives are obtained through mixed halogenation. Chlorinated green copper phthalocyanines can also be prepared from already perchlorinated starting materials. Tetrachloro phthalonitrile, for instance, may be condensed in nitrobenzene with copper(II)chloride to form copper perchloro phthalocyanine. Costly starting materials, however, make this an industrially unattractive route. Tetrachloro phthalonitrile may be synthesized from phthalonitrile by gas-phase chlorination, a process which is quite demanding.

The phthalic anhydride/urea process may also be employed to convert tetra-chloro phthalic anhydride to green copper hexadecachloro phthalocyanine by condensation. In this case, titanium or zirconium dioxides, particularly in the form of hydrated gels, are used instead of the molybdenum salts which are used in the phthalic anhydride process [18]. There is a certain disadvantage to the fact that the products lack brilliance and require additional purification.

The exact hue which is produced by a green copper phthalocyanine is defined only by the degree of halogenation and by the type of halogen used (chloro or bromo substitution) and not by different crystal phases. Only one crystal modification of copper polyhalophthalocyanine has been reported. A higher degree of chlorination is accompanied by a shift of color from greenish blue to bluish green. A considerable difference is observed with the introduction of approximately the tenth chloro atom. Bromination affords yellowish shades of green.

Introduced in 1959 by GAF, a copper polybromochloro phthalocyanine was the first commercially used mixed halogenated copper phthalocyanine pigment. It was manufactured, as described above, in molten $NaCl/AlCl_3$ with $CuCl_2$ as a catalyst and bromine besides chlorine to afford a sequence of bromination followed by a chlorination reaction. The thus prepared pigment contains approximately 11 bromine and 3 chlorine atoms per molecule. In the most yellowish pigments, the degree of halogenation is 11 to 12 bromo and 4 to 5 chloro substituents.

Copper Perbromo Phthalocyanine Green may also be obtained from tetrabromo phthalic anhydride by the phthalic anhydride/urea process in the presence of titanium or zirconium catalysts. This route has not yet been introduced on a commercial scale.

3.1.2.6 Metal-free Phthalocyanine Blue

There are several pathways to metal-free Phthalocyanine Blue:
- by synthesizing suitable unstable ionic metal phthalocyanine salts, such as alkali or alkaline earth salts, with subsequent demetallization by alcohol or acid,
- by direct synthesis from phthalonitrile or amino-iminoisoindolenine,
- by fusing phthalic anhydride with urea.

The first route to preparing metal-free Phthalocyanine Blue involves treating phthalonitrile with the sodium salt of a higher-boiling alcohol, for instance with sodium amylate. The resulting phthalocyanine disodium salt is demetallized by stirring in cold methanol:

$$PcM_2 + 2\,H_3O^{\oplus} \longrightarrow PcH_2 + 2\,M^{\oplus} + 2\,H_2O$$

$$M : Na, \frac{Ca}{2}, \frac{Mg}{2}$$

Phthalocyanine is likewise obtained by hydrolyzing a corresponding calcium or magnesium salt in an acidic medium.

The route which proceeds via phthalonitrile has recently stimulated interest in connection with semiconductor technology for photoelectric copying, especially with respect to the pure Phthalocyanine Blue which is prepared by this path. An excellent pure product is obtained by heating phthalonitrile in 2-dimethylamino-ethanol while passing ammonia through the reaction mixture. A second scheme involves heating 1,3-diiminoisoindolenine in the same solvent to 135 °C, a method which affords the product not only in equal purity but also in high yield (approx. 90%). A third route proceeds via phthalonitrile, which is heated with hydrogen under pressure in an inert solvent [19].

The similarly blue and equally polymorphous metal-free phthalocyanine is chemically somewhat less stable than its copper complex: it decomposes slowly in a sulfuric acid solution. On the other hand, it can be chlorinated to afford metal-free Phthalocyanine Green.

3.1.3   Properties

While copper phthalocyanine blue pigments range in shade from greenish to reddish blue, the copper phthalocyanine green series covers the bluish to yellowish green portion of the spectrum. All copper phthalocyanines exhibit excellent fastness properties, especially regarding lightfastness and weatherfastness, and most types show very good heat stability. The pigments do not melt. Copper Phthalocyanine Blue may be sublimed in an inert gas atmosphere at approximately 550 °C and ambient pressure.

Three dimensional X-ray diffraction analysis has been employed to elucidate the molecular and crystal structure of Copper Phthalocyanine Blue (β-modification). In all modifications, the planar and almost square phthalocyanine molecules are arranged like rolls of coins, i.e., in one dimensional stacks. The modifications vary only in terms of how these stacks are arranged relative to each other. One important aspect is the angle between staple axis and molecular plane. The α-phase features an angle of 26.5°, while the stacks in the β-modification deviate by as much as 45.8° [9].

Fig. 94 shows cross sections through the molecular staples of both the α- and the β-modification. Sandwiched on top of each other, the molecules form acicular crystals. The orientation of the molecules within the crystal structure is such that the surfaces of the prisms which are arranged parallel to the longitudinal axis of the stacks and mainly carry H atoms and substituents of the benzene rings impart largely nonpolar properties on these surfaces. The basal planes of the stacks, on the other hand, which incorporate the π-electronic systems, the nitrogen atoms, and the copper atom, exhibit a relatively polar character. On the whole, the surface of the acicular substituted phthalocyanine pigment is largely nonpolar, irrespective of the crystal modification. There is consequently very little chance for specific interaction with the surrounding medium. This also explains the very low hydrophilicity of the compounds discussed thus far, a facet which has some impact on the disper-

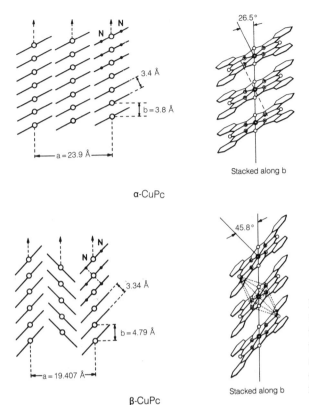

α-CuPc

Stacked along b

β-CuPc

Stacked along b

Fig. 94: Arrangement of the CuPc molecules of the α- and the β-modification. The data for the α-modification relate to the structure of 4-mono-chloro copper phthalocyanine, which is isomorphous with α-CuPc [9].

sibility of a pigment, particularly regarding surface wetting and stabilization of a dispersed pigment.

It is therefore not surprising that phthalocyanine pigments show a considerable tendency to flocculate. Types which are stabilized towards flocculation, for instance through introduction of polar groups, exhibit much improved properties in commercially important media. This is true for all modifications. Thus achieved improvement in pigment performance is a particular asset in various types of gravure and flexographic printing inks and in oven drying paints. This stabilization, however, will often adversely affect the stability of a pigment to organic solvents and its fastness to overcoating.

Copper phthalocyanine pigments also demonstrate good overall stability to organic solvents. A number of solvents, however, especially aromatics, may cause a change of modificaton in unstable types or "overcrystallization" in stable varieties. This phenomenon is largely due to the tendency of the stable phase to nucleate. The particle size of the resulting cystals decreases as the number of nuclei rises. β-Copper Phthalocyanine Blue is the thermodynamically stable modification.

Only one cystalline modification has been reported of polyhalogenated copper phthalocyanine green pigments. This phase is closely related to α-Copper Phthalocyanine Blue. The degree and type of halogenation and also the ratio between chloro and bromo atoms in a molecule defines not only the coloristics but also, if only marginally, a series of fastness properties.

Phthalocyanine pigments, which show high tinctorial strength, provide an excellent ratio of strength versus price. The strongest member is α-Copper Phthalocyanine Blue, while yellowish Copper Phthalocyanine Green is the weakest representative.

3.1.4 Application

By far the largest portion of the worldwide phthalocyanine production is targeted for use as organic pigment. The compound is employed almost exclusively in the form of copper phthalocyanine or as one of its halogenated derivatives. Commercial dyes are produced by introducing solubilizing groups, such as one or more sulfonic acid functions, into the phthalocyanine and especially into the copper phthalocyanine host structure. If desired, the system may undergo further chemical modification. The thus prepared dyes find extensive use in various areas of textile dyeing (they are utilized as direct dyes for cotton), for spin dyeing, and in the paper industry.

The five crystal modifications of copper phthalocyanine blue pigments which have been discovered thus far differ in terms of coloristics. The β-modification provides the most greenish and cleanest shades of blue, the α-form is distinctly redder than the β-modification, and the ϵ-phase is even redder than the α-form. γ- and δ-Copper Phthalocyanine Blue have not yet been commercialized.

Thermal treatment converts the thermodynamically unstable modifications into the stable β-form. The same change of modification may be effected in a high boiling inert solvent, especially in an aromatic one. This should be considered if unstable modifications of Copper Phthalocyanine Blue are to be applied in plastics or in oven drying paints containing aromatics. The phase transitions have been studied extensively [9]. Minor chemical changes to the copper phthalocyanine structure, for instance through partial chlorination, increase the activation energy for the change of crystal modification, which is a particular concern with the α-phase. If this energy barrier becomes large enough, sufficient stability is imparted on the crystal modification to make it suitable for application in a variety of areas, even in aromatic systems or in polymers. Besides, there are various methods to also improve the stability of these pigments to flocculation (Sec. 3.1.2.4). Other additives are available which enhance the somewhat poor rheological properties of copper phthalocyanine pigments.

Although low-cost green shades may be produced by combining Copper Phthalocyanine Blue with yellow pigments, the fastness properties of the resulting mixtures are usually inferior to those of green polyhalogenated copper phthalocyanine pigments.

3.1.5 Commercially Available Pigments

General

The α- and β-types of Copper Phthalocyanine Blue reign supreme among commercially available phthalocyanine pigments. There is also an increasing amount of interest in the phase-stabilized form of the α-crystal modification. Both modifications are also supplied as flocculation resistant types.

In view of the fact that phthalocyanine blue pigments are utilized in a host of applications, the requirements for their use vary considerably. A large number of special-purpose types are therefore supplied which are custom designed to exhibit optimized properties in certain media to suit particular demands.

The ε-modification has gained some interest in recent years. Commercial types of metal-free phthalocyanine blue are now also available.

Nonpolymorphous copper phthalocyanine green pigments are primarily offered as partially chlorinated products containing 14 to 15 chlorine atoms per pigment molecule (Pigment Green 7). Various copper polybromochloro phthalocyanine derivatives featuring 4 to 9 bromine and 8 to 2 chlorine atoms per molecule (Pigment Green 36) play a role as yellowish green pigments.

Table 27 lists the types of copper phthalocyanine pigments which are currently supplied:

Tab. 27: Commercially available copper phthalocyanine pigments

C.I. Name	C.I. Constitution No.	Stabilized toward change of crystal modification	Crystal modification	Range of shades	Number of halogen atoms	Comments
P.Bl.15	74160	no	α	reddish blue	–	
P.Bl.15	74160:1	yes	α	greener than P.Bl.15	0.5-1 Cl	
P.Bl.15	74160:2	yes	α	reddish blue	0.5-1 Cl	non-flocculating
P.Bl.15	74160:3	–	β	grenish blue	0*	
P.Bl.15	74160:4	–	β	like P.Bl.15:3	0*	non-flocculating
P.Bl.15	74160:6	yes	ε	very reddish blue	0*	
P.Bl.16	74100	metal-free phthalocyanine				
P.Gr.7	74260	–		bluish green	14-15 Cl	
P.Gr.36	74265	–		yellowish green	4-9 Br, 8-2 Cl	

* Depending on the synthetic route, small amounts of chlorine may be present.

Individual Pigments

Pigment Blue 15

Under the name of Pigment Blue 15, the Colour Index lists types of α-Copper Phthalocyanine Blue which are not stabilized towards phase-transfer. Compared to other phthalocyanine blue pigments, Pigment Blue 15 types are reddish blue in shade, tinctorially strong, and provide high color yield and economy in use. Their impact on the market, however, is inferior to that of the corresponding stabilized types, by which they are increasingly being replaced. In many media, P.Bl.15 types are redder and frequently cleaner than stabilized types, an advantage which is often compromised by less tinctorial strength.

Nonstabilized α-Copper Phthalocyanine Blue is utilized in the printing industry, to a certain extent in oil-based binder systems, such as offset printing inks for packaging and metal deco printing. Under standard conditions, P.Bl.15 is stable to a variety of organic solvents, such as alcohols, esters, ketones, aliphatic hydrocarbons, toluene, and plasticizers, such as dioctyl phthalate (Sec. 1.6.2). Ethylglycol or the standard DIN 16 524 solvent mixture, however, are slightly colored. The prints are equally fast to acids and alkali and show perfect fastness to soap, butter, and sterilization. P.Bl.15 is too reddish to be used in process colors, i.e., to produce the standard cyan for three and four color printing. A strong tendency to change its cystal modification in the presence of aromatics makes P.Bl.15 a doubtful candidate for packaging gravure inks. For the same reason, the pigment is only rarely found in paints.

The fact that unstable P.Bl.15 types tolerate less than 200 °C is a disadvantage in the coloration of plastics. With increasing temperature, the pigment frequently converts to the β-modification, a tendency which is especially pronounced in polymer materials which contain aromatic rings, such as polystyrene, ABS, or PETP.

As a result of their reddish shade, P.Bl.15 types are used to a certain extent to color polyethylene, which is processed below 200 °C, to produce items such as films. Another suitable medium is PVC, which is commonly treated at moderate temperature. There is a certain disadvantage to the fact that the tinctorial strength of such systems is frequently inferior to that of stabilized Copper Phthalocyanine Blue varieties.

Incorporated in plasticized PVC, P.Bl.15, like other phthalocyanine pigments, is usually entirely fast to migration. Moreover, it provides excellent lightfastness. P.Bl.15 also finds use in various types of PUR foam materials as well as in rubber. Its redder and frequently cleaner shade compared to corresponding stabilized types makes it an equally useful pigment for other media. This applies especially for water-based systems. Textile printing, paper mass coloration, paper surface treatment, and paper pulp are areas of application as suitable for the use of P.Bl.15 as office articles, including colored pencils, blackboard chalks for schools, and water colors.

Using P.Bl.15 in aqueous media may be problematic in view of the fact that the highly nonpolar character (Sec. 3.1.3) of the pigment may make it necessary to use a considerable amount of dispersants such as oxyethylated phenols, alcohols, or alkylsulfonates. Adequate dispersion relies as much on sufficient wetting of the pigment particles as it does on vigorous mechanical shear. Thus, adiabatic heat is frequently developed during pigment processing. Unsubstituted types of α-Copper Phthalocyanine Blue may convert to the β-modification if they are treated in insufficiently cooled dispersion units. This process is associated with a color change to a greener blue as well as with loss of cleanness and tinctorial strength. P.Bl.15 is thus supplied mainly in the form of highly pigmented aqueous pastes. Pigment manufacturers may either provide such products themselves or leave this task to specialized manufacturers of pigment preparations.

Introducing suitable substituents into the Copper Phthalocyanine Blue structure or covering the crystal surface with appropriate substances may specifically influence the hydrophilicity or the polarity of the pigment (Sec. 3.1.2.4). The ease of dispersion and wetting may thus be improved or optimized for certain applications, such as for aqueous media.

Pigment Blue 15:1

In the Colour Index, the designation P.Bl.15:1 refers to a phase-stabilized α-Copper Phthalocyanine Blue. This pigment has gained great commercial importance in almost all areas. Stabilization, however, for instance through partial chlorination, usually causes loss of tinctorial strength and cleanness as well as a color shift towards a greener blue. Despite this disadvantage, however, stabilized P.Bl.15:1 types reign supreme among Copper Phthalocyanine Blue types as colorants for coatings and paints, packaging printing inks, and plastics. They show good resistance to organic solvents. Excellent lightfastness and weatherfastness, superior migration fastness, and high heat stability, along with a reasonable price make these pigments attractive products. There is almost no limitation to pigment use in the above-mentioned areas. Moreover, P.Bl.15:1 types are often employed in combination with other pigments. Blends with Dioxazine Violet, for instance, provide shades of navy blue, while P.Bl.15:1 in conjunction with titanium dioxide produces good hiding power. Besides, P.Bl.15:1 blends are also utilized to brighten white paints, printing inks, and other media.

The paint industry employs P.Bl.15:1 for its excellent fastness properties to color all types of paints, including automotive refinishes and original automotive finishes, in both solid and metallic shades.

The so-called flop effect is a primary concern in connection with metallic finishes. Products may exhibit a different color and/or brightness, depending on the direction of incident light and the angle of observation. Flop is an important consideration in automotive finishes because different auto parts, although finished with the same product, may appear in a different color, depending on the angle at which they are viewed. Differences are particularly visible in locations such as between hood and fender. This phenomenon is attributed to a considerable extent of scattering in the colored pigment, thus sufficiently transparent pigments present somewhat of advantage. More extensive chlorination of α-Copper Phthalocyanine

Blue makes it possible to afford products which exhibit very little flop. Suitable special-purpose products are commercially available.

In practical application, most types of P.Bl.15:1 are completely fast to over-coating in oven drying systems. The pigments are suitable candidates for powder coatings, for instance for acrylate or polyurethane-based systems, in which they do not exhibit plate-out (Sec. 1.6.4.1).

The use of P.Bl.15:1 in various types of paints has for many years been hampered by unsatisfactory flow behavior. Newer methods involving surface treatment of the pigment with suitable agents have made it possible to improve the rheology in certain media enough to afford pigments which satisfy the use requirements.

Many P.Bl.15:1 type types demonstrate very good stability to change of modification, a feature which largely eliminates application problems. A sample may be tested for its stability in this regard by refluxing the pigment in toluene or xylene, filtering it, dispersing it in its target paint system, and evaluating the resulting product by coloristic comparison with an untreated but equally dispersed reference sample. The tendency of a product to convert to a different modification may similarly be assessed by storing the pigmented paint at elevated temperature, for instance at 50 °C, for two weeks. The thus prepared dispersion is then compared with a reference system to assess possible color change.

Incorporated in printing inks, phase-stabilized α-Copper Phthalocyanine Blue, like nonstabilized types, is too reddish to be employed as a process color for three and four color printing. It is used, however, to a considerable extent in all types of printing inks for special and packaging purposes. The prints are stable to common organic solvents and exhibit perfect fastness properties in special application (Sec. 1.6.2.3). Metal deco prints withstand up to 200 °C for 10 minutes or 170 to 180 °C for 30 minutes. They may safely be sterilized.

Many types of P.Bl.15:1 are initially very strong in printing inks. In order to produce 1/1 SD letterpress proof prints at standard conditions, a pigment concentration of 16 to 19% will suffice, depending on the type. 1/3 SD prints are produced using 7.5 to 10% pigment. The resulting prints are entirely fast to light: they score as high as step 8 and step 7 on the Blue Scale, respectively.

The list of applications also includes decorative printing inks for melamine based laminated plastic sheets. It should be noted, however, that P.Bl.15:1 may not be used in decorative printing inks for polyester-based sheets, because the pigment reacts with accelerators and hardening agents. Moreover, most types of P.Bl.15:1 are not entirely fast to styrene monomer, which is a frequently used solvent for polyesters.

Incorporated in aqueous systems, the main areas of application for stabilized α-Copper Phthalocyanine Blue are in emulsion paints, pastes for textile printing, and aqueous or water soluble gravure and flexographic printing inks for packaging and wallpaper. The textiles printing industry frequently prefers stabilized over nonstabilized α-types, especially if the resulting product must be fast to dry cleaning, i.e., to exposure to halogenated hydrocarbons. Even stabilized Copper Phthalocyanine Blue is frequently worked into aqueous application media in the form of an aqueous paste. Such pastes are concentrated pigment formulations which contain predispersed pigment. In contrast to the handling of nonstabilized types

(P.Bl.15), it is not necessary to exercise particular caution in dispersing P.Bl.15:1 in the presence of dispersion or wetting agents or particular solvents.

Likewise, P.Bl.15:1 also lends blue shades to plastics, in which it is incorporated either individually or in combination with inorganic or organic pigments. The stabilized type, like its nonstabilized counterpart, is a pigment with high tinctorial strength in polyolefins. 1/3 SD colorations (1% TiO$_2$) are formulated at 0.08 to 0.1% pigment, depending on the type. Most grades resist exposure to approximately 300 °C and satisfy almost all demands. Pigment dispersibility, however, is somewhat of a problem. Only very few commercial products meet the stringent requirements particularly for use in thin films and similar products. Like other types of Copper Phthalocyanine Blue, P.Bl.15:1 varieties also affect the shrinkage of partially crystalline thermoplastics to an appreciable extent. They are therefore only marginally suited to application in nonrotation-symmetrical injection-molded articles, such as bottle crates. Many grades of P.Bl.15:1 are more or less fast to bleeding in plasticized PVC, depending on the type of stabilization. In this medium, the pigment is also very lightfast and weatherfast. Incorporated in rigid or impact resistant PVC, stabilized α-types are less fast to long-term outdoor exposure than β-Copper Phthalocyanine Blue or Green.

P.Bl.15:1 is also applied in polystyrene, polyamide, polycarbonate (in which it is heat stable up to 340 °C), PUR foam materials, and cast resins. It should be noted, however, that the hardening of cast resins which are based on unsaturated polyesters is usually much retarded.

Problems may arise if P.Bl.15:1 is to be used in natural rubber, because the presence of free copper affects not only the vulcanization process, but also considerably affects the fastness of the product to aging. P.Bl.15:1 is thus considered a rubber poison. The free copper content in a pigment should therefore never exceed 0.015%. Commercial types are available which have been tested accordingly.

P.Bl.15:1, like other types of Copper Phthalocyanine, find extensive use in the spin dyeing of polypropylene, polyester, polyamide, secondary acetate, viscose rayon, and spun rayon. In these, as in other media, P.Bl.15:1 is very lightfast, and its textile fastness properties are almost entirely if not entirely satisfactory.

Pigment Blue 15:2

A number of α-Copper Phthalocyanine types which are stabilized towards flocculation and change of modification are registered in the Colour Index as P.Bl.15:2. Their main area of application is in paints. They are employed wherever P.Bl.15:1 types show too much of a tendency to flocculate in application or where economical considerations make P.Bl.15:2 appear more attractive. In accordance with its good fastness to flocculation, P.Bl.15:2 demonstrates good resistance to rub-out [20] in most common binder systems. Gradual improvement of the flocculation behavior has been achieved by suitably adjusting the particle size or particle size distribution as well as the shape of the pigment particles. However, stability to flocculation is imparted on these types primarily through chemical modification or through adsorption of suitable agents. The chemically induced resistance to flocculation, however, is frequently associated with a certain loss of fastness to overcoating, especially in systems containing aromatics. For this reason, flocculation resistant types have recently been introduced which largely eliminate this problem.

Other fastness properties in application largely equal those of P.Bl.15:1. In printing inks, P.Bl.15:2 is employed mostly in special gravure and flexographic inks. This is an area in which lack of fastness to overcoating is frequently of no consequence. P.Bl.15:2, like other α-Copper Phthalocyanine Blue types, is too reddish to be used as a standard cyan for three or four color printing.

Pigment Blue 15:3

P.Bl.15:3, the β-modification of Copper Phthalocyanine Blue, affords a clean shade of turquoise. Pigments of this type are used primarily in graphical printing as well as in finishes and paints, plastics and rubber, textile printing, and other areas, such as office articles.

The commercial importance of P.Bl.15:3 is reflected in the wide variety of available grades which cover a range of coloristic and application properties. Differences among application media are particularly noticeable in aspects such as dispersibility, tinctorial strength, shade, and transparency.

The printing inks industry uses P.Bl.15:3 especially as a blue component on different color scales for three and four color printing. The pigment corresponds to the CIE 12-66 standard shade of cyan on the European Color Scale for offset and letterpress prints (Sec. 1.8.1.1).

Resinated types are supplied for use in so-called oily binder systems for offset printing inks. However, these are much fewer in number than corresponding P.R.57:1 types which are used to produce the standard shade of ruby, or azo yellow pigments, which are used for standard yellow.

β-Copper Phthalocyanine Blue is one of the tinctorially strong pigments, although its strength is approximately 15 to 20% below that of α-Copper Phthalocyanine Blue. At standard conditions, 1/1 SD letterpress proof prints are formulated at 16% of a tinctorially strong type of P.Bl.15:1, but require approximately 21% β-Copper Phthalocyanine Blue. 1/3 SD prints contain, respectively, 7.5% and 9% pigment. In order to produce the standard shade of cyan, inks containing 14 to 15% of a tinctorially strong grade are used.

The prints exhibit excellent application properties. They are, for instance, entirely fast to organic solvents, soap, alkali, and acids. They are also fast to sterilization. Metal deco prints demonstrate very good heat stability. The products withstand exposure to 200 °C for 10 minutes or to 180 °C for 30 minutes. Although not quite as fast to heat as halogenated types of Copper Phthalocyanine Green, P.Bl.15:3 is thus somewhat more heat stable than stabilized α-Copper Phthalocyanine Blue.

The fact that P.Bl.15:3 provides good dispersibility has met with considerable appreciation in recent years. This development results from using more economical but less effective dispersion units (such as agitated bead mills) to produce web offset inks today. Excellent solvent fastness renders P.Bl.15:3 largely insensitive to high processing temperatures, which affect neither the tinctorial strength nor the transparency of the system, the latter being of considerable concern. This is in contrast to the yellow pigments which are employed in three color printing and which are accordingly processed at lower temperature. On the other hand, there is some danger that flocculation may occur in systems which are rich in mineral oil, a phenomenon which resembles pigment recrystallization.

In Europe, Copper Phthalocyanine Blue is usually supplied as a powder or as a granulate. The granulated product is somewhat less dusty but also more difficult to disperse. In the USA, P.Bl.15:3 also continues to be offered in the form of flushed pastes to be incorporated into oil-based printing inks. These pastes offer improved pigment dispersion and frequently afford more glossy and transparent prints.

As in coatings and paints, β-Copper Phthalocyanine Blue tends to flocculate if employed in publication gravure printing inks which contain a high content of aromatics and very little resin. Therefore nonflocculating P.Bl.15:4 grades are becoming increasingly important in this area. Sometimes also predispersed, solvent-containing pastes (toluene) or concentrates, nonaqueous dispersions (NAD, Sec. 1.8.2.2) are used. These systems confer better stability toward flocculation and tinctorial strength on the consequently improved pigment.

P.Bl.15:3 is usually also offered as a powder, to be used in packaging gravure printing inks. β-Copper Phthalocyanine Blue, like other pigments, is also supplied in the form of nitrocellulose chips or other preparations, for instance on vinyl chloride/vinyl acetate mixed polymer basis, ethyl cellulose, or polyvinylbutyral. These pigment preparations are used to advantage where superior transparency, high cleanness of shade, and optimized gloss are prime considerations.

Pigment preparations are more commonly used in aqueous/alcohol flexographic printing inks than in other types of inks. Aqueous/alcohol systems present somewhat of a problem, since hydrophobic phthalocyanine pigments are very difficult to adequately wet and disperse. Stabilized α-Copper Phthalocyanine Blue types, which provide a redder shade, dominate throughout the coatings and paints industry. Despite this fact, β-Copper Phthalocyanine Blue types are also useful, although often in flocculation resistant form (see P.Bl.15:4). They color industrial paints, architectural paints, and emulsion paints. The β-phase, like the stabilized α-crystal modification, is also excellently lightfast and weatherfast in these media.

P.Bl.15:3, the stable crystal modification of Copper Phthalocyanine Blue, demonstrates excellent heat stability. Most commercial types are therefore entirely suitable candidates for the pigmentation of plastics. β-Copper Phthalocyanine Blue, however, like the α-modification, often presents dispersion problems, especially in polyolefins. As a result of these diffculties, P.Bl.15:3 is usually employed in the form of pigment preparations which contain predispersed pigment. Moreover, β-Copper Phthalocyanine Blue types, like α-types, tend to nucleate in polyolefins, a problem which may lead to distortion and stress cracking in injection-molded parts. Good stability to organic solvents, plasticizers, etc., makes β-Copper Phthalocyanine Blue also a useful pigment for plasticized PVC. The resulting pigmented systems are fast to bleeding, although somewhat less so than halogenated Copper Phthalocyanine Green types. The pigment has equally excellent lightfastness and weatherfastness in plastics. Incorporated in rigid PVC, for instance, β-Copper Phthalocyanine Blue is one of the most stable organic pigments known. It is only slightly less fast than halogenated Copper Phthalocyanine Green types. Besides, P.Bl.15:3 is also found in polystyrene, impact resistant PS types, ABS, and similar polymers. Transparent P.Bl.15:3 systems, like those containing α-Copper Phthalocyanine Blue and Copper Phthalocyanine Green types, are thermally stable up to 300 °C. In very light white reductions (0.01% pigment/0.5% TiO$_2$), however, the samples only withstand up to 250 °C, while corresponding green types (P.Gr.7) are stable up to 300 °C.

β-Copper Phthalocyanine grades are also supplied as special-purpose types with a limited or defined free copper content, targeted for the coloration of rubber (see p. 439).

P.Bl.15:3, like stabilized α-Copper Phthalocyanine Blue, markedly affects the hardening of unsaturated polyester cast resins. The list of applications also includes PUR foam materials, office articles, such as colored pencils, wax crayons, and water colors, as well as spin dyeing of polypropylene, polyacrylonitrile, secondary acetate, polyamide, polyester, and viscose. Used in polyester spin dyeing, P.Bl.15:3 satisfies the thermal requirements of the condensation process (Sec. 1.8.3.8). 1/3 and 1/25 SD samples equal step 7–8 on the Blue Scale for lightfastness. Textile fastnesses, such as stability to wet and dry crocking are perfect.

Pigment Blue 15:4

Under the designation P.Bl.15:4, the Colour Index lists β-Copper Phthalocyanine Blue types which are stabilized towards flocculation. These products show largely the same coloristic and fastness properties as P.Bl.15:3 types, but often exhibit much better rheology. As with stabilized α-Copper Phthalocyanine Blue types, stabilization through surface treatment has proven to decrease the solvent fastness of β-Copper Phthalocyanine Blue, sometimes considerably so, making the pigment more sensitive to aromatics, alcohols, ethylene glycol, and ketones.

Stabilized types of β-Copper Phthalocyanine Blue are becoming increasingly important throughout the printing inks and paints field. Improved stability to flocculation and other rub-out phenomena [20] make P.Bl.15:4 an attractive choice, especially for oven drying paints. The pigments have also gained importance in gravure printing inks, particularly in toluene-based publication gravure printing inks and in various types of flexographic printing inks. They now dominate in these areas. Often, P.Bl.15:4 types are optimized for use in specific media, such as publication gravure printing inks. In these media the pigment exhibits high tinctorial strength and good flow. Other products have been developed for processing in particular dispersion units.

Pigment Blue 15:6

P.Bl.15:6, which is the ε-modification of copper phthalocyanine, affords the reddest blue shade provided by any copper phthalocyanine blue pigment. In recent years it has become possible to sufficiently stabilize these pigments towards change of modification in order to largely eliminate undesirable coloristic effects in application. Sec. 1.6.5 illustrates the different processes which simultaneously occur during pigment dispersion if an insufficiently stabilized grade of this modification is used. Oven drying paints containing aromatics clearly undergo a color change towards greener shades as the dispersion time increases. This phenomenon is attributed to a conversion to β-Copper Phthalocyanine Blue. Recrystallization effects, which accompany this phase transition, markedly decrease the tinctorial strength of the system. Although very interesting coloristically, the ε-modification is too expensive to find more extensive application. Corresponding reddish shades of blue may also be produced by combining α-Copper Phthalocyanine Blue types with small amounts of Dioxazine Violet (P.V.23). The shade of ε-Copper Phthalocyanine Blue demonstrates noticeably higher cleanness than the equally reddish blue shades provided by P.Bl.60.

In terms of fastness properties, P.Bl.15:6 performs like other modifications. In baking enamels, stabilized types are completely fast to overcoating, but they are not entirely fast to bleeding in plasticized PVC.

Pigment Blue 16

To date, metal-free phthalocyanine blue pigment has little commercial importance. Used to an appreciable extent as a greenish blue pigment until the 1950s, it was later replaced largely by β-Copper Phthalocyanine Blue. Of the different crystal modifications, only the α-form continues to play a certain commercial role. The pigment is supplied as a nonstabilized type and also as a type which is stabilized towards change of modification. Regular grades, like the unstabilized α-Copper Phthalocyanine Blue P.Bl.15, lose tinctorial strength if they are exposed to certain solvents and undergo a color shift towards a greener blue. In a number of systems these types may present problems due to lack of dispersibility and flocculation stability.

P.Bl.16 is used especially to produce metallic finishes. Incorporated in acrylate resin systems for this purpose, the pigment is weatherfaster than types of Copper Phthalocyanine Blue. P.Bl.16 also lends color to plastics, although to a more limited extent. It is not as heat stable as stabilized α- or β-Copper Phthalocyanine Blue types. P.Bl.16 is also used for artists' paints.

Pigment Green 7

P.Gr.7 type pigments provide a bluish green shade. The fact that P.Gr.7 only exists in one crystal modification which resembles that of α-Copper Phthalocyanine Blue eliminates the problems which are associated with the possibility of phase transition.

P.Gr.7, like copper phthalocyanine blue pigments, demonstrates good overall fastness properties. Its lightfastness and weatherfastness, heat stability, and solvent stability are superior even to those of the corresponding blue types. Therefore P.Gr.7 is broad enough in scope to be found in all areas of pigment application. P.Gr.7 grades, together with the yellower chlorinated/brominated green types of P.Gr.36, dominate wherever green shades are required which cannot be produced by combining α- or especially β-Copper Phthalocyanine Blue with suitable organic yellow pigments. Introducing halogen substituents into the copper phthalocyanine molecule reduces the tinctorial strength of the product, a tendency which becomes more noticeable as the halogen content and thus the molecular weight of the pigment increases.

The main area of application for P.Gr.7 is in paints.

P.Gr.7 is lightfast and durable enough to satisfy almost all requirements. It is almost entirely fast to overcoating. Moreover, the pigment performs very well in terms of other properties which are of interest in this type of application. As with other copper phthalocyanine pigments, problems may occasionally occur in attempting to disperse the pigment in its medium or to stabilize the pigment dispersion to flocculation. Improved versions, however, have been introduced in recent years. Untreated green pigments frequently show a greater tendency to flocculate than grades of Copper Phthalocyanine Blue which have not been stabilized to flocculation.

P.Gr.7 is used in all types of paints, including high grade original automotive finishes. Paints containing P.Gr.7 are completely suitable for exterior application. Incorporated in powder coatings, the pigment performs better than β-Copper Phthalocyanine Blue. Plate-out is not observed (Sec. 1.6.4.1).

The printing ink industry utilizes P.Gr.7 particularly for packaging printing inks. In this area, however, green shades are frequently also produced by combining the less expensive β-Copper Phthalocyanine Blue with suitable organic yellow pigments.

Prints made of P.Gr.7 are tinctorially weaker than those containing Copper Phthalocyanine Blue types. At standard conditions, for instance, 1/3 SD letterpress proof prints are made of inks formulated at approximately 17% pigment. Only approx. 8 to 9% β-Copper Phthalocyanine Blue are needed for the same purpose. Phthalocyanine Green demonstrates excellent overall fastness properties in application. Metal deco prints are thermally stable up to 220 °C for 10 minutes, they are fast to clear lacquer coating and may safely be sterilized. Incorporated in special gravure printing inks, some types of P.Gr.7 may flocculate, a tendency which is also observed in paints and which results in loss of tinctorial strength, reduced gloss, and other problems. In contrast to the different types of Copper Phthalocyanine Blue, P.Gr.7 is suited to use in decorative printing for laminated plastic sheets based on polyester. Melamine resin systems, on the other hand, render the pigment phototropic in daylight, a phenomenon which is in contrast to the behavior of Copper Phthalocyanine Blue. Chemical compounds which undergo a reversible shade change by exposure to light are referred to as being phototropic. The occurrence of phototropicity in a material is limited to a particular portion of the spectrum and may be reversed by irradiating the sample with a source which emits different radiation, usually in the more long-wave region. Irradiating a decorative P.Gr.7 melamine resin-based print with a xenon arc lamp rapidly destroys the pigment, resulting in a dull shade. The lightfastness of such samples equals step 5 on the Blue Scale.

In plastics, phthalocyanine green pigments are also tinctorially much weaker than corresponding blue pigments. More than 0.2% Pigment Green 7, for instance, is required to produce 1/3 SD HDPE samples (1% TiO_2). Less than 0.1% β-Copper Phthalocyanine Blue and approximately 0.08% α-Copper Phthalocyanine Blue are needed for the same purpose. The pigment withstands more than 300°C. Some types of P.Gr.7 influence the shrinkage of HDPE and other partially crystalline thermoplastics at elevated processing temperature much less than do Copper Phthalocyanine Blue types or brominated derivatives. This is an asset in the pigmentation of larger injection-molded parts.

P.Gr.7 is also difficult to disperse, especially in polyolefins. Easily dispersible products, however, are available. In plasticized PVC, as in other polymers, the pigment is completely fast to bleeding and exhibits excellent lightfastness and durability. P.Gr.7 is one of the most durable organic pigments in rigid PVC and may even be considered for long-term exposure. It is equally suitable polystyrene, impact resistant polystyrene, and ABS. Its heat stability in polystyrene, even in white reduction, is up to 300°C, while blue grades only withstand up to 240°C. Incorporated in unsaturated polyester-based cast resins, P.Gr.7, in contrast to Copper Phthalocyanine Blue types, does not affect the hardening of the medium.

Grades with a limited free copper content are supplied for the pigmentation of rubber. These pigments do not disturb caoutchouc vulcanization and do not affect the resistance of rubber to aging.

Utilized in spin dyeing, P.Gr.7 lends color to all types of commercially important fibers. The products demonstrate excellent lightfastness and weatherfastness. Used in polyacrylonitrile, for instance, P.Gr.7 satisfies the stringent requirements for use in outdoor textiles such as canvasses. Its textile fastness properties are almost, if not completely satisfactory. This textiles field is another area in which Copper Phthalocyanine Blue types are more than twice as strong as P.Gr.7.

Moreover, P.Gr.7 is also found in a variety of applications outside the previously discussed areas, for instance in surface covering colorations for leather and in furniture stains.

Pigment Green 36

Pigments of this type provide very yellowish shades of green. They are much yellower than P.Gr.7 types. The color of P.Gr.36 becomes yellower as more chlorine atoms are replaced by bromine. In the USA, these products are therefore classified according to the yellowness of shade and the bromine content. The bromine content may range from 25 to 30 wt% for less yellowish to 50 to 53 wt% for considerably yellower varieties. Again, only one crystal modification has been found.

All types of P.Gr.36 provide excellent lightfastness and weatherfastness, heat stability, and solvent fastness. Their main area of application is in paints. The pigment covers the entire range of applications, including various types of high grade automotive finishes. P.Gr.36 types are weaker and more expensive than P.Gr.7 or

Copper Phthalocyanine Blue. Yellowish shades of green are thus also produced by combining more inexpensive chlorinated copper phthalocyanine green pigments with suitable organic yellow pigments, although some of the excellent fastness properties are lost in the process.

The list of applications also includes printing inks, in which P.Gr.36 is tinctorially weaker than its copper phthalocyanine counterparts. For comparison, 1/3 SD letterpress proof prints containing P.Gr.36, printed at standardized film thickness, require approximately 26% pigment, while only 17% of P.Gr.7 suffice for the same purpose. The fastness properties exhibited by P.Gr.36 in prints, as in paints, closely resemble those of P.Gr.7, including the phototropic behavior of decorative printing inks for melamine-based laminated plastics sheets.

More than 0.3% pigment is required to formulate 1/3 SD HDPE samples. P.Gr.36 is thus much weaker than P.Gr.7 types, although it withstands more than 300°C and is thus equally heat stable. Considerable influence on the shrinkage of injection-molded polyethylene parts necessitates some caution in using P.Gr.36 for this purpose. Many types are difficult to disperse in plastics, especially in polyolefins.

References for Section 3.1

[1] F.H. Moser and A.L. Thomas: The Phthalocyanines, Vol. I and II, CRC Press, Boca Raton, Florida, 1983.
[2] F.H. Moser, A.L. Thomas: Phthalocyanine Compounds, Reinhold, New York, 1963.
[3] G. Booth in: Venkataraman: The Chemistry of Synthetic Dyes, Vol. V, p. 241, Academic Press, New York, London, 1971.
[4] Pigment Handbook, ed. by P.A. Lewis, Vol. I, p. 679 and further Vols., J. Wiley, New York, 1988.
[5] K. Weissermel and H.-J. Arpe, Industrial Organic Chemistry: 3. Ed. VCH, Weinheim, 1988.
[6] US-PS 2 964 532 (Du Pont) 1957.
[7] DE-AS 2 256 485 (Bayer) 1972; CH-PS 471 865 (BASF) 1966.
[8] DE-OS 2 432 564 (Bayer) 1974; DE-AS 1 569 650 (BASF) 1975.
[9] D. Horn and B. Honigmann, XII. Congr. FATIPEC, Garmisch 1974, Congress book p. 181.
[10] W. Herbst and K. Merkle, DEFAZET Dtsch. Farben Z. 24 (1970) 365.
[11] B. Honigmann, H.-U. Lenné, and R. Schrödel, Z. Kristallogr. 122 (1965) 185.
[12] G. Booth, Chimia 19 (1965) 201.
[13] DE-OS 3 023 722 (BASF) 1980.
[14] DE-AS 2 841 244 (BASF) 1978.
[15] S. Suzuki, Y. Bansho, and Y. Tanabe, Kogyo Kagaku Zasshi 72 (1969) 720.
[16] GB-PS 912 526 (ICI) 1960; DE-PS 1 161 532 (ICI) 1960.
[17] DE-AS 2 210 072 (BASF) 1972.
[18] DE-AS 1 172 389 (Allied) 1954; CH-PS 431 774 (Geigy) 1963.
[19] DE-PS 1 234 342 (BASF) 1963; US-PS 3 492 308 (Xerox) 1970; US-PS 3 509 146 (Xerox) 1970.
[20] U. Kaluza: Physikalisch-chemische Grundlagen der Pigmentverarbeitung für Lacke und Druckfarben, BASF 1979, 90–122.

3.2 Quinacridone Pigments

Strictly speaking, quinacridone pigments are dioxotetrahydroquinolinoacridine compounds. The typical quinacridone structure is a five-ring polycyclic system. The molecule consists of a central benzene ring which is bridged to two peripheral six-membered aromatic rings by two 4-pyridone rings [1]. Quinacridone molecules have been prepared in an angular form, such as in (**50, 51**), as well as in a linear arrangement, as in (**52, 53**).

50 (cis)   **51** (trans)

52 (cis)   **53** (trans)

While compounds **50** through **52** are only slightly yellowish, **53** provides an intensely bluish red shade.

Among the four constitutions, only **53** has stimulated commercial interest. Systematically referred to as a 7,14-dioxo-5,7,12,14-tetrahydro(2,3-b)quinolinoacridine, it is commonly referred to as linear trans-quinacridone. The pigment has gained commercial recognition especially for its high tinctorial strength. Unless mentioned otherwise, the term "quinacridone" is therefore used in this context exclusively to refer to the parent ring system **53**.

The first literary evidence of an angular quinacridone (structure **51**) dates to 1896, but it was not until 1935 that the quinacridone **53** was first synthesized by H. Liebermann [2].

Some 20 years later, W.S. Struve (1955) at Du Pont was the first to recognize the impact of linear quinacridone on the pigment industry. In due course, the first industrially useful synthetic pathway was found. Three commercial types of unsubstituted quinacridone consisting of two crystal modifications were introduced in 1958 [3]. From then on, quinacridone pigments have been one of the most recent classes of pigments to experience rapid development, especially throughout the USA and Western Europe.

3.2.1   Manufacture, Starting Materials

Several synthetic pathways for the commercial manufacture of quinacridone pigments have been published. In this context, only those routes are mentioned which were developed for industrial scale production. There are four options, the first two of which are preferred by the pigment industry. It is surprising to note that these are the methods which involve total synthesis of the central aromatic ring. On the other hand, routes which start from ready-made aromatic systems and thus might be expected to be more important actually enjoy only limited recognition.

The synthesis found by Liebermann involves heating 2,5-diaminoterephthalic acid with boric acid to 200 to 250 °C. Ring closure through condensation affords linear quinacridone [2]. A variation of this route, known as acidic ring closure (Sec. 3.2.1.2), currently plays a major role in pigment manufacture.

3.2.1.1   Thermal Ring Closure

The oldest known industrially used synthesis is carried out in a solvent and involves several intermediate steps. It is not necessary to isolate the intermediates, and the reaction can proceed in one vessel [4].

The starting material is succinic dialkylester*, which is easily obtained from maleic anhydride. Cyclization is accomplished with sodium alcoholate in a high boiling solvent or solvent mixture (such as diphenylether/diphenyl) to afford the succinylosuccinic dialkylester **54**:

R: alkyl group at $C_1 - C_4$, primarily C_1, C_2

A newer synthetic pathway, found by Lonza, involves chlorinating diketene to form the γ-chloroacetoacetic ethylester, followed by condensation to afford the succinylosuccinic diethylester [5]:

The typical quinacridone synthesis may be exemplified by the manufacture of unsubstituted quinacridone.

Succinylosuccinic dialkylester is treated with two equivalents of aniline to yield the 2,5-dianilino-3,6-dihydroterephthalic dialkylester **55**. Without separating the intermediate, the ring is closed **thermally** (at 250 °C) in the same reaction medium to afford the α-modification of dihydroquinacridone **56** [4,6]:

---

* These are esters or mixtures of esters of lower alcohols (C_1, C_2).

55   56

According to the patent literature, this synthesis proceeds in one vessel. The yield is approximately 75% relative to the starting material succinic dialkylester.

The condensation reaction with aniline may also be accomplished in two separate steps. It is possible to separate succinylosuccinic ester, followed by condensation in boiling ethanol in the presence of an acid (acetic acid, hydrochloric acid, phosphoric acid).

Oxidation of dihydroquinacridone to quinacridone may be achieved, for instance, with the sodium salt of m-nitrobenzene sulfonic acid in aqueous ethanol in the presence of sodium hydroxide solution [7]. A distinction is made between heterogeneous and homogeneous oxidation. The reaction is referred to as a "solid state oxidation" if the solvent contains approximately 2% sodium hydroxide solution. A content of approximately 30% sodium hydroxide solution relative to the solvent mixture, on the other hand, converts the reaction into a so-called "solution oxidation". The type of ring closure defines the crystal modification of the resulting dihydroquinacridone, while the oxidation technique defines the crystal phase of the quinacridone pigment.

Solid state oxidation, both of the α- and the β-phase [8] of dihydroquinacridone, affords crude α-quinacridone. Subsequent milling with salt in the presence of dimethylformamide produces the γ-modification, while the β-form evolves in the presence of xylene. Solution oxidation of dihydroquinacridone, possibly performed as air oxidation in the presence of 2-chloroanthraquinone [9], forms crude β-quinacridone. Milling with xylene likewise affords β-quinacridone pigment (see tables of chemical structures on p. 613).

Other patents describing variations of this method have been issued to Du Pont. However, no new insight has yet been gained from these publications of the described processes.

3.2.1.2   Acidic Ring Closure

Acidic ring closure resembles the Du Pont process in that the central benzene ring of the quinacridone structure is synthesized totally. The process bears resemblance to the Liebermann method [2, see also 1.3]. Condensation of succinylosuccinic ester with two equivalents of arylamine affords 2,5-diarylamino-3,6-dihydroterephthalic diester. Subsequent oxidation with suitable agents yields 2,5-diarylamino-terephthalic diester **57**. Hydrolysis and cyclization in polyphosphoric acid or other acidic condensation agents produces crude quinacridone [10]. This product already consists of very small particles.

Performing the last step (thermal aftertreatment) in an appropriate solvent provides a simple and convenient method of producing the particular target crystal modification.

PPA: Polyphosphoric acid

The β-crystal modification is prepared by pretreating the crude quinacridone product with alkali base before finishing with the solvent. Immediate solvent treatment, on the other hand, produces the γ-modification of the quinacridone pigment.

Among syntheses which start from preformed aromatic systems, the Sandoz process in particular has stimulated interest. It is the only route which allows the manufacture of asymmetrically substituted quinacridones. A typical synthesis follows.

3.2.1.3   Dihalo Terephthalic Acid Process

2,5-Dibromo-1,4-xylene or its 2,5-dichloro derivative is obtained by bromination or, correspondingly, chlorination of 1,4-xylene. It is oxidized to form 2,5-dibromo-terephthalic acid or its dichloro derivative **59**. Subsequent reaction with arylamine, for instance in the presence of copper acetate, affords 2,5-diarylaminoterephthalic acid **60**. It is also possible to replace the halogen atoms stepwise by arylamino moieties [11]. Cyclization to form linear trans-quinacridones, as in the above-mentioned method, is achieved by using acidic condensation agents:

Hal: Br, Cl; R,R': e.g. H, Cl, CH₃ (R≠R' or R=R')

Although the synthesis is completed in very few steps, oxidation of 1,4-xylene to the corresponding terephthalic acid does not afford a uniform product. Partial dihalogenation gives rise to side products. The condensation reaction requires two equivalents of arylamine per halogen atom. One equivalent is needed to neutralize the generated hydrohalogen acid, which is subsequently separated as arylamine-hydrohalide and recycled as hydrohalide and arylamine.

3.2.1.4 Hydroquinone Process

The hydroquinone process was developed by BASF [12]. Hydroquinone-2,5-dicarboxylic acid is prepared by a modified Kolbe-Schmidt synthesis from hydroquinone and carbon dioxide. Subsequent reaction with arylamine in an aqueous-methanolic suspension in the presence of an aqueous sodium chlorate solution and a vanadium salt affords the product in good yield:

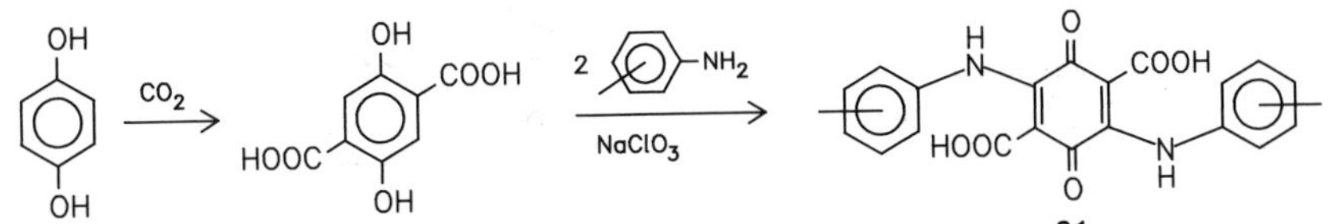

Subsequent ring closure of the 2,5-diarylamino-1,4-benzoquinone-3,6-dicarboxylic acid (61) is performed in concentrated sulfuric acid (or with thionyl chloride/nitrobenzene) to afford the linear trans-quinacridone quinone 62.

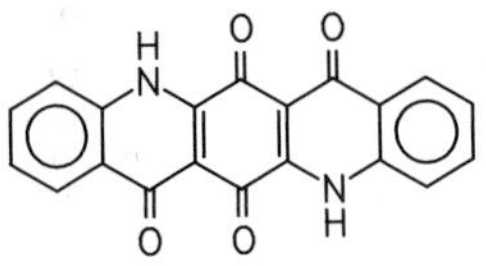

62

62 is also a component in mixed crystal phases with quinacridones, types of which are commercially available (Sec. 3.2.4). Quinacridone quinone is reduced with zinc or aluminum powder in dilute sodium hydroxide solution, in an aluminum chloride/urea melt, or in sulfuric/phosphoric acid under pressure to form quinacridone. There are certain disadvantages to this method in view of the fact that hydroquinone is relatively expensive, quinacridone is not easily reduced, and ecological problems prevail (wastewater pollution).

To summarize this discussion, the two latter methods have failed to gain the commercial recognition which is enjoyed by the first two processes (Sec. 3.2.1.1 and 3.2.1.2).

3.2.1.5 Substituted Quinacridones

There are also certain disubstituted quinacridones which are used commercially. The two peripheral rings in such quinacridone systems (53) are partially chlorinated or methylated. In the sequence **57** → **58** → **53** (p. 450), possible substitution sites on the peripheral rings are indicated by dashes (see also tables of chemical structures in the appendix).

Succinylosuccinic diester is cyclized according to the common route by using chlorinated or methylated aniline. Starting from anilines which are substituted in 2- or 4-position, the reaction affords the symmetrical 4,11- or 2,9-disubstitution products. Reaction with 3-substituted anilines, on the other hand, produces a mixture of both of the symmetrical 1,8- and 3,10-disubstitution products, as well as unsymmetrical 1,10-disubstituted compound. The following chart illustrates this point.

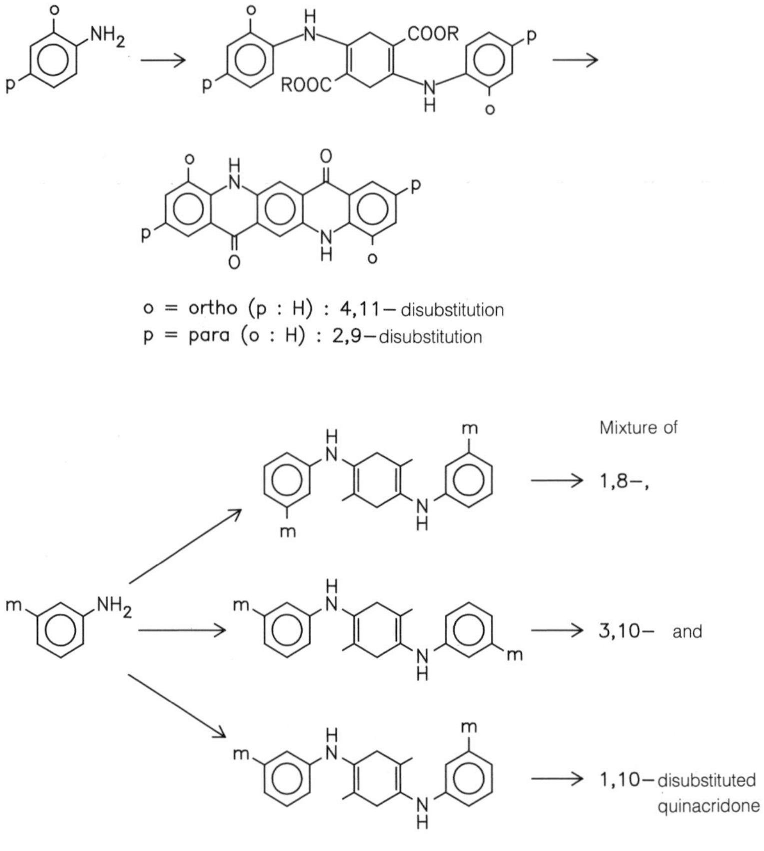

o = ortho (p : H) : 4,11−disubstitution
p = para (o : H) : 2,9−disubstitution

3.2.1.6   Quinacridone Quinone

Quinacridones are not the only industrially significant products. The list may be extended to include the derivative mentioned in Section 3.2.1.4, the linear trans-quinacridone quinone. There are two other synthetic pathways besides the hydroquinone method. The older method involves cyclization of the 2,5-bis-(2'-carboxyanilino-)-1,4-benzoquinone **63** with concentrated sulfuric acid or polyphosphoric acid at 150 to 200 °C. The starting material **63** is obtained through condensation of 1,4-benzoquinone with anthranilic acid:

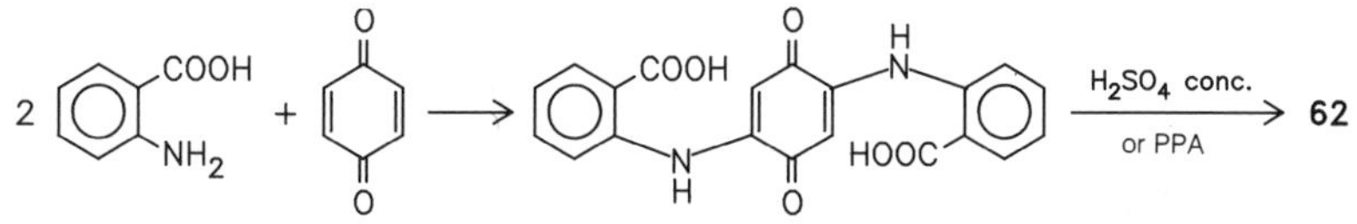

COOH
NH$_2$

+

→

COOH
HOOC

H
N
N
H

H$_2$SO$_4$ conc.
or PPA

→ **62**

63

A more recent synthesis proceeds via 2,5-diarylamino-3,6-dicarbethoxy-1,4-hydroquinone **65**, which is obtained by reducing 2,5-diarylamino-3,6-dicarbethoxy-1,4-benzoquinone **64** with sodium hydrogensulfite or sodium in ethanol:

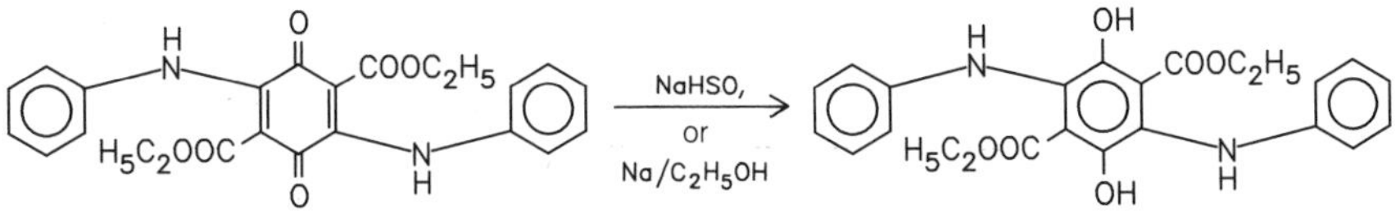

COOC$_2$H$_5$
H$_5$C$_2$OOC

NaHSO$_3$
or
Na/C$_2$H$_5$OH

COOC$_2$H$_5$
H$_5$C$_2$OOC

64   **65**

The hydroquinone **65** is cyclized in an organic solvent at 230 to 270 °C to yield 6,13-dihydroxy-quinacridone **66**. Oxidation with suitable agents (such as nitrobenzene, chromic acid, nitric acid) provides the linear trans-quinacridone quinone (**62**):

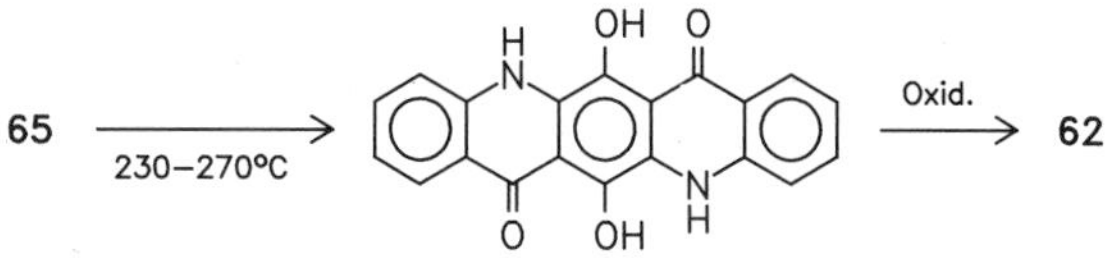

65 → 230–270°C

OH O
O OH

Oxid. → **62**

66

3.2.1.7 Polymorphism

Linear trans-quinacridones exhibit multiple crystal modifications. This phenomenon is reflected in the different reflection angles which are observed in the X-ray diffraction spectra of the pigment powders. This effect may be exemplified by desribing the polymorphous phases of unsubstituted quinacridone.

Most synthetic methods afford crude α-quinacridone [13]. This modification, however, has not been commercialized because of its lack of fastness properties. α-Quinacridone may be converted to the β- or γ-phase by organic solvents, especially at elevated temperature. Moreover, the α-phase is much less lightfast and weatherfast than its β- and γ-counterparts. It is by employing suitable solvents, therefore, that the β- and γ-phases are commonly obtained. The list of options includes milling the crude pigment with salt in the presence of a solvent, by first dissolving and then precipitating crude quinacridone, or by heating the crude product. Grinding with salt in a ball mill in the presence of xylene or o-dichlorobenzene, for instance, yields the β-modification [14]. The more stable γ-form [15] evolves in the presence of DMF [16].

The β-form is obtained by dissolving crude quinacridone in any one of a variety of solvents (such as concentrated sulfuric acid/toluene or methylated sulfuric acid), followed by precipitation with water. The same end is achieved by dissolving the product in polyphosphoric acid, followed by rapid precipitation with ethanol at 45 °C. The β-product, however, is not pure and usually contains some α-crystal modification aswell.

The γ-modification is produced by heating crude quinacridone in ethanol (under pressure) or in dimethylformamide or dimethylsulfoxide at 150 °C.

Nonpolar solvents generally aid the formation of the β-phase, while polar solvents assist in producing the γ-modification.

So far, neither the δ- [17] nor the γ'-crystal modification [18] has any commercial value.

Fig. 95 shows the reflection angles of the α-, β-, and γ-crystal modifications. The measurements were carried out on powder diagrams derived from X-ray diffraction spectra.

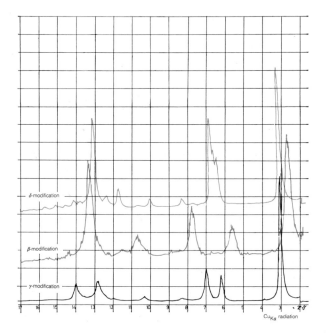

Fig. 95: Pigment Violet 19, 46500 – powder diagrams of X-ray diffraction spectra: reflection angles of the α-, the β-, and the γ-modification, measured with $Cu_{K\alpha}$ radiation.

3.2.2 Properties

All industrially significant quinacridone pigments are deeply colored products. They cover the spectral range from yellowish red to violet shades. Quinacridone pigments do not melt but decompose at high temperature.

Color and properties may be influenced by:
- changing the particle size,
- changing the crystal modification,
- introducing substituents into the ring system,
- forming mixed crystal phases of quinacridone with substituted quinacridones or with quinacridone quinone.

The pigments are practically insoluble in most common solvents and provide extensive migration resistance in all application media. They are very lightfast and weatherfast.

The exact cystal structure of quinacridone pigments has been published recently. So far, models indicated a planar arrangement of the molecules within the crystal lattice [19]. In fact, it has now been proved very recently by three-dimensional X-ray analysis that the pigment exists in two different crystal modifications [20]:

- a layer structure whereby a quinacridone molecule forms two hydrogen bonds each connected to two atoms of neighbour molecules. The resulting layers are arranged in staples.
- a "hunter's fence" structure, whereby one quinacridone molecule forms hydrogen bonds each bridged to four neighbour atoms. The planes of the molecule form an angle.

The strong intermolecular hydrogen bonds between the CO and the NH-groups combined with the interaction of the π-orbitals and the van der Waals forces lead to a very stable structure of the crystal lattice. This all together causes the very good thermal stability and solvent resistance of these pigments. The fact that quinacridone pigments furnish only weak orange shades in solution (solvents: phenol, dimethylformamide) shows that the color of a quinacridone pigment is controlled by its crystal lattice. The solutions, in contrast to the pigment itself, are only poorly lightfast.

The exact shade of a quinacridone pigment is determined not only by the crystal modification but also by the substitution pattern. Substitution, especially disubstitution, generally lightens the shade. This is observed in the sequence 2,9-substituted quinacridone to 3,10-substituted compound and to the 4,11-derivative. The fact that substitution in 4,11- positions also reduces the lightfastness of the pigment is attributed to steric considerations. Substituents approaching the NH moiety are likely to hinder the formation of hydrogen bonds. This assumption is confirmed by the fact that 5,12-dimethylquinacridone is least lightfast and dissolves even in ethanol.

Methods of changing the shade of a quinacridone pigment, other than changing the chemical modification and the substitution pattern, include the formation of mixed crystal phases, also referred to as "solid solutions". Mixed phases supplement chemically uniform quinacridone pigments in extending the commercially available range of shades (Sec. 3.2.4). The second component in a solid solution is usually either quinacridone quinone or 4,11-dichloroquinacridone.

3.2.3 Application

Good to excellent fastness properties justify the comparatively high price of quinacridone pigments. They are thus suitable candidates for the pigmentation of high
grade industrial finishes. Systems containing quinacridone pigments include original automotive finishes and automotive refinishes, weatherfast emulsion paints
such as house paints, plastics, high grade printing inks for purposes such as metal
deco and poster printing, and weatherfast textile printings, as well as spin dyeing.

3.2.4 Commercially Available Quinacridone Pigments

General

Although quinacridone pigments were first prepared in 1935, their technical value
initially went unrecognized. It was only when their polymorphism was detected that
quinacridones gained importance. The systematic synthesis of different crystal
modifications by appropriate aftertreatment of crude quinacridone made it possible to obtain a variety of highly stable pigments which rapidly gained commercial
interest as a new class of pigments.

A considerable number of mono- to tetra-substituted quinacridones has been
described so far. A publication which includes the literature up to the year 1965
already lists more than 120 compounds, and many exhibit more than one crystal
modification [21]. However, this diversity of compounds does not reflect commercial variety. Only very few species have been offered, and even less command a
major share of the pigments market.

Heading the list are two crystal modifications of Pigment Violet 19 (the parent
compound of the quinacridone series). Both the red-violet β- and the red γ-modification are encountered in the market, while the pure α-form is not lightfast and
durable enough to have any commercial value. However, traces of the α-modification are found in a variety of types.

Of the substituted quinacridone systems, it is primarily the 2,9-dimethyl derivative (Pigment Red 122) which enjoys appreciation for its excellent fastness properties and its pure bluish red shade. Besides, the list also includes 2,9-dichloro, 3,10-
dichloro, 4,11-dichloro, and 4,11-dimethyl quinacridones. Some of these are available as mixed phases with other quinacridone derivatives [22].

Other mixed phases are made from quinacridone quinone with unsubstituted
quinacridone or 4,11-dichloroquinacridone as a second component.

Quinacridone quinone itself (Sec. 3.2.1.6) is a tinctorially relatively weak yellow
compound with poor lightfastness. Oxidation of dihydroquinacridone with less
than molar amounts of chromate affords a quinacridone/quinacridone quinone
mixed phase which offers an interesting shade of gold.

Representative commercial quinacridone pigments are listed in Table 28. Each specimen is offered in only one crystal modification. Unsubstituted quinacridone is somewhat of an exception in that two crystal phases are commercially available.

Tab. 28: Commercially available quinacridone pigments.

C.I. Name	C.I. Constitution No.	R^2	R^3	R^4	R^9	R^{10}	R^{11}	Range of shades	Total number of known modifications
Quinacridone pigments									
P.V.19	73900	H	H	H	H	H	H	red-violet to bluish red	5
P.R.122	73915	CH$_3$	H	H	CH$_3$	H	H	bluish red (magenta)	4
P.R.192	–	–	–	–	–	–	–	bluish red	1
P.R.202	73907	Cl	H	H	Cl	H	H	bluish red to violet	3
P.R.207	73906	H	H	Cl	H	H	Cl	yellowish red	1 (4,11-dichloro:3)
	mixed crystal phase with unsubstituted quinacridone								
P.R.209	73905	H	Cl	H	H	Cl	H	red	1
	(mixed with 1,8- and 1,10-dichloroquinacridone)								
Quinacridone quinone pigments (Quinacridone quinone, C.I. 73920)									
P.R.206	–	H	H	H	H	H	H	maroon (golden yellow)	1
	mixed crystal phase with quinacridone quinone								
–	–	H	H	Cl	H	H	C	scarlet	1 (4,11-dichloro:3)
	mixed crystal phase with quinacridone quinone								
P.O.48	–	unsubstituted quinacridone + quinacridone quinone						maroon (golden yellow)	1
P.O.49	–	unsubstituted quinacridone + quinacridone quinone						maroon (golden yellow)	1
P.V.42	–							maroon	1

Individual Pigments

Pigment Violet 19, β-Modification

The β-modification affords a reddish violet shade. These pigments are frequently used in combination with inorganic pigments, especially with iron oxide, or to a decreasing extent with Molybdate Red pigments, to color industrial paints. Such pigment combinations provide comparatively dull shades of red to bordeaux. The excellent hiding power of the resulting systems is attributed to the content of highly scattering inorganic pigments with higher refractive indices. Combinations with opaque organic pigments, especially the P.O.36 type, are equally common. The products are used mostly in original automotive and automotive refinishes. The shade of the β-modification of P.V.19 is close to that of P.R.88, a tetrachloro-thioindigo pigment. The tinctorial strength of the reddish violet quinacridone grades, however, is considerably higher in combination with these inorganic pigments, especially the Molybdate Red types. They also produce cleaner shades than thioindigo pigments. Quinacridone pigments, despite their higher price, have the advantage of being more economical in use. Moreover, the β-modification of P.V.19 is also more lightfast and weatherfast than its thioindigo counterparts. Thioindigo pigments and quinacridone pigments, however, demonstrate different metameric effects (also in combination with inorganic pigments) and may thus not be exchanged directly. The higher tinctorial strength of quinacridone pigments is less noticeable in white reductions, but the pigment is appreciably more lightfast and weatherfast. Full shades tend to darken. The β-modification of P.V.19, like other quinacridone pigments, shows good resistance to the common organic solvents. Incorporated in baked enamels, the pigment is correspondingly fast to over-coating. The systems show excellent fastness up to 140 °C, while some bleeding is observed at 160 °C. In this respect, the β-crystal modification of P.V.19 is inferior to the γ-form. Paints made of β-P.V.19 are fast to acid and alkali.

Like a number of other quinacridone pigments, P.V.19 types of the β-modification are prone to cause problems, especially through pigment flocculation, while being dispersed in paint systems.

The different grades exhibit a wide range of particle size distributions. As with other pigments, a decreasing average particle size is accompanied by enhanced transparency in full shade, increased tinctorial strength in white reduction, and a general color shift towards more intense reddish violet shades. There is no noticeable change in lightfastness and weatherfastness if the particle size is changed. The transparent types are also suited to application in metallic finishes. Besides, they lend color to a variety of other media throughout the paint industry wherever high lightfastness and weatherfastness, chemical fastness, or heat stability are required. The list includes coil coating and various types of powder coatings, although it is recommended to adjust the pigment to the hardener before it is applied in a powder coating. β-P.V.19 does not show plate-out in most powder coatings. The β-modification of P.V.19 is also used as a colorant for plastics. In the coloration of PVC and PUR coatings, β-P.V.19 is also often used in combination with inorganic iron oxide and Molybdate Red pigments. The pigment is especially suitable for applications which require high hiding power as well as good lightfastness and weather-fastness, such as for automobile interiors, purses, and canvasses for tarps, awnings, etc. It is important to note that β-P.V.19 is more lightfast and weatherfast in combination with Iron Oxide or Molybdate Red pigments than in combination with TiO_2.

The β-modification of P.V.19 is very fast to bleeding in plasticized PVC. Its lightfastness in this medium satisfies the requirements for long-term weathering, as long as the demands are not exceptionally high. The pigment is thermally stable in polyolefins, it withstands exposure up to 300 °C. Its tinctorial strength in polyolefins, like in other media, is average. Approximately 0.2% pigment is needed to formulate 1/3 SD HDPE samples containing 1% TiO_2, if a tinctorially strong type with a high specific surface area is used. However, the pigment noticeably affects the shrinkage of PE and other partially crystalline polymers. High heat stability makes β-P.V.19 a suitable colorant for polystyrene and other industrial synthetics. Likewise, β-P.V.19 is also used in spin dyeing, for instance of PP. The printing ink industry utilizes β-P.V.19 only where high lightfastness or other special fastness properties are required. It is found, for instance, in inks for printing on PVC and for posters.

Metal deco prints containing β-P.V.19 withstand up to 180°C for 10 minutes and 170°C for 30 minutes, respectively. The prints are fast to sterilization and calendering and to a number of special-purpose media, such as particular organic solvents.

Pigment Violet 19, γ-Modification

The γ-crystal modification of P.V.19 affords bluish red shades which are much yellower than those obtained by the β-modification. Commercial types of γ-P.V.19 show a wide range of particle size distributions. The specific surface areas vary accordingly, they range from approximately 30 to 70 m^2/g.

The particle size of γ-P.V.19 has more of an effect on the lightfastness and weatherfastness of the resulting pigmented system than can be said of the β-modification. The grades of γ-P.V.19 with finer particle sizes, although less lightfast and weatherfast, have the advantage of being more transparent, tinctorially stronger, and bluer than varieties with coarser particle sizes.

High lightfastness and durability makes γ-P.V.19 a useful product for a variety of media throughout the coatings and paints industry, including high grade industrial paints and automotive finishes. γ-P.V.19 is used both in solid shades and as a shading pigment. Applied in combination with Molybdate Red, γ-P.V.19 affords comparatively brilliant, opaque shades, which are not accessible by using other quinacridone pigments. Moreover, γ-P.V.19 is also utilized in combination with other opaque pigments. Full shades and similarly deep shades darken slightly as they are exposed to light or weather. On the whole, γ-P.V.19 grades with fine grains are less weatherfast than the β-crystal modification and also various other types, such as P.R.122. The products of γ-P.V.19 with coarser particle sizes, however, frequently perform better than these quinacridone pigments. Excellent weatherfastness makes γ-P.V.19 a suitable candidate for exterior house paints. Its use in these areas, however, is somewhat limited by the fact that long-term exposure to alkali, for instance on freshly applied plaster, may considerably reduce the lightfastness and weatherfastness of the product as well as its tinctorial strength.

The stronger and more economical types with fine particle sizes are also found in transparent paints. If it is to be used in two-coat metallics, however, the application of the pigment must be combined with UV absorbants in topcoats to render it weatherfast enough to be applied in original automotive finishes.

The printing ink industry mostly prefers the type with finer particle sizes. These are incorporated into high grade printing inks for metal deco and poster printing.

Combination with 2,9-dimethylquinacridone (P.R.122) affords a bluish shade of red, which closely approaches the standard magenta for three and four color printing. Compared to other types of pigments covering the same portion of the spectrum, γ-P.V.19 types lack tinctorial strength. Inks formulated at almost 20% pigment, for instance, are needed to produce 1/3 SD letterpress proof prints at standard conditions. The γ-crystal modification is thus utilized primarily in areas where other pigments are not fast enough to satisfy the use requirements. In terms of lightfastness, 1/3 SD letterpress proof prints equal step 6–7 on the Blue Scale. The prints are thermally stable up to 190°C for 10 minutes or to 170°C for 30 minutes. They are fast to sterilization and calendering. Excellent fastness properties make γ-P.V.19 an equally suitable pigment for decorative printing inks for melamine and polyester based laminated plastic sheets. The samples equal step 8 on the Blue Scale for lightfastness.

γ-P.V.19 is frequently used for its high heat stability to replace more expensive bluish Cadmium Red pigments (cadmium selenide / cadmium sulfide mixed crystals) in lightfast thermoplastic systems. 0.2% pigment is required to formulate 1/3 SD HDPE samples containing 1% TiO_2. Some types exhibit a very moderate tendency to nucleate, which is why there is no limitation to pigment use in HDPE.

γ-P.V.19 as a colorant for plastics is available both as a powder and as a pigment preparation containing predispersed pigment. The list of applications includes PVC films, technical articles and toys made of PVC or PO, as well as spin dyeing products of PP fibers and PP monofilaments. Employed in polyester spin dyeing, the pigment is heat stable enough to be used in the condensation process at 240 to 290°C for 5 to 6 hours. A certain color change at low concentrations is attributed to the fact that γ-P.V.19 dissolves in polyester, a feature which makes it unsuitable as a shading pigment. Like in other media, the pigment is often incorporated in polymers together with grades of Molybdate Red.

The γ-form of P.V.19 is also applicable in injection-molded and extrusion-made polyamide. It satisfies not only the high thermal requirements in connection with these purposes but has the added advantage of being, like P.R.122 and 209, chemically inert to the slightly alkaline and reducing plastic melt.

Moreover, γ-P.V.19 is also found in a variety of other media, such as powder coatings and cast resins. This includes systems based on unsaturated polyester resins whose hardening is not affected by the pigment. The list of application media includes plastics which are processed at very high temperature (such as polycarbonates), in which the pigment is thermally stable up to 320°C. PUR foams and polyacetals should also be mentioned, although γ-P.V.19, like other quinacridone pigments, tends to migrate in polyacetals if it is used at concentrations below 0.1%. The pigment is also found in water colors for artists.

Pigment Red 122

2,9-Dimethylquinacridone is more weatherfast than most other unsubstituted types. It possesses excellent fastness to migration and outstanding heat stability. P.R.122 offers a very clean bluish shade of red, which is usually referred to as pink or magenta. Its main areas of application are in high grade paints, printing inks, and plastics, which is also true for the γ-modification of unsubstituted quinacridone.

P.R.122, which is more durable than unsubstituted types of quinacridone with fine particle sizes, may safely be used in automotive metallic finishes. Highly transparent types are available for this important purpose. However, such finishes occasionally present problems in terms of viscosity and flow. Its clean bluish shade makes P.R.122 an equally suitable pigment for use in combination with Molybdate Orange. The resulting colorations are particularly lightfast and weatherfast in automotive finishes. The thus prepared shades are distinctly cleaner than combinations of Molybdate Red with β-P.V.19. Combinations of Molybdate Red with the γ-modification are less economical than those with P.R.122 and afford somewhat cleaner and more bluish shades. The pigment is used primarily as a shading pigment. Excellent stability to common organic solvents renders the pigment completely fast to overcoating in oven drying systems up to 160°C. In this respect, P.R.122 performs better than the β-modification of unsubstituted quinacridone. Moreover, P.R.122 is recommended for use in architectural paints: its excellent weatherfastness is of interest for exterior application. P.R.122, like P.V.19, is also used in powder coatings.

2,9-Dimethylquinacridone has a slightly nucleating effect on partially crystalline polymers such as HDPE. The plastics industry therefore uses P.R.122, like the γ-modification of unsubstituted quinacridone, primarily in systems such as PVC films and coatings made from PVC or from PUR, in technical products or toys made from PVC or polyolefins, or in PO films. Besides, P.R.122 is also utilized in PP spin dyeing. In polyester spin dyeing, P.R.122 satisfies the high thermal requirements of the condensation process (exposure to 280°C for 5 to 6 hours). At low concentrations, however, the pigment dissolves to a certain extent in its medium and accordingly changes color. Its lightfastness is also affected. Plastics are advantageously colored with pigment powders or preparations containing predispersed pigment. P.R.122 grades are considerably stronger than γ-P.V.19, they match the β-modification in tinctorial strength. 0.21% pigment is required to produce a 1/3 SD HDPE sample containing 1% TiO_2. Comparative systems containing γ-P.V.19 are formulated at 0.26% pigment. P.R.122 demonstrates excellent heat stability, which makes it a suitable colorant for polystyrene, impact-resistant polystyrene, ABS, and polycarbonate. Besides, it is also used in a variety of other media wherever high lightfastness, good heat stability, and other special fastnesses are a prime consideration. Examples are systems such as unsaturated polyester resins and PUR foams.

P.R.122, like other quinacridone pigments, shows excellent application properties in high grade printing inks. It is fast to sterilization and to calendering. However, P.R.122 is somewhat weaker in these media than P.V.19 types. At standard film thickness, 1/3 SD letterpress proof prints are made from printing inks which contain approximately 25% pigment. In order to compare values, it should be noted that approximately 13% are needed of the β-modification of P.V.19 and about 19% of the γ-modification. The standard magenta for three and four color printing on cans, posters, and packaging may be approached by combining

P.R.122 with pigments of the γ-P.V.19 type. P.R.122 is also found in decorative printing inks for laminated plastic sheets. Complete fastness to the solvents which are commonly used in this area, such as styrene monomer and acetone, makes P.R.122 a useful colorant for polyester-based sheets, as well as for melamine resin systems. The lightfastness of such products equals step 8 on the Blue Scale.

Pigment Red 192

Production of P.R.192, a quinacridone pigment, has recently been discontinued. The pigment never gained considerable commercial recognition. It provides a shade somewhere between the shades of the quinacridone pigments P.R.122 and γ-P.V.19. The commercial type was distinguished by good transparency, but inadequate flow behavior. In some systems, the pigment was almost as durable as P.R.122. It was recommended for use in high grade industrial paints, especially for automotive finishes, plastics, and special-purpose printing inks.

Pigment Red 202

P.R.202, a very lightfast and weatherfast mixed phase quinacridone pigment, provides more bluish and considerably duller shades than 2,9-dimethylquinacridone. (2,9-dichloroquinacridone, which is listed in the Colour Index as Pigment Violet 30, has been discontinued). P.R.202 is primarily applied in automotive finishes. Even highly transparent types with fine particle sizes provide two-coat metallic finishes which offer excellent rheological properties. P.R.202 is accordingly superior to 2,9-dimethylquinacridone (P.R.122). On the other hand, metallic finishes made from P.R.202 are duller and more bluish. P.R.202 types exhibit more apparent flop (Sec. 2.3.1.5 under P.Bl.15:1) than grades of P.R.122. At similar transparency, the mixed phase pigment is more durable than the individual components, especially more so than unsubstituted quinacridone pigment. Its response to weathering corresponds to that of 2,9-dimethylquinacridone. Two-coat finishes made of P.R.202, however, are frequently less weatherfast than single-layer products. In terms of fastness to overcoating and other aspects of pigment performance and properties, P.R.202 largely behaves like P.R.122.

Pigment Red 207

P.R.207 is a mixed phase pigment made from unsubstituted and 4,11-substituted dichloroquinacridone. Pure 4,11-dichloroquinacridone is not commercially available. P.R.207 affords a yellowish red shade, which is very yellowish compared to other nonoxidized quinacridone pigments. P.R.207 is still somewhat yellower, but also duller than P.R.209. The commercial type of P.R.207 is very opaque. In fact, it has better hiding power and is considerably weaker in white reductions than the commercially available version of 3,10-dichloroquinacridone. P.R.207 demonstrates excellent weatherfastness, and its response to weathering approximately equals that of P.R.209. This is also true for other aspects of pigment performance, including resistance to organic solvents, fastness to overcoating, tendency to flocculate, and other properties and fastnesses. P.R.207 is used primarily in automotive finishes but is also utilized as a colorant for plastics and artists' colors.

Pigment Red 209

P.R.209 is a specialty product. It is used in much less volume than the β- and γ-modifications of unsubstituted quinacridone and also than 2,9-dimethylquinacridone. P.R.209 has stimulated a certain amount of interest, not only for its excellent lightfastness and weatherfastness, migration fastness and heat stability, which are comparable to those of P.R.122, but also for its very yellowish shade of red, which is unusual for a quinacridone pigment.

P.R.209 is applied in paints, especially in various types of automotive finishes. Exceptionally clean shades of red are furnished if P.R.209 is combined with Molybdate Orange. There is a certain disadvantage to the fact that such combinations are formulated at a relatively high P.R.209 content. The need for a high pigment content considerably compromises the economy of such systems in contrast to pigment combinations with other quinacridones. Moreover, P.R.209 is found in metallic automotive finishes, for instance in conjunction with transparent Iron Oxide pigments. Comparatively poor tinctorial strength, however, restricts the use of P.R.209 in this area to occasions where the desired shade is not accessible through pigments derived from 2,9-dimethylquinacridone or 2,9-dichloroquinacridone.

Other areas of application include a number of plastics which are processed at elevated temperature. P.R.209 lends color to polyolefins (in which it only slightly affects the shrinkage), to ABS, to polyacetal resins, and other industrial polymers.

Quinacridone Quinone Pigments

In contrast to most quinacridone pigments, commercial products based on quinacridone quinone afford relatively dull, slightly less weatherfast red and reddish to yellowish shades of orange. These pigments are used as special-purpose products to create metallic effects in paints. It is possible, particularly in combination with transparent Iron Oxide Yellow pigments, to produce interesting shades of gold for fashionable automotive finishes, but these products exhibit more of a flop than transparent grades of quinacridone pigments. These mixed phase pigments are used particularly throughout the USA.

Pigment Red 206

P.R.206 is a mixed crystal type and consists of unsubstituted quinacridone and quinacridone quinone. The ratio between the two components as well as the crystal modification is not yet known. P.R.206 affords a very dull, yellowish shade of red, referred to as maroon. The pigment is considerably weaker than perylene pigments. All commercially available types of P.R.206 are more or less transparent and are used mostly in metallic finishes for automobiles, to which they lend reddish shades of copper. The pigment is often found to be difficult to disperse. The finishes frequently exhibit rheological problems, especially at high pigment concentration.

Pigment Orange 48

P.O.48, like P.R.206, is a mixed phase pigment, made of unsubstituted quinacridone and quinacridone quinone. The ratio of the two components within the pigment crystal and the modification of the commercial types are not known. The full shade is a deep maroon which turns into a yellowish brown as TiO_2 is added, such as in 1:5 TiO_2 reductions. The main area of application for P.O.48 is in metallic finishes, to which the pigment lends interesting shades of copper and gold. P.O.48 produces considerably redder shades than P.O.49, a member of the same class of pigments. P.O.60, a benzimidazolone pigment, affords colors which at equal conditions are intermediate between those of the two quinacridone quinone pigments. Newly developed grades of P.O.48 and 49 have recently been introduced offering appreciably higher weatherfastness, which now corresponds to that of other representatives of the same series which are found in original automotive finishes. Moreover, these novel types also exhibit improved flow behavior. P.O.48 is also found in plastics and is recommended for spin dyeing. It dissolves in polyamide and in various other polymers.

Pigment Orange 49

The exact physical properties of P.O.49, i.e., mainly the crystal modification of this quinacridone/quinacridone quinone pigment remain to be published. P.O.49, like P.R.209, is a specialty product for metallic shades. It is used to produce shades of gold in finishes, which are considerably more yellowish than those of P.O.48.

Pigment Violet 42

The chemical constitution of P.V.42, a mixed phase pigment, remains to be published. P.V.42 has little commercial value. Its shade is a maroon, which is used primarily to afford metallic shades in original automotive finishes.

Other Quinacridone Pigments

A number of other members of this class are also available, although not all are listed in the Colour Index. Their exact chemical and physical properties remain to be published. The list includes mixed phase pigments made of unsubstituted quinacridone with dimethylquinacridone or other substituted quinacridones. These pigments are considered specialty products for automotive finishes, especially for metallic finishes.

References for Section 3.2

[1] S.S. Labana and L.L. Labana, Chem. Rev. 67 (1967) 1–18; W.F. Spengeman, Paint Varn. Prod. 60 (1970) 37–42.
[2] H. Liebermann, Ann. 518 (1935) 245–259.
[3] T.B. Reeve and E.C. Botti, Official Digest 31 (1958) 991–1002.
[4] US-PS 2 821 529 (Du Pont) 1958.
[5] P. Pollak, Chimia 30 (1977) 357–361 and Progr. Org. Coatings 5 (1977) 245; DE-OS 2 313 329 (Lonza) 1972.
[6] US-PS 2 821 541 (Du Pont) 1958; US-PS 3 009 916 (Du Pont) 1961.
[7] US-PS 2 969 366 (Du Pont) 1961.
[8] US-PS 3 007 930 (Du Pont) 1961.
[9] US-PS 3 475 436 (Du Pont) 1969.
[10] US-PS 3 342 823 (Allied Chem.) 1967.
[11] F. Kehrer, Chimia 20 (1974) 174–176; DE-PS 1 200 457 (Sandoz) 1958; DE-PS 1 250 033 (Sandoz) 1962.
[12] DAS 1 195 425 (BASF) 1963.
[13] US-PS 2 844 484 (Du Pont) 1958.
[14] US-PS 2 844 485 (Du Pont) 1958.
[15] F. Jones, N. Okui and D. Patterson, J. Soc. Dyers Col. 91 (1975) 361–365.
[16] US-PS 2 844 581 (Du Pont) 1958.
[17] US-PS 3 272 821 (Eastman Kodak) 1966.
[18] DE-PS 1 183 884 (BASF) 1959.
[19] G. Lincke, Farbe + Lack 86 (1980) 966–972.
[20] E. Dietz, 11th International Color Symposium, Montreux 1991.
[21] S.S. Labana and L.L. Labana, Chem. Rev. 67 (1967) 1–18.
[22] US-PS 3 160 510 (Du Pont) 1964; US-PS 3 148 075 (Du Pont) 1964.

3.3 Vat Dyes Prepared as Pigments

In this section a number of polycyclic pigments are discussed which have been used for a long time as vat dyes for textile fibers. Heading the list are perylene, perinone, and thioindigo pigments, as well as pigments derived from anthraquinone.

Original attempts at using these vat dyes as pigments afforded only dull shades with insufficient tinctorial strength. As pigment technology developed into a well-established industry and chemical and physical methods were found to increase the quality of pigments, vat dyes stimulated more and more interest throughout the pigment industry. In order to employ a vat dye as a pigment, it was necessary to improve the chemical purity of the product, to optimize the particle size and the particle size distribution, and to synthesize certain target crystal modifications.

The polycyclic colorants which are discussed in this context (and this is also true for copper phthalocyanine and quinacridone pigments) as a rule evolve from the manufacturing process as products with relatively coarse particles, which is in contrast to azo pigments (Sec. 2.2.3).

Although a crude pigment may be converted to a material with finer particle sizes by milling, there is a certain disadvantage to the fact that the frequently wide particle size distributions (approx. 0.5 to 200 μm) make these compounds unsuitable for use as pigments. The crystallinity and deagglomeration is commonly improved by treating the pigment in an organic solvent or water in the presence of a dispersing agent. This type of aftertreatment is referred to as finishing a pigment.

Various methods of aftertreatment have been developed to transform vat dyes into easily dispersible pigments. Heading the list are the following methods:

- Reduction to form a leuco base, removal of impurities, and subsequent reoxidation.
- Dissolving the crude product, for instance in sulfuric acid, precipitation in water and/or organic solvents, possibly in the presence of surfactants at exactly defined conditions.
- Formation of a pigment sulfate in the presence of 70 to 100% sulfuric acid, isolation of the intermediate, and subsequent conversion to the pure pigment by treatment with water, possibly in the presence of surfactants, and heating.
- Thermal aftertreatment in water and/or organic solvents.
- There is a variety of dispersion and milling techniques. Particle size reduction may be achieved, for instance, through fine dispersion in kneaders, fast rotating stirrers with milling effect, or in grinding units based on the impact principle. Some crush the material through mutual friction between particles or by rotation or vibration. Vibration mills, roll mills, or ball mills may be used. The crude pigment is usually prepurified, for instance by reprecipitation, before being dispersed.
- Milling the pigment in the presence of salts or solvents, possibly in the presence of surfactants, or in the presence of strong, nonoxidizing acids (with a pK_α below 2.5).

Often two or more of these processes are combined as necessary to optimize the results.

The methods which are used to convert individual vat dyes into pigments are described in the respective chapters.

Despite the extensive technology that has developed in this field, only very few of the originally large number of useful vat dyes continue to play a role as pigments. The somewhat narrow selection is also made on the basis of the cost/performance considerations for each of these relatively complicated compounds. Only very few of these synthetically demanding pigments match the fastness standards of phthalocyanine pigments.

3.4   Perylene and Perinone Pigments

Perylene and perinone pigments are chemically related. The group of perylene pigments is derived from perylene-3,4,9,10-tetracarboxylic acid **67**, while perinone pigments are derivatives of napthalene-1,4,5,8-tetracarboxylic acid **68**:

67   68

Members of both groups are manufactured by essentially the same route. Starting from the anhydrides of **67** and **68**, formation of bisimido or, respectively, bisimidazole compounds is accomplished with amines or diamines. The compounds have been known as vat dyes since 1913 (perylenes) and 1924 (perinones).

3.4.1   Perylene Pigments

Perylene pigments are diimides of perylene tetracarboxylic acid. The unsubstituted compound, although never used as a vat dye, is the oldest known member of this class. It was described as early as 1912. In due course, a number of perylene tetracarboxylic acid diimides emerged (dimethylimide in 1913) which were initially used exclusively as vat dyes. It was not until 1950 that perylene compounds found use as pigments. This shift in emphasis is credited to investigations by Harmon Colors which broke new ground in describing the conversion of vat dyes to pigments. Several members of the perylene pigment series have thus found their way towards industrial scale production.

The customary method of preparing perylene pigments is by reaction of perylene tetracarboxylic dianhydride with primary aliphatic or aromatic amines in a high boiling solvent. The dianhydride itself is also used as a pigment. Dimethylperylimide may also be obtained by treating the diimide with methyl chloride or dimethyl sulfate.

3.4.1.1   Preparation of the Starting Materials

The primary starting material for the synthesis of perylene tetracarboxylic acid pigments is the dianhydride **71**. It is prepared by fusing 1,8-naphthalene dicarboxylic acid imide (naphthalic acid imide **69**) with caustic alkali, for instance in sodium hydroxide/potassium hydroxide/sodium acetate at 190 to 220 °C, followed by air oxidation of the molten reaction mixture or of the aqueous hydrolysate. The reaction initially affords the bisimide (peryldiimide) **70**, which is subsequently hydrolyzed with concentrated sulfuric acid at 220 °C to form the dianhydride:

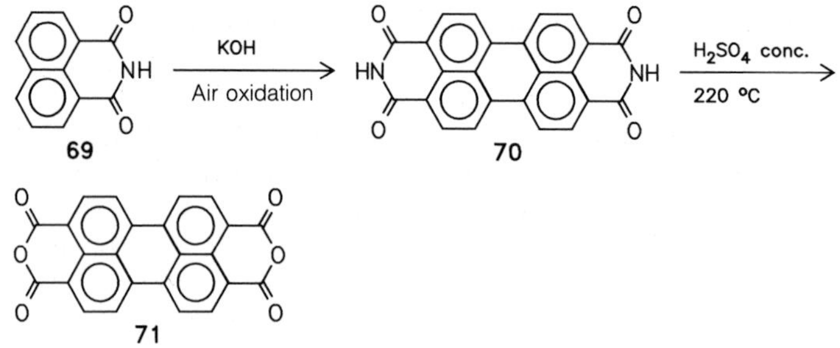

Naphthalic acid imide **69** is obtained through air oxidation of acenaphthene **72** with vanadium peroxide as a catalyst. The intermediate, naphthalic anhydride **73**, is subsequently reacted with ammonia:

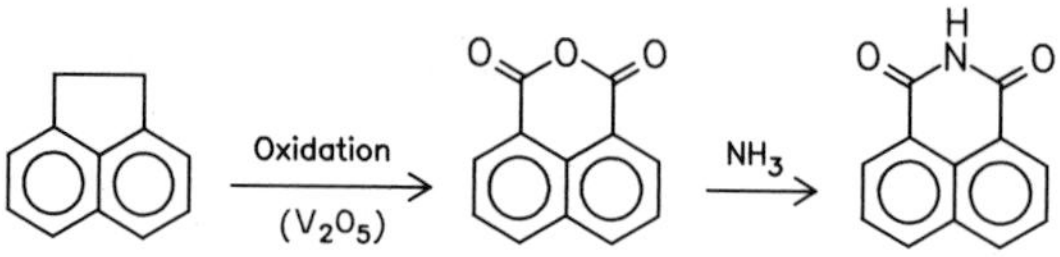

3.4.1.2 Chemistry, Manufacture

Perylene pigments are derivatives of the general structure **74**:

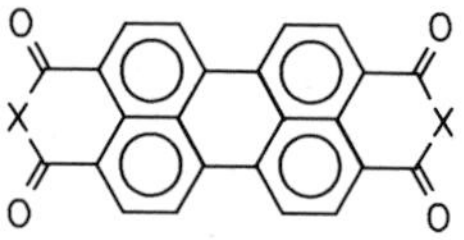

In industrially interesting pigments, X usually stands for O or N—R, and R represents H, CH$_3$, or possibly a substituted phenyl moiety. The phenyl ring may possess a methyl, a methoxy, an ethoxy, or an $-N{=}N{-}$⟨◯⟩ substituent.

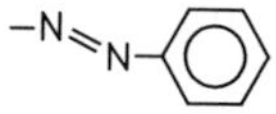

Heating perylene tetracarboxylic dianhydride **71**, as described in Section 3.4.1.1, to 150 to 250 °C with a primary aliphatic or aromatic amine in a high boiling solvent affords the desired crude pigment. The reaction may be accelerated by adding agents such as sulfuric acid, phosphoric acid, or zinc salts [1]. Pigment synthesis through condensation in aqueous medium has also been described.

A different route proceeds by alkali fusion of the corrspondingly substituted naphthalic acid imide. This pathway parallels the synthesis of perylene tetracarboxylic acid diimide **70** (Sec. 3.4.1.1). The method is particularly suited to aliphatic amines.

Unsymmetrically substituted perylene pigments are a comparatively recent novelty. Selective protonation of the tetra sodium salt of perylene tetracarboxylic acid affords the monosodium salt of perylene tetracarboxylic monoanhydride in high yield. Stepwise reaction with amines produces unsymmetrically substituted perylene pigments [2].

There are various techniques of aftertreating the crude product which convert it into an industrially useful pigment. Important methods include reprecipitation from sulfuric acid, milling, and recrystallization from solvents. It is also customary to combine these methods in order to optimize the results [3].

The dianhydride of perylene tetracarboxylic acid is converted into the pigment form by preparing the corresponding alkali salt and then reprecipitating the compound with an acid. The dianhydride is formed after separating the acid by thermal aftertreatment at 100 to 200 °C, possibly under pressure, with an organic solvent. The list of suitable media includes alcohols, ketones, carboxylic acid esters, hydrocarbons, and dipolar aprotic solvents.

Opaque perylene pigment varieties are treated in ball mills, either in the presence or absence of milling auxiliaries. The pigments are typically heated to 80 to 150 °C in solvents such as methyl ethyl ketone, isobutanol, diethylene glycol, N-methylpyrrolidone, possibly in the presence of water.

Perylene tetracarboxylic acid dimethylimide may also be prepared by methylating the corresponding diimide.

3.4.1.3 Properties

Perylene pigments exist in a wide range of hues, they provide red, bordeaux, violet, and brown shades. The pigments exhibit excellent solvent stability, good to very good migration stability in plastics, fastness to overcoating in paints, high chemical inertness, and superior thermal stability. Only the basic compound, the dianhydride of tetracarboxylic acid, is not fast to alkali.

As a group, perylene pigments offer high tinctorial strength. They are frequently found to be appreciably stronger than quinacridone pigments. Moreover, perylene pigments provide excellent lightfastness and weatherfastness. In this respect, they approximately equal the performance of quinacridone pigments.

There are recent publications describing black perylene pigments [4]. These compounds are chemically closely related to the corresponding red species, from which they are derived through slight changes in the substitution pattern on the moiety X which is not part of the conjugated system (see structure **74**).

The perylene pigment in which $X = \!\!>\!\!N-CH_2-CH_2-O-CH_2-CH_3$, for instance, affords a red shade, while $X = \!\!>\!\!N-CH_2-CH_2-CH_2-OCH_3$ produces a black shade. The difference in shade is attributed to a structurally controlled special arrangement of the pigment molecules within the crystal lattice.

3.4.1.4 Application

Most perylene pigments, like quinacridone pigments, are used primarily in high grade industrial paints, especially in original automotive finishes and in automotive refinishes. Different types are available of some pigments, ranging from varieties with very fine particle sizes and high specific surface areas and correspondingly high transparency to versions with coarse particle sizes and low specific surface areas as well as high hiding power. Types with fine particle sizes are used especially in metallic and transparent paints, while opaque types provide full shades, frequently in combination with inorganic or other organic pigments. Some types are particularly suited to use in plastics and in spin dyeing products, in which they demonstrate excellent heat stability. However, perylene pigments are rarely used to color HALS (Hindered Amine Light Stabilizer) stabilized polyolefins, i.e., polyolefins that are stabilized by steric amines. At medium to high pigment concentrations, these stabilizers may be inactivated or even destroyed by exposure to light, rapidly converting the polyolefin system into a brittle material.

3.4.1.5   Commercially Available Perylene Pigments

General

Representative commercial perylene pigments are listed in Table 29:

Tab. 29: Commercially available perylene pigments.

C.I. Name	C.I. Constitution No.	X	Shade
P.R.123	71145	H_5C_2O—⟨◯⟩—N	scarlet to red
P.R.149	71137	H_3C / ⟨◯⟩—N / H_3C	red
P.R.178	71155	⟨◯⟩—N=N—⟨◯⟩—N	red [5]
P.R.179	71130	H_3C–N	red to maroon
P.R.190	71140	H_3CO—⟨◯⟩—N	bluish red
P.R.224	71127	O	bluish red
P.V.29	71129	HN	red to bordeaux
P.Bl.31	71132	⟨◯⟩—CH_2CH_2–N	black
P.Bl.32	71133	H_3CO—⟨◯⟩—CH_2–N	black

Individual Pigments

Pigment Red 123

P.R.123 is not produced in large volume, although it has stimulated more commercial interest than the chemically closely related P.R.190. P.R.123 is primarily used in paints. The pigment provides a medium red shade of much higher cleanness than that of P.R.190. P.R.123 is somewhat duller and slightly yellower than P.R.178. Transparent types are employed especially in shading systems for emulsion paints. P.R.123 offers high weatherfastness. In full shades and similarly deep shades, most grades darken slightly upon exposure to weather. In concentration ranges where high weatherfastness is a primary concern, P.R.123 is replaced by opaque types of the more durable P.R.178. Opaque varieties of P.R.123 are commonly used in combination with other organic or inorganic pigments, including TiO_2, to suppress or cover up this darkening.

P.R.123 shows good resistance to organic solvents. At common baking temperatures, however, oven drying systems containing P.R.123 are not entirely fast to overcoating. Opaque versions of P.R.123 are used not only in industrial paints, including automotive finishes, but also in emulsion paints. Good weatherfastness makes these grades suitable colorants for exterior house paints. P.R.123 is alkali proof.

In plastics, P.R.123 retains its color value up to 220 °C. At higher temperature, the color changes towards bluer and duller shades. No further change is observed, however, between 240 and 300 °C. Incorporated in polyolefins, P.R.123 considerably affects the shrinkage of injection-molded products, which excludes it from being recommended for use in these media. Although not entirely fast to bleeding in plasticized PVC, P.R.123 is used to a certain extent in PVC coatings. Plasticized PVC systems, like other media containing P.R.123, perform much more poorly than systems containing the somewhat yellower and considerably stronger, more heatstable and more lightfast P.R.149, which is a member of the same class of pigments. P.R.123 is also used in spin dyeing. The pigment is even fast to the considerable heat which accompanies the condensation process of polyester spin dyeing (Sec. 1.8.3.8). Besides, P.R.123 is also used to color PUR foam.

Pigment Red 149

P.R.149, a clean, medium to slightly bluish red pigment, is used primarily in the coloration of plastics. With a melting point above 450 °C, the pigment is exceptionally heat stable. Polyolefin systems containing P.R.149 may be processed up to 300 °C. Like other members of its class, P.R.149 affects the shrinkage of injection-molded articles. Its influence, however, decreases with increasing temperature. P.R.149 exhibits high tinctorial strength. Less than 0.15% pigment is needed to formulate 1/3 SD HDPE systems containing 1% TiO_2. For comparison, approximately 20% more of the somewhat bluer P.R.123 are required for the same purpose, although the two pigments have the same average particle size. Like other members of its class, P.R.123 destroys any HALS stabilizers present in plastics by exposure to light. Good heat stability makes the pigment a suitable colorant for polystyrene, impact-resistant polystyrene, ABS, and other plastics which are processed at high temperature. P.R.149, especially in white reductions with TiO_2, is much more heat stable than most quinacridone pigments.

In plasticized PVC, P.R.149 is very fast to migration. It also shows excellent tinctorial strength. 1/3 SD samples in plasticized PVC containing 5% TiO_2 are formulated at only 0.63% pigment. Comparative values are: 0.75% P.R.123, 1.2% of the γ-modification of P.V.19, a quinacridone pigment, and 1.0% of P.R.122.

P.R.149 furnishes high lightfastness. Transparent colorations at 1/25 SD and opaque colorations at 1/3 to 1/25 SD equal step 8 on the Blue Scale for lightfastness. The pigment is not durable enough to be used in long-term exposure. The commercially available products are extremely transparent, a feature which is exploited in a variety of applications.

P.R.149 also lends color to cast resins made from materials such as unsaturated polyester or methacrylic acid methylester, which are polymerized with peroxide catalysts. P.R.149 is equally lightfast in these media. In polycarbonate, the pigment tolerates exposure to more than 320°C. This is an asset in view of the fact that polycarbonate shows high melt viscosity and is thus processed at up to 340°C. The list of applications also includes other media, such as PUR foams and elastomers, for which P.R.149 is recommended because of its good heat stability and its coloristic properties.

Besides, P.R.149 is also used to a considerable extent in the spin dyeing of polyacrylonitrile and polypropylene. Its high lightfastness presents an advantage in these media. In the concentration range between 0.1 and 3%, P.R.149 equals step 7 to step 7–8 on the Blue Scale. High pigment concentrations are to be avoided, however, in systems containing HALS stabilizers. P.R.149 performs perfectly with regard to the more important textile fastnesses, such as fastness to perspiration, dry and wet crocking, and resistance to perchloroethylene and similar solvents. It is often found in polyamide. Partial reduction of the pigment can lead to a change in shade towards a dull brown color. On the fiber, the original color of the pigment is quickly restored through oxidation. P.R.149 is an equally interesting candidate for the spin dyeing of polyester, although it does dissolve at low concentrations and consequently undergoes a color shift towards orange. High thermal stability makes it resistant to the extreme temperatures to which it is exposed during the condensation process (Sec. 1.8.3.8).

The list of media wich are colored with P.R.149 includes not only polymers but also paints and printing inks, but these only to a lesser degree. The pigment is used, for instance, in industrial finishes, where its high transparency is particularly useful in creating transparent and metallic effects in coatings. Not only is P.R.149 a pigment with high tinctorial strength, but it also offers high cleanness of shade and average lightfastness and weatherfastness. Full shades darken upon exposure to weather or light. In baked enamels, P.R.149 is not perfectly fast to overcoating. Special-purpose printing inks are available which are designed for substrates such as PVC films and metal deco prints. Such inks contain P.R.149 because of its fastness to organic solvents and the packaged products, as well as for its high heat stability (up to 220 °C for 10 minutes). Fastness to sterilization is another asset in this type of application.

Pigment Red 178

P.R.178 has been used for some time throughout the paint industry, where it is found especially in industrial finishes, including various types of automotive finishes. Two crystal modifications are known, only one of which possesses some commercial value. Only one version, a highly opaque type, is available. Full shades and similarly deep shades provide good lightfastness and weatherfastness, although some darkening is observed as the samples are exposed. The weatherfastness deteriorates rapidly in systems which are reduced with TiO_2. P.R.178 covers the same range of shades as P.R.123, the shades are only slightly yellower and somewhat duller. Full shades of P.R.178 are more weatherfast than those of P.R.123, but the reverse is true for white reductions. P.R.178 offers good stability to the solvents which are commonly found in oven drying systems. In this respect, it performs approximately like other members of its class. The pigment is not completely fast to overcoating, but is thermally stable in excess of 200 °C.

As a colorant for plastics, P.R.178 exhibits good heat stability. The pigment shows an exceptionally strong effect on the shrinkage of injection-molded polyolefin articles. P.R.178 is less lightfast than the traditionally used, somewhat yellower and noticeably stronger P.R.149. Like other perylene tetracarboxylic acid pigments, P.R.178 is not recommended for use in combination with HALS stabilizers. Its fastness to bleeding in plasticized PVC is very good and is equal to that of P.R.149. The commercial type of P.R.178 provides very opaque colorations, which are not accessible with other pigments covering the same portion of the spectrum.

Pigment Red 179

P.R.179, a dimethylperylimide compound, is probably the most significant member of its class. The pigment is primarily used in industrial coatings, especially for high grade original automotive (O.E.M) finishes and automotive refinishes. Coloristically, the pigment covers two different portions of the spectrum. On the one hand, it provides clean shades of red, with newer types producing comparatively yellowish shades. On the other hand, the pigment also affords maroon and bordeaux shades. Transparent types are available for metallic finishes, sometimes also in combination with other organic or inorganic pigments. These products extend the range of shades provided by quinacridone pigments towards the yellowish portion of the spectrum.

P.R.179 types demonstrate excellent weatherfastness. They are as weatherfast or even more weatherfast than substituted quinacridone pigments. The various types differ considerably in their flop behavior. Some of the commercially available types provide high hiding power and are used especially in combination with

Molybdate Red pigments to produce opaque dark red shades of high gloss and excellent weatherfastness. In paints, the pigments are heat stable up to 180 °C, sometimes even up to 200 °C. They demonstrate excellent fastness towards the organic solvents which are typically found in baking enamels. The systems are accordingly fast to overcoating. P.R.179 types, in contrast to those of P.R.224, are entirely fast to alkali. Dispersibility, fastness to rub-out, and especially resistance to flocculation are frequently less than perfect.

Although they are used to a considerable extent in paints, P.R.179 types have stimulated little interest throughout the plastics industry. The pigments are used in PVC and in PVC plastisols. Besides, P.R.179 is heat stable enough to be recommended for use in polyolefins. Like P.R.149 and other members of this class, P.R.179 undergoes considerable color change if it is used in spin dyeing polyamide. This effect is attributed to the reducing effect of the melt. On the fiber, however, the pigment rapidly returns to its original color.

Pigment Red 190

P.R.190 has a relatively small impact on the market. It is a specialty product for industrial paints, especially for automotive finishes, but offers no advantage over other members of its class. Its shade is dull and referred to as scarlet. In white reductions, the commercial type is very bluish and equally dull. The particle sizes of this product are too coarse for it to be used in metallic finishes. P.R.190 is very durable and very fast to organic solvents and to migration. The pigment is heat stable up to processing temperatures of 200 °C. Despite its good heat stability, the pigment possesses only limited commercial value as a colorant for plastics.

Pigment Red 224

Commercial types of perylene tetracarboxylic dianhydride offer a wide range of coloristic properties in application. The varieties range from highly transparent to very opaque grades. Both versions are used primarily in industrial paints, especially for automotive finishes. The transparent types are particularly interesting for metallic finishes. Since its chemical constitution renders the pigment sensitive to alkali, it fails to satisfy the stringent demands of the standardized test (Sec. 1.6.2.2). Other tests show that P.R.190 is not suited to single-layer metallic finishes. It is thus used primarily in base coat/clear coat metallic systems. The pigmented base coat, in Europe commonly a polyester lacquer, is protected against alkali, such as in car wash detergents, by a clear acrylic resin based varnish.

Commercial types exhibit a surprisingly wide range of shades. A new version has been introduced recently which, although performing like other versions of P.R.224 in terms of weatherfastness, solvent fastness, and fastness to overcoating, is substantially yellower and noticeably cleaner than traditional types. Opaque varieties also provide brilliant shades of red. P.R.224 is used primarily in full shades, sometimes in combination with other organic or inorganic red pigments, such as Molybdate Red.

Besides, P.R.224 types are found in a number of other high grade industrial finishes. The pigments exhibit high tinctorial strength, which makes them suitable shading components for inorganic pigments. The resulting brilliant shades of red are considerably more yellowish than those furnished by quinacridone pigments. Because of the risk of chemical interaction, the pigment is not recommended for use in amine hardening binders. The same is true for exterior house paints which are applied onto stucco plaster. However, P.R.224 is very fast to organic solvents and overcoating. It is thermally stable up to 200 °C.

Apart from paints, P.R.224 is also used in polyacrylonitrile spin dyeing. Application in the spin dyeing of polypropylene is compromised by the fact that medium to high pigment concentrations accelerate the degradative action of light on HALS stabilizers (Sec. 3.4.1.4).

Pigment Violet 29

P.V.29 types demonstrate excellent weatherfastness, much more so than other perylene pigments. Their impact on the market, however, is limited by their very dull shade of maroon. Full shades are deep brown, almost black. The pigment is very fast to organic solvents and overcoating. Commercial types are utilized especially in metallic finishes. Full shades frequently bronze upon exposure to weather.

P.V.29 is also highly heat stable, which makes it a suitable colorant for plastics which are processed at high temperature. However, this advantage is somewhat compromised by its lack of cleanness.

P.V.29 is also used in polyester spin dyeing. It meets the high thermal demands of the condensation process, during which the pigment is exposed to 290 °C for 5 to 6 hours. At 1/3 and 1/9 SD, the systems equal step 7 to step 8 on the Blue Scale for lightfastness. P.V.29 performs excellently in terms of the more important textile fastnesses, such as fastness to wet and dry crocking.

3.4.1.6   Various Other Perylene Tetracarboxylic Acid Pigments

As a class, these compounds also include other pigments which although registered in the Colour Index have not yet been found on the market. In full shade, these pigments provide shades of black or very dark olive or violet hues, which satisfy certain spectral requirements in the IR region. Examples are Pigment Black 31,71132 and 32,71133. Their properties include resistance to certain chemicals and good weatherfastness.

3.4.2   Perinone Pigments

Section 3.3 also applies to perinone pigments. The earliest commercial product in this group was a yellowish red vat dye which had been discovered in 1924 by Eckert and Greune at Hoechst AG. The original product was a mixture of the cis and trans isomers; however, it later became possible to separate the two isomers into the clean orange trans isomer and the dull red cis isomer. Both were used for a long time exclusively as vat dyes for cotton. It was not until 1950 that these compounds found recognition as pigments.

3.4.2.1 Preparation of the Starting Materials

Perinone pigments are obtained from naphthalene-1,4,5,8-tetracarboxylic acid or its monoanhydride.

The acid, referred to as "tetra acid", is prepared as follows:

In a Friedel–Crafts reaction, acenaphthene **72** is reacted with malonic dinitrile and aluminum chloride. The resulting condensation product **75** is oxidized with sodium chlorate/hydrochloric acid to form the dichloroacenaphthindandione **76**. Oxidation with sodium hypochlorite solution/sodium permanganate affords naphthalene tetracarboxylic acid **68**, mostly existing as the monoanhydride **68a**. The dianhydride, on the other hand, evolves only after drying at approx. 150 °C.

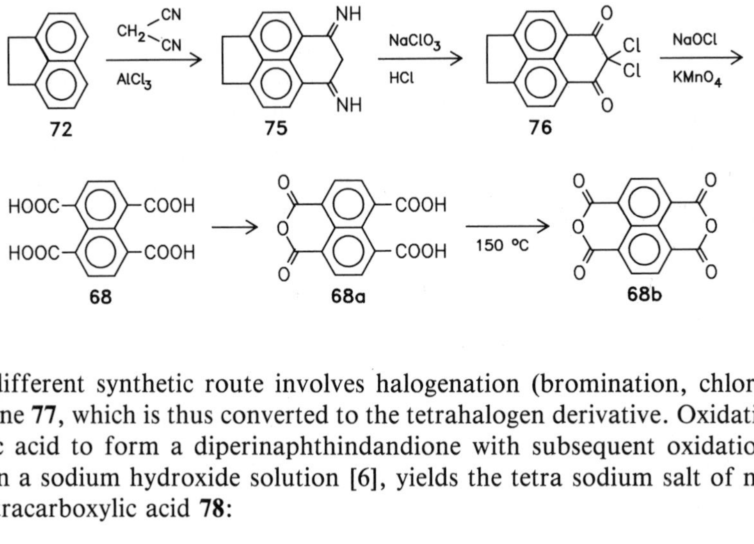

A different synthetic route involves halogenation (bromination, chlorination) of pyrene **77**, which is thus converted to the tetrahalogen derivative. Oxidation with sulfuric acid to form a diperinaphthindandione with subsequent oxidation, once again in a sodium hydroxide solution [6], yields the tetra sodium salt of naphthalene tetracarboxylic acid **78**:

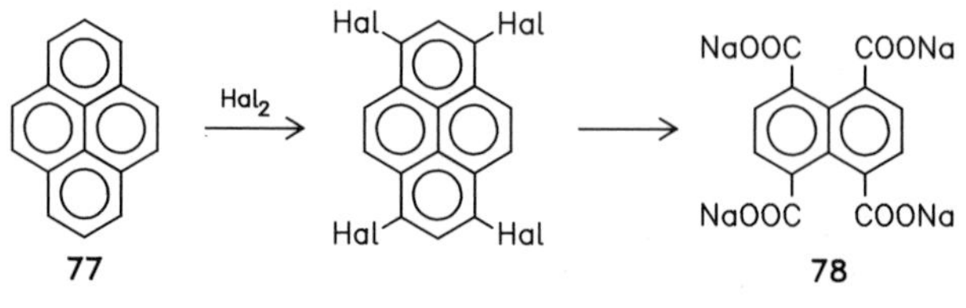

3.4.2.2 Chemistry, Manufacture of the Pigments

The synthetic route to perinone pigments, as to perylene pigments, starts from an anhydride, in this case from the monoanhydride of naphthalene tetracarboxylic acid.

The perinone ring system is thus formed by condensation with aromatic o-diamines. Reaction of o-phenylenediamine with the monoanhydride is typically achieved in glacial acetic acid at 120 °C. A mixture of the cis and trans isomers evolves, which appear as mixed crystals:

cis   trans

The isomers are separated by taking advantage of the solubility characteristics of their respective salts. The trans isomer is precipitated as a sparingly soluble colorless potassium hydroxide addition compound by heating the mixture in ethanol/potassium hydroxide. Satisfactory results are also achieved through fractionation with concentrated sulfuric acid. Either one of these techniques is followed by traditional methods of converting the product into a commercially useful pigment. The options include milling, acid treatment, and solvent treatment at elevated temperature.

3.4.2.3 Properties

Perinone pigments perform very much like perylene pigments. They exhibit shades in the range from orange to bordeaux. Only two members of this group, however, have gained commercial importance. Perinone pigments demonstrate high heat stability and are very lightfast and weatherfast.

3.4.2.4 Commercially Available Perinone Pigments and Their Application

At present, only the two isomers which constitute the original cis/trans mixture (Vat Red 74) and, to a limited extent, also the mixture itself are commercially used as pigments.

Tab. 30: Commercially available perinone pigments.

C.I. Name	C.I. Consti- tution No.	Structure (see p. 477)	Shade
P.O.43	71105	trans isomer	reddish orange
P.R.194	71100	cis isomer	bluish red
Vat Red 74		isomer mixture of P.O.43 and P.R.194	scarlet

Pigment Orange 43

P.O.43, the trans isomer, affords a clean reddish shade of orange. Of the two iso-mers, this is the more important one in its role as a pigment. High cost limits its use to areas which involve high fastness requirements. This is especially true as a col-orant for spin dyeing and plastic products, which is the main area of application for P.O.43. Its high lightfastness and weatherfastness is an asset especially in po-lyacrylonitrile spin dyeing. The pigments are used to color awnings, tents, can-vasses, and other items. Samples formulated at 0.3 to 3% pigment are rated step 7–8 on the Blue Scale for lightfastness, while 0.1% specimen match step 7. The pigment meets high standards in terms of other important fastness properties, such as resistance to dry and wet crocking as well as fastness to chlorohydrocarbons. In the spin dyeing of polyester, P.O.43 withstands the harsh conditions of the conden-sation process (Sec. 1.8.3.8). The pigment is also used for the spin dyeing of cellu-lose acetate and polypropylene fibers and threads. It also lends itself to textile printing.

P.O.43 exhibits good fastness to bleeding in plasticized PVC. It should be noted, however, that bleeding may be observed at extremely low pigment concen-trations and at a high plasticizer content. Although P.O.43 is excellently fast to organic solvents and plasticizers; high processing temperatures and low pigment concentrations may lead to blooming (Sec. 1.6.3.1). In plasticized PVC, like in oth-er media, P.O.43 offers average tinctorial strength. Approximately 0.9% pigment is needed, for instance, to afford 1/3 SD samples containing 5% TiO_2. P.O.43 demonstrates excellent lightfastness, even more reduced shades equal step 8 on the Blue Scale. Deep shades, however, darken to a certain extent. No such darkening is observed in corresponding transparent colorations. P.O.43 is also very durable, al-though it does not satisfy the requirements for long-term exposure.

Transparent and reduced 1/3 SD polyolefin systems are stable up to 300 to 320 °C. 0.25% pigment is required in order to formulate 1/3 SD HDPE samples containing 1% TiO_2. P.O.43, incorporated in injection-molded partially crystalline plastics such as HDPE, affects the shrinkage of the medium, which precludes its use in nonrotation symmetrical injection-molded articles (Sec. 1.6.4.3).

The lightfastness of transparent systems matches step 7 to 8 on the Blue Scale, depending on the type of plastic and on the pigment content. In white reductions, at 1/3 to 1/100 SD, the pigment scores between step 8 and step 6.

P.O.43 is useful for transparent colorations of polystyrene. At low concentrations (below 0.1%), the pigment dissolves in thermoplastic polyester to produce yellow shades. Even in its dissolved state, the pigment is very lightfast. P.O.43 is inert to peroxides which are typically used to polymerize cast resins based on methacrylic acid methylesters or unsaturated polyesters. The pigment does not affect the hardening of these polymers.

The paint industry uses P.O.43 primarily as a shading pigment or in white reductions, together with large amounts of TiO_2. The pigment is also frequently combined with Molybdate Red pigments. Full shades and similarly deep shades, such as 1:10 reductions with TiO_2, darken by exposure to light. Considerably high concentrations of TiO_2, however, still show excellent lightfastness. Although they do darken to a certain extent upon exposure to weather, this effect is not so noticeable as to affect the applicability of the pigment.

P.O.43 is used especially in industrial paints, including original automotive (O.E.M) and automotive refinishes, although its significance is somewhat declining. The pigment is used to create metallic effects in paints, especially shades of copper, which are produced by combination with aluminum. There is a certain disadvantage to the fact that the pigment darkens by weathering, even in the presence of UV absorbants. Its heat stability satisfies all types of requirements throughout the paint industry. The pigment is very fast to overcoating and fast to acid, alkali, and lime. P.O.43 is also found in emulsion paints. Its excellent weatherfastness makes it a suitable colorant for exterior house paints based on dispersions of synthetic resins. It is mostly used in very strong white reductions.

The printing inks industry uses P.O.43 to color various types of special-purpose printing inks. The pigment exhibits good heat stability (220°C for 30 minutes). Prints containing P.O.43 are fast to sterilization and to a large number of organic solvents and chemicals, as well as to packaged goods. 1/3 SD letterpress proof prints equal step 6 on the Blue Scale for lightfastness, although the shade becomes slightly dull by exposure to light.

Besides, P.O.43 is found in a variety of other media. In wood stains, P.O.43 is frequently used in combination with yellow pigments such as P.Y.83 and carbon black to produce shades of brown. It is also found in leather seasoning systems.

Pigment Red 194

P.R.194, which is the corresponding cis isomer, has much less of a commercial impact than the trans isomer. Its shade is a very dull, bluish red. The pigment is mainly used in paints, especially in architectural paints. It is very lightfast and weatherfast. In contrast to the trans isomer, P.R.194 does not darken by exposure to light and weather, not even in the full shade range. Although its stability to organic solvents is high, the pigment is not entirely fast to overcoating. This limits its use in certain areas, such as in automotive finishes. P.R.194 is also found in emulsion paints, especially since its excellent weatherfastness makes it a useful product for exterior application. The pigment is fast to acid, alkali, and lime.

P.R.194 is also used in textile printing, for instance in combination with carbon black, to produce shades of brown. It is both very lightfast and durable in these media. The coloration equal step 7 on the 8 step weatherfastness scale [7].

P.R.194 is less important in plastics. It shows some bleeding in plasticized PVC. Moreover, the pigment frequently dissolves in its medium as the temperature is increased, even in rigid PVC. It thus exhibits a temperature-dependent color shift. P.R.194 affords an orange shade as it dissolves in polystyrene, a fact which may present problems in matching defined colors within a certain concentration/temperature range. In polyolefins, the pigment is heat stable up to 270 °C. In some countries, P.R.194 is also an important colorant for polypropylene spin dyeing. Dispersed in polyester during spin dyeing, P.R.194 initially dissolves but recrystallizes as the fiber is thermoset, a process which is associated with an appreciable color change.

Vat Red 74

The mixture of P.O.43 and P.R.194, which is a mixed phase comprising both isomers, is also used as a pigment. Its main market is in spin dyeing, especially of polypropylene. Almost as fast as P.O.43, its color is referred to as scarlet. The product is thermally stable up to 300 °C.

References for Section 3.4

[1] DE-AS 1 071 280 (Sandoz) 1957.
[2] DE-OS 3 008 420 (Hoechst) 1980; H. Tröster, Dyes and Pigments 4 (1983) 171–177.
[3] DE-OS 2 545 701 (BASF) 1975; DE-OS 2 546 266 (BASF) 1975.
[4] G. Graser and E. Hädicke, Liebigs Ann. Chem. 1980, 1994–2011.
[5] B.K. Manukian and A. Mangini, Helv. Chim. Acta 54 (1971) 2924.
[6] DE-PS 602 445 (IG Farben) 1932; Friedländer XX, 1433.
[7] DIN 54 071: Bestimmung der Wetterechtheit von Färbungen und Drucken in Apparaten (künstliche Bewetterung, Xenonbogenlicht).

3.5   Thioindigo Pigments

As a class, these compounds are related to indigo, which is the oldest of all vat dyes. Indigo itself, during its comparatively short history as a commercial pigment, was used especially in rubber. Its structure was first described in 1883 by A.V. Bayer. Knowledge of this structure, together with the development of an industrial scale synthesis, facilitated the development of a series of indigo-based colorants.

The parent structure of all thioindigo pigments is the unsubstituted thioindigo molecule (Friedländer, 1905), which is of little technical importance. More important products are chlorinated and/or methylated thioindigo derivatives, which were developed in subsequent years. During the 1950s, several thioindigo derivatives gained commercial recognition as pigments. This was only after it became possible to appropriately finish the crude products after manufacture (Sec. 3.3). Manufacture of most of these pigments has been discontinued.

3.5.1   Chemistry, Manufacture

Thioindigo pigments have the following general structure:

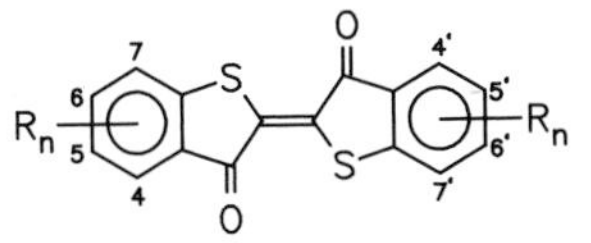

In commercially important products, R typically represents a chlorine atom or a methyl group, and n stands for 2 or 3. The compounds assume the trans structure.

The synthesis is performed in two steps [1]:
- attaching the 5-membered ring to the benzene ring, and
- oxidatively linking two molecules of the resulting thionaphthenone-3.

The most important route involves cyclization of appropriately substituted phenylthioglycolic acids:

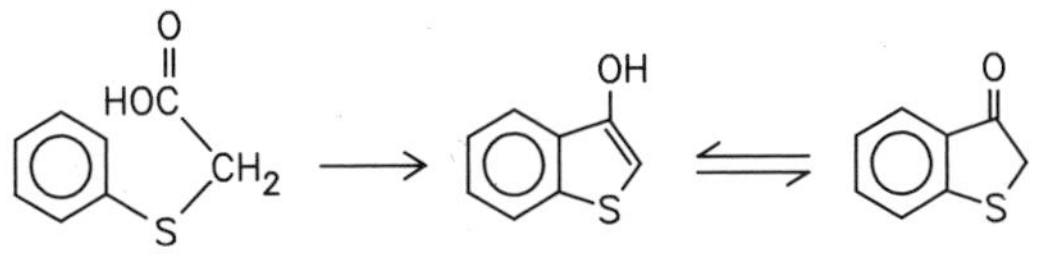

Suitable agents are chlorosulfonic acid, monohydrate, or concentrated sulfuric acid. Other methods include cyclization of phenylthioglycolic acid chlorides, which is afforded by Friedel–Crafts reaction with aluminum chloride, possibly in the presence of sulfuric acid/sodium chloride. The acid chloride is obtained from phenylthioglycolic acid with thionyl chloride or phosphorus trichloride.

There are more routes to cyclization. Phenylmercaptoacetic acid may be replaced by its o-carboxy or o-amino derivative to form the thionaphthenone-3. The first reaction is performed in molten alkali, the second proceeds via diazotization, Sandmeyer reaction with sodiumcyanide/copper cyanide, alkaline hydrolysis, and acid treatment.

Oxidation may be achieved in the presence of oxygen or air. Other suitable oxidants include sulfur, sodium polysulfide, iron (III) chloride, potassium ferrocyanide (III) or potassium dichromate, peroxydisulfate or salts of aromatic nitrosulfonic acids. An aqueous/alkaline medium is used in the presence of a high boiling organic solvent which is not miscible with water or which is almost immiscible with water. Cyclization with chlorosulfonic acid can be followed directly by oxidation with bromine to afford the thioindigo system, without separation of the intermediate.

The synthetic route to tetrachlorothioindigo may serve as an example for an industrial scale synthesis. Starting material is 2,5-dichlorothiophenol, which is obtained by reduction of 2,5-dichlorobenzene sulfochloride or by reaction of the 2,5-dichlorobenzene diazonium salt with potassium-o-ethyldithiocarbonate and hydrolysis or with disodium sulfide and subsequent reduction:

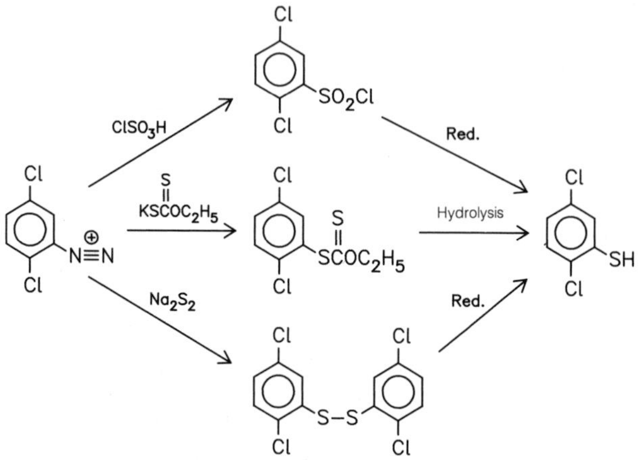

2,5-Dichlorothiophenol is treated with chloroacetic acid to form 2,5-dichlorophenylmercaptoacetic acid. Cyclization followed by oxidation with chlorosulfonic acid at 35 °C affords 4,7-dichlorothionaphthenone-3. Direct addition of bromine in chlorosulfonic acid without separating the intermediate yields the 4,4',7,7'-tetrachlorothioindigo derivative through oxidative dimerization. It is also possible to separate the monoheterocycle and to oxidize it with oxygen in an alkaline medium.

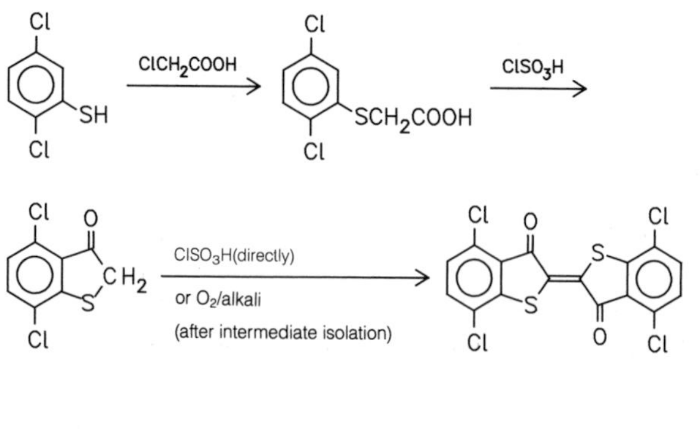

Aftertreatment

Tetrachlorothioindigo, by far the most important member of this pigment series, will serve as a frame of reference.

Although there is one route [2] which describes the direct synthesis of the tetrachlorothioindigo pigment by oxidation of 3-hydroxy-4,7-dichlorothionaphthenone with oxygen in an aqueous alkaline medium, this is somewhat of an exception. In most cases, it is necessary to modify the crude thioindigo derivative by appropriate aftertreatment in order to develop the desired pigment properties.

The list of options includes milling the crude product with salt or dispersing agents or reprecipitation from sulfuric acid or chlorosulfonic acid, followed by aftertreatment with organic solvents [3]. Tinctorially strong transparent pigment grades, for instance, are obtained by milling the crude pigment suspension in the presence of an aqueous base [4]. The same result is achieved by oxidizing the leuco form of tetrachlorothioindigo in the presence of sodium dithionite with air while applying shearing forces (for instance in a pearl mill) [5].

3.5.2 Properties

Thioindigo pigments provide a range of hues from red violet and maroon to brown shades. The type and position of the substituents affects the shade:

Electron donating substituents (methyl groups) in 5-position cause the most drastic bathochromic shift, which decreases in the following order: 4- > 7- > 6-position. The reverse is also true in the case of electron acceptors (such as chlorine): they cause more of a bathochromic shift in 6-position than in 5-position.

As a class, tetrachlorothioindigo pigments exhibit good to excellent lightfastness and weatherfastness as well as solvent stability and migration fastness. 4,4',7,7'-substituted derivatives demonstrate highest fastness to solvents. Chlorine substitution has a better effect in this respect than methyl substitution. Shifting only one of the substituents on 4,4',7,7'-tetrachlorothioindigo, for instance, will adversely affect the migration resistance of the parent compound (fastness to bleeding).

3.5.3 Commercially Available Types and Their Application

General

Of the two remaining thioindigo pigments, only Pigment Red 88, the 4,4',7,7'-tetrachlorothioindigo, has been able to maintain interest throughout the pigments industry. It is broad in scope. The other type, Pigment Red 181, is a special-purpose type for a limited number of applications. Table 31 lists the chemical constitutions of both pigments.

Tab. 31: Commercially available thioindigo pigments.

C.I. Name	C.I. Constitution No.	R^4	R^5	R^6	R^7	Shade
P.R.88	73312	Cl	H	H	Cl	red-violet
P.R.181	73360	CH_3	H	Cl	H	bluish red

Unsubstituted thioindigo, which is marketed as Vat Red 41, 73300, is used in rigid PVC, polystyrene, and a number of other plastics. Dissolved in its medium, the pigment affords fluorescent bluish red shades.

Unsubstituted indigo is marketed as Pigment Blue 66, 73000. It is a suitable pigment for spin dyeing of synthetic viscose rayon and spun rayon, to which it lends navy blue shades. The pigment provides good lightfastness: the samples match step 5–6 to step 7 on the Blue Scale, depending on the standard depth of shade. However, Pigment Blue 66 performs poorly regarding a number of textile fastnesses, especially in terms of bleaching with chlorite, and vat resistance.

Individual Pigments

Pigment Red 88

P.R.88 has stimulated considerable interest throughout the paint industry. It is a suitable colorant especially for industrial coatings, including automotive (O.E.M) finishes and automotive refinishes. Its red-violet shade is used frequently to produce opaque dark shades of red, bordeaux, and maroon, for which purpose the pigment is typically used in combination with inorganic pigments, such as Iron Oxide Red or Molybdate Red. P.R.88 is both very lightfast and weatherfast, although in white reductions and as a shading pigment it fails to reach the standards of P.V.19, the β-modification of unsubstituted quinacridone, which covers the same range of shades. Moreover, β-P.V.19 performs much better in pigment combinations than P.R.88, and it provides much cleaner shades in such combinations.

Deep shades of maroon tend to form water spots in certain binder systems, especially in media which are based on acrylic resin. More or less distinctive light spots appear on the coating. The effects that cause this phenomenon remain to be elucidated. Factors such as long-term weathering at elevated temperature, UV radiation, and the presence of demineralized water probably cause reduction and solvation effects within the coating. Products are available which are much less susceptible to these agents. Rub-out effects, especially flocculation, may also present problems in various binder systems. Special-purpose grades are therefore available which are more stable to flocculation.

P.R.88 shows good stability to organic solvents. Its fastness to overcoating is good, but not reliable at baking temperatures above 140 °C. Paints containing P.R.88 are acid and alkali resistant.

P.R.88 is also found in other systems throughout the paint industry, such as in various types of air drying systems and powder coatings.

P.R.88 is one of the standard red-violet pigments for use in plastics. It is particularly suitable for use in PVC, including PVC plastisols and PUR coatings. The pigment shows a more or less pronounced tendency to migrate, depending on the type. New types have been introduced recently which are much faster to bleeding. As a pigment for plasticized PVC, P.R.88 is in direct competition with Pigment Violet 32, which affords a very similar shade, has the advantage of being fast to bleeding, and demonstrates considerably higher tinctorial strength. In light tint, however, P.V.32 is somewhat less lightfast than P.R.88. P.R.88 shows excellent lightfastness: transparent and white reductions at 1/1 to 1/25 SD equal step 8 to step 6 on the Blue Scale, depending on the method of stabilization. The pigment is also very weatherfast, but fails to satisfy the requirements for long-term exposure.

Transparent polyolefin systems containing P.R.88 are stable up to 260 to 300 °C, depending on the type and on the pigment content. 1/3 SD systems are heat stable up to approximately 240 to 260 °C. Some types are recommended only for use in LDPE at low processing temperatures. The lightfastness of such specimens is between step 6 and step 7 on the Blue Scale. P.R.88 considerably affects the shrinkage of injection-molded articles, a feature which somewhat restricts its application in such systems.

P.R.88 is recommended for use in PP spin dyeing, special-purpose grades are used for polyacrylonitrile spin dyeing. The resulting fibers are completely fast to perspiration, dry and wet crocking, perchloroethylene, and other important solvents, as well as to dry heat fixation. In the concentration range between 0.1 and 3.0%, the pigment equals step 7 and step 7–8 on the Blue Scale, respectively. P.R.88 meets the performance standards for use in canvasses, tents, and similar articles. Used in the spin dyeing of polyester, the pigment dissolves in the fiber. Both sublimation and migration fastness are excellent in this medium. P.R.88 is not always suitable for use in polystyrene. Depending on the temperature, the pigment dissolves in this polymer and accordingly changes its color and fastness properties. This presents major difficulties in trying to match a defined shade.

Incorporated in methacrylate and unsaturated polyester cast resins, P.R.88 not only withstands several hours of thermal exposure during processing but is also resistant to the peroxides which are used as catalysts. Some types accelerate the polymerization process, i.e., the hardening of the plastic.

P.R.88 is also used for printing inks. Its red-violet shade is especially used for printing inks which are to be targeted for packaging, posters, and other special purposes. The prints are entirely fast to organic solvents, plasticizers, and packaged goods, such as butter and soap. P.R.88 prints are fast to alkali and acid, heat stable up to 200 °C, and fast to sterilization.

The list of applications also includes decorative printing inks for laminated plastic sheets based on melamine or polyester resin. The required fastness and performance properties are very good: the pigment does not, for instance, color the clear sheets. 8% gravure prints (20 μm cell depth) equal step 8 on the Blue Scale for lightfastness.

P.R.88 is also used in wood stains, for instance in combination with yellow pigments, such as P.Y.83, or carbon black to produce shades of brown.

Pigment Red 181

P.R.181 is a special-purpose product for polystyrene and similar polymers (Sec. 1.8.3.3). P.R.181 largely dissolves in its medium at the high temperatures at which these polymers are processed. The pigment affords brilliant, bluish shades of red. It possesses excellent lightfastness and satisfies the application requirements.

Another important market is in toothpaste, for which the pigment has been approved worldwide. It also lends color to lipstick and other decorative cosmetics. In the US, it is registered by FDA as D&C Red 30, in Germany as C-Rot 28 according to the DFG Catalog for Cosmetic Colorants.

References for Section 3.5

[1] H. Zollinger, Color Chemistry, 2. Ed., VCH-Verlag 1991.
[2] DE-OS 2 457 703 (Bayer) 1974.
[3] DE-OS 2 504 962 (Hoechst) 1975.
[4] DE-OS 2 043 820 (Hoechst) 1975.
[5] DE-OS 2 916 400 (BASF) 1979.

3.6   Various Polycyclic Pigments

The term polycyclic pigments refers to pigments which are derived from the anthraquinone skeleton, either by chemical structure or by synthesis. The complex fused-ring systems which are discussed in this context are all at least remotely related to the parent anthraquinone structure.

This chapter classifies polycyclic pigments by chemical constitution. The resulting classes include aminoanthraquinones, hydroxyanthraquinones, heterocyclic and polycarbocyclic anthraquinone pigments.

With the exception of the anthraquinone-azo series and the salts of hydroxyanthraquinone sulfonic acids, all of the listed compounds have known a long history as vat dyes before their pigment properties gained commercial recognition (Sec. 3.3).

3.6.1   Aminoanthraquinone Pigments

As a group, these pigments include derivatives of 1-aminoanthraquinone, which is the only derivative of the possible parent compounds which has stimulated commercial interest. While the synthetic routes to the pigment derivatives are found under the respective pigments, the pathway for 1-aminoanthraquinone will be described separately.

1-Aminoanthraquinone

The traditionally most important route to 1-aminoanthraquinone [1] proceeds via nucleophilic exchange of anthraquinone-1-sulfonic acid or 1-chloroanthraquinone with ammonia. Replacing ammonia by amines affords the corresponding alkyl or arylaminoanthraquinones.

Sulfonation of anthraquinone to form the 1-sulfonic acid is achieved at approximately 120 °C with 20% oleum in the presence of mercury or a mercury salt as a catalyst [2]. Without this catalyst, the reaction produces the 2-sulfonic acid. Exchange with aqueous ammonia (30%) at about 175 °C under pressure converts the potassium salt of 1-sulfonic acid to 1-aminoanthraquinone in 70 to 80% yield. To avoid sulfite formation, the reaction is performed in the presence of an oxidant, such as m-nitrobenzosulfonic acid, which destroys sulfite.

A less important pathway for 1-aminoanthraquinone involves replacing the chlorine atom of 1-chloroanthraquinone by an amino group. The displacement is achieved with an excess amount of aqueous ammonia at about 200 to 250 °C in the presence of acid-binding agents.

Nitration of anthraquinone with a stoichiometric amount of nitric acid in sulfuric acid affords 1-nitroanthraquinone.

This reaction, which has been known for approximately 100 years, has recently been reinvestigated extensively in an attempt to furnish a purer reaction product and to improve the yield. Interrupting the nitration at approximately 80% completion by removing 1-nitroanthraquinone by distillation makes it possible to separate the side products, which remain in the distillation residue.

The subsequent reaction of the pure 1-nitroanthraquinone to 1-aminoanthraquinone used to be carried out mainly by reaction with sodium sulfides in an alkaline medium and is now performed by nucleophilic replacement of the nitro group with ammonia in organic solvents, affording up to 98% yield. The overall reaction affords approximately 70% yield, in comparison to the roughly 50% yielded by the "classical method" which proceeds via the 1-sulfonic acid. Moreover, the newer method is also superior to the old one because recyclization of the solvent makes it ecologically more attractive.

3.6.1.1 Anthraquinone-Azo Pigments

1-Aminoanthraquinone has only little importance as a diazo component for corresponding azo pigments. Diazotization of 1-aminoanthraquinone with subsequent coupling onto different coupling components, which, especially by heterocyclic groups with additional carbonamide groups, leads to highly insoluble, migration fast pigments, has been described in several patents. This synthesis produces yellow to red monoazo pigments, which are recommended primarily for use in plastics.

   1-Aminoanthraquinone is diazotized advantageously with nitrosylsulfuric acid after being dissolved in sulfuric acid. Coupling the diazo compound onto barbituric acid, for instance, affords a yellow pigment [3] with the structure **79**:

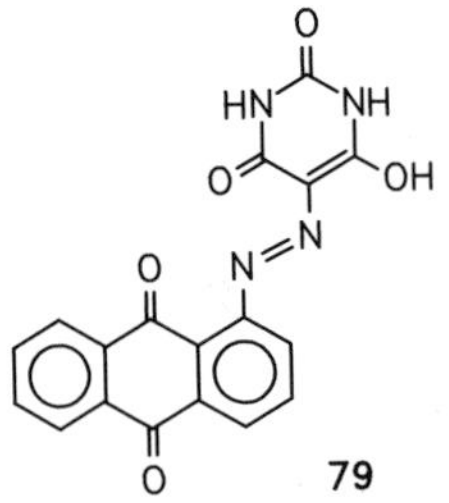

Coupling 1-aminoanthraquinone with certain Naphthol AS derivatives produces red pigments [4] with the structure **80**, which are suitable colorants for plastics:

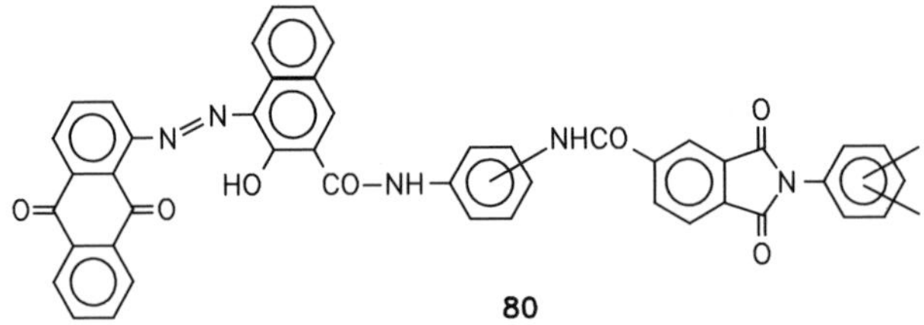

   The compounds are obtained by coupling diazotized 1-aminoanthraquinone onto 2-hydroxy-3-naphthoic acid, followed by separation, drying, and conversion into the azo dye acid chloride. Condensation with amines (structure **81)** is achieved in an aprotic organic solvent.

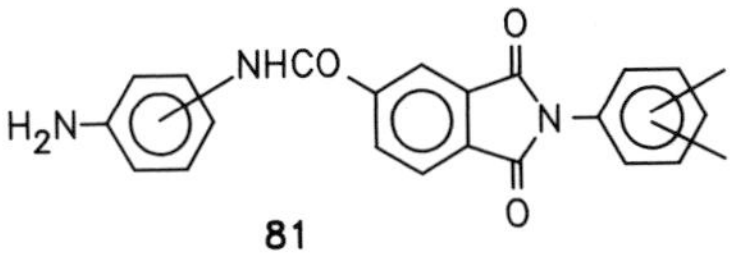

81

Coupling 1-aminoanthraquinone onto pyrazolo-(5,1b)-quinazolone with the general structure

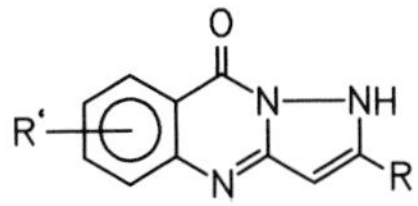

primarily affords red pigments with very good fastness properties. R represents a methyl or phenyl group and R' stands for various substituents, but primarily for hydrogen, chlorine, or bromine. This point is illustrated by Pigment Red 251, 12925 [5] as follows (see p. 561):

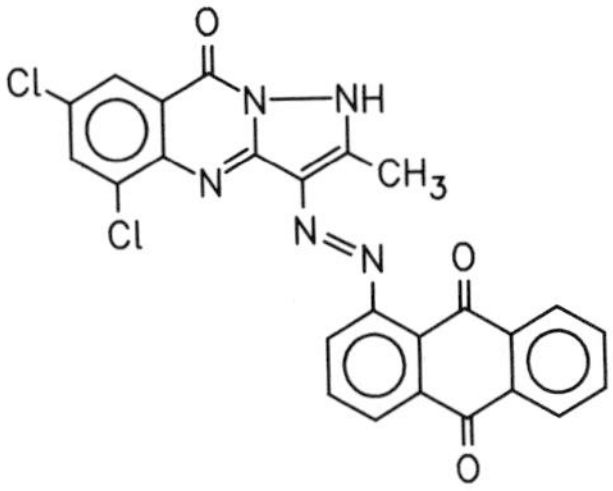

Other anthraquinone/azo pigments are obtained by coupling 1-aminoanthra-quinone derivatives onto bis-quinazolinone methanes with the following chemical structure **82**:

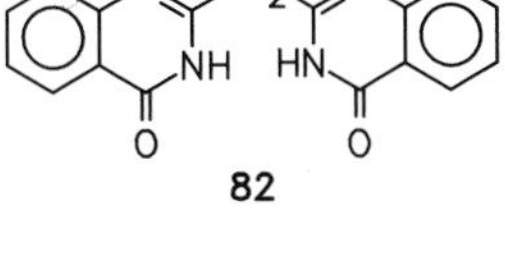

82

82 is prepared through condensation of possibly substituted anthranilic acid amides with malonic diethyl esters in the presence of pyridine. Diazotization of 1-amino-5-benzoylaminoanthraquinone, which is then coupled onto **82**, for instance, affords an orange pigment with the structure **83** [6]:

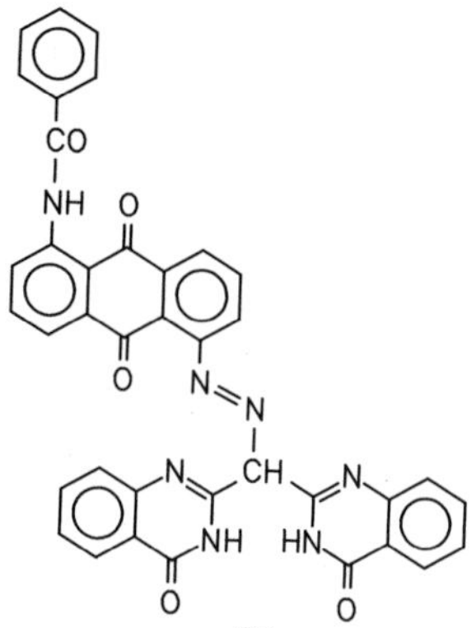

83

3.6.1.2 Other Aminoanthraquinone Pigments

As a class, these pigments include 1-aminoanthraquinone compounds which bear free amino groups, and also derivatives in which the primary amino group is substituted by an aryl or a heteroaryl moiety. Only very few representatives are industrially important.

Formation of C—C links in the anthraquinone molecule, especially substitutions with aryl moieties, proceed via copper-catalyzed nucleophilic exchange of halogenated anthraquinone compounds. These methods include dimerization of 1-aminoanthraquinone.

Pigment Red 177, a red pigment which is listed in the Colour Index under Constitution No. 65300, has the chemical structure of 4,4'-diamino-1,1'-dianthraquinonyl (**84**) [7]:

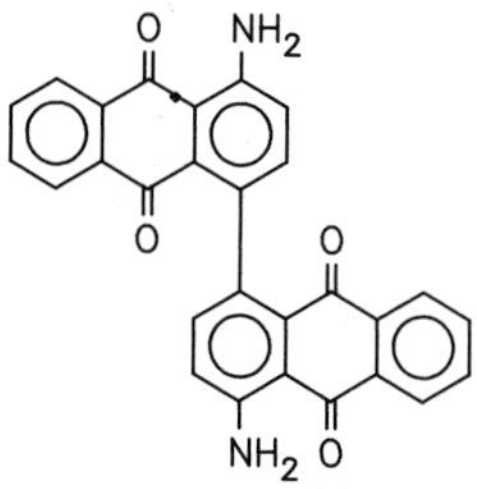

84

Simultaneously with Hansa Yellow G, compound **84** was first described as early as 1909 by Farbenfabriken Bayer. Its preparation starts from 1-amino-4-bromo-anthraquinone-2-sulfonic acid (bromamine acid) **85**. Dimerization is achieved through the Ullmann reaction, i.e., treatment with fine-grain copper powder in dilute sulfuric acid at 75 °C. The separated intermediate, the disodium salt of 4,4'-diamino-1,1'-dianthraquinonyl-3,3'-disulfonic acid **86**, is heated to 135 to 140 °C in the presence of 80% sulfuric acid in order to cleave the sulfonic acid groups [8]:

The result is a bluish red pigment, the only one among the commercial products which possesses free amino groups.

X-ray diffraction analysis of P.R.177 disclosed a twisting of the two anthraquinone units by 75° relative to each other. This allows optimum formation of intermolecular hydrogen bonds [8a].

Acylation of 1-aminoanthraquinone affords 1-acylaminoanthraquinones. The carbonamide function renders these compounds very fast to migration. The migration fastness may be improved further by introducing substituents which carry additional carbonamide groups, or by molecular dimerization using aromatic dichlorides.

The most important route to 1-acylaminoanthraquinones involves reaction of 1-aminoanthraquinone with acid chlorides in an organic solvent. Reaction of 1-aminoanthraquinone with benzoylchloride in nitrobenzene at 100 to 150 °C affords 1-benzoylaminoanthraquinone, a yellow pigment which is registered as Colour Index Constitution No. 60515. The reaction may also be performed in the presence of a tertiary amine, which acts as a proton acceptor:

Benzoylation of 1-amino-4-hydroxyanthraquinone affords a bluish red pigment, which is known as Pigment Red 89, 60745:

1-Amino-4-hydroxyanthraquinone may be obtained from 1-nitro-4-hydroxyanthraquinone by reduction with sodium sulfide.

Condensation of 1-aminoanthraquinone with phthalic acid chloride in o-dichlorobenzene at 145 °C also yields a yellow pigment, registered as Pigment Yellow 123, 65049 [9]:

It is important at this point to mention a structural isomer of P.Y.123, Pigment Yellow 193, which has recently been added to the Colour Index. P.Y.193 provides a reddish yellow shade.

The three pigments Pigment Red 89, Pigment Yellow 193, and 123, however, currently lack significant commercial use.

Among heterocyclic substituted 1-aminoanthraquinone derivatives, the 2:1 reaction product with 1-phenyl-2,4,6-triazine, also referred to as Pigment Yellow 147, 60645, should be mentioned as an example [10]. It is a reddish yellow pigment with the chemical structure **87**:

87

This pigment is prepared by condensation of two equivalents of 1-aminoanthraquinone with 1-phenyl-3,5-dichloro-2,4,6-triazine in the presence of a base in an organic medium. It is also possible and even preferable to treat one equivalent of phenylguanamine with two equivalents of 1-halogenanthraquinone in the presence of an additional tertiary base with copper(I)iodide as a catalyst. In contrast to copper iodide itself, the corresponding addition compounds have the advantage of dissolving easily in organic solvents, which facilitates their separation.

Phenylguanamine (2,4-diamino-6-phenyl-1,3,5-triazine) is synthesized from benzonitrile and dicyanamide.

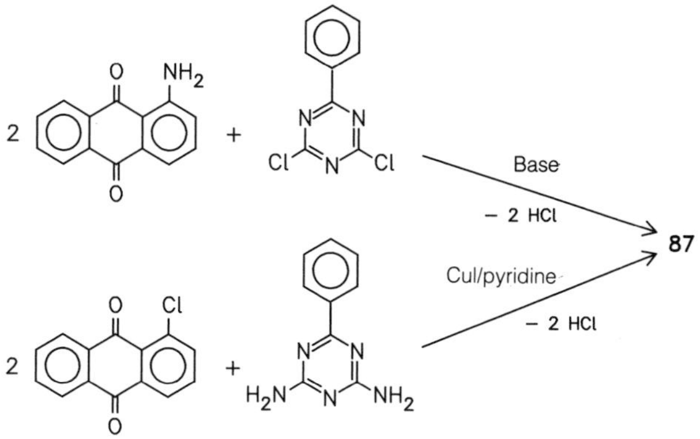

3.6.1.3 Properties and Application

Pigment Yellow 147

P.Y.147 provides a medium to reddish shade of yellow. It is a specialty product for use in polystyrene. Colorations up to 1/3 SD are heat stable up to 300 °C. In terms of lightfastness, these samples match step 7 on the Blue Scale. Incorporated in polyolefins, P.Y.147 is also stable to temperatures up to 300 °C. The pigment is comparatively weak. 0.35% P.Y.147 is required to afford 1/3 SD HDPE samples (1% TiO$_2$). The pigment is also recommended for use in PP spin dyeing.

Pigment Red 89

P.R.89 affords a very bluish shade of red, referred to as pink. It exhibits good tinctorial strength and good lightfastness. Full shades in a combination nitro lacquer, for instance, equal step 7–8 on the Blue Scale. Somewhat poor overall fastness restricts the applicability of P.R.89. The pigment is considered a special-purpose product for artists' colors. It used to be employed also in spin dyeing vicose products.

Pigment Red 177

P.R.177 is mainly applied in industrial paints, in spin dyeing, and in polyolefin and PVC coloration.

The paint industry uses P.R.177 primarily in combination with inorganic pigments, especially with Molybdate Red Pigments. The resulting colorations have the advantage of showing high brilliance and cleanness in shades which are not accessible with other organic pigments. Combined with inorganic pigments, P.R.177 exhibits high lightfastness and weatherfastness. Combinations with Molybdate Red are also used in automotive finishes, especially for original automotive (O.E.M) finishes and for automotive refinishes. The types which have been marketed thus far are highly transparent, which makes them suitable colorants for transparent paints. In metallic effect coatings, P.R.177 is not weatherfast enough to meet higher standards of application. In an improperly formulated paint system, the pigment may react with aluminum or other reducing agents. Its weatherfastness decreases rapidly in white reductions. At typical processing temperatures for oven drying systems, P.R.177 is entirely fast to overcoating.

A very opaque version has recently been introduced to the market in the USA. This type, which is also yellower than traditional varieties, demonstrates not only excellent rheological behavior but also high flocculation stability and therefore high gloss. It is slightly more weatherfast than more transparent types. The main application is in lead-free automotive finishes for full shades.

P.R.177 shows excellent heat stability in plastics. 1/3 SD HDPE samples containing 1% TiO_2, for instance, are fast to up to 300 °C. Nucleation is not observed, i.e., the pigment does not affect the shrinkage of injection-molded polymer articles. In terms of tinctorial strength, P.R.177 is comparatively weak in these media: 0.3% pigment is needed to afford 1/3 SD colorations containing 1% TiO_2. P.R.177 systems are less lightfast than those containing other pigments of this class. P.R.177 is not entirely fast to bleeding in plasticized PVC.

Good transparency makes the grades with fine particle sizes suitable colorants for transparent films. The products exhibit excellent lightfastness. P.R.177, in combination with Molybdate Red pigments, provides better properties than other organic red pigments. Suitable media, also for such pigment combinations, are PUR and PVC coatings. Besides, P.R.177 is also used for spin dyeing polypropylene, polyacrylonitrile, and polyamide.

The printing ink industry uses P.R.177 primarily to print securities, especially bank notes.

3.6.2 Hydroxyanthraquinone Pigments

The class of hydroxyanthraquinone pigments comprises two different groups of compounds: metal complexes of hydroxyanthraquinones on the one hand and metal salts of hydroxyanthraquinone sulfonic acids on the other hand. Some of the products are metal chelates.

The first group includes 1,2-dihydroxyanthraquinone, commonly known as alizarin, 1,4-dihydroxyanthraquinone (quinizarin), and 1,2,4-trihydroxyanthraquinone (purpurin). Alizarin in particular has been known and appreciated for thousands of years in the form of its "lake", i.e., the coordination complex of 1,2-hydroxyanthraquinone **88** with aluminum and calcium (Madder Lake, Turkey Red).

The chemical structure of **88** was first described in 1963 [11]. Aluminum and calcium are equally important constituents of other hydroxyanthraquinone compounds, although iron salts are also known.

The calcium lake of 1,2-dihydroxyanthraquinone is marketed as Pigment Red 83:1, 58000:1. It is produced commercially by treating a slightly basic alizarin solution with aqueous calcium chloride.

All of these compounds were originally precipitated onto aluminum oxide hydrate, which served as a carrier. This applies accordingly to hydroxyanthraquinone sulfonic acids.

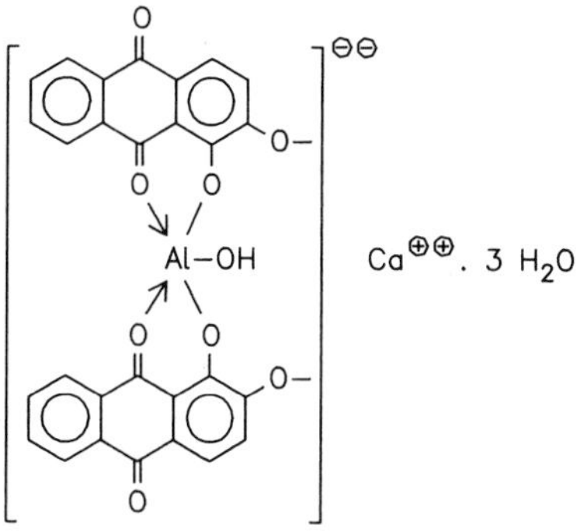

88

Both groups have lost most of their commercial importance as pigments.

There is one derivative of hydroxyanthraquinone sulfonic acid which is used to some extent. This is the reddish violet Pigment Violet 5:1, 58055:1, which has the basic structure **89**:

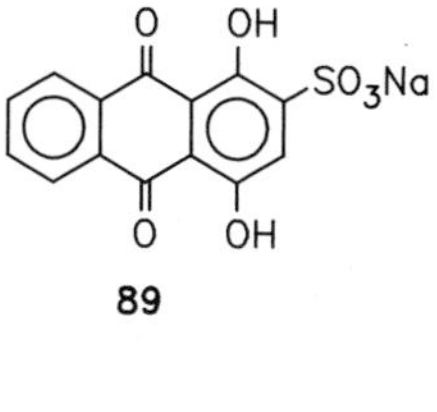

89

This compound is obtained from phthalic acid and p-chlorophenol with sulfuric acid in the presence of boric acid. The intermediate product is quinizarin, which is sulfonated in oleum or with sodium hydrogen sulfite and oxidants to form **89**:

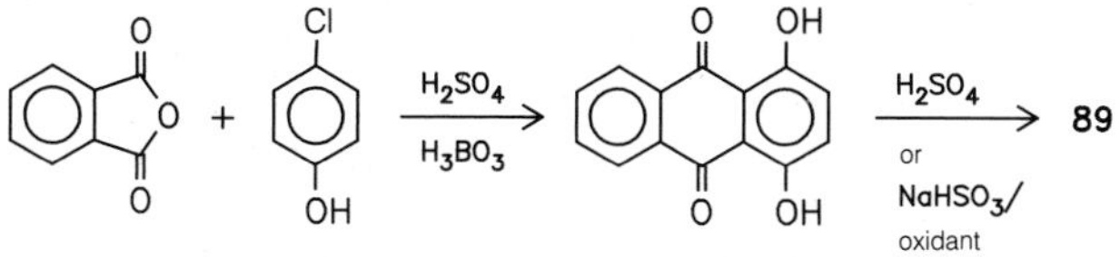

Pigment Violet 5:1 is an aluminum complex of **89**.

3.6.2.1   Properties and Application

Pigment Red 83

P.R.83, listed under Constitution No. 58000:1, continues to be used only in the USA. According to the US International Trade Commission, 150 t were produced nationwide in 1983, but only 19 t in 1988. The pigment affords brilliant, bluish shades of red. Traces of iron as an impurity adversely affect the full shades and shift the color towards duller and bluer shades. The pigment is not fast to common organic solvents, especially to esters and ketones. It therefore lacks stability to overcoating. Its lightfastness, particularly in tint, is poor. P.R.83 is used in paints for toys, in packaging printing inks, especially for soap and butter, and in artists' colors.

Pigment Violet 5:1

Commercial interest in P.V.5:1, Constitution No. 58055:1, has declined considerably. The pigment continues to be used in industrial paints, especially throughout the USA. According to the US International Trade Commission, only 28 t of P.V.5:1 were produced nationwide in 1988. Its full shade is a brilliant, deep bluish maroon. In white reductions, the pigment produces clean, reddish violet shades. It lacks tinctorial strength and the coatings are fast to neither acid nor alkali. P.V.5:1 is also not very lightfast, which practically precludes its use in products for exterior application, particularly in reduced shades.

P.V.5:1 is used to a certain extent in PVC. Plasticized PVC systems are fast to bleeding and reasonably lightfast in full shades. Addition of TiO_2, however, markedly affects its lightfastness. The pigment is heat stable up to 170 °C.

3.6.3   Heterocyclic Anthraquinone Pigments

As a group, these pigments include compounds whose heterocyclic system is formally derived from the anthraquinone nucleus.

3.6.3.1   Anthrapyrimidine Pigments

1,9-Anthrapyrimidine **90**, which is not used commercially, provides the parent structure for an important yellow pigment.

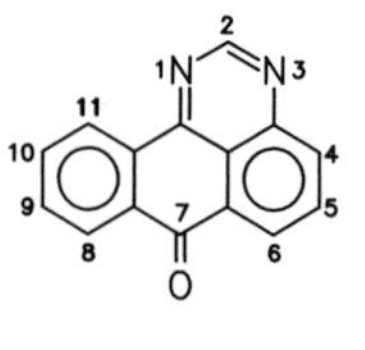

Anthrapyrimidine and its substituted derivatives are obtained by condensation of 1-aminoanthraquinone (or its derivatives) with formamide or aqueous formaldehyde/ammonia in the presence of an oxidant, such as ammonium vanadate or m-nitrobenzosulfonic acid. A newly developed, more simple route proceeds via formamidinium chloride, which is prepared from 1-aminoanthraquinone with dimethylformamide and thionyl chloride or phosphorus oxychloride. Cyclization in a solvent in the presence of ammonium acetate affords the desired product:

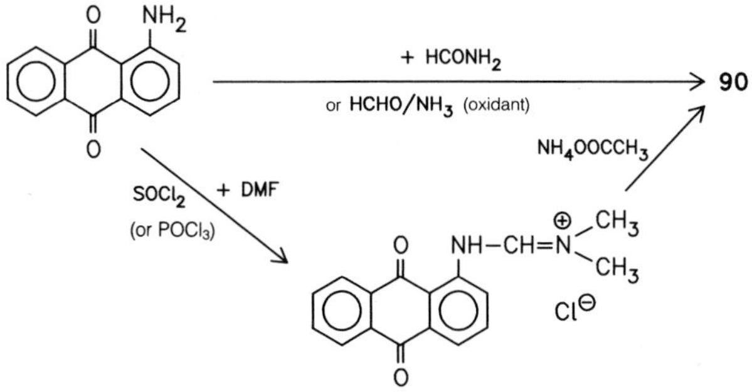

Vat Yellow 20, a product known for a long time and patented as early as 1935, has the following chemical constitution, derived from anthrapyrimidine:

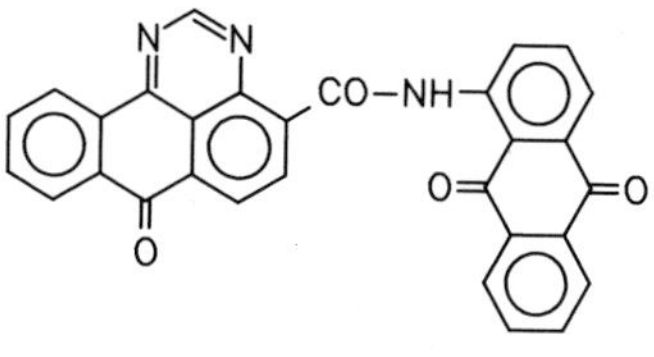

The compound has gained commercial recognition primarily as an organic pigment (Pigment Yellow 108, 68420).

P.Y.108 is obtained under condensation from 1-aminoanthraquinone, 1,9-anthrapyrimidine-2-carboxylic acid, and a chlorinating agent, or directly with 1,9-anthrapyrimidine-2-carboxylic acid chloride in an organic solvent in the presence of an acid trap.

Thus, P.Y.108 is commercially produced by heating 1,9-anthrapyrimidine-2-carboxylic acid with 1-aminoanthraquinone and thionyl chloride in a high boiling solvent, such as o-dichlorobenzene or nitrobenzene, to 140 to 160 °C. The product is separated, washed with methanol, and residual solvent removed by steam distillation. The aqueous suspension is then boiled down with sodium hypochlorite solution.

Finer particle sizes are obtained if 1,9-anthrapyrimidine-2-carboxylic acid chloride is condensed with 1-aminoanthraquinone in a dipolar aprotic solvent (such as N-methylpyrrolidone) at a temperature between 70 and 110 °C. The reaction may be accelerated by using a proton acceptor such as triethylamine or tert-butanol, which reacts with hydrochloric acid. A particularly useful route proceeds via the acid chloride, which is prepared by reacting anthrapyrimidine carboxylic acid with thionyl chloride at 40 °C. A condensation reaction follows without separation of the intermediate [12].

Properties and Application of Pigment Yellow 108

Anthrapyrimidine Yellow is primarily applied in paints. In lighter tints, the pigment provides a medium, dull yellow color, while full shades and similarly deep shades are distinctly redder and even duller than tints. P.Y.108 is not fast to the solvents which are commonly used in paints. It bleeds more or less into solvents such as aromatic and aliphatic hydrocarbons, alcohols, ketones, and esters (Sec. 1.6.2.1). P.Y.108 accordingly lacks entire fastness to overcoating. As a colorant for coatings, it is heat stable up to 160 °C. P.Y.108 is both very lightfast and weatherfast in lighter tints, but its fastness to light and weather deteriorates rapidly as more white pigment is added. Some darkening is observed. P.Y.108 is therefore used particularly in medium to very light shades, as a shading pigment, and as a source of cream colors. For a long time, such P.Y.108 systems were considered the most weatherfast products within the medium yellow portion of the spectrum.

P.Y.108 is applied in various types of industrial finishes, especially in original automotive (O.E.M) and in automotive refinishes. It is also recommended for metallic finishes, although it is much less weatherfast in such systems. The pigment tends to seed, i.e., it forms specks upon storage. The mechanism behind this phenomenon remains to be elucidated. Besides, P.Y.108 also lends color to emulsion paints, in which it is durable enough to satisfy the requirements for exterior paints based on synthetic resin dispersions. It is also fast to acids, alkali, and plaster.

The printing inks industry utilizes P.Y.108 only in special-purpose media, including packaging inks, metal decorating inks, posters, and similar applications. The prints are comparatively dull, but very lightfast. They may safely be sterilized.

3.6.3.2   Indanthrone and Flavanthrone Pigments

In 1901, R. Bohn synthesized indanthrone and flavanthrone. Both compounds are thus among the oldest synthetic vat dyes known.

Indanthrone

Vat dyes of the highest quality have derived their name from indanthrone, a blue compound, which was originally known as Indanthrene Blue. The compound has been used for a long time as a pigment and is registered in the Colour Index as Pigment Blue 60, 69800 (**91**).

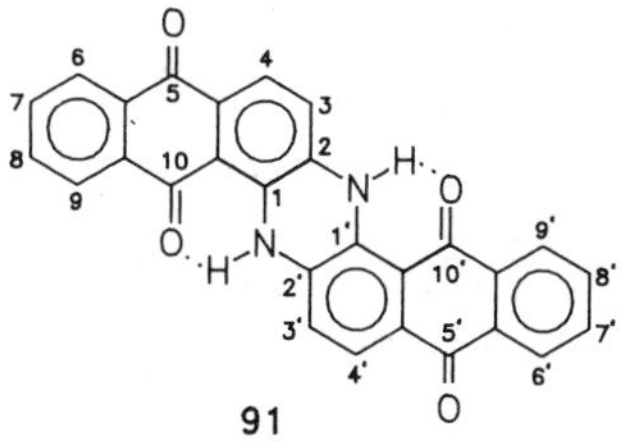

91

The primary synthetic route proceeds via oxidative dimerization of 2-aminoanthraquinone in the presence of an alkali hydroxide. 2-aminoanthraquinone, for instance, is fused with potassium hydroxide/sodium hydroxide at 220 to 225 °C in the presence of sodium nitrate as an oxidant. New techniques involve air oxidation of 1-aminoanthraquinone at 210 to 220 °C in a potassium phenolate/sodium acetate melt or in the presence of small amounts of dimethylsulfoxide. A certain amount of water which is formed during the reaction may be removed by distillation in order to improve both efficiency and yield.

One out of several possible options for the synthesis of indanthrone from 2-aminoanthraquinone is illustrated in the following scheme:

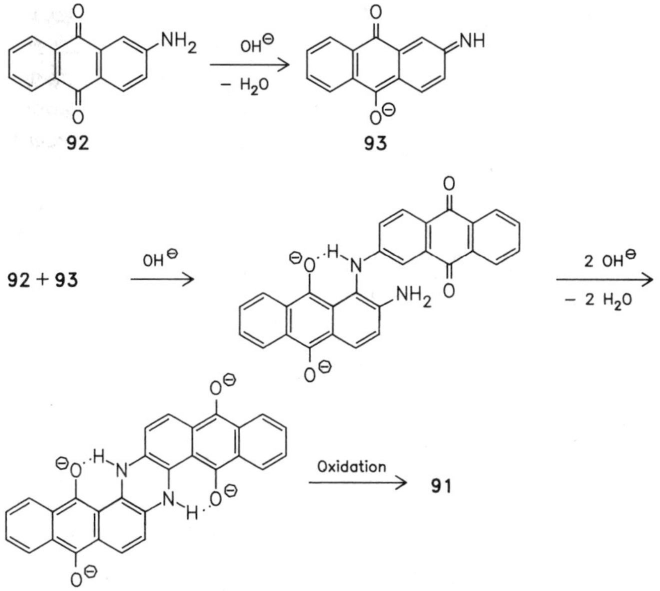

92   93

92 + 93

Oxidation
91

The reactions involve a variety of preliminary equilibria, which open up a number of different pathways. Therefore, the reaction conditions during indanthrone synthesis must be followed exactly [13, 14].

The pigment form [15] is obtained from the leuco form, which in turn is prepared by oxidation with alkali fusion, followed by treatment with sodium hydrogensulfite solution or sodium dithionite (vatting).

Heating the leuco form in an aqueous alkaline medium in the presence of air, possibly also in the presence of a surfactant, affords the pigment form. Oxidation may also be achieved with sodium-m-nitrobenzenesulfonate. Another alternative is to precipitate the vat acid from the salt of the leuco form and to subsequently oxidize the product. It is also possible to mill the suspension of the leuco form together with aqueous sodium hydroxide/sodium dithionite in air, using a pearl mill.

Other methods start from crude indanthrone, which is dissolved in sulfuric acid or oleum (sometimes mixed with organic solvents). The product may be precipitated with water. It is more advantageous to separate the precipitated crude pigment, to reslurry it in water, and to heat the resulting aqueous suspension in the presence of a cationic surfactant. Subsequent treatment of the sulfuric acid solution with nitric acid, manganese dioxide, or chromium trioxide is followed by transfer into a solution containing sodium sulfite or iron(II)sulfate.

Kneading or milling the crude indanthrone in the presence of finishing agents, such as polyols, or milling it with salt also affords a product which provides useful pigment properties.

Indanthrone exists in four crystal modifications, of which the α- and the β-modification afford greenish and reddish blue shades, respectively, while the γ-form provides reddish hues. The δ-modification has no coloristic importance. Being the most stable thermodynamically, the α-form is most suitable for use as a pigment. It affords reddish and greenish blue grades. The more greenish type is prepared by precipitating a crude indanthrone solution in sulfuric acid, or it is obtained from the leuco form with a surfactant by air oxidation. The only commercially available reddish form is also obtained through air oxidation, although it is necessary to simultaneously apply shearing forces [16].

In the past, not only unsubstituted indanthrone but also some chloro derivatives were commercially available. 3,3'-Dichloroindanthrone, also known as Pigment Blue 64, 69825, has had a certain impact on the market.

No current manufacturers are known of this and other indanthrone derivatives which are preferably chlorinated in α-position (4,4'-, 5,5'-, or 8,8'-) and used as pigments.

Flavanthrone

Flavanthrone has also been used for a long time as a vat dye. It gained recognition as a pigment when more and more lightfast and durable paints were required.

The yellow pigment is registered in the Colour Index as Pigment Yellow 24, 70600. It has the structure **94**:

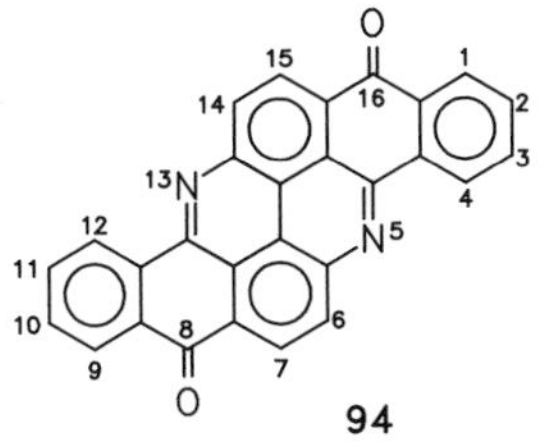

94

The synthetic route involves treating 1-chloro-2-aminoanthraquinone with phthalic anhydride (PA), which initially affords 1-chloro-2-phthalimidoanthraquinone **95**. Subsequent Ullmann reaction with copper powder in refluxing trichlorobenzene yields 2,2'-diphthalimido-1,1'-dianthraquinonyl **96** through dimerization. This product is cyclized with 5% aqueous sodium hydroxide solution at 100 °C, cleaving phthalic acid units, to yield **94**:

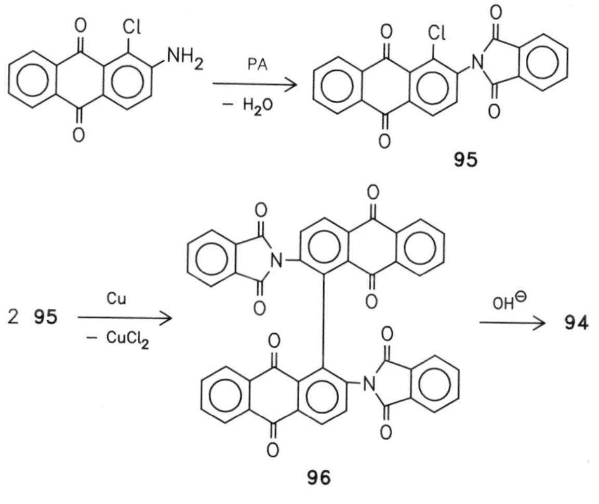

95

96

This three-step synthesis may be performed in one step by heating 1-halogeno-2-aminoanthraquinone with copper in a highly polar aprotic solvent [17].

An older, equally interesting industrial route involves condensing 2-aminoanthraquinone in nitrobenzene in the presence of antimony pentachloride or titanium tetrachloride. Complex **97** prevents any undesirable formation of anthrimide (**98**).

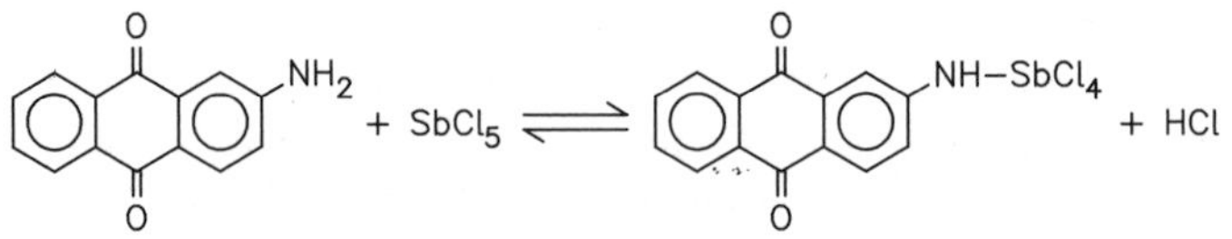

97

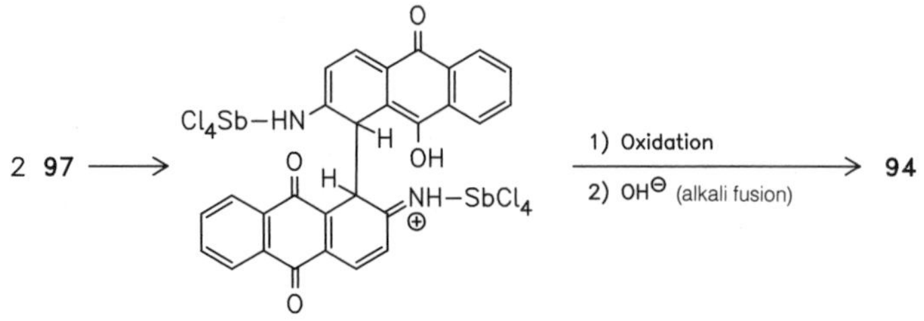

2 97 ⟶

1) Oxidation

2) OH⊖ (alkali fusion)

94

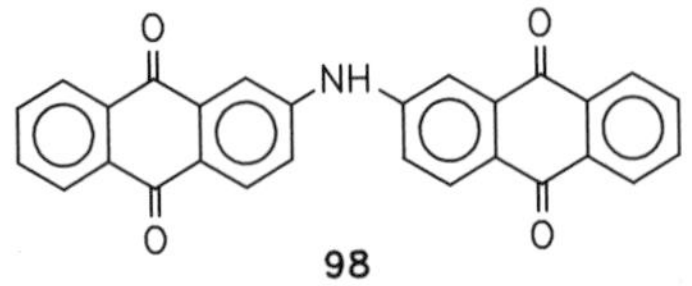

98

The comparatively low yield, together with the high price of antimony penta-chloride, make this process economically unattractive.

A more recent method starts from 2,2'-diacetylamino-1,1'-dianthraquinonyl, which is cyclized in a phase-transfer reaction in chlorobenzene using tetrabutylam-monium bromide and a 30% aqueous sodium hydroxide solution [18].

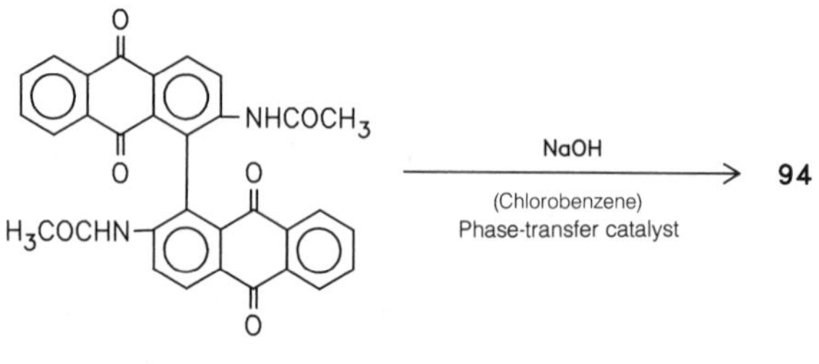

NaOH

(Chlorobenzene)
Phase-transfer catalyst

94

Flavanthrone, like indanthrone, must be extremely pure in order to develop useful pigment properties. Subsequent finishing converts the thus prepared materi-al into an appropriate product for use in paints or plastics.

The crude product may be purified by one of the following methods:
- Dissolving the material in concentrated sulfuric acid at 50 to 70 °C and hydro-lyzing the thus prepared sulfate with water.
- Converting the crude product into its leuco form, separating the intermediate, and reoxidizing the compound to form pure flavanthrone.
- Extracting the crude product with a dipolar aprotic solvent, such as dimethyl-formamide or dimethylsulfoxide.

A variety of milling techniques are available to develop the desired pigment properties, performing the finishing process either in an aqueous suspension or in the presence of organic solvents or milling agents. A pigment which is prepared by this route typically evolves in an opaque, reddish yellow form.

A product which provides the same shade but higher transparency is obtained by converting crude flavanthrone into its leuco form (for instance with sodium dithionite/aqueous sodium hydroxide). The resulting intermediate material is separated and resuspended in water. It undergoes facile reoxidation through shearing forces and/or in the presence of surfactants, affording the pigment [19].

The crude pigment may also be treated with an aromatic sulfonic acid (such as toluene sulfonic acid, xylene sulfonic acids, m-nitrobenzene sulfonic acid) in sulfuric acid or with nitric acid at 80 °C to yield a somewhat redder yellow transparent modification of flavanthrone [20].

Properties and Application of Pigment Blue 60

Indanthrone demonstrates excellent weatherfastness. Most grades of P.Bl.60, like other commercial vat type pigments, provide good transparency. They are primarily applied in automotive finishes, especially in metallic finishes. In these media, P.Bl.60 is even more weatherfast than copper phthalocyanine pigments, especially in light tints. This makes P.Bl.60 particularly attractive in areas where copper phthalocyanine pigments are not weatherfast enough to satisfy the application requirements. P.Bl.60 is highly durable even in light white reductions. Its shade is noticeably redder and duller than α-Copper Phthalocyanine Blue. Likewise, P.Bl.60 is also duller than the similarly redder ε-Copper Phthalocyanine Blue. Although a pigment with high tinctorial strength, P.Bl.60 is weaker than α-Copper Phthalocyanine Blue types. P.Bl.60 shows very good fastness to organic solvents, including alcohols, esters, and ketones, aromatic and aliphatic hydrocarbons, and various types of plasticizers. Incorporated in baked enamels, the pigment is fast to overcoating. P.Bl.60 is also fast to acids and alkali and heat stable up to 180 °C. The pigment is not only used in automotive finishes but also in general industrial paints wherever the high demands on pigment fastness properties or the very reddish shade warrant its use. It should be noted, however, that a very similar shade is accessible through shading of the more brilliant α-Copper Phthalocyanine Blue with Dioxazine Violet. The cleanness of the product may be reduced if needed. The resulting mixture, however, is less weatherfast in light tints than Indanthrone Blue, especially in metallic finishes. Adding suitable UV absorbers eliminates the difference in weatherfastness.

Besides, P.Bl.60 is employed in plastics wherever its shade is required, or for its excellent fastness properties. The pigment has the advantage of being highly heat stable. Dispersed in polyolefins, it is fast up to 300 °C for 5 minutes. The difference in color between full shades and 1/3 SD HDPE samples at 300 °C as opposed to 200 °C is only approximately 1.5 CIELAB units. 1/25 SD specimens are thermally stable up to 280 °C. Indanthrone Blue exhibits average tinctorial strength. 0.15% pigment is necessary to produce 1/3 SD colorations with 1% TiO$_2$. The same result is achieved with as little as approximately 0.08% α-Copper Phthalocyanine Blue. P.Bl.60 does not nucleate, i.e., it does not affect the shrinkage of injection-molded plastic articles to a noticeable extent. As in PVC, P.Bl.60 is frequently used in these materials as an alternative to phthalocyanine blue pigments for its redder shade. It is almost entirely fast to bleeding in plasticized PVC. The pigment shows excellent lightfastness: 1/3 SD colorations equal step 8 on the Blue Scale. Likewise, P.Bl.60 is also highly weatherfast, fast enough to be recommended for long-term exposure. This pigment is a valuable colorant for a wide variety of plastic media. Used in PP spin dyeing, for instance, P.Bl.60 exhibits greater fastness than copper phthalocyanine pigments. It is a suitable pigment for the coloration of rubber and other elastomers. Polyamide is an unsuitable medium for P.Bl.60.

Printing inks made of P.Bl.60 are used to print banknotes.

Properties and Application of Pigment Yellow 24

Flavanthrone Yellow, together with its chemical structure, is listed in the Colour Index under Constitution No. 70600. It was temporarily known as Pigment Yellow 112, but now it is exclusively referred to as Pigment Yellow 24.

The commercially available types are reddish yellow. They are specialty products for use in paints. Besides, P.Y.24 is also employed for polyacrylonitrile spin dyeing. The pigment equals the considerably more greenish and distinctly weaker Anthrapyrimidine Yellow in terms of fastness to organic solvents and chemicals. P.Y.24 is almost entirely fast to overcoating. Full shades and similarly deep shades darken considerably upon exposure to light and weather. This effect, however, is not noticeable in very light tints; corresponding samples show very good lightfastness and weatherfastness. P.Y.24 is, however, not quite as durable as anthrapyrimidine yellow pigments, such as P.Y.108.

High transparency makes P.Y.24 a valuable pigment for metallic finishes. It is used in relatively light shades, typically at a ratio of one part of color pigment to three parts of aluminum pigment. Thus prepared systems demonstrate excellent weatherfastness. Flavanthrone Yellow, like P.Y.108, tends to seed (Sec. 3.6.3.1). The pigment is heat stable up to 200 °C and thus satisfies all possible heat stability requirements in this area. Flavanthrone Yellow is used in various industrial paints, especially in original automobile (O.E.M) finishes and in automotive refinishes.

Although highly reduced P.Y.24 plastic systems are also very lightfast, such products are used only to a limited extent. P.Y.24 shows average tinctorial strength. 1/3 SD colorations in HDPE (1% TiO_2), for instance, are formulated at approximately 0.25% pigment. Such products may be safely exposed to temperatures up to 270 °C, while addition of TiO_2 (1/25 SD) reduces the heat stability of the pigmented systems to 230 °C. The tinctorial strength increases at higher temperature and the color turns greener, a phenomenon which points to pigment dissolution. Polyacrylonitrile spin dyed fibers are lightfast and weatherfast enough to satisfy stringent requirements, which qualifies the pigment for use in canvasses and similar media. In this respect, P.Y.24 is somewhat of an exception among organic yellow pigments.

P.Y.24 is also sold for use in other areas in which high lightfastness is a prime concern. It is thus used in media such as solvent-based wood stains and in artists' colors.

3.6.4 Polycarbocyclic Anthraquinone Pigments

This class includes polycarbocyclic compounds which are at least formally derived from the anthraquinone structure. The products are considered members of the higher condensed carbocyclic quinone series, which even in the absence of additional substituents provide yellow to red shades. Halogenation is frequently found to afford cleaner shades and improved fastness properties. Heading the list of such derivatives are pyranthrone, anthanthrone, and isoviolanthrone pigments.

3.6.4.1 Pyranthrone Pigments

Pyranthrone pigments are related to the pyranthrone structure (**99**), which is formally derived from flavanthrone (Sec. 3.6.3.2) in that the nitrogen atoms are replaced by CH groups.

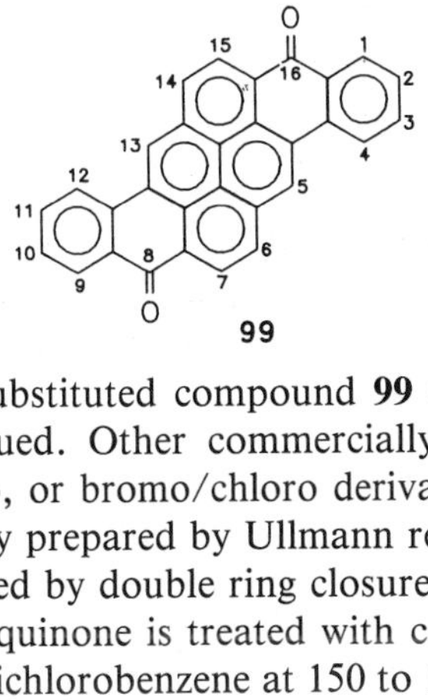

99

Manufacture of the unsubstituted compound **99** (Pigment Orange 40, 59700) has recently been discontinued. Other commercially available pyranthrone pigments include bromo, chloro, or bromo/chloro derivatives of the parent structure.

Pyranthrone is commonly prepared by Ullmann reaction of 1-chloro-2-methylanthraquinone (**100**), followed by double ring closure.

1-Chloro-2-methylanthraquinone is treated with copper powder, pyridine, and dry sodium carbonate in o-dichlorobenzene at 150 to 180 °C to form 2,2'-dimethyl-1,1'-dianthraquinonyl (**101**). Subsequent condensation is achieved after several hours of boiling in a sodium hydroxide/isobutanol solution at 105 °C, which affords the leuco form. The product is then oxidized to form pyranthrone by blowing air through the reaction mixture.

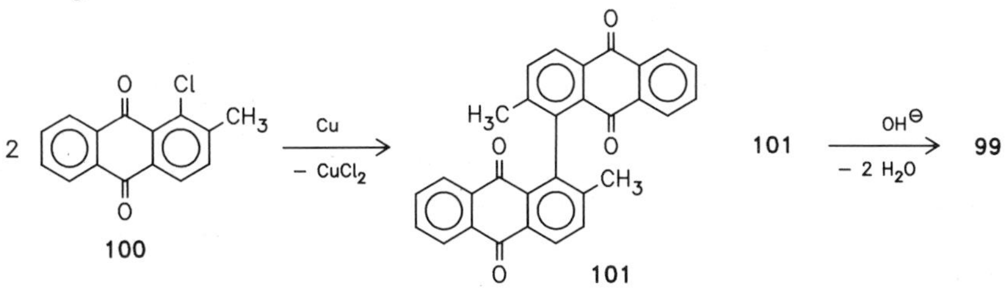

100   **101**

Heating **101** with aqueous sodium hydroxide in diethyleneglycol monomethyl-ether or with alkali acetate, a reaction which may also be performed in other polar organic solvents, such as dimethylformamide, N-methylpyrrolidone, or dimethyl-acetamide at 150 °C to 210 °C, also provides pyranthrone. As an alternative, **101** may be cyclized to form pyranthrone in a two-phase reaction, carried out in the presence of quarternary ammonium salts in a phase-transfer system comprising an aqueous and an organic phase [18].

Halogenated pyranthrones may be obtained by two different routes. One route proceeds via ring closure of halogenated 2,2'-dialkyl-1,1'-dianthraquinonyl, in accordance with the synthesis of pyranthrone. The same end may be accomplished by halogenation of ready-made unsubstituted pyranthrone.

101, for instance, may be chlorinated in o-dichlorobenzene, followed by ring closure through phase-transfer reaction with aqueous sodium hydroxide/water and dichlorobenzene. 3,3'-dichloro-2,2'-dimethyl-1,1'-dianthraquinonyl is also accessible through ring closure by traditional alkaline condensation or by adding polar solvents in the presence of sodium acetate at 110 to 150 °C [21].

3,3'-Dichloro-2,2'-dimethyl-1,1'-dianthraquinonyl can be synthesized by Friedel-Crafts reaction from phthalic anhydride and 2,6-dichlorotoluene. Subsequent cyclization of the resulting substituted benzoylbenzoic acid **102** in sulfuric acid affords the corresponding anthraquinone derivative **103**, which is dimerized by Ullmann reaction:

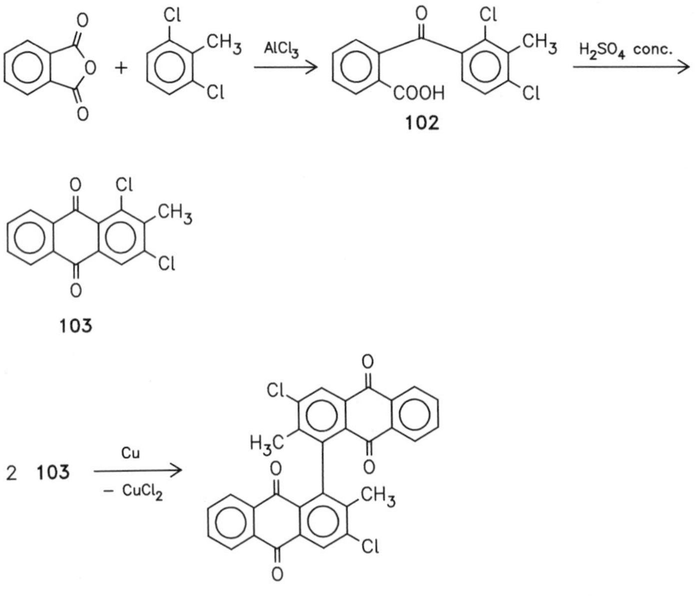

Pyranthrone may be halogenated, for instance, in chlorosulfonic acid in the presence of small amounts of sulfur, iodine, or antimony as a catalyst. This procedure necessitates intermediate separation and purification of pyranthrone after manufacture, because the products, unless purified, fail to furnish the solvent fastness which is characteristic of a typical pigment.

While 3,3'-dichloro-2,2'-dimethyl-1,1'-dianthraquinonyl is easily converted into well-defined 6,14-dichloropyranthrone, direct halogenation affords changing amounts of chloro and bromo derivatives with undefined substitution patterns on the pyranthrone ring. Depending on the amount of added halogen, an average of two to four halogen atoms per molecule is expected.

Bromination of 6,14-dichloropyranthrone in chlorosulfonic acid in the presence of a catalyst provides 0.1 to 2.2 bromine atoms per molecule, depending on the reaction conditions. The exact outcome is controlled not only by the choice of solvent but also by the type of catalyst and the amount of bromine, as well as by the reaction time and reaction temperature.

The halogenated pyranthrones dissolved in chlorosulfonic acid may be isolated by precipitation into ice water. Residual acid is removed by washing. The thus prepared products, such as 6,14-dichloropyranthrone, do not require further treatment before bromination. An opaque version of 6,14-dichloropyranthrone as well as transparent types of chlorobromopyranthrones are accessible by initially dissolving the crude pigment in sulfuric acid, reprecipitating it with ice/water, and separating the product. The aqueous presscake is then treated with C_4–C_{10}-alkanols or alkanones, or with C_6–C_8-cycloalkanols or alkanones [22].

An opaque form of pyranthrone may also be prepared by treating the corresponding pigment in a polar organic solvent (such as isobutanol) at elevated temperature in the presence of some (0.5 to 10%) halogenated anthraquinone or another anthraquinone derivative.

While 7,15- and 3,11-dichloropyranthrone demonstrate such poor solvent fastness as to be unsuitable for use as pigments, 6,14-dichloropyranthrone possesses excellent solvent resistance.

In addition to 6,14-dibromopyranthrone, there are some mixed halogenated pyranthrone derivatives which exhibit equally useful pigment properties.

Commercially Available Pyranthrone Pigments and Their Application

Only very few halogenated types are currently available. Pyranthrone pigments are supplied primarily to the paint industry.

Pigment Orange 40

P.O.40, unsubstituted pyranthrone, is registered under Constitution No. 59700. Its production has recently been discontinued. The pigment used to be employed in spin dyed viscose rayon and viscose cellulose. The resulting products demonstrate good lightfastness. 1/1 to 1/25 SD specimens equal step 6–7 to step 6 on the Blue Scale. The pigment exhibits average tinctorial strength. Its textile fastnesses are excellent. As a result of its chemical constitution, P.Y.40 is not vat resistant, i.e., it is only moderately fast to an alkaline sodium dithionite solution (60 °C for 30 minutes). Spin dyed fibers undergo noticeable color change under such conditions, they bleed and distinctly color white staple fiber and cotton.

Pigment Orange 51

P.O.51, which is the 2,10-dichloropyranthrone derivative, affords a medium orange shade. It is not only slightly redder than the halogen-free variety, but also exhibits higher cleanness and is tinctorially stronger in tints. The pigment is used especially to provide reddish shades of yellow. P.O.51 provides good fastness to organic solvents, it may safely be overcoated under normal conditions if incorporated in baking enamels. Heat stable up to more than 200 °C, P.O.51 satisfies practically all possible requirements for use in this area. Its full shade and similarly deep shades, down to approximately 1/9 SD, show excellent lightfastness and weatherfastness. The commercial version possesses only poor transparency, despite its high specific surface area of about 60 m^2/g. This brand is also recommended for use in metallic shades and is used extensively in two-coat metallic finishes. Although highly weatherfast, medium colors do not perform as well as the other halogenated, redder pyranthrone pigments.

Pigment Red 216

P.R.216, Constitution No. 59710, a tribromopyranthrone derivative, is distinctly bluer than P.R.226, and its shade is comparatively dull. As a result of a high specific surface area of about 80 m^2/g, the commercial version is highly transparent. The pigment is specifically recommended for use in metallic finishes. P.R.216, like P.R.226, is frequently used in two-coat metallic finishes for automobiles, especially if very little colored pigment is applied. Combination with UV absorbants in the top coat is also common. Pigment durability in medium tints, like 1/3 SD, is good. However, the weatherfastness decreases rapidly as more TiO$_2$ is added. This effect is more noticeable in P.R.216 than in P.R.226. P.R.216 is also employed in pigment combinations with colored inorganic pigments, especially with Molybdate Red derivatives. It shows good fastness to solvents and performs even better in this respect than P.R.226. This is also true for the fastness to overcoating, although baking temperatures above 140 °C should be avoided.

P.R.216 lends color to all types of industrial paints, it is heat stable up to 200 °C. Like other pyranthrone pigments, it is suitable for use in unsaturated polyester systems, in which it is resistant to peroxides.

P.R.216 demonstrates excellent lightfastness in plastics, especially in PVC. 1/3 SD plasticized PVC systems (5% TiO$_2$) equal step 8 on the Blue Scale for lightfastness. Pigment weatherfastness is similarly excellent, but fails to meet the standards for long-term exposure. P.R.216 is not entirely fast to bleeding, and its tinctorial strength is poor: 1/3 SD (5% TiO$_2$) is formulated at 1.85% pigment. 1/3 SD colorations in HDPE with 1% TiO$_2$ withstand exposure to 250 °C. In terms of lightfastness, these specimens equal step 6 on the Blue Scale.

Pigment Red 226

Pigment Red 226 is a dibromo-4,6-dichloro-pyranthrone. It provides a medium, somewhat dull red shade, which is considerably more bluish than that of P.O.51. The shade, although noticeably duller, corresponds to that of the less fast toluidine red pigments. P.R.226 is distinctly yellower than P.R.216, a pyranthrone derivative. Despite the fact that P.R.226 demonstrates good fastness to solvents, it is much less resistant than other members of its class. The difference is most noticeable in toluene and xylene. P.R.226 is fast to acids and alkali. Like P.O.51, it is thermally stable up to more than 200 °C, even in long-term exposure for several weeks. The commercial grade is more transparent than P.O.51 and provides excellent lightfastness and weatherfastness. P.R.226 is a special-purpose type which is used for metallic finishes. It is a suitable pigment for automotive finishes, especially for two-coat metallic finishes, quite often in combination with UV absorbants in the clean top coat.

3.6.4.2 Anthanthrone Pigments

Anthanthrone pigments are characterized by the basic structure **104**:

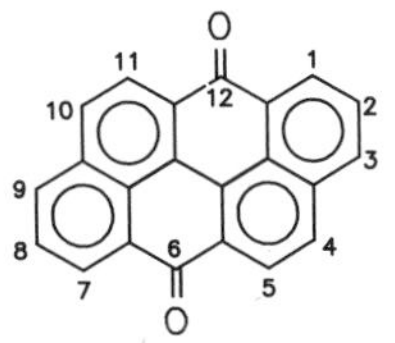

104

The unsubstituted ring system, although exhibiting an orange shade, apart from other deficiencies is tinctorially not strong enough to stimulate interest. Only halogenated derivatives have gained some interest throughout the pigment industry.

Anthanthrone is synthesized from naphthostyril (**105**), which is saponified to form 1-aminonaphthalene-8-carboxylic acid (**106**). Naphthostyril itself is prepared from 1-naphthylamine with phosgene in the presence of dry aluminum chloride.

Diazotizing **106** and boiling the solution in the presence of copper powder affords 1,1'-dinaphthyl-8,8'-dicarboxylic acid (**107**), which is cyclized with aluminum chloride, but preferably with concentrated sulfuric acid at 30 to 40 °C to produce anthanthrone.

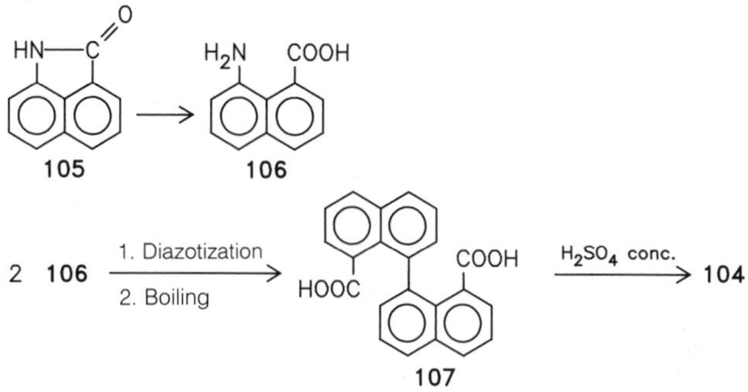

The Friedel–Crafts reaction, which proceeds via electrophilic aromatic substitution, as illustrated in the following scheme, is unique to the manufacture of anthanthrone pigments. Most other polycyclic anthraquinone pigments are synthesized via nucleophilic ring closure.

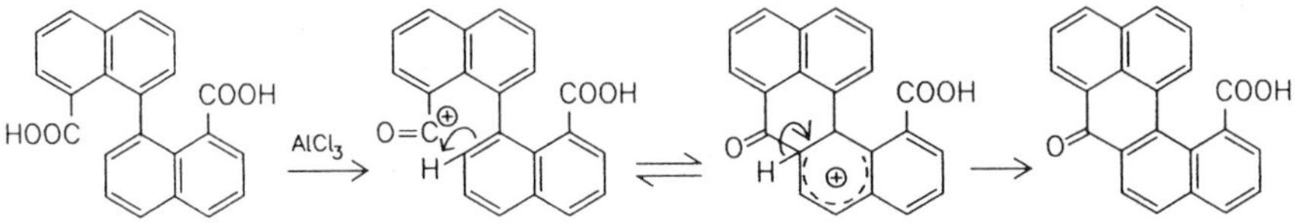

4,10-Dibromoanthanthrone (**108**) is registered as Pigment Red 168, 59300. First synthesized as early as 1913 as a vat dye, this compound is the commercially most interesting halogenated anthanthrone derivative.

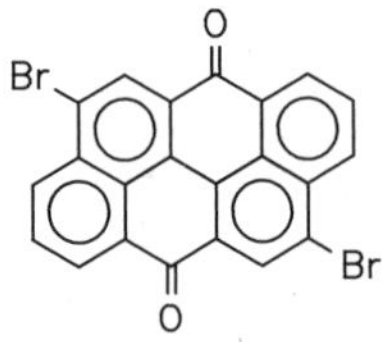

108

Compound **108** may be prepared directly from **104** without intermediate isolation of **104** by treating the dicarboxylic acid **107** with monohydrate or with concentrated sulfuric acid at 35 °C, followed by bromination in the presence of iodine as a catalyst.

In contrast to this reddish orange color, which is referred to as scarlet, the yellower 4,10-dichloroanthanthrone has no commercial interest as a pigment and is therefore not produced.

Properties and Application of Pigment Red 168

Dibromoanthanthrone is a vat type pigment which demonstrates excellent fastness properties, a feature which is an asset in high grade paints. It provides a clean yellowish shade of scarlet, somewhere between that of P.O.43, a naphthalenetetra-carboxylic acid derivative, and those of yellowish perylenetetracarboxylic acid type pigments.

P.R.168 is entirely or almost entirely resistant to most of the organic solvents which are commonly found in typical binder systems. P.R.168, like other vat type pigments, is not completely fast to overcoating in oven drying systems which are baked at 120 to 160 °C. The degree of bleeding depends on the individual system. The pigment is thermally stable up to 180 °C. It is one of the most lightfast and weatherfast organic pigments known. Excellent performance standards make it a suitable candidate for all types of coatings and paints, even at very low pigment concentrations. P.R.168 is used both as a shading pigment and in mixed systems. It exhibits comparatively low tinctorial strength. The commercial products, like those of other vat type pigments, are more or less transparent, which makes them useful products for metallic finishes. P.R.168 is used, for instance, in one- and two-coat automotive finishes. Moreover, it is weatherfast enough to be used in low concentrations. In such systems, P.R.168 is also employed in combination with highly fast, typically reddish yellow pigments in order to produce shades of bronze and copper. The pigment is equally useful in conjunction with more reddish organic pigments, such as with perylenetetracarboxylic acid pigments. High weatherfastness makes it a suitable candidate for use in architectural paints and in emulsion paints, including those which are targeted for outdoor use. The fact that P.R.168 is only used in very light tint allows its comparatively high price in this market. Specialized pigment preparations are also available for these purposes. P.R.168 is fast to alkali and plaster. The pigment is also recommended for use in coil coating, including PVC coatings (Sec. 1.8.2.2).

P.R.168 is found to a lesser extent in printing inks and plastics. The printing ink industry utilizes P.R.168 to produce special-purpose printing inks, which may be applied to substrates such as posters or metal deco prints. The pigment demonstrates equally excellent fastness in these materials. 1/1 SD systems equal step 8 on the Blue Scale for lightfastness, while 1/3 to 1/25 SD formulations match step 7. The prints are resistant to common organic solvents and chemicals. The pigment is thermally stable up to 220 °C for 10 minutes, and its prints may safely be sterilized.

3.6.4.3 Isoviolanthrone Pigments

Isoviolanthrone (**109**) is an highly anellated polycyclic quinone system. It is derived from the chemical structure of isodibenzanthrone, which may be visualized as being obtained by unsymmetrical condensation of two benzanthrone molecules. The compound itself affords an intense blue shade.

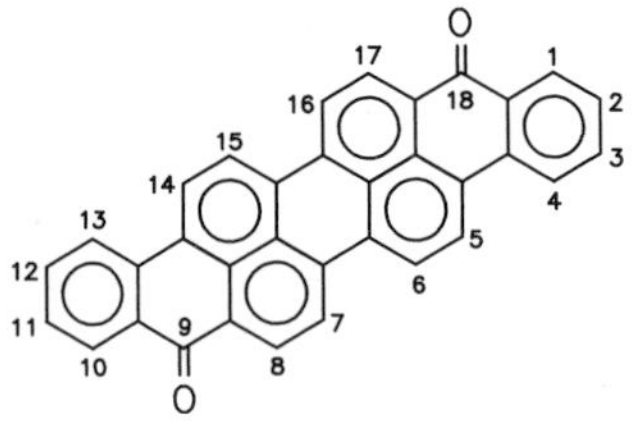

109

Isoviolanthrone is currently prepared by boiling 3,3'-dibenzanthronyl sulfide (**110**) with potassium hydroxide in alcohol (ethanol or isobutanol):

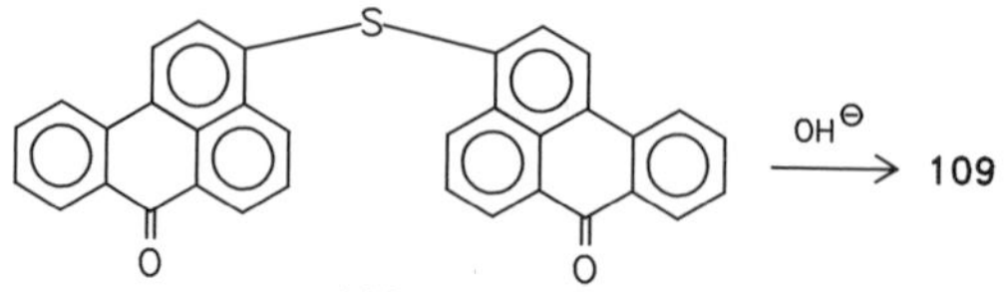

110

110 is made from benzanthrone, which is either chlorinated or brominated in 3-position (**111**). Subsequent treatment with sodium disulfide at 135 to 150 °C under pressure leads to **110**.

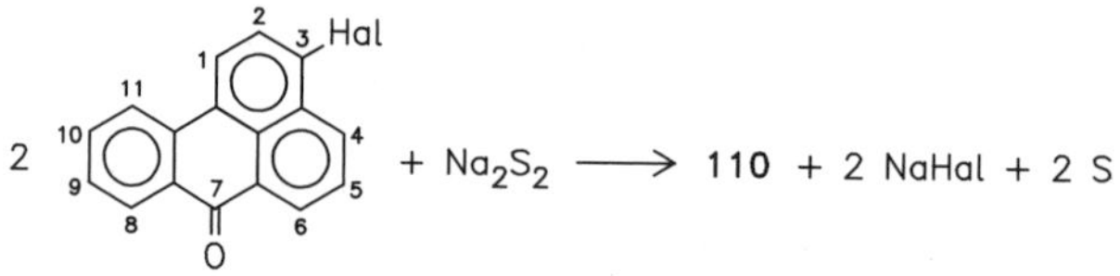

111   Hal: Cl, Br

In contrast to earlier methods which involved alcoholic potassium hydroxide fusion of 3-chloro-benzanthrone, this route affords pure isoviolanthrone, free of isomers.

Benzanthrone, in turn, is obtained from anthraquinone by reaction with glycerine and sulfuric acid in the presence of a reducing agent such as iron. Anthraquinone is initially reduced to anthrone, which is condensed with acrolein. Acrolein, on the other hand, is an intermediate of the reaction between glycerine and sulfuric acid. The synthesis proceeds by oxidative cyclization with sulfuric acid. Finally, halogenation of benzanthrone in sulfuric acid or in chlorosulfonic acid in the presence of a catalyst (sulfur, iodine, or iron) affords 3-chloro or 3-bromobenzanthrone.

Isoviolanthrone was first synthesized in 1907 and has long been known as a vat dye. However, in contrast to its halogenated derivatives, unsubstituted isoviolanthrone has failed to gain technical recognition as a pigment.

Pigment Violet 31, 60010, a dichloroisoviolanthrone (**112**), is commercially available.

Dichloroisoviolanthrone is prepared by chlorination of isoviolanthrone with sulfuryl chloride in nitrobenzene. Chlorine substitution occurs in positions 6 and 15.

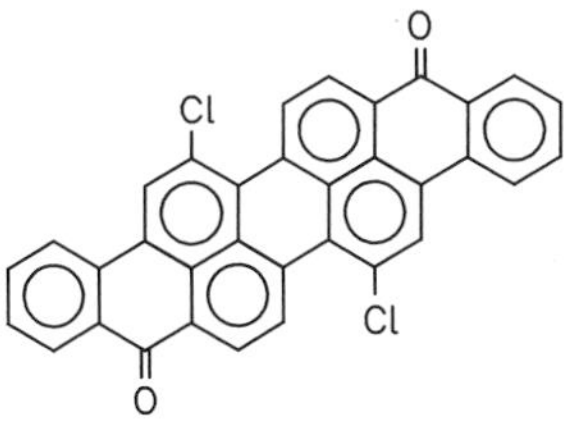

112

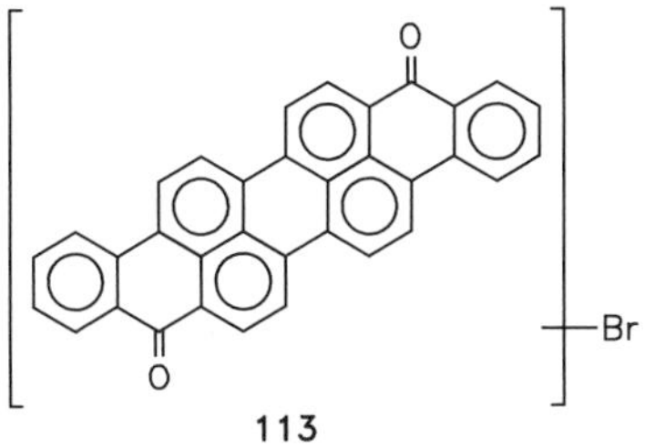

113

Compounds containing 12 to 17% bromine (and up to 1% chlorine) are referred to as bromoisoviolanthrone (**113**). The product has only recently been discontinued in the USA. It is obtained by bromination of isoviolanthrone at 80°C in chlorosulfonic acid, using iodine as a catalyst. Two crystal modifications are obtained: the α-modification is produced by treating crude isoviolanthrone in 90% sulfuric acid and discharging it into an aqueous dispersion solution. The β-modification evolves if the aqueous α-paste is heated in the presence of N-methylpyrrolidone and a surfactant, or generally by stirring the α-modification with organic solvents [23].

A higher quality α-modification with enhanced tinctorial strength and transparency is prepared from the leuco form of bromoisoviolanthrone. This intermediate in turn is manufactured by vatting crude bromoisoviolanthrone with sodium dithionite/aqueous sodium hydroxide. The product is separated and oxidized in an aqueous alkaline medium in the presence of surfactants. Application of shearing forces, preferably by means of a sand or pearl mill, and maintaining a temperature of 50°C produces improved pigment quality [24].

Properties and Application of Pigment Violet 31

Dichloroisoviolanthrone is supplied by the pigment industry in the form of a pigment preparation, which is a specialty product for use in spin dyed viscose fibers. Its shade is a reddish violet, which is distinctly redder than that of the known shade of P.V.23. The pigment exhibits very high quality fastness properties. Depending on the depth of shade, P.V.31 systems equal step 7 to step 8 on the Blue Scale for lightfastness. Their fastness properties in application, especially their fastness to perspiration, dry cleaning, and crocking are excellent. P.V.31 shows only poor resistance to vatting. Although the shade of spin dyed articles changes only slightly in an alkaline sodium dithionite solution (60°C/30 minutes), color is clearly transferred onto white spun rayon and cotton.

References for Section 3.6

[1] N.N. Woroshzow, Grundlagen der Synthese von Zwischenprodukten und Farbstoffen, Akademie Verlag, Berlin 1966; FIAT Final Report 1313 II, p. 22.

[2] H.-S. Bien et al., in Ullmann's Encyclopedia of Industrial Chemistry, 5. Ed., Vol. A2 (1985), p. 355–417.

[3] DE-AS 1 544 372 (BASF) 1965; DE-OS 1 544 374 (BASF) 1965.

[4] DE-OS 2 059 677 (BASF) 1970.

[5] DE-OS 2 843 873 (BASF) 1978.

[6] DE-OS 2 644 265 (Bayer) 1976.

[7] B.K. Manukian and A. Mangini, Helv. Chim. Acta 54 (1971) 2093–97.

[8] CH-PS 396 264 (Ciba) 1960.

[8a] K. Ogawa, H.J. Scheel, F. Laves, Naturwissenschaften **53** (24) 700–701 (1966).

[9] US-PS 2 727 044 (Interchemical) 1953.

[10] DE-PS 1 283 542 (Ciba) 1962; DE-AS 1 795 102 (Ciba) 1967.

[11] E.G. Kiel and P.M. Heertjes, J. Soc. Dyers Col. 79 (1963) 21–27.

[12] DE-OS 2 300 019 (BASF) 1973.

[13] P. Rys and H. Zollinger, Farbstoffchemie, 3. Ed., Verlag Chemie, Weinheim, 1982, S. 146 ff; H. Zollinger, Color Chemistry, 2. Ed., VCH-Verlag 1991.

[14] K. Venkataraman, The Chemistry of Synthetic Dyes Vol. V, Academic Press, New York, 1971, 182 ff.

[15] e.g. DE-AS 2 854 190 (BASF) 1978.

[16] DE-AS 2 705 107 (BASF) 1977.

[17] DE-OS 2 105 286 (Ciba-Geigy) 1970.

[18] DE-OS 2 812 192 (Ciba-Geigy) 1977.

[19] DE-AS 2 748 860 (BASF) 1977.

[20] DE-OS 2 428 121 (Allied Chem. Corp.) 1973.

[21] DE-AS 2 007 848 (BASF) 1970.

[22] DE-AS 2 702 596 (BASF) 1977.

[23] DE-AS 1 284 092 (BASF) 1961.

[24] DE-AS 2 909 568 (BASF) 1979.

3.7   Dioxazine Pigments

Pigments which are derived from the triphenedioxazine structure (**114**)

114

have been known since 1928. The compound provides an orange color, but has no technical importance as a colorant. It was in 1928 that Kränzlein and coworkers (Farbwerke Hoechst) found that sulfonated derivatives of this basic structure pro-

vide dyes which can be used as direct dyes on cotton. In the synthesis, sulfonation
follows the formation of the heterocyclic ring system.

It was not until 25 years later that a tinctorially strong violet pigment was obtained by formation and finishing of a water insoluble 9,10-dichlorotriphenedioxazine derivative [1].

At present, dioxazines no longer enjoy commercial significance as dyes. The
pigments industry shows little interest in substituted derivatives of the described
parent structure other than this violet pigment.

3.7.1 Preparation of the Starting Materials

One of the primary starting materials for all members of this group is chloranil
(tetrachloro-p-benzoquinone). Today, it is customarily prepared by oxidative chlorination of hydroquinone.

As an example, a mixture of hydroquinone and concentrated hydrochloric acid
is chlorinated initially at 10 °C. After adding water, the reaction mixture is heated
and the chlorination continued. Similar but slightly modified routes involve using
hydrochloric acid/H_2O_2 or chlorine/water. Hydroquinone may be replaced by benzoquinone [2].

Another important starting material is 3-amino-N-ethylcarbazol. Carbazol is
obtained from hard coal tar and is usually ethylated with ethylbromide or with
ethylchloride. Subsequent nitration and reduction affords the required compound.

3.7.2 Chemistry, Preparation of the Pigments

Dioxazine Pigments have the general structure **115**:

115

Industrially important pigments typically show the following substitution pattern: A represents an ethoxy group, B represents an acetylamino or a benzoylamino moiety, and X is a chlorine atom or an NHCOCH$_3$ function. The commercially most important compound **115**, however, is somewhat of an exception in that A and B represent the heterocyclic moiety 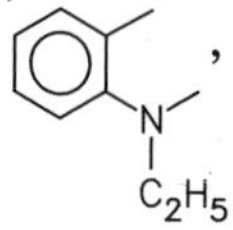 , which results in the following structure:

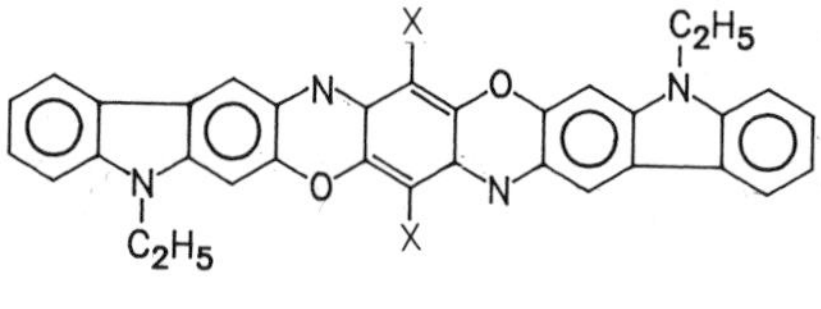

The synthetic route proceeds via a two-step pathway:
- Linking the central ring via NH bridges to aromatic units, i.e., formation of the 2,5-diarylamino-1,4-benzoquinones (**116**), and
- Cyclization of **116** to afford the triphenoxazine ring system.

The first step, i.e., linkage via NH bridges, may start from o-substituted anilines or o-alkoxyanilines. The reaction typically proceeds in an organic solvent (such as boiling ethanol) below 100 °C and in the presence of an acid trap such as sodium acetate:

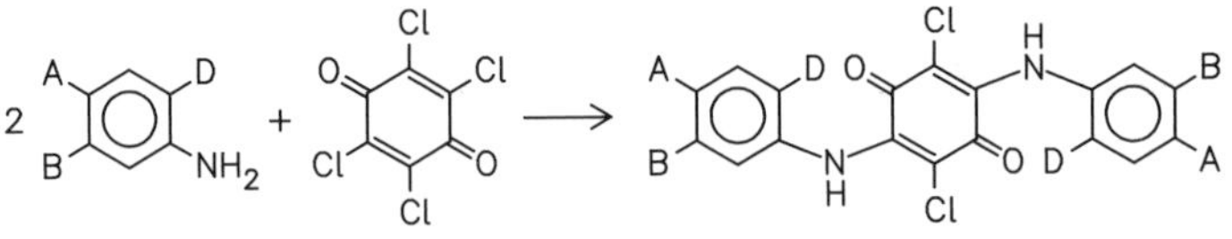

116

D: hydrogen or alkoxy group (OC$_2$H$_5$)

116 featuring hydrogen in D-position is cyclized by oxidative condensation under comparatively harsh conditions. The material is treated at 180 to 260 °C in a high boiling organic solvent such as chloronaphthalene in the presence of a condensation agent in the form of an acidic catalyst. Suitable agents include benzenesulfochloride, p-tosylchloride, m-nitrobenzenesulfochloride, and aluminum chloride.

Compounds with the structure **116** carrying ethoxy or methoxy groups in D-position are prepared from o-alkoxyanilines. Not only are cyclization and condensation achieved at comparatively lower temperature, between 170 and 175 °C, but the reaction also proceeds at a higher rate than for compounds with D = H. One of the above-mentioned condensation agents is similarly necessary in this case. o-Dichlorobenzene, for instance, is a suitable organic solvent. The list of solvents for condensation and cyclization also includes trichlorobenzene and nitrobenzene. The reaction conditions are not markedly affected by the type of substitution in positions A and B.

Both 3-amino-N-ethylcarbazole and 1,4-diethoxy-2-amino-5-benzoylaminobenzene are suitable aromatic amines. Chloranil may also be replaced by 2,5-dichloro-3,6-bisacetylamino-1,4-benzoquinone to react with appropriately substituted o-alkoxyanilines.

Pigment Violet 23 may serve as an example to illustrate the synthesis of dioxazine pigments in general. The technique has not changed much since it was first developed [3].

An excess of tetrachloro-p-benzoquinone (chloranil) in o-dichlorobenzene is added to 2 moles of 3-amino-N-ethylcarbazol and dry sodium acetate, which acts as an acid trap. The mixture is stirred for 6 hours at 60 °C, then within 5 hours heated to 115 °C under vacuum, after which benzenesulfochloride is added at the same temperature. Cyclization is achieved by increasing the temperature to 175–180 °C. The reaction mixture is agitated until no more acetic acid appears in the distillation receiver (for 4 to 8 hours). The reaction product is vacuum filtered, residual o-dichlorobenzene removed by steam distillation, the product washed, and dried. A patent which was issued in 1980 claims a much improved yield if the reaction is performed in the presence of only a slight excess of chloranil, provided 0.15 to 1.8 wt% water is added to the reaction mixture [4].

Production of the crude pigment is followed by finishing. The options include milling the material with salt in a ball mill in the presence of an organic solvent, or treating it similarly in a kneader, or producing a 60 to 90% sulfuric acid slurry, or using aromatic sulfonic acids.

3.7.3   Properties

Those pigments of the dioxazine series which are of importance to the pigment industry provide a clean violet (considerably bluish red) shade. They demonstrate excellent lightfastness and weatherfastness, even in light shades. There is a certain disadvantage to the fact that these pigments, and especially the most important one, tend to bleed if incorporated in plastics.

Dioxazine pigments are found in more than one crystal modification [5].

3.7.4   Commercially Available Dioxazine Pigments and Their Application

General

At present, this class produces only two representatives which continue to be used industrially: Pigment Violet 23, 51319 (**117**):

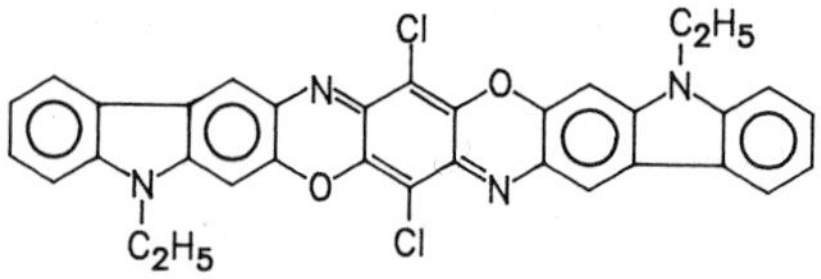

and Pigment Violet 37, 51345 (**117a**):

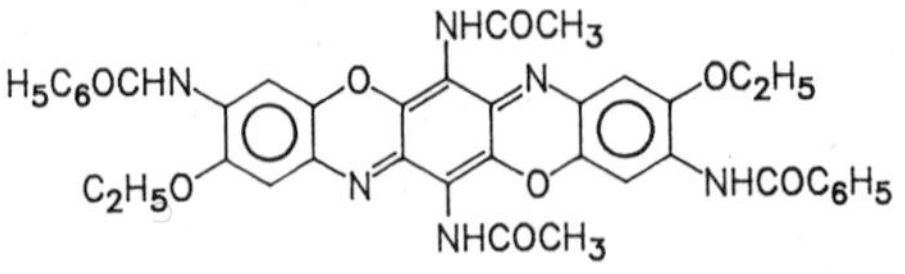

While P.V.23 is used in considerable volume, P.V.37 is a specialty product. Pigments of the P.V.34 and 35 type, which are much inferior in terms of tinctorial strength and sometimes also migration fastness, especially in plastics, have been discontinued several years ago. These compounds typically contained ethoxyaniline with acetylamino or benzoylamino moieties instead of the carbazole group of P.V.23 [6,7,8].

Individual Pigments

Pigment Violet 23

P.V.23, also referred to as Carbazole Violet, is a universally useful product. Its color, a bluish violet shade, is not accessible with other Pigments. P.V.23 is used in almost all media which are typically colored with pigments. The list of suitable systems ranges from coatings and paints to plastics, printing inks, and other spe-cial-purpose media. P.V.23 is entirely fast to many organic solvents. At standar-dized conditions (Sec. 1.6.2.1), it is fast to alcohols, esters, and aliphatic hydrocar-bons as well as to plasticizers such as dibutyl and dioctyl phthalate. Other solvents, such as ketones, are colored slightly (step 4).

Tinctorially, P.V.23 is an uncommonly strong pigment in almost all media, which qualifies it, even in very small amounts, for use as a shading pigment. Used to a considerable extent as a shading component for paints, P.V.23 adds a reddish tinge to the shade of copper phthalocyanine blue pigments. Although P.V.23 is not quite as lightfast and weatherfast as phthalocyanine blue pigments, it does satisfy most requirements, even very stringent ones. Moreover, P.V.23 is also a useful ton-ing pigment for white enamels. The pigment is particularly important in systems based on TiO_2/rutile with their yellowish undertone. Trace amounts of P.V.23 are used to tone the system: only 0.0005 to 0.05 parts of P.V.23 are required per 100 parts of TiO_2. White enamels may also be toned by mixtures of P.V.23 and α-Copper Phthalocyanine Blue. P.V.23 is both completely lightfast and weather-fast, even in very light shades (for instance at 1:3000 reduction with TiO_2). Apart from this excellent fastness, P.V.23 has the added advantage of being entirely fast to overcoating. It is therefore used in all types of media throughout the paint in-dustry. The list ranges from air drying house and trade sales paints, i.e., decorative (architectural) paints, to general and high grade industrial finishes, such as original automotive (O.E.M) and automotive refinishes. P.V.23 is used in uniformly col-ored finishes as well as in metallic finishes. Baking enamels containing P.V.23 are thermally stable up to 160 °C.

Most grades exhibit a high specific surface area, sometimes above 100 m^2/g. For good dispersion and in order to avoid flocculation, it is therefore necessary to maintain an adequate pigment/binder ratio. It should be noted that the required amount of binder is higher in this case than for most other organic pigments.

P.V.23 is a favorite shading pigment for use in emulsion paints, where it lends a reddish tinge to Phthalocyanine Blue shades. Excellent weatherfastness makes it a suitable candidate for exterior application in media based on synthetic resin disper-sions. The systems are fast to alkali and plaster.

Plate-out is observed with various hardeners if P.V.23 is incorporated in most types of powder coatings, for instance in epoxy systems. This phenomenon, howev-er, is of no consequence for the applicability of the pigment: only minor amounts of P.V.23 are required in these media to lend a blue shade to white enamels.

The list of suitable applications also includes a variety of special-purpose sys-tems, especially where high lightfastness and durability, but also high heat stability or other perfect fastness properties are a prime concern. This is true, for instance, for coatings to be applied onto aluminum window blinds. Bilaterally coated, usual-ly in pastel colors, these aluminum strips are baked at 250 °C or less and subse-quently subjected to considerable mechanical stress.

P.V.23 is also frequently encountered in plastics. Although plasticized PVC sys-tems containing P.V.23 are not entirely migration resistant, the pigment exhibits unusually high tinctorial strength. Less than 0.3% pigment is needed, for instance, to produce 1/3 SD systems (5% TiO_2). The systems demonstrate excellent lightfast-ness: 1/3 SD specimens equal step 8 on the Blue Scale, while 1/25 SD samples match step 7–8. Moreover, P.V.23 provides excellent weatherfastness and is suited to long-term exposure. It is an important colorant for PVC and PUR plastisols.

In terms of heat stability, 1/3 SD polyolefin systems containing P.V.23 with-stand exposure to 280 °C. In lighter tints, this value is appreciably lower: 1/25 SD samples only tolerate up to 200 °C. Dissolution effects shift the shade considerably towards the reddish region, while the pigment maintains its shade at higher temper-atures. P.V.23 is an equally tinctorially strong product in these media. 1/3 SD HDPE samples (1% TiO_2) are formulated at less than 0.07% pigment. P.V.23 is a typical polycyclic pigment in that it affects the shrinkage of injection-molded arti-cles made of HDPE and other partially crystalline polymers. This is a tendency which somewhat restricts its use in such media. The pigment concentration in a transparent HDPE system should not be less than 0.05%. P.V.23 systems rapidly lose lightfastness as more white pigment is added. 1/3 SD samples equal step 8 on the Blue Scale, while 1/25 SD specimens only score as high as step 2.

Carbazol Violet is a suitable pigment for transparent polystyrene products. A certain degree of dissolution in this medium at elevated temperatures makes it necessary to limit the use to processes up to approximately 220°C. P.V.23 is also used in polyester which is processed at very high temperature. The pigment is equally suitable for use in polyester spin dyeing, it is stable to the harsh conditions of the condensation process, during which it is exposed to 240 to 280°C for 5 to 6 hours. At low pigment concentrations there is also a color shift towards redder shades, due to pigment dissolution. P.V.23 is highly lightfast. Although it generally offers high textile fastness properties, the pigment is not entirely fast to dry and wet crocking. P.V.23 also satisfies the thermal and other requirements for use in spin dyed polyacrylonitrile. In this context, the pigment is completely fast to dry and wet crocking. In PP spin dyeing, P.V.23 should not be applied at low concentrations in order to prevent dissolution effects. It provides excellent lightfastness in medium and deep shades. P.V.23 is also recommended for use in spin dyed viscose, where it exhibits good lightfastness and generally excellent fastness properties in application.

Worked into cast resins based on methacrylate or unsaturated polyester, P.V.23 has the advantage of being fast to peroxides, which act as catalysts in these media. The lightfastness of such systems is between step 7 and step 8 on the Blue Scale, both for transparent and opaque colorations.

P.V.23 has stimulated interest throughout the printing ink industry. It is frequently utilized in combination with copper phthalocyanine blue pigments to produce reddish shades of blue. P.V.23 is equally strong in these media. At standardized film thickness, 1/1 SD letterpress proof prints are prepared from inks formulated at only 8.7% pigment, while 1/3 SD specimens require inks containing 5.6%. In terms of lightfastness, these prints equal step 6 to step 7 on the Blue Scale. The prints are very fast to organic solvents and also exhibit excellent application properties, such as fastness to soap, alkali, and fat. The prints may be exposed to 220°C for 10 minutes or to 200°C for 30 minutes and may safely be calendered or sterilized. P.V.23 is also used in textile printing. It is very lightfast and weatherfast in this application and also performs well in terms of other important fastness properties.

There are a number of other media which are also pigmented with P.V.23. The list includes office articles and artists' colors, such as drawing inks and fiber-tip pen inks, wax crayons, oil paints, and high quality water colors, water- or solvent-based pigmented wood stains, cleaning agents, and mass colored paper.

Pigment Violet 37

P.V.37, a dioxazine pigment, provides a much more reddish shade than P.V.23 but is equally fast to organic solvents. P.V.37, although weaker in some media than P.V.23, is broad in scope. Printing inks based on nitrocellulose are unique in that P.V.37 provides high tinctorial strength and excellent gloss in these media, although the products are somewhat opaque. Despite the fact that, incorporated in oil-based printing inks for purposes such as offset printing, it is somewhat weaker than P.V.23, P.V.37 exhibits high gloss and good flow behavior. Its main field of application within the printing ink industry is in metal deco printing. Its fastness properties in application equal those of P.V.23.

In paints, P.V.37 also exhibits a considerably more reddish shade and reduced tinctorial strength. Similar observations apply to its coloristic properties in polymers. In plasticized PVC, P.V.37 has the advantage of being considerably more migration resistant than P.V.23; in fact, it is almost entirely fast to migration. Since P.V.37, like P.V.23, dissolves to some extent in polymers, it loses much of its heat stability as the white pigment concentration in a system is increased. Similar observations apply to polyolefin systems: the shade provided by P.V.37 is considerably more reddish and the tinctorial strength much inferior to that of P.V.23. 0.09% P.V.37 is required to afford 1/3 SD in HDPE systems (1% TiO_2). Such systems are heat stable up to 290 °C, but it should be noted that the heat stability declines rapidly in certain concentration ranges, a tendency which is also observed with P.V.23. Like P.V.23, P.V.37 also considerably affects the shrinkage of HDPE and other partially crystalline injection-molded plastics. P.V.37 is also recommended for use in PP spin dyeing. As with P.V.23, systems formulated at low concentrations tend to undergo a color shift due to pigment dissolution.

References for Section 3.7

[1] DE-PS 946 560 (Hoechst) 1952.
[2] DE-OS 3 707 148 (Hoechst) 1988; E-PS 326 455 and E-PS 326 456 (Rhone-Poulence) 1988.
[3] BIOS Final Report No. 960, p. 75.
[4] DE-OS 3 010 949 (Cassella) (1980).
[5] A. Pugin, Offic. Dig. Fed. Soc. Paint Technol. 37 (1965) 782–802.
[6] DE-AS 1 243 303 (Ciba) 1962.
[7] DE-AS 1 142 212 (Ciba) 1959.
[8] DE-AS 1 174 927 (Ciba) 1959; DE-AS 1 185 005 (Ciba) 1961.

3.8 Triarylcarbonium Pigments

As a class, these pigments share the triarylcarbonium structure. At least two of the aryl moieties carry amino groups, which act as electron donating substituents.

The compounds are therefore basic and thus capable of combining with acids to form insoluble salts, which is the form in which they can be applied as pigments.

Triarylcarbonium compounds can be described by several mesomeric structures, either in quinoid form with an ammonium ion, or by a benzenoid structure with a carbonium ion.

For reasons of molecular symmetry, only the carbonium structure is used in this book to describe triarylcarbonium pigments.

Two types of insoluble triarylcarbonium compounds are used industrially as pigments. Both are salts of these basic dyes. So-called Alkali Blue type triarylcarbonium pigments are inner salts of sulfonic acids, while the second group comprises salts of complex inorganic anions of heteropolyacids.

For a long time, triarylcarbonium compounds were used exclusively as textile dyes. Wool was colored using acidic dyes, i.e., compounds which contained several sulfonic acid functions, while silk was dyed with basic products. Cotton was prepared by "mordanting" the fiber with amine/tartar emetic (potassium antimonyl tartrate), the dye was fixed in the form of an insoluble compound. "Lakes" were prepared by precipitating aqueous solutions of basic dyes with tannine/tartar emetic onto a mineral carrier (aluminum oxide, barium sulfate, calcium sulfate). Although the resulting "pigments" provided brilliant shades, they offered insufficient lightfastness and are thus no longer of importance.

3.8.1   Inner Salts of Sulfonic Acids (Alkali Blue Types)

These compounds are triphenylmethane pigments. Strictly speaking, they should be referred to as triaminophenylmethane derivatives. The parent structure is known as reddish violet parafuchsin (**118**) or its anhydro base pararosaniline (**119**). The parent compound may bear between one and three methyl substituents (fuchsin, new fuchsin).

118   119

The entire group of these compounds dates back to the very beginnings of organic dye chemistry. In 1858, E. Verguin in France oxidized a material which he named "aniline" but which was in actual fact a mixture of aniline, o-toluidine, and p-toluidine. He performed the reaction in nitrobenzene in the presence of tin(IV)chloride or iron(III)chloride and received bluish red fuchsin (**120**). The process has been industrially exploited since 1859. The central carbon atom is furnished by the CH_3 group of p-toluidine, which is initially oxidized to its aldehyde.

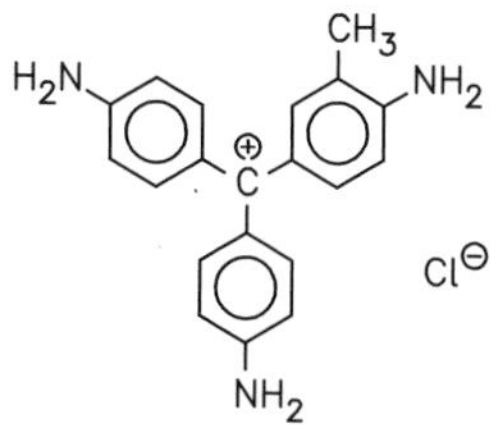

120

Shortly afterwards, a route to acidic triphenylmethane compounds, referred to as Alkali Blue types, was developed.

In 1860, Girard and de Laire heated fuchsin with aniline and found triphenylfuchsin, known as Lyon Blue. Nicholson, in an attempt to make this structure water soluble, in 1862 introduced free sulfonic acid groups into the molecule through sulfonation.

The thus prepared product, however, being the inner salt of a sulfonic acid, turned out to be entirely insoluble in water. The compound only became a suitable textile dye after it had been converted into its sodium salt, i.e., through application in a slightly alkaline medium (= Alkali Blue).

The principle of forming the important inner salts of sulfonated triphenylmethane derivatives had thus been discovered. All of the commercially important products within this group are derived from phenylated rosaniline corresponding to the structure **121.**

3.8.1.1   Chemistry, Manufacture

Alkali Blue type pigments are based on the following general structure:

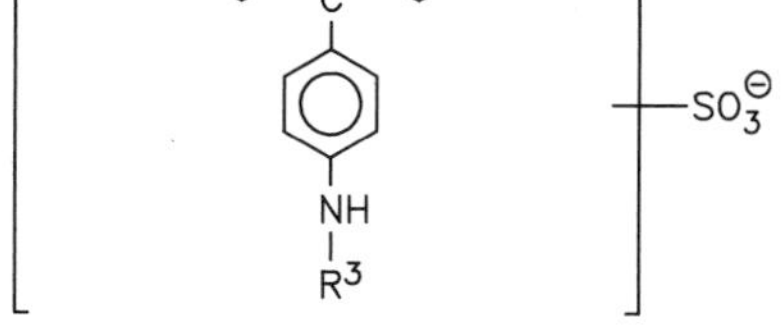

121

Manipulation of the number of phenyl groups R^1 to R^3 on the rosaniline nucleus and their substitution pattern has proven a useful tool in producing a variety of commercially important derivatives. Currently, compounds with two (R^1, R^2 = $C_6H_4CH_3/C_6H_5$; R^3 = H) and especially three phenyl and/or toluene moieties (R^1, R^2, R^3 = $C_6H_5/C_6H_4CH_3$) are technically important. The CH_3 groups are primarily located in m-position relative to the secondary amino group.

It should be mentioned that all triarylcarbonium pigments are described by structures which are no more than a useful approximation of reality. These products in actual fact represent mixtures of various compounds which are obtained through an intricate reaction pattern, the respective written structure being the main component.

The substituted triphenylmethyl (trityl) system **121** may basically be synthesized by two different routes, both of which are in use.

Option A involves preparation of triaminotriarylmethane derivatives and their subsequent reaction with aromatic amines in the presence of acidic catalysts.

Route B proceeds via trihalogentriarylcarbonium compounds, which are reacted with aromatic amines.

The most important starting materials for process A are 4,4',4''-triaminotriphenylmethane, pararosaniline (**119**), and parafuchsin (**118**). Aniline and formaldehyde are treated at 170 °C to form, apart from some formaldehyde-aniline intermediates, 1,3,5-triphenylhexahydrotriazine as the main component. Subsequent treatment with an acidic catalyst, for instance with hydrochloric acid, in excess aniline as a solvent initially affords 4,4'-diaminodiphenylmethane, which is finally oxidized to yield parafuchsin (**118**). Iron(III)chloride and nitrobenzene, which in the past were used as oxidants, are no longer used. The reaction is now performed by air oxidation in the presence of vanadium pentoxide as a catalyst.

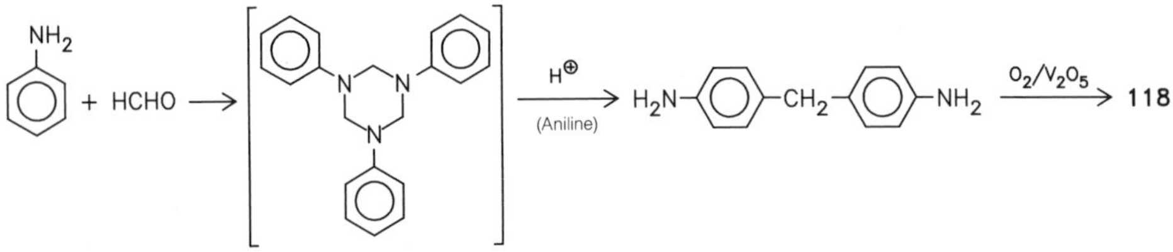

Other intermediate products are [1]:

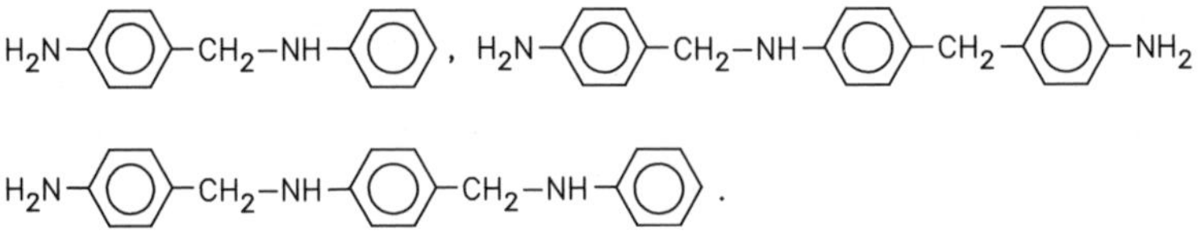

Pararosaniline, after being formed from parafuchsin with sodium hydroxide, is fused at 175 °C with aniline and/or m-toluidine, for instance in the presence of catalytic amounts of benzoic acid, and thus arylated.

The aniline or toluidine residue in the melt is removed by vacuum distillation at 150 °C. The melt is allowed to cool to room temperature and then broken.

Various degrees of arylation may be achieved, depending on the fusion time and on the amount of benzoic acid used. Longer fusion times afford more greenish Alkali Blue types, corresponding to a higher yield of phenylated triaryl product, while the amounts of the more reddish diaryl and monoaryl pararosanilines decrease with the time. Careful reaction control makes it possible to convert the materials into the corresponding water insoluble monosulfonates through sulfonation with concentrated sulfuric acid:

R^1, R^2, R^3: C_6H_5 and/or $C_6H_4CH_3$ and/or H

The crude pigment provides only limited tinctorial strength. This necessitates the following aftertreatment: dissolution as a sodium salt, precipitation with mineral acid, possibly in the presence of surface active agents, followed by drying or flushing in order to transfer the material from the aqueous to the oily phase (mineral oil, linseed oil).

Flushing involves kneading the aqueous pigment presscake with varnish until a complete exchange has taken place between the two liquid phases. The aqueous phase is then separated and residual water removed by vacuum distillation at elevated temperature. Flushing affords a product which can be incorporated directly into a printing ink (Sec. 1.6.5.8).

Route A is generally unacceptable, due to a poor parafuchsin yield (in the past: 35%, now: approx. 60%). The arylation step, on the other hand, proceeds with good yield. The formation of a comparatively large amount of by-products makes it necessary to purify the product extensively before it is processed further. Pathway B starts from p-chlorobenzotrichloride and chlorobenzene, which undergo a Friedel–Crafts reaction, forming the tetrachloroaluminum complex of trichlorotriphenylmethyl chloride (122). 122 reacts with aromatic amines only if it is first condensed with at least one meta- or para-substituted aniline. Unsubstituted aniline, deactivating the aluminum complex, does not afford arylation products. Meta- or para-substituted anilines, however, produce arylated pararosanilines in almost theoretical yield.

R : H or (m- or p-)CH$_3$

Subsequent treatment with alkali provides a free base solution in excess amine, which can be separated from the aqueous/alkaline aluminate layer. The thus prepared product is precipitated with acid, possibly converted into the free base with aqueous alkaline solution, and isolated. The dried base or its salt is then sulfonated with concentrated sulfuric acid to form the monosulfonic acid 121.

Route B of this process may be substantially improved in terms of yield and product quality (purity) of the resulting triarylaminoarylcarbonium pigment. To this end, the solution of the free dye base is treated with an excess of aqueous sulfuric acid (20 to 40%) in a solvent such as chlorobenzene or an aromatic amine. This method produces the sulfate of the basic dye, which is insoluble in this medium, together with the soluble sulfates of the primary aromatic amines, which can therefore easily be separated. The isolated sulfate of the basic dye is then washed and in dry or wet condition monosulfonated with 85 to 100% sulfuric acid. Based on the dye base sulfate, this step affords 96 to 98% yield, compared to only 83 to 89% achieved by the previously described method. The entire synthesis, including the intermediate isolation of the triarylaminoarylmethane sulfate, may also be performed by continuous process [2].

Route B has the added advantage over route A of making it possible to perform a stepwise reaction of trichlorophenylmethyl-tetrachloroaluminate **122** (i.e., the complex compound formed from tri-(chlorophenyl)methylchloride and AlCl$_3$) with various amines. It is thus possible to systematically synthesize triphenylmethane pigments with two or three differently substituted arylamino groups.

3.8.1.2 Properties

As a result of the nonuniform reaction process, all commercial products within the Alkali Blue series represent mixtures of various products. The respective structure which is listed in the Colour Index only reflects the main component of a differently arylated mixture. Moreover, the aromatic moieties not only represent differently substituted compounds but also mixtures of various degrees of sulfonation.

Special-purpose triphenylmethane pigments are frequently even "tailor-made", designed to suit the individual needs of different areas of application.

Alkali Blue types cover a wide range of shades from the reddish blue to the violet portion of the spectrum. The shade becomes greener as the number of phenyl groups in the molecule increases. Both the tinctorial strength and the solubility of the pigment are controlled by manipulating the degree of sulfonation: one sulfonic acid function per molecule affords an optimum in tinctorial strength, while a higher degree of sulfonation adversely affects the strength but simultaneously improves the solubility of the pigment in water. This explains why compounds with two to three sulfonic acid groups (such as "Water Blue" for paper mass coloration) are only infrequently used as pigments (by forming an insoluble aluminum lake by adding alum). Products featuring four to five sulfonic acid groups (such as "Ink Blue" for inks) are completely unsuitable for use as pigments.

3.8.1.3 Commercially Available Types and Their Application

The list of significant Alkali Blue type triarylcarbonium pigments includes compounds which are derived from diarylated and triarylated rosanilines. Table 32 shows triarylcarbonium pigments which are listed in the Colour Index.

In this chapter, relevant fastness properties and aspects of pigment performance as well as various facets of pigment application are discussed. The entire class is treated comprehensively instead of focussing on individual types. In Europe, Alkali Blue pigments are also referred to as Reflex Blue pigments, a trade name which was created by the local manufacturer.

The main area of application for these products is in printing inks, especially in offset and letterpress application and to a lesser extent also in aqueous flexographic inks. The pigments have less of an impact on the office articles sector. The printing ink industry employs triarylcarbonium pigments less as independent self colors but rather as shading pigments to tone black inks. Used at low concentrations, these pigments correct the brown touch of carbon black and provide a coloristically neutral, particularly deep black.

Tab. 32: Triaryl carbonium pigments.

C.I. Name	C.I. Constitution No.	R	R′	R^1	R^2	R^3
P.Bl.18	42770:1	C$_6$H$_5$NH	H	H	H	H*
P.Bl.19	42750	NH$_2$	CH$_3$	H	H	H
P.Bl.56	42800	(see structure) —NH	H	CH$_3$	CH$_3$	H
P.Bl.61	42765:1	C$_6$H$_5$NH	H	H	H	H

* The constitution which is listed in the Colour Index erroneously contains another SO$_3$H group in the R^2 and R^3 carrying ring. In fact, however, P.Bl.18 and P.Bl.61 are chemically identical.

Alkali Blue types cover the reddish blue portion of the spectrum. Even the very greenish specimens are still considerably redder than α- or ε-Copper Phthalocyanine Blue. Alkali Blue pigments are uncommonly strong. At standardized conditions concerning printed film thickness, 1/3 SD letterpress proof prints are prepared from inks containing only approx. 3.5% pigment. More than twice as much pigment is needed to prepare corresponding α-Copper Phthalocyanine Blue prints.

As a consequence of their polarity, triarylcarbonium pigments show a considerable tendency to agglomerate. Compared to other organic pigments, they are very difficult to disperse. Normal powder pigments are therefore rarely used in actual practice, the pigments are supplied predominantly in predispersed form as flushed pastes (Sec. 1.6.5.8). Pigments may be flushed into very different types of binders, which makes it possible to optimize the products for specific printing purposes, such as heat set inks. The fact that a flushed paste consists of predispersed pigment makes it possible to add any amount of paste, as needed, to a ready-made dispersion of carbon black.

The rheological and coloristic differences between flushed pastes are partly attributed to the pigment concentration. It should be noted, however, that tinctorial strength is a minor concern in a shading pigment: these products are selected primarily for their shading behavior. This depends not only on the type of carbon black, for instance its color cast, but it is also a function of the shade of Alkali Blue. Black prints made from inks containing Alkali Blue tend to bronze, a feature which equally contributes to the shading behavior of the product. This effect, which is difficult to assess and measure, results from a variety of factors [3]. Studies have indicated that, depending on the pigment concentration, practically all blue pigments bronze to a certain extent. The minimum extent of bronzing is given by the optical constants of the pigment. Stronger bronzing is attributed to floating of the blue pigment or to the penetration of a considerable amount of binder into the surface of the substrate, consequently increasing the pigment concentration on the surface of the dry printed film, a phenomenon which makes the reflection of light at the surface sensitive to the wavelength of the incident light.

Commercial flushed pastes commonly contain about 40% pigment; however, these products are standardized not in terms of pigment concentration but regarding their tinctorial strength. The ratio between carbon black and Alkali Blue pigment in a toned product may range between approximately 2:1 and 4:1, depending on the nature of the components and on the desired effect.

For several years, specialized Alkali Blue pigments, referred to as easily dispersible, have been available in powder form. Already during synthesis, these pigments are prepared with suitable resins or other agents in order to reduce the problem of pigment agglomeration during drying and milling. Besides, thus prepared pigments also have the advantage of facilitating the deagglomeration of agglomerated units within a printing ink. These types, however, do not disperse as readily as the easily dispersible members of other classes of organic pigments, for instance, of the diarylide yellow pigment series.

Powder grades are commonly dispersed using agitated ball mills, often directly in combination with the carbon black. Theoretically, one should expect pigments which are used in offset and letterpress inks to dissolve in their medium or to more or less recrystallize. Such effects are normally expected in view of the considerable thermal demands placed upon pigments, especially carbon black, by modern dispersion technology. In practice, however, no coloristic evidence of dissolution or recrystallization has been found. The primary components in the respective oil-based binders, i.e., offset and letterpress printing inks, are largely nonaromatic mineral oils as well as linseed oil and similar solvents. Alkali Blue pigments, tested for solvent fastness in pigment powders at standardized conditions (ambient temperature) (Sec. 1.6.2) prove to be completely resistant to these solvents, but not to alcohols and ketones. Insufficient fastness to alcohols is one of the main factors in precluding Alkali Blue from use in nitrocellulose-based printing inks. The printing ink often loses tinctorial strength and gloss, an effect which, due to pigment recrystallization, cannot be controlled.

Although Alkali Blue pigments are considerably faster to aromatic hydrocarbons than to alcohols and ketones, they are by no means entirely fast to these solvents. Alkali Blue is not applied in toluene publication gravure printing inks. Such black systems are typically shaded with Milori Blue, although sometimes in combination with Alkali Blue. Milori Blue exhibits a shading effect which is different from that of Alkali Blue types. At the concentrations which are needed to afford a satisfactory shading effect in carbon black, Milori Blue, incorporated in toluene-based publication gravure printing inks, unlike Alkali Blue, does not adversely affect the rheological properties of the black ink.

Prints containing Alkali Blue are not fast to the standard DIN 16524 solvent mixture, but they are fast to acid, paraffin, butter, and other materials. Tested in accordance with normative testing standards (Sec. 1.6.2.2), the prints unexpectedly also show fastness to alkali. It should be noted, however, that at higher alkali concentrations the tinctorial strength of the system declines and the shade becomes duller. This is a result of the fact that the pigment reacts with alkali.

Alkali Blue pigments are used to a considerable extent in aqueous flexographic inks, despite a slightly alkaline pH value. The resulting printing inks, however, may show a viscosity increase. Such inks are prepared from powder grades, because flushed pastes contain binders which are mostly incompatible with aqueous printing inks. Special-purpose pigment preparations have also recently been introduced to the market for this application.

Alkali Blue pigments may safely be exposed to 140 to 160 °C for 30 minutes. They are not fast to clear lacquer coatings and to sterilization. The pigment is therefore normally excluded from use in metal deco printing, although there are exceptions. Alkali Blue is employed, for instance, for blue cream containers which are not exposed to direct sunlight. Pigment blends of Copper Phthalocyanine Blue and Pigment Red 57:1 are frequently utilized as suitable shading pigments for black metal deco printing inks or for other black printing inks which are fast to overlacquering.

In terms of lightfastness, Alkali Blue performs moderately. 1/1 SD letterpress proof prints equal step 3 on the Blue Scale, while 1/3 to 1/25 SD formulations equal only step 2. However, as shading components their lightfastness is excellent and satisfies the requirements for all types of applications. This is due to the fact that a large portion of the incident light is absorbed by carbon black.

Alkali Blue pigments are used in large volume to color office articles, especially ribbons for typewriters and computers, as well as blue copy paper. Other areas of application, such as the plastics industry, do not employ Alkali Blue pigments because of their lack of fastness.

3.8.2   Dye Salts with Complex Anions

This class of pigments consist of salts of the anions of complex inorganic acids with dye cations, primarily with triarylcarbonium cations.

As early as 1913, BASF research succeeded in precipitating cationic ("basic") dyes with a heteropolyacid. The specific acid used in this case was phosphotungstic acid (which at the time was still applied on aluminum oxide hydrate as a carrier). The new products were patented. Using these and other heteropolyacids proved a useful tool in improving the lightfastness of the dye salts, and the importance of the pigments was enhanced even further as it became apparent that eliminating the mineral carrier led to a drastic increase in the tinctorial strength. As a consequence of these advances in pigment technology, complex salts became very important in the time between the First and the Second World War. Other heteropolyacids, es-

pecially phosphomolybdic acid and finally also a combined phosphotungstomolybdic acid, likewise stimulated considerable interest.

The shortage of tungsten and molybdenum during the 1930s in Germany neces-
sitated the use of copper ferrocyanide as anion. Some of these salts have maintained their commercial position to the present time. Besides, silicomolybdates are
also used to produce this type of pigments.

3.8.2.1   Chemistry, Manufacture

The list of dye cations which are used to prepare these complex salts includes primarily two types of triarylcarbonium compounds, namely:

-   triphenyl or diphenylnaphthyl derivatives with the chemical structure **123**, and
-   phenylxanthene derivatives with the structure **124**.

The most important group in this series comprises compounds with the structure **123** (A$^-$ = acid anion)

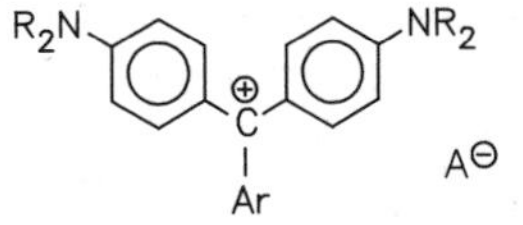

123

with R = methyl or ethyl and Ar = phenyl, 4-dimethylaminophenyl, or 4-ethyl-
aminonaphthyl. It is the substitution pattern that defines whether a product will
provide violet, blue, or green shades.

Compounds with the structure **124**

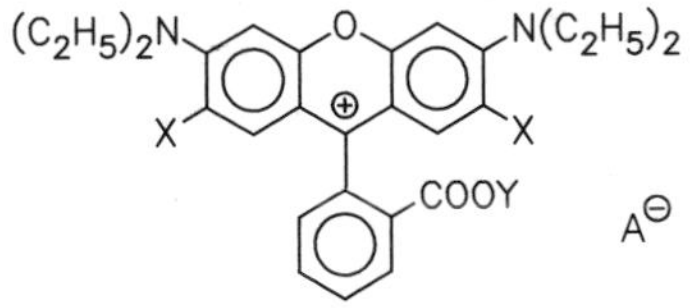

124

with X = hydrogen or methyl and Y = hydrogen or ethyl primarily afford bluish
red ("pink") colors.

Another member of this group is a yellowish green pigment, structurally derived from a benzothiazolium system (**125**)

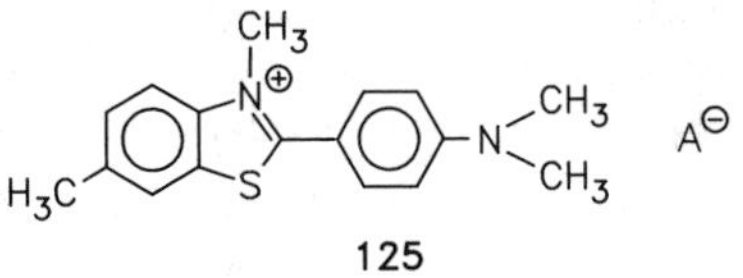

125

125, supplied in the form of a complex salt with a heteropolyacid, is used in combination with Pigment Green 1.

Heading the list of possible acid anions, A^-, are complex phosphoric acids, particularly those with $Mo_3O_{10}^{3-}$ or $W_3O_{10}^{3-}$ ligands. Formally, these acids are derived from phosphoric acid H_3PO_4 by replacing all oxygen atoms by molybdate and/or tungstate groups. The resulting heteropolyacids may be written as follows:

$$H_3[P(W_3O_{10})_4].aq \qquad H_3[P(Mo_3O_{10})_4].aq$$

The same structures may also be represented as

$$H_3H_4[P(W_2O_7)_6].aq \qquad H_3H_4[P(Mo_2O_7)_6].aq$$

Alternatively, it is possible to replace the phosphorus atom by silicon, which affords products among which silicomolybdic acid $H_4[Si(W_3O_{10})_4].aq$ and $H_4H_4[Si(Mo_2O_7)_6].aq$ are commercially of some significance. The second kind of formula emphasizes the fact that only 3 (or 4) out of 7 (or 8) hydrogen atoms provide possible substitution sites.

The dye salts are therefore represented as follows (F = dye group):

$$F_3[P(Mo_3O_{10})_4], \quad F_3[P(Mo_3O_{10})_3(W_3O_{10})] \quad \text{or} \quad F_4[Si(Mo_3O_{10})_4].$$

However, it should be noted that the stoichiometric aspect is idealized. The ratios between the different components can vary widely and are in actual fact controlled by the pH value and the precipitation temperature.

Reducing agents, such as zinc dust or sodium dithionite, convert heteropolyacids into deeply colored blue compounds consisting of heteropolyacids with four hydrogen atoms to be substituted. This makes it possible to precipitate more dye with the same heteropolyacid to produce an insoluble pigment.

Copper ferrocyanide complex salts, which are occasionally used, are derived from copper-1-hexacyano-iron-2-acid $HCu_3[Fe(CN)_6]$. Three equivalents of copper are required for each unit of ferrocyanide, which furnishes the following general pigment structure:

$$FCu_3[Fe(CN)_6].$$

Dye Synthesis

This chapter outlines the different synthetic routes to the most significant cationic dyes that these pigments are derived from.

Basically, compounds with a central C atom, which are considered electrophilic reaction centers, are treated with a nucleophilic aromatic compound. Subsequent oxidation affords the desired carbonium compound:

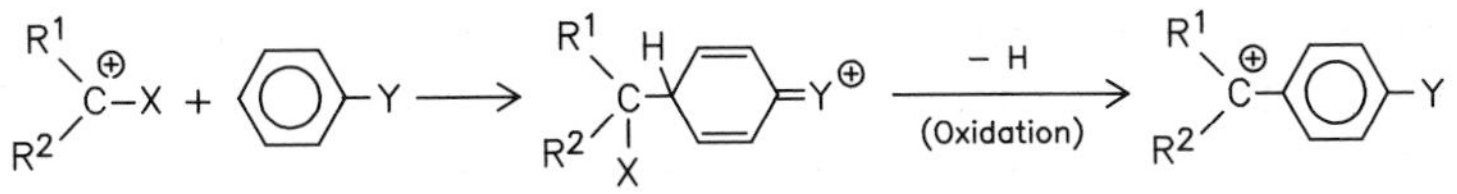

X : OH, SH;   R^1,R^2: aryl groups;
Y : N(CH$_3$)$_2$, N(C$_2$H$_5$)$_2$

Dyes With the Structure 123

Methyl Violet consists of the basic structure of Pigment Violet 3 (**126**).

It is prepared by air oxidation of dimethylaniline in the presence of phenol and copper salts as well as sodium chloride. The reaction product consists of tetra- to hexamethylated pararosanilines (see **119**, p. 522):

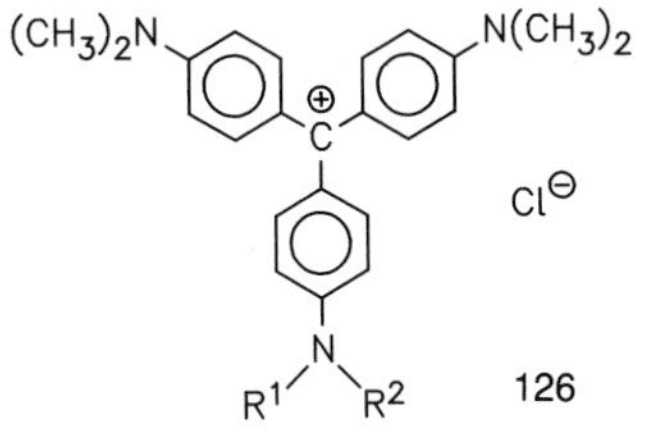

R^1, R^2: H and CH$_3$

126

In contrast to Methyl Violet, Crystal Violet (**127**) is a uniform compound. This is the dye which constitutes Pigment Violet 39.

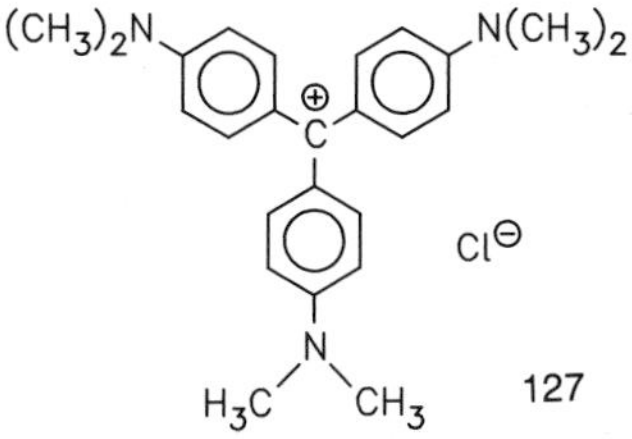

127

The compound is obtained from "Michler ketone" (**128**) and dimethylaniline in the presence of POCl$_3$. Michler ketone in turn is synthesized from dimethylaniline (DMA) and phosgene with zinc chloride:

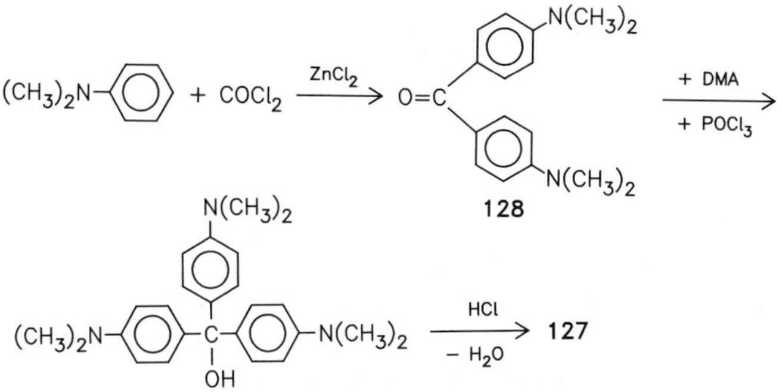

128

Another route proceeds via bis(dimethylaminophenyl)methane, which is synthesized from dimethylaniline and formaldehyde. The thus prepared intermediate may be oxidized to form the hydrol and then reacted again with dimethylaniline to afford the leuco base, which is finally oxidized to yield **127**:

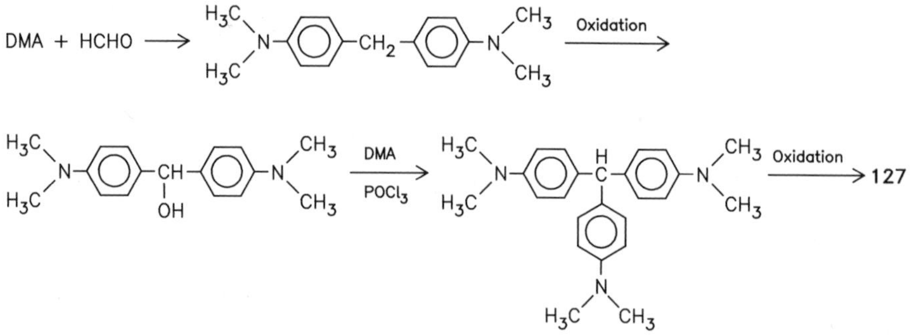

A blue shift is observed if the phenyl ring Ar in structure **123** is replaced by a naphthyl ring. The product is known as "Victoria Blue" (**129**):

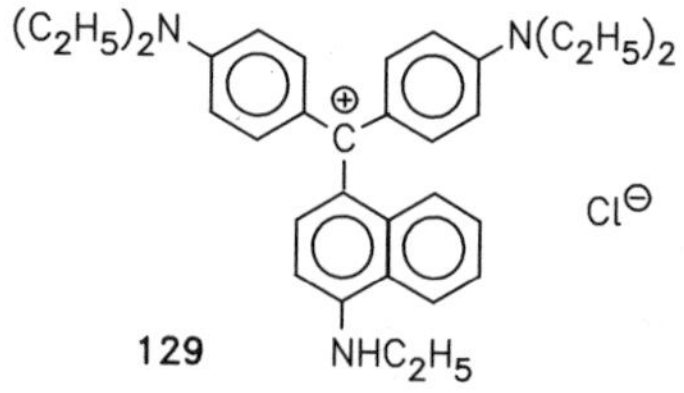

The product may be obtained from tetraethyldiaminobenzophenone and N-ethylnaphthylamine with phosphorus oxychloride or with phosphorus trichloride, the route parallels the synthesis of Crystal Violet. The resulting colorant is the basic dye for Pigment Blue 1.

Substituting the Ar moiety in structure **123** by a phenyl group without alkylamino groups affords green shades. A typical product is "Brilliant Green" (**130**), which is the basis for Pigment Green 1. With the chemical substituent $N(C_2H_5)_2$, this compound is homologous to Malachite Green with $N(CH_3)_2$:

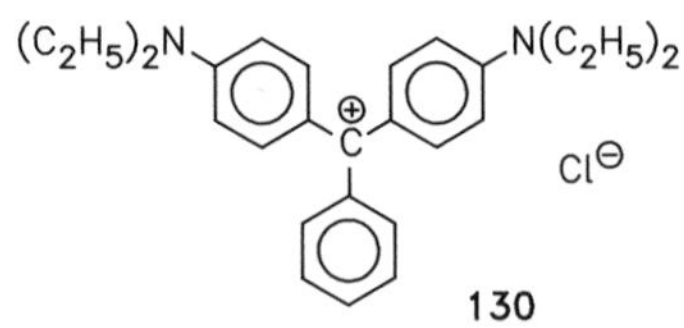

130 is manufactured by condensing one equivalent of benzaldehyde with two equivalents of diethylaniline in the presence of zinc chloride or hydrochloric acid. Proceeding via the carbinole, the reaction ultimately affords the leuco base, which is oxidized with agents such as PbO_2 to afford the dye:

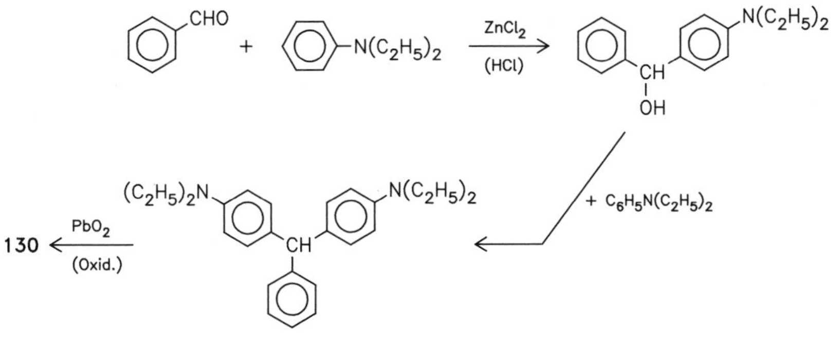

Dyes With the Structure 124

As a group, these typically pink compounds are derived from the xanthene structure **(131)**:

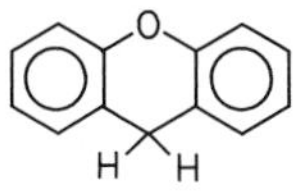

131

Condensation of a m-dialkylaminophenol with one equivalent of aldehyde in the presence of sulfuric acid or zinc chloride, followed by oxidation (for instance with $FeCl_3$) provides the bis(dialkylamino)-phenylxanthenium skeleton, which is the parent compound for dyes with the structure **124**:

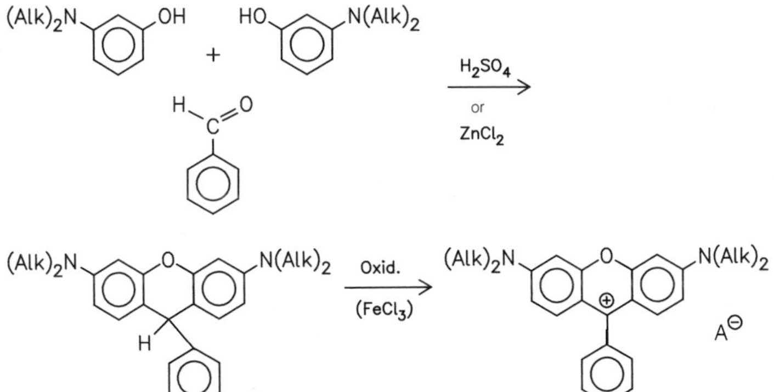

$A^{\ominus}$: acid anion

The most significant parent structure for pigments of this group is obtained by a slightly modified route by simply using phthalic anhydride instead of aldehyde. Reaction with m-diethylaminophenol at 180 °C in the presence of sulfuric acid or zinc chloride and subsequent oxidation with iron(III)chloride thus affords a dye known as "Rhodamine B" **(132)**, the basis of Pigment Violet 1:

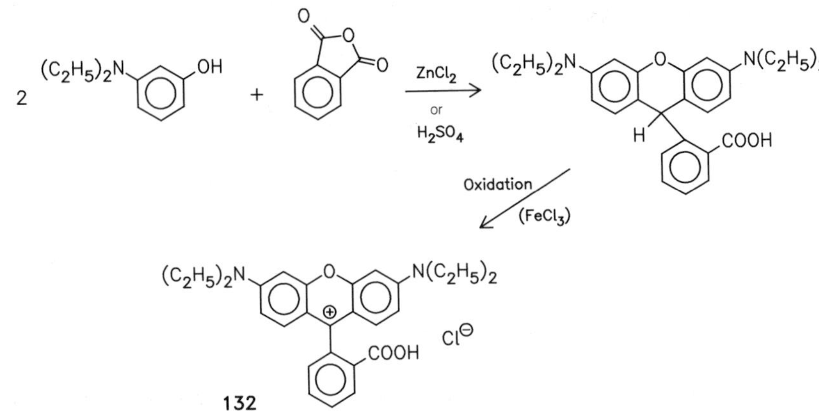

132

The ethyl ester of **(132)** is the dye component of Pigment Violet 2.

Starting from 3-ethylamino-4-methylphenol, it is possible to synthesize the cation **(133)** by using phthalic anhydride, as described above, and forming the ethylester. The thus prepared product is the dye cation of Pigment Red 81.

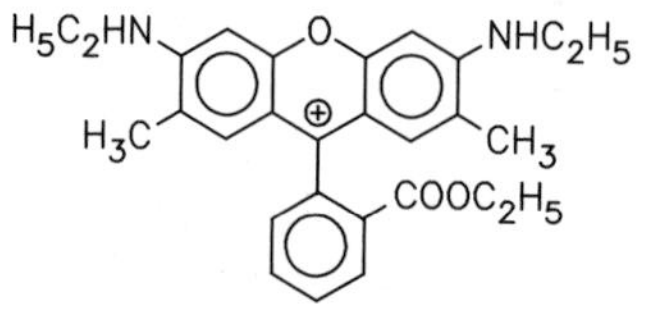

133

The dye component with the structure **125** is obtained through so-called primuline fusion: p-toluidine is heated with sulfur to 200 to 280 °C. Subsequent distillation affords not only the "primuline base"

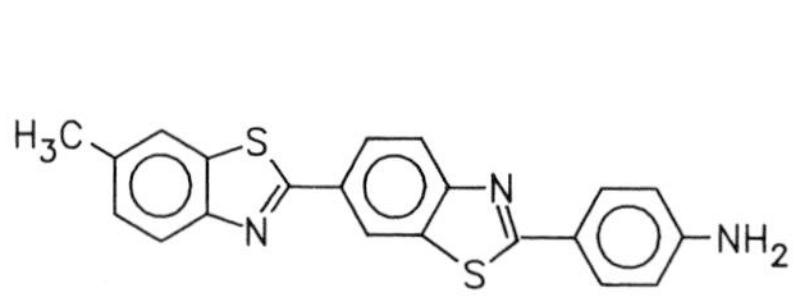

but also a product referred to as dehydrothio-p-toluidine, i.e., 2-(4'-aminophenyl)-5-methylbenzothiazole **(134)**. Permethylation of the nitrogen atoms with agents such as methanol and hydrochloric acid and simultaneous quarternization affords the thiazolium salt **135**:

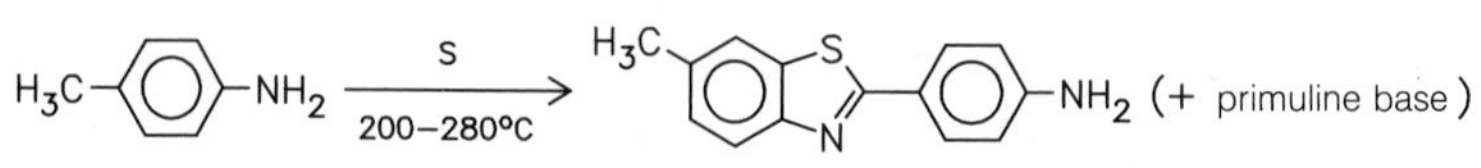

134

134 should first be removed from the primuline base by distillation:

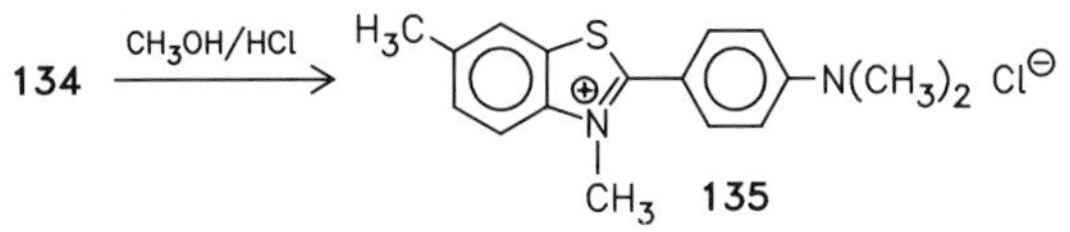

Manufacture of the Heteropolyacids and the Pigments

Mixing aqueous disodium hydrogenphosphate solutions with sodium molybdate and/or sodium tungstate and subsequently acidifying the mixture with mineral acids yields the desired heteropolyacids. While phosphotungstic acid is used only to a minor extent, phosphomolybdic acid and phosphotungstomolybdic acid reign supreme in the industrial manufacture of pigments of this class. Adding molybdate to phosphotungstic acid confers much higher brilliance and improved lightfastness on the resulting pigment complexes.

The pigments are obtained by preparing an aqueous solution of disodium hydrogenphosphate and adding a sodium molybdate solution (containing molybdenum trioxide and aqueous sodium hydroxide). Acidified with hydrochloric acid or with sulfuric acid, the reaction mixture is then added to an aqueous solution of the cationic dye at 65 °C.

The product is aftertreated at refluxing temperature, possibly in the presence of a surfactant. The pigment is then isolated by filtration, washed, dried, and milled. Instead of drying and milling, it is also useful to flush the aqueous pigment press-cake.

The corresponding copper ferrocyanide salts of basic dyes are obtained by treating potassium ferrocyanide $K_4[Fe(CN)_6]$ with sodium sulfite. Dissolved together, these two constituents are transferred to a solution of the cationic dye. A copper sulfate solution is finally added at 70 °C.

3.8.2.2 Properties

As a class, these pigments are characterized by uncommonly clean and brilliant shades. They provide a number of shades, particularly red and violet ones, that can be duplicated by other organic pigments, but without their brilliance and cleanness. These pigments fail to satisfy more stringent fastness requirements. They are not stable to polar solvents, such as alcohols, ketones, and ethylene glycol. Moreover, they also decompose if exposed to alkali.

It is for economical reasons that these complex pigments are declining in spite of their excellent coloristic properties. Many attempts have been made to replace more expensive types, such as those based on phosphotungstomolybdic acid, by less costly ones. As a consequence of these efforts, the properties of the pigments have improved greatly in recent decades, especially in terms of tinctorial strength and lightfastness. Complex pigments typically demonstrate average lightfastness. This property is a function both of the type of heteropolyacid and of the depth of shade. Copper ferrocyanide types, for instance, score an average of 2 steps lower on the Blue Scale than phosphomolybdic acid, phosphotungstomolybdic acid, and silicomolybdic acid types.

3.8.2.3   Application

These complex pigments are utilized almost exclusively in printing inks, especially in packaging and special printing inks. Of the large number of pigments falling within this class, only the most significant ones are surveyed in this chapter. The exact chemical compositions of the individual species, especially the respective dye/ heteropolyacid ratio, remain to be published.

3.8.2.4   Important Representatives

General

The pigments are commonly referred to not only by their Colour Index name but also by an added abbreviation for the respective heteropolyacid. For instance:

PM or PMA stands for **phosphomolybdic acid,**
PT or PTA for **phosphotungstic acid,**
PTM or PTMA for **phosphotungstomolybdic acid,**
SM or SMA for **silicomolybdic acid,** and
CF represents **c**opper **f**errocyanide (copper-1-hexacyano-iron-II-acid).

Table 33 lists the predominant species among these pigments (dyes with complex anions).

Individual Pigments

Pigments with the General Structure 123

Pigment Violet 3

P.V.3 provides a very bluish violet shade, which is considerably bluer than that of its counterparts P.V.1 and P.V.2. P.V.3 exhibits higher fastness properties than P.V.27. This is particularly true with respect to the lightfastness: 1/3 SD P.V.3 letterpress proof prints score approximately 1 1/2 steps higher on the Blue Scale than comparative P.V.27 prints. Similarly, P.V.3 is used primarily in printing inks, especially in flexographic and packaging gravure printing inks. The list of suitable media also includes oil-based vehicle systems. Possible processing problems are re-

Tab. 33: Commercial products of cationic dyes with inorganic heteropolyacids.

Pigments of the general structure 123:

$$R_2N \;\overset{\oplus}{\underset{Ar}{C}}\; NR_2 \qquad A^{\ominus}$$

C.I. Name	C.I. Constitution No.	$A^{\ominus}$	Ar	R	Common name of the corresponding dye
P.V.3	42535:2	PTM PM	$-\!\!\bigcirc\!\!-N\big\langle{}^{CH_3}_{H(CH_3)}$	CH_3	Methyl Violet
P.V.27	42535:3	CF	$-\!\!\bigcirc\!\!-N\big\langle{}^{CH_3}_{H(CH_3)}$	CH_3	Methyl Violet
P.V.39	42555:2	PM PTM	$-\!\!\bigcirc\!\!-N(CH_3)_2$	CH_3	Crystal Violet
P.Bl.1	42595:2	PTM PM	(naphthyl) HNC_2H_5	CH_3	Victoria Pure Blue B
P.Bl.2	44045:2	PTM PM	(naphthyl) HNC_6H_5	CH_3	Victoria Blue 4R
P.Bl.9	42025:1	PTM PM	Cl $-\!\!\bigcirc$	CH_3	–
P.Bl.10	44040:2	PTM PM PT	(naphthyl) HNC_2H_5	CH_3	–
P.Bl.14	42600:1	PTM PM	$-\!\!\bigcirc\!\!-N(C_2H_5)_2$	CH_3	Ethyl Violet
P.Bl.62	44084	CF	(naphthyl) HNC_2H_5	C_2H_5	Victoria Blue R
P.Gr.1	42040:1	PTM PM	$-\!\!\bigcirc$	C_2H_5	Diamond Green G
P.Gr.4	42000:2	PTM PM	$-\!\!\bigcirc$	CH_3	Malachite Green
P.Gr.45	–	P.Gr.1/P.Gr.2 cations with $CF^{\ominus}$ [4]			

Pigments of the general structure 124:

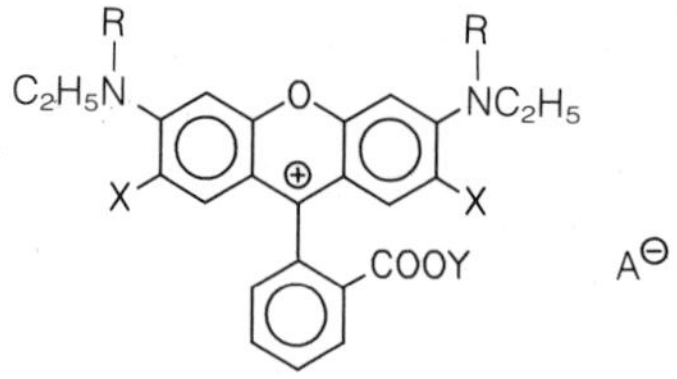

$$\text{C}_2\text{H}_5\overset{\overset{\text{R}}{|}}{\text{N}} \quad \text{O} \quad \overset{\overset{\text{R}}{|}}{\text{N}}\text{C}_2\text{H}_5$$

X ... X

COOY   A$^{\ominus}$

C.I. Name	C.I. Consti- tution No.	A$^{\ominus}$	R	X	Y	Common name of the corre- sponding dye
P.R.81:1	45160:1	PTM	H	CH$_3$	C$_2$H$_5$	Rhodamine 6G
P.R.81:x	45160:x	PM	H	CH$_3$	C$_2$H$_5$	(Basic Red 1)
P.R.81:3	45160:3	SM	H	CH$_3$	C$_2$H$_5$	Rhodamine 6G
P.R.169	45160:2	CF	H	CH$_3$	C$_2$H$_5$	Rhodamine 6G
P.V.1	45170:2	PTM	C$_2$H$_5$	H	H	Rhodamine B (Basic Violet 10)
P.V.2	45175:1	PTM	C$_2$H$_5$	H	C$_2$H$_5$	Rhodamine 3B ethylester

Pigments of the general structure 125:

P.Gr.2 is a mixture of

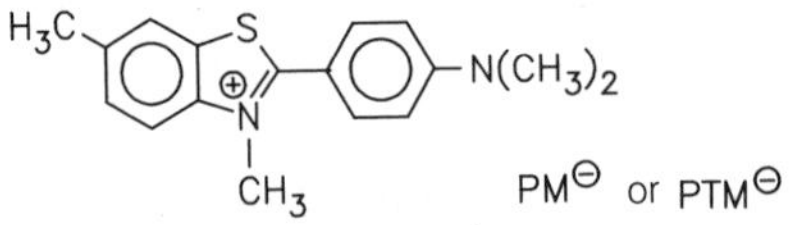

H$_3$C ... S ... N(CH$_3$)$_2$

N

CH$_3$ PM$^{\ominus}$ or PTM$^{\ominus}$

P.Y.18, 49005:1
("Thioflavine")

and P.Gr.1, 42040:1

ferred to under P.V.2. According to the US International Trade Commission, 190 t of PM type P.V.3 and 10 t of PTM type were produced nationwide in 1983. US manufacturers also supply flushed pastes in which P.V.3 is dispersed in systems such as offset and letterpress vehicles or in mineral oil.

Pigment Violet 27

P.V.27 is found on the product line of several European manufacturers. The pigment affords a very bluish, clean violet shade, similar to that of P.V.39. P.V.27, however, is tinctorially stronger and can therefore be used more economically than P.V.39. It is, on the other hand, less fast than P.V.39, and the difference is occasionally considerable. This is partly attributed to the complex acid component. P.V.27, for instance, is less resistant to a large number of organic solvents than

P.V.39, the difference being at least one step on the 5 step scale (Sec. 1.6.2.1). Similar observations have been recorded for the fastness of prints. They showed much less stability than prints containing the considerably redder P.V.1 and 2 types. Prints containing P.V.1 and 2, for instance, are entirely fast to water, while similar P.V.27 samples distinctly color moist filter paper (Sec. 1.6.2.2). Moreover, P.V.27 specimens, in contrast to P.V.1 prints, are not acid proof. P.V.27 is very fast to aliphatic and aromatic hydrocarbons. The pigment is used to shade and tone black inks. Insufficient fastness properties preclude P.V.27 from being used in packaging gravure and flexographic printing inks based on NC, but the pigment is used in aqueous flexographic inks. Copper ferrocyanide as a complex acid in oil-based systems, such as for letterpress or offset printing inks, catalytically accelerates the drying process. This drying effect largely restricts the use of P.V.27, since it is responsible for phenomena such as film formation on the ink surface, thickening of the ink, and yellowing of the print.

Pigment Violet 39

Types containing phosphotungstomolybdic acid and types based on phosphomolybdic acid are commercially available. The tinctorially somewhat stronger PM version is a more economical alternative. Both types afford the same shade, a very clean, bluish violet, which cannot be duplicated by other types of pigments. Compared to Dioxazine Violet, which is the standard pigment in this range of shades, P.V.39 exhibits a still considerably bluer and noticeably cleaner shade. The pigment darkens upon exposure to light. 1/3 SD letterpress proof prints equal step 3–4 on the Blue Scale for lightfastness.

P.V.39 is primarily applied in special printing inks. Besides, it is also used as a shading pigment in other colors. Possible disadvantages to pigment processing in different binders are listed under P.V.2. Incorporated in NC printing inks, P.V.39 does not change color upon storage.

Pigment Blue 1

P.Bl.1, which is supplied by several manufacturers, affords clean reddish shades of blue. Although its shade is also accessible by using chemically different pigments, for instance by combining α- or ε-Phthalocyanine Blue and Dioxazine Violet, these shades are not as brilliant. P.Bl.1 is available in the form of PM, PT, PTM, and SM types. The PM and SM varieties are tinctorially stronger, but offer less clean shades than PT and PTM types. The higher the PT content in the pigment, the cleaner the shade, an advantage which is somewhat compromised by the fact that the tinctorial strength decreases in the same order. P.Bl.1 is largely as fast as other types of pigments within the same class. The products, for instance, are very unstable to polar solvents such as alcohols, ketones, and esters. P.Bl.1 is comparatively lightfast. 1/3 SD letterpress proof prints equal approximately step 4 on the Blue Scale, although the pigment darkens distinctly.

Its coloristic properties make P.Bl.1 a suitable and widely used colorant for special printing inks. The pigment may be incorporated without problems in oily binders, i.e., in letterpress and offset inks. It is equally often used in publication gravure printing inks based on toluene and in packaging printing inks based on NC. Problems which may arise as P.Bl.1 is incorporated in these vehicles are mentioned under P.V.2. Besides, P.Bl.1 lends color to paper, wallpaper, typewriter ribbons, and other media, although it is often difficult to disperse.

Pigment Blue 2

At present, P.Bl.2 is only produced in the USA, where both PM and PT types are available. They are greener than grades of P.Bl.1. P.Bl.2 has much less impact on the market than P.Bl.1.

Pigment Blue 9

P.Bl.9 is similarly restricted to the North American market, which supplies both PM and PTM types. The PM varieties are stronger than PTM types but furnish shades that are not as clean.

P.Bl.9 matches the standard cyan for three and four color printing (Sec. 1.8.1.1). The pigment is continually losing significance as it is being displaced by the similarly shaded β-modification of Copper Phthalocyanine Blue. The latter offers a number of applicational and economical advantages. P.Bl.9 continues to be used in mass colored paper, textile printing, and in colored pencils.

Pigment Blue 10

P.Bl.10 is only utilized locally in East Asia and has no commercial significance.

Pigment Blue 14

Pigment Blue 14, like P.Bl.2 and 9, is produced exclusively in the USA, where it has only gained limited commercial importance. Both PM and PTM types are available. They provide very reddish shades of blue.

Pigment Blue 62

In terms of shade, P.Bl.62 parallels the PTM and PM types of P.Bl.1. P.Bl.62 exhibits high tinctorial strength. Its shade is not as clean as those of similar pigments, but it may be applied more economically. P.Bl.62 is less lightfast: 1/3 SD letterpress proof prints do not quite reach step 3 on the Blue Scale, and the samples darken upon exposure to light. The same applies to stability to organic solvents: compared to the two P.Bl.1 types, P.Bl.62 is somewhat less fast to a number of solvents. P.Bl.62 is primarily used as a shading pigment for publication gravure printing inks, as well as for aqueous flexographic inks. It is not recommended for NC-based printing inks. Incorporated in oily inks for applications such as offset printing inks, P.Bl.62 tends to catalytically accelerate the drying process, which restricts its suitability for these media. The pigment is also used to color office articles.

Pigment Green 1

P.Gr.1 is less brilliant than other triarylcarbonium pigments. Its shade may also be obtained by combining pigments of other chemical classes, such as copper phthalocyanine blue pigments or Copper Phthalocyanine Green types with monoazo yellow pigments. The resulting shades are even more brilliant than those of P.Gr.1. P.Gr.1 has an advantage over other pigments in providing higher tinctorial strength, which makes it an attractive product despite its lack of brillance.

In terms of lightfastness, P.Gr.1 parallels the blue types. Irradiated 1/3 SD letterpress proof prints equal step 3 on the Blue Scale, although the shade darkens considerably. P.Gr.1 is recommended for use in oily binders, publication gravure printing inks, and NC-based printing inks. Possible problems are referred to under P.V.2.

Pigment Green 4

P.Gr.4 is manufactured in the USA, but has only attracted very limited attention. Both PM and PTM types are available. P.Gr.4 provides an intermediate shade between those of P.Gr.1 and 2. The pigment performs much like other green pigments of its class.

Pigment Green 45

Both the chemical constitution of this basic dye and the complex acid of P.Gr.45 remain to be published. P.Gr.45 affords yellowish shades of green, which are within the range of shades provided by P.Gr.36, the brominated copper phthalocyanine pigment. The shade produced by P.Gr.45 may be matched by combining pigments of other classes. Suitable blends are provided by greenish P.Y.74 type monoazo yellow pigments with the β-modification of Copper Phthalocyanine Blue. The resulting shade equals P.Gr.45 shades in terms of brilliance. P.Gr.45 is appreciated for its high tinctorial strength. High gloss and transparency add to the value of the pigment. P.Gr.45 is used primarily in offset and aqueous printing inks, but also in solvent-based printing inks, such as those based on NC.

Pigments With the General Structure 124

Pigment Red 81

P.R.81 affords a very clean bluish red shade, which matches the purple-red on the DIN 16 508 color scale for letterpress application and also on the DIN 16 509 offset scale. The shade is not accessible by using other pigments (Sec. 1.8.1.1). Pigments which are closely related by shade, such as P.R.147, provide shades which are not even remotely as clean. The standard magenta on the European Color Scale for letterpress (DIN 16 538) and offset printing (DIN 16 539) is somewhat yellower. For such printing inks, certain high solvent fastness properties are required in order to safely overlacquer the prints. These conditions cannot be met with P.R.81. The pigment in particular lacks fastness to polar solvents, such as alcohols, ketones, and esters, as well as to the DIN 16 524 solvent mixture. On the other hand, P.R.81 prints are very fast to aliphatic and aromatic hydrocarbons, paraffin, butter, and many other fats, although they are not entirely stable to sterilization. Like other members of its class, P.R.81 affords prints which are not fast to alkali and soap but largely fast to acid.

Tinctorially, P.R.81 is a strong pigment, which, as with other representatives, is especially true for the PTM types. The pigment commonly darkens noticeably as it is exposed to light. 1/3 SD letterpress proof prints equal step 4 on the Blue Scale for lightfastness. The lightfastness of P.R.81 systems may be adversely affected if the surface of the pigment particles is attacked by polar solvents, such as esters, ketones, low molecular weight alcohols, or glycol ether. Alkali has a similar effect.

P.R.81 is used especially in three and four color printing and lends itself to various printing processes, therefore pigments of this type are referred to as "Process Red" in the USA. (According to the US International Trade Commission, 165 t of P.R.81 types were produced nationwide in 1988.) Used as a colorant for NC-based printing inks, SM types of P.R.81 may present problems as they are dispersed with steel balls or stored in steel containers as well as at elevated temperature. Catalytic decomposition of the binder and damage to the pigment may induce a color shift and increase the viscosity.

Pigment Red 169

P.R.169 parallels the P.R.81 types in terms of shade and fastness properties. Thus, P.R.169 types are equally suited to letterpress and offset printing inks, where they provide the standard purple red for three and four color printing in accordance with the so-called DIN scale (Sec. 1.8.1.1). Moreover, P.R.169 is frequently used for toluene-based publication gravure printing inks, where it is sometimes applied in combination with P.R.57:1. A number of grades, however, are not as fast to toluene as comparable P.R.81 varieties. One of the major fields of application for P.R.169 is in aqueous flexographic printing inks.

There are various systems in which the salt character of these pigments adversely affects the ease of processing in an ink or even the finished print. Most of these problems are attributed to the copper ferrocyanide content, which is used as a complex acid in P.R.169 and CF types of other basic dyes. In offset inks and other oxidatively drying printing inks, catalytic acceleration of the drying process may occur, resulting in a thickening of the ink, film formation on the surface of the ink, and yellowing of the prints. Incorporation in polyamide vehicles has the disadvantage that P.R.169 catalytically decomposes the resin and thus causes the prints to stick and develop an odor. An unsuitable or shifting pH value in an aqueous ink may thicken the vehicle and affect the coloristic properties of the product. Highly acidic binder components may chemically react with the pigment or with the dye component and thus adversely affect the coloristic properties, an effect which is often accompanied by considerable flocculation. Deficiencies also tend to appear in NC-based printing inks. These are a consequence of dye nitrosation through formation of nitrous gases. As a result, gas develops, the shade demonstrates an often considerable shift, and the printing ink loses some of its strength. Pigment manufacturers supply their products with detailed information on how to obviate these difficulties.

Pigment Violet 1

P.V.1 is broad in scope; it is closely related to P.V.2, both in terms of coloristics and fastness properties. Although the shades are very similar, P.V.1 usually provides not only higher tinctorial strength and greater brilliance but also higher cleanness of shade. In terms of lightfastness, P.V.1 scores about 1/2 step on the Blue Scale less than P.V.2 types. Moreover, P.V.1 prints are also less fast to organic solvents and to agents such as soap and butter than prints made from P.V.2. The areas of application are the same as for the other violet type.

Pigment Violet 2

One of the predominant features of P.V.2 varieties is the fact that they are comparatively lightfast. At standardized conditions regarding printed film thickness and substrate, 1/3 SD letterpress proof prints equal step 5 on the Blue Scale. Like other representatives of this class of pigments, P.V.2 loses much of its lightfastness through exposure to polar solvents, such as low molecular weight alcohols, esters, and ketones. Alkali has a similar effect. Moreover, such solvents (and also alkali) enhance the tendency of the pigment to migrate and reduce its fastness to overlacquering. PTM types are noticeably less fast to polar solvents than pigments based on other complex acids. P.V.2 is acid fast in accordance with standard test conditions (Sec. 1.6.2.2).

The pigment provides a very bluish shade of red, referred to as violet. It furnishes a very clean shade. P.V.2 systems show a brilliance which is not reproduced by using chemically different pigments. Tinctorially, P.V.2 is a particularly strong product.

The printing ink industry utilizes P.V.2 particularly for special printing inks and as a shading pigment. It is much too blue to be used in process colors. As is the case with other PTM types and with types containing phosphomolybdic acid or silicomolybdic acid, there is some difficulty in dispersing P.V.2 in certain binders. Both the processing of the ink and the printed products may be affected. Inks containing polyamide, for instance, if printed onto polyolefin films may exhibit dye migration through the film. This phenomenon is attributed to the presence of small amounts of dye within the pigmented system which have not been converted into salts. NC-based printing inks containing P.V.2 are affected through contact with steel. This is likely to occur, for instance, as printing inks are prepared using steel balls or as they are transported in steel containers. Consequences include catalytic degradation of nitrocellulose and damage to the pigment, visibly resulting in increased viscosity and coloristic changes involving transparency, tinctorial strength, and shade. Elevated temperature has a similar effect or may even enhance the action of steel. Pigments decompose in strongly basic aqueous systems and suffer coloristic and rheological changes. Interaction between pigment and polar solvents also frequently affects the shade and the rheology of the system, thickening the material or flocculating the pigment. Finally, the different components of a blend of this type of pigment, incorporated in a liquid printing ink, may interact with one another, not only forming a sediment on the bottom of the container but also affecting the coloristic properties of the ink. Pigment manufacturers provide their products with information on how to meet these problems.

Pigments With the General Structure 125

Pigment Green 2

P.Gr.2 provides yellowish shades of green which are considerably more yellowish than those furnished by P.Gr.1. P.Gr.2, like P.Gr.1, lacks brilliance, a deficiency which within this class of pigments is unique to these two species. The shade produced by P.Gr.2 is also obtained with blends comprising other organic pigments, such as phthalocyanine green pigments and monoazo yellow pigments or diarylide yellow pigments. A prominent characteristic of P.Gr.2 is its high tinctorial strength. In recent years, P.Gr.2 has lost most of its commercial importance and continues to be used only in the USA. Both PTM and PM types are available. The pigment parallels other representatives of its series in terms of fastness properties. It shows good lightfastness, but darkens slightly upon exposure.

References for Section 3.8

[1] Kh. Ringel et al., Zh. Org. Chim. 18 (1982) 1018–1022.
[2] DE-AS 1 919 724 (Hoechst) 1969.
[3] H. Schmelzer, XIII. Congr. FATIPEC, Juan les Pins 1976, Congress book p. 572–574.
[4] J.O. Sanders, Pigments for Inkmakers, SITA Technology, London 1989.

4 Miscellaneous Pigments

This chapter lists pigments which, either for reasons of different chemical structure or for lack of knowledge thereof, are not included in any other chapter of this book.

4.1 Quinophthalone Pigments

4.1.1 Chemistry and Manufacture

E. Jacobson in 1882 fused phthalic anhydride with quinoline bases obtained from coal tar, which also contained quinaldine (136). He thus received quinophthalone (137). Quinophthalone derivatives bearing sulfonic or carboxylic acid functions represent suitable anionic dyes. Derivatives carrying basic side chains containing quarternary nitrogen, on the other hand, provide cationic dyes. The compounds are used especially as disperse dyes [1].

137, the parent structure to all pigments which are described in this chapter, continues to be prepared by fusion or, even better, by treating quinaldine with phthalic anhydride in an inert high boiling solvent at 200 °C to 220 °C.

Although Eibner elucidated the structure between 1904 and 1906, it was only through IR and nuclear magnetic resonance spectroscopy (NMR) that the chromaticity of these molecules could be attributed to keto–enol tautomerism and simultaneous hydrogen bond formation (structures 137a = 137b) [2].

Quinophthalone molecules are too soluble in various media to be used under normal application conditions. Both solvent and migration resistance may be enhanced by enlarging the molecule, and the options are as follows:

- introducing suitable substituents,
- condensing quinaldines with pyromellitic dianhydride instead of phthalic anhydride,
- doubling the molecule via diamide bridges.

These methods produce yellow to red compounds which exhibit satisfactory pigment properties.

The list of suitable substituents includes acylamino groups, but especially halogen atoms such as chlorine or bromine. Halogenated derivatives are obtained from tetrahalogen phthalic anhydride or naphthalene-2,3-dicarboxylic anhydride, by reaction with quinaldine (or one of its derivatives). A patent has been issued describing the reaction of 8-aminoquinaldine with twice the stoichiometric amount of tetrahalogen phthalic anhydride in an inert solvent in the presence of zinc chloride [3]. This route makes it possible to introduce eight halogen atoms into the molecule (**138**), resulting in greenish yellow pigments:

Pigment Yellow 138, 56300, has the chemical structure **138** with X = Cl [4].

Another publication describes the conversion of 8-chloro-5-aminoquinaldine with possibly halogenated phthalic anhydride [5]. These publications also discuss the fact that halogenation of the phthalimide ring improves the lightfastness of the product, but on the other hand that lightfastness is diminished if substitution is made to the benzene ring of the quinaldine system. Apart from compounds bearing an OH group in 3'-position, 4'-OH and 4'-acylamino derivatives also show enhanced lightfastness. Condensation of 3-hydroxyquinaldine or its derivatives with pyromellitic dianhydride affords bis-quinophthalones (**139**). These are very lightfast pigments affording yellow, red, or brown shades.

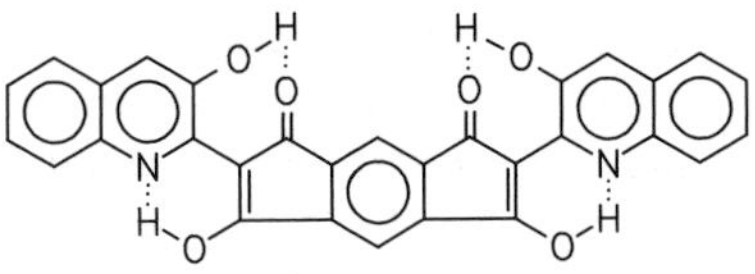

139

Doubling the quinophthalone structure may also be achieved through condensation of two equivalents of 3-hydroxy-quinophthalone carboxylic acid chloride with aromatic diamines [6]. The reaction affords orange pigments (140). The acid chloride is prepared from 3-hydroxyquinaldine-4-carboxylic acid and benzene-1,2,4-tricarboxylic acid.

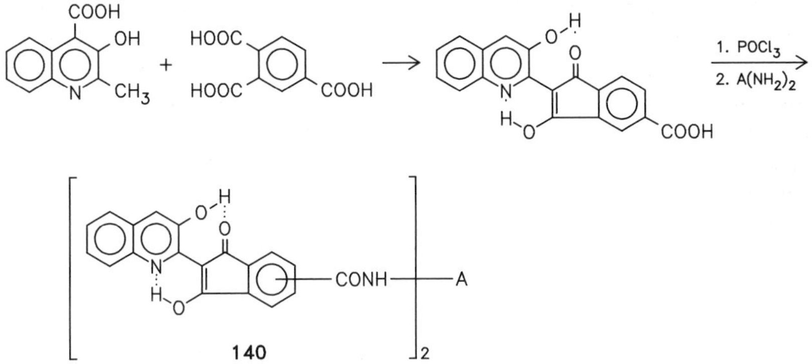

140

A: bifunctional aromatic group

4.1.2 Properties and Application

Quinophthalone pigments are marketed in very limited number. The pigments provide yellow to red shades and are used primarily to color paints and plastics.

Pigment Yellow 138

P.Y.138 type quinophthalone pigments afford exceedingly lightfast and weatherfast greenish yellow shades with good heat stability. Their main fields of application are in paints and in plastics.

Most commercial P.Y.138 types have low specific surface areas of approximately 25 m^2/g and consequently provide good hiding power, a feature which qualifies them for use in opaque systems. In this field, P.Y.138 competes with opaque yellow pigments of other classes, which, although tinctorially weaker, sometimes offer better hiding power and cleaner shades. Paints formulated at higher pigment concentrations show good flow behavior. P.Y.138 is entirely or at least largely resistant to a large number of organic solvents, such as alcohols, esters, ketones, and aliphatic as well as aromatic hydrocarbons. Oven drying systems containing P.Y.138 are fast to overcoating at common baking temperatures; the pigment is thermostable up to 200 °C.

Full shades and similar deep shades exhibit excellent weatherfastness, but rapidly decrease in tints made by adding TiO_2. The cleanness of full shades increases noticeably by exposure to weather. There are certain binders in which P.Y.138 tends to chalk if employed at high concentration.

P.Y.138 is primarily applied in industrial finishes for commercial vehicles, automotive refinishes, and original automotive finishes. Incorporated in these systems, P.Y.138 is acid and alkali fast. However, exterior house paints which are applied onto a basic substrate show some sensitivity to alkali.

Types which feature slightly higher specific surface areas are more transparent and demonstrate somewhat different coloristic properties. Such types are not only tinctorially stronger and more greenish but also somewhat less fast to light and weather than opaque varieties, their flow behavior is less favorable.

Another main field of application for P.Y.138 is in plastics, in which it demonstrates average to good tinctorial strength. 1/3 HDPE samples (1% TiO_2) are formulated at approx. 0.2% pigment. Such systems are heat stable up to 290 °C, as are full shades. 1/25 SD specimens only withstand 250 °C. The pigment nucleates in this polymer, as in other partially crystalline plastics, i.e., it affects the shrinkage of injection-molded articles. This effect becomes markedly less noticeable as the processing temperature rises. In terms of lightfastness, full shades equal step 7–8 on the Blue Scale, while white reductions down to 1/25 SD match step 6–7. P.Y.138 is an equally suitable colorant for polystyrene, ABS, and various other plastics, such as polyurethane foam. A wide variety of pigment preparations are supplied for various areas of application within the paint and plastics industry. These preparations have the advantage of containing already dispersed pigment.

4.2   Diketopyrrolo-Pyrrole (DPP) Pigments

4.2.1   Chemistry and Manufacture

In the early eighties, Ciba-Geigy discovered a new type of heterocyclic pigments, based on a symmetric chromophor, the 1,4-diketopyrrolo(3,4c)pyrrole system (**141**):

141

In 1974, at a different place, an attempt to synthesize a 2-acetinone [7] by reacting benzonitrile with bromoacetic acid in the presence of zinc dust failed and instead led to the preparation of **141** (R = C_6H_5) in small yield:

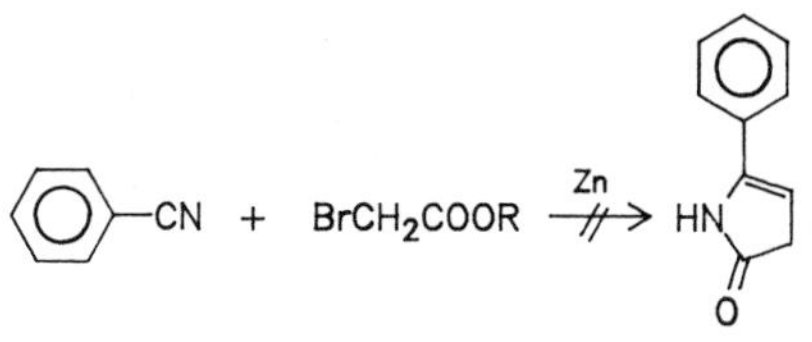

$$\text{C}_6\text{H}_5\text{-CN} \ + \ \text{BrCH}_2\text{COOR} \ \xrightarrow[\ ]{\text{Zn}} \!\!\!\!/\!\!\!\!\to \ \text{141}$$

These studies were followed up by Ciba-Geigy and systematically developed into a synthetic route to a group of very fast red pigments, known as diketopyrrolo-pyrrole pigments (DPP pigments) [8,9], since the brilliant red crystals of **141** turned out to be extremely insoluble.

DPP pigments are synthesized by reacting succinic ester with benzonitriles in the presence of sodium methylate in methanol:

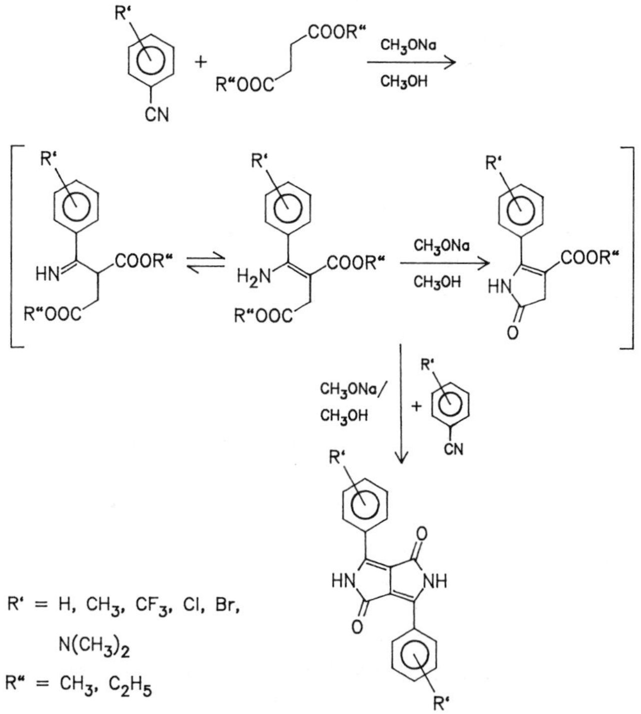

R' = H, CH_3, CF_3, Cl, Br,

$N(CH_3)_2$

R" = CH_3, C_2H_5

The intermediates which are formed during the reaction are not isolated; the synthesis proceeds in one step and affords a very good yield.

Depending on the importance of R^1 (CH_3, CF_3, Cl, Br, $N(CH_3)_2$), the resulting reddish yellow to bluish violet pigments show excellent lightfastness and weatherfastness and very good migration fastness. If two differently substituted benzonitriles are used, the product is a mixture of the symmetrically substituted and the two unsymmetrically substituted diphenyl-DDP pigments (mixed synthesis).

It is possible to manufacture either the transparent or the opaque pigment grade by controlling the particle size during or after synthesis. Small-grained pigment, for instance, can be prepared by adding small amounts of m-phthalonitrile. Other methods of affecting the particle size involve thermal aftertreatment after hydrolysis with optional variation of solvent and pH.

4.2.2   Properties and Application

At present, all commercially available representatives of this new class of pigments have the same chemical constitution **141** ($R = C_6H_5$).

Pigment Red 254, 56110

P.R.254, which was introduced into the market in recent years, shows good coloristic and fastness properties and has therefore within a short period of time developed into a widely used pigment for high quality industrial paints, especially in original automotive finishes and automotive refinishes. The commercially available type affords medium shades of red in full shades, while white reductions are somewhat bluish red. The pigment shows good hiding power; it is used primarily in automotive finishes wherever lead-free formulations are required. For economical reasons, P.R.254 is frequently used in combination with the coloristically similar but less weatherfast opaque type of P.R.170. Combination, for instance, with quinacridone pigments affords opaque shades of bluish red.

P.R.254 shows very good fastness to organic solvents. It is therefore fast to blooming and to bleeding in baking enamels. Full shades and similar deep shades are lightfast enough to reach step 8 on the Blue Scale. The pigment also shows very good lightfastness—a reason for its primary use in original automotive finishes. Its fastness to flocculation can be improved by employing suitable additives.

P.R.254 is also used to color plastics which are processed at high temperature. A specialty type has recently been introduced to the market which is used for this purpose. 1/3 SD HDPE colorations of this type are stable up to 300 °C for 5 minutes. The colorations exhibit high tinctorial strength; their shade is considerably yellower and cleaner than that of the type which is used in paints. 1/3 SD colorations (1% TiO_2) require 0.16% pigment. The pigment only affects the shrinkage of the plastic to a minor extent. The lightfastness of 1/3 SD colorations (1% TiO_2 and transparent) and of full shades equals step 7–8 on the Blue Scale.

In plasticized PVC, P.R.254 reaches step 8 on the Blue Scale for lightfastness. It shows high tinctorial strength and bleeding fastness.

Pigment Red 255

According to the manufacturer, P.R.255 is considerably yellower than the chemically similar P.R.254. It has not yet been encountered in the market.

4.3 Aluminum Pigment Lakes

This class includes the aluminum salts of some carboxylic acids or sulfonic acids of polycyclic dyes.

Pigment Red 172, 45430:1

P.R.172 is the aluminum salt of tetraiodofluorescein (**142**):

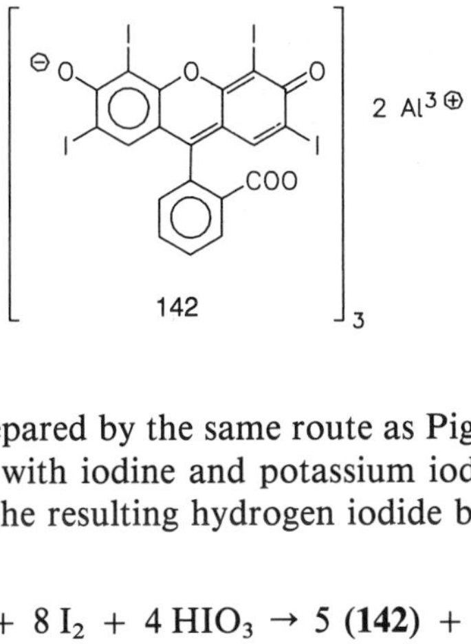

142

Fluorescein (**147**) is prepared by the same route as Pigment Red 90 (see p. 559). The compound is iodized with iodine and potassium iodate in an acidic medium. The iodic acid reoxidizes the resulting hydrogen iodide back to iodine:

$$5 \ (\mathbf{147}) + 8 \ I_2 + 4 \ HIO_3 \rightarrow 5 \ (\mathbf{142}) + 12 \ H_2O$$

Aluminum chloride is used to convert tetraiodofluorescein, also known as erythrosine, to the P.R.172 lake.

Provided it meets certain purity criteria, the pigment may be used in foodstuffs, pharmaceuticals, and cosmetics. It is registered as E 127 throughout the EC and as FD&C Red No.3 in the USA. The bluish red pigment P.R.172 provides little tinctorial strength. The pigment generally shows poor application and fastness properties, including fastness to organic solvents, alkali, and acids, as well as heat stability and lightfastness. Consequently, there are limits to its technical applicability.

Pigment Blue 24:x, 42090:2 (Al) and Pigment Blue 24:1, 42090:1 (Ba)

The parent structure is a trisulfonated triphenylmethane dye (**143**):

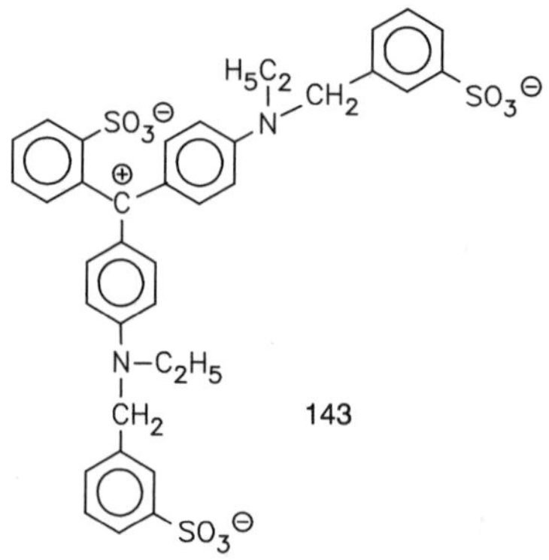

143

It is obtained by condensing benzaldehyde-o-sulfonic acid with 2 equivalents of N-ethylbenzylaniline, followed by sulfonation, oxidation, and conversion to the ammonium salt. The dissolved dye is then immersed in a dispersion of aluminum hydroxide and converted to the corresponding salt using aluminum chloride or barium chloride solution:

$$[143]_3 \cdot 2 \text{ Al as well as } [143] \text{ Ba}$$

Both types of lakes are available in the USA.

Pigment Blue 24:x

P.Bl.24:x, a greenish blue compound, is registered in the USA as FD&C Blue 1 for use in foodstuffs, pharmaceuticals, and cosmetics, provided certain purity standards are met. The pigment possesses poor lightfastness. It largely parallels P.Bl.24:1, also in terms of tinctorial strength and stability to organic solvents.

Pigment Blue 24:1

P.Bl.24:1 provides brilliant greenish shades of blue. Although a pigment with very good tinctorial strength, P.Bl.24:1 has largely been displaced by Copper Phthalocyanine Blue. In the past, P.Bl.24:1 used to be applied in large volume as a process color blue for three and four color printing. The pigment lacks fastness to acid, alkali, and soap, and its lightfastness is poor. P.Bl.24:1 is found in inexpensive letterpress printing inks as well as in office articles such as colored pencils and less costly water colors.

Pigment Blue 63, 73015:x

Likewise, P.Bl.63 is an aluminum lake. It is derived from the indigo structure, which is sulfonated to afford indigo-5,5'-disulfonic acid. For indigo syntheses, see [10].

Reaction with aluminum trichloride yields the insoluble pigment:

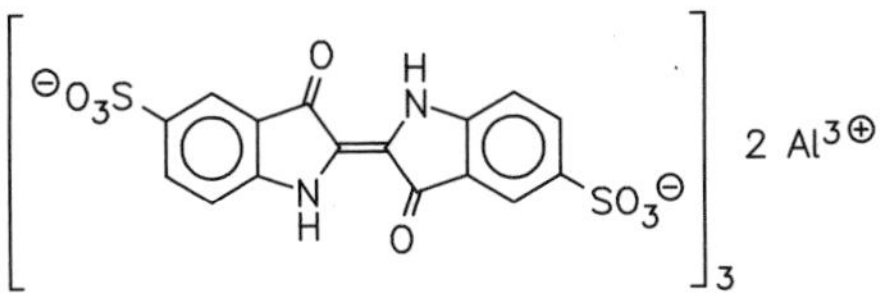

The aluminum lake is registered in the USA as FD&C Blue 2 as a colorant for foodstuffs and pharmaceuticals, provided certain purity conditions are met. Its shade is a bluish red. The pigment is somewhat sensitive to chemicals and to over-coating as well as to light. It is a tinctorially weak product.

4.4   Pigments With Known Chemical Structure Which Cannot Be Assigned To Other Chapters

These products are classified according to their shade.

Pigment Yellow 101, 48052

P.Y.101 has the chemical structure of a disazomethine compound (**144**):

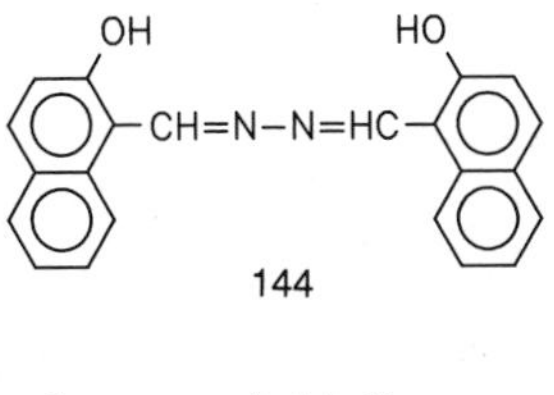

144

Known since 1899, this product was initially patented as a fluorescent dye and was used later as a pigment for the mass coloration of viscose.

It is obtained by condensing 2 moles of 2-hydroxy-1-naphthaldehyde with one mole of hydrazine:

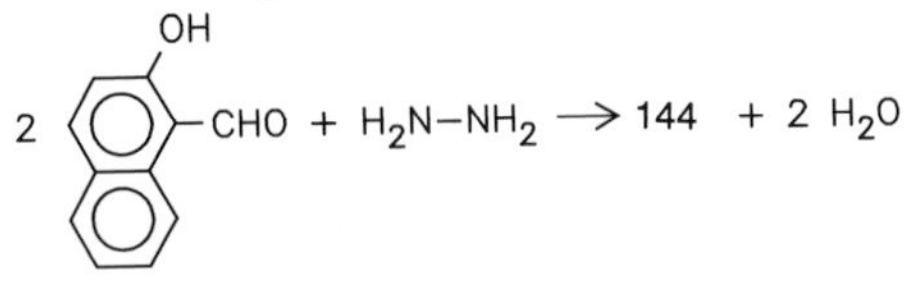

$$2 \quad \text{[naphthol-CHO]} + H_2N-NH_2 \longrightarrow 144 \ + \ 2 \ H_2O$$

P.Y.101 produces uncommonly brilliant greenish yellow shades. It is referred to as a fluorescent pigment. In this respect, P.Y.101 is unique among the currently available organic pigments. All other industrial fluorescent pigments, which are not discussed in this book, represent fluorescent dyes which are dissolved in suitable media (resins). Some of these provide good migration resistance through chemical interaction between dye and resin. P.Y.101, on the other hand, is a crystalline pigment, although it shows somewhat less fluorescence than resin-based dyes. The optical effect of P.Y.101 and other fluorescent colorants is a consequence of selective light absorption and simultaneous luminescence, initiated by high-energy radiation, i.e., UV radiation and/or light with a short wavelength.

P.Y.101 is fast to neither acids nor bases and is not entirely fast to important organic solvents, either. It is moderately lightfast.

The commercial product is highly transparent and is considered a special-purpose product, particularly for printing inks. The color intensity reaches an optimum when applied over light, if possible white substrates. P.Y.101 is frequently applied in combination with nonhiding fillers, such as barium sulfate, but particularly with other yellow, green, or red pigments. Even minor amounts of P.Y.101 noticeably increase the brilliance of prints containing nonfluorescent pigments.

Other areas of application are in office articles and in the arts field, where P.Y.101 lends color to pencils, chalks, watercolors, and other articles. It is also used for fluorescent markers.

Pigment Yellow 148, 59020

The chemical structure of P.Y.148, a greenish yellow pigment is listed in the Color Index under Constitution Number 59020:

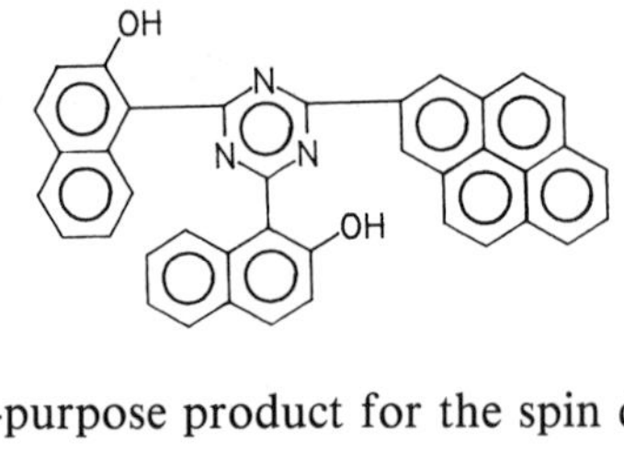

This pigment is a special-purpose product for the spin dyeing of polyamide.

Pigment Yellow 182

Diazotization of terephthalic acid methylester and coupling onto **(146)** affords P.Y.182 [11]:

The corresponding coupling component is synthesized by treating s-trichlorotria-zine with acetoacetic arylamides in an alkaline medium. Upon addition of sulfuric acid at room temperature the resulting compound (145) releases the acetyl group and hydrolyzes to afford 146. The yield is more than 90% [12, 13]:

P.Y.182, which has been available for some years, provides somewhat reddish shades of yellow and is tinctorially strong. It is sensitive to a variety of organic solvents, especially to ketones such as methylethylketone and cyclohexanone, as well as to aromatic solvents such as toluene or xylene. In this respect, the pigment equals step 2 on the 5-step scale (Sec. 1.6.2.1). P.Y.182 is targeted for the paint and the plastics industry.

The paint industry utilizes P.Y.182 especially in various types of industrial fin-ishes wherever the weatherfastness requirements are not too stringent. A high TiO_2 content adversely affects both the lightfastness and the weatherfastness of P.Y.182 systems. At typical baking temperatures (150°C), such systems are not entirely fast to overcoating. The list of recommended applications for P.Y.182 includes emul-sion paints, although a certain sensitivity of the pigment to alkaline agents prevents P.Y.182 containing paints from being applied onto alkaline substrates, such as fresh plaster or cement.

Used as a colorant for plastics, P.Y.182 exhibits medium to high tinctorial strength. 0.8% pigment is required to produce a 1/3 SD PVC sample containing 5% TiO_2. There is some blooming at low pigment concentrations, and bleeding is observed over the entire concentration range. P.Y.182 is therefore only recom-mended for use in rigid PVC.

1/3 SD HDPE systems (1% TiO_2) are thermally stable up to 250°C, while 1/25 SD specimens withstand 280°C. P.Y.182 does not affect the shrinkage of the plas-tic. 0.37% pigment is needed to formulate a 1/3 SD HDPE sample. Combining P.Y.182 with nickel stabilizers in polypropylene is to be avoided. The pigment dissolves in polystyrene as the temperature reaches 200 °C, a process which is accompanied by considerable color change. P.Y.182 is also an unsuitable candidate for ABS.

Pigment Yellow 192

This polycyclic pigment is synthesized by condensation of 5,6-diaminobenzimidazolone with a naphthalic anhydride in a high boiling solvent [12]:

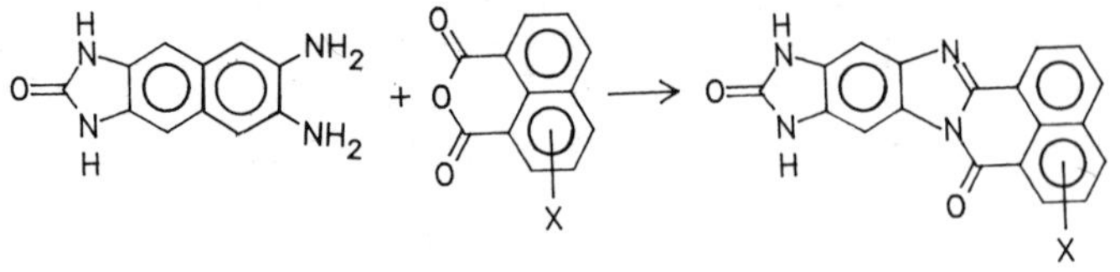

The pigment has only recently been introduced to the market. It is a specialty product for the spin dyeing of polyamide. 1/3 SD colorations with and without TiO_2 are thermally stable up to 300 °C. Even samples which are formulated at very low pigment concentrations completely maintain their color. The pigment offers excellent textile fastnesses, such as resistance to dry cleaning and dry heat at 200 °C (Sec. 1.6.2.4). Besides, P.Y.192 is also used in other polymers which are processed at high temperature. According to the manufacturer, no coloristic changes are observed if P.Y.192 is incorporated in polyester; the pigment maintains its coloristic properties even during the condensation process (Sec. 1.8.3).

Pigment Orange 64, 12760

The chemical structure of P.O.64 is listed in the Colour Index under Constitution Number 12760:

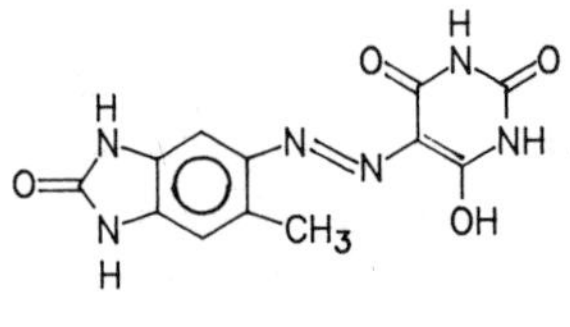

P.O.64 is synthesized by diazotization of 5-amino-6-methylbenzimidazolone and subsequent coupling of the diazonium component onto barbituric acid.

Its main field of application is in plastics. 1/3 SD HDPE samples, both transparent and opaque, withstand exposure to 300 °C for 5 minutes, while 1/25 SD specimens are heat stable up to 250 °C. Temperatures in excess of this value shift the shade towards more yellowish hues. P.O.64 does not have a nucleating effect on its medium, i.e., it does not affect the shrinkage of partially crystalline injection-molded polymers.

P.O.64 shows good migration resistance in plasticized PVC and exhibits medium tinctorial strength. The pigment is also used in polystyrene and similar polymers and is also recommended as a colorant for rubber.

The printing ink industry employs P.O.64 in metal deco printing inks in view of the fact that the pigment is thermally stable up to 200 °C. The prints may safely be overcoated.

Pigment Orange 67, 12915

P.O.67 is a member of the pyrazolo-quinazolone pigment series. Its constitution is listed in the Colour Index under Constitution No. 12915. Coupling component is the heterocycle shown on p. 489. The diazonium component is 2-nitro-4-chloroaniline. Coupling affords the pigment, which was found to exhibit the hydrazone structure:

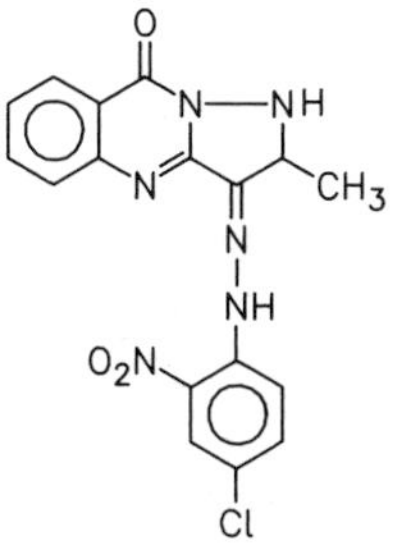

P.O.67 is a comparatively recent product. The commercial grade is very opaque. Its main area of application is in paints, to which it lends full shades and relatively deep shades. P.O.67 is employed wherever use of Molybdate Orange pigments is to be avoided. P.O.67 affords brilliant yellowish shades of orange, which are close to the standard shade RAL 2004. The pigment is not sufficiently fast to important organic solvents. Incorporated in oven drying systems, the pigment is not entirely fast to overcoating at common baking temperatures (140 °C). Systems which are processed at low temperatures between 80 and 100 °C, however, may safely be overcoated. P.O.67 is particularly recommended for use in long-oil and medium-oil alkyd resin systems, especially for decorative paints and emulsion paints. In these media, P.O.67 exhibits very good lightfastness and weatherability. The pigment should not be used in epoxy resins.

Pigment Red 90, 45380:1

P.R.90 is derived from the fluorescein structure (**147**). Fluorescein is a yellow dye with intensely green fluorescence which was discovered by A.v.Bayer in 1871. It is prepared by heating resorcin and phthalic anhydride with zinc chloride or concentrated sulfuric acid:

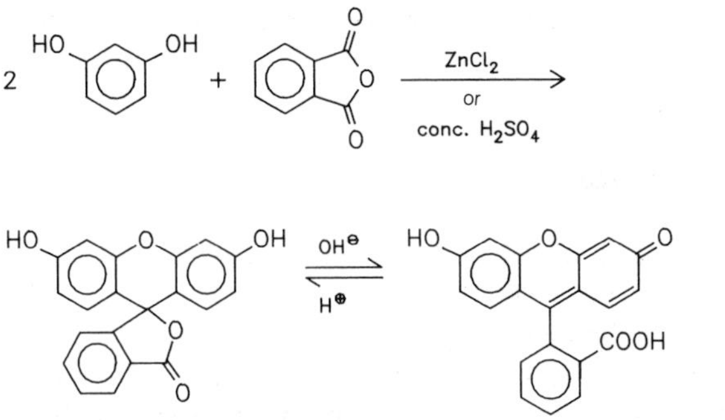

147

Brominating fluorescein in the presence of sodium chlorate affords the red tetrabromo derivative, whose sodium salt is known as eosine (Caro 1871). Sodium chlorate reoxidizes, evolving hydrogen bromide, to bromine which can react again:

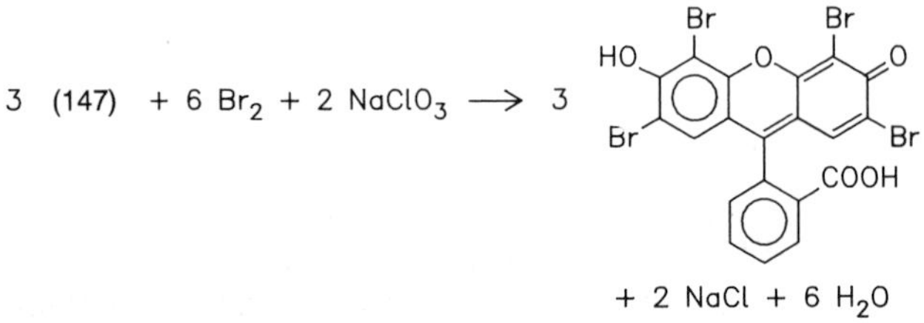

$$3 \ (147) \ + 6 \ Br_2 + 2 \ NaClO_3 \longrightarrow 3$$

+ 2 NaCl + 6 H$_2$O

The bromination product is actually contaminated with a certain percentage of mono to tribromo derivative. Although impurities, these side products apparently furnish the good tinctorial strength of P.R.90. Eosine itself is only a tinctorially weak pigment.

Treating the aqueous eosine solution with lead nitrate or lead acetate affords bluish red P.R.90, referred to as phloxine:

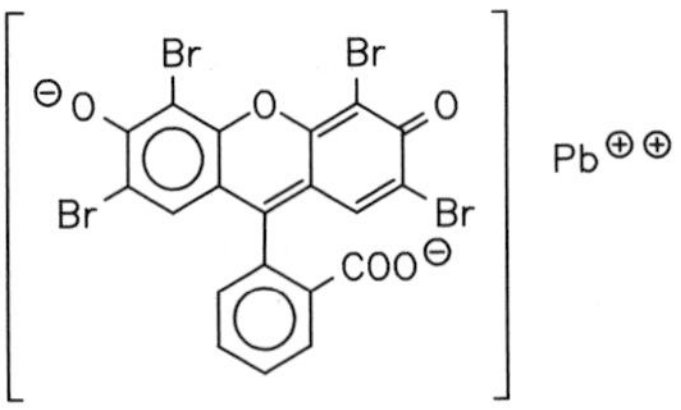

Phloxine, the lead precipitate, is manufactured in Japan and the USA, although its significance even in these countries is declining rapidly as a result of its lead content. The pigment provides a brilliant medium red shade and high tinctorial strength. P.R.90 is used in inexpensive letterpress and offset inks, for which it used to be supplied in the form of flushed pastes. The prints lack fastness to organic solvents, especially to alcohols, esters, and ketones, as well as to various chemicals such as alkali, acid, and soap. Likewise, P.R.90 shows poor lightfastness and heat stability.

Pigment Red 251, 12925

The chemical constitution of P.R.251, a pyrazoloquinazolone pigment, is described in the Colour Index [14]. See also p. 489.

A yellowish red pigment, P.R.251 has been known for some years. It is marketed as a variety with a very coarse particle size, which exhibits good hiding power and is recommended particularly as an alternative for Molybdate Red pigments in paints. Systems containing P.R.251 are only fast to bleeding up to 80 to 100 °C. The pigment is frequently found in long-oil alkyd resin paints (house paints), in medium-oil alkyd resin systems (air drying industrial paints, including systems which are dried at elevated temperature up to about 80 °C, such as automotive refinishes), as well as in emulsion paints. P.R.251 exhibits excellent lightfastness and weatherfastness.

Pigment Red 252

P.R.252 is a similarly recent product. It is a member of the pyrazoloquinazolone series and chemically differs from P.R.251 only in its substitution pattern.

P.R.252 provides yellowish to medium red shades and is recommended particularly for use in architectural paints. The pigment shows very poor fastness to a number of organic solvents which are commonly used in paints, a deficiency which largely precludes it from being used in oven drying systems. Regarding lightfastness and weatherfastness, the only available type with coarse particle sizes performs somewhat better than the much more yellowish P.O.5.

Pigment Brown 22, 10407

The reddish brown pigment is characterized by the following chemical structure:

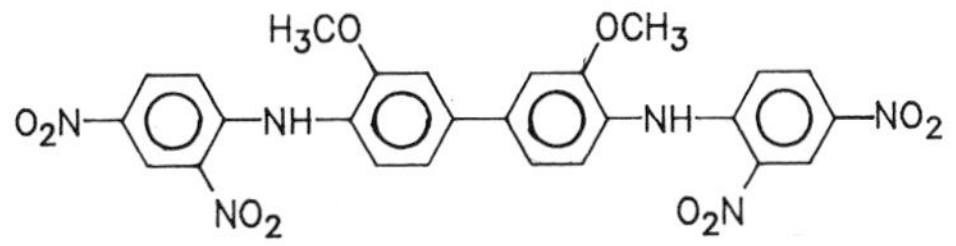

P.Br.22 presents medium to slightly reddish shades. It is used for the spin dyeing especially of polyacrylonitrile and viscose.

Pigment preparations of P.Br.22 are supplied for these purposes. The pigment demonstrates good fastness properties, it is fast to perspiration, dry cleaning in perchloroethylene, and dry heat. In terms of lightfastness, P.Br.22 systems equal step 5 to step 7 on the Blue Scale, depending on the depth of shade.

P.Br.22 is offered in the form of a variety of pigment preparations, designed for industrial finishes, packaging gravure printing inks, or wood stains. The pigment is not supplied as a powder.

Pigment Black 1, 50440

P.Bl.1, referred to as Aniline Black, is an indazine derivative:

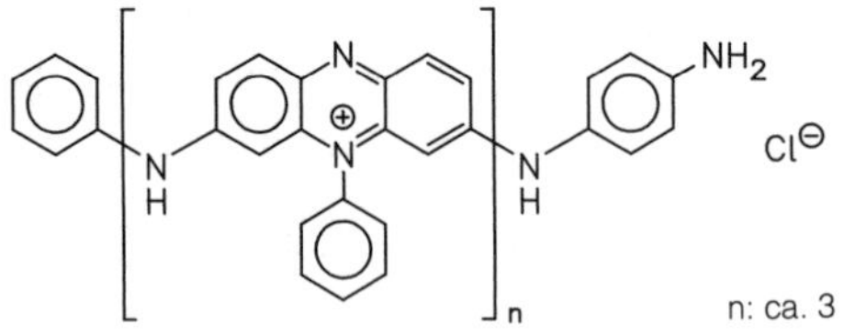

n: ca. 3

A technique of developing Aniline Black directly on the fiber was found by Lightfoot in the period between 1860 and 1863. In accordance with this process, the fiber is soaked with aniline, aniline hydrochloride, and sodium chlorate in the presence of an oxidation catalyst (e.g., ammonium vanadate, potassium hexacyanoferrate(II)). The compound is "developed" at 60 to 100 °C and then oxidized further with sodium chromate. It should be noted, however, that Perkin had already synthesized a black compound which he called Aniline Black as early as 1856. He oxidized aniline (containing toluidine) with potassium dichromate and separated Aniline Violet from the resulting black mixture (Aniline Black).

Modern methods of manufacturing this probably oldest representative among synthetic organic pigments involve dissolving aniline in strong sulfuric acid. Oxidation is achieved with sodium dichromate in the presence of a copper salt or one of the above-mentioned oxidation catalysts. Oxidation with sodium chlorate initially affords an indamine polymer (pernigraniline):

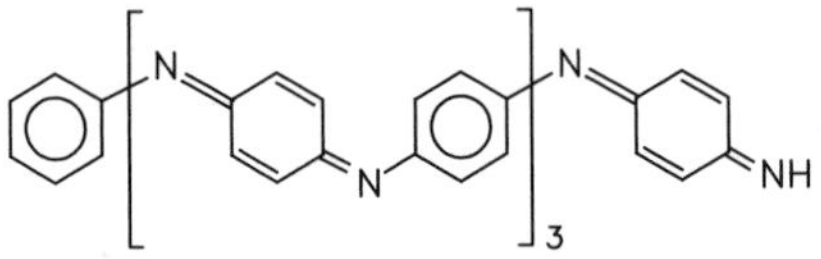

This intermediate must be oxidized further to afford the azine pigment (Green, Willstätter, 1907 to 1909).

Aniline Black provides a deep, neutral shade of black. Extensive absorption and little scattering make for good hiding power. The commercial grades cover a comparatively wide range of particle size distributions. The types with fine particle sizes in particular provide characteristically dull, velvety effects in finishes and prints. Even types with fine particle sizes show only a very slight tendency to agglomerate, which makes them easy to disperse. The pigment is not an electrical conductor.

P.Bl.1 is used in a variety of media. Incorporated in paints, full shades show excellent lightfastness and weatherfastness, qualities which deteriorate rapidly as more TiO_2 is added. The pigment is tinctorially weaker than carbon blacks. Some types are not entirely fast to overcoating, a property which extends to acid and alkali, oxidants and reducing agents. The paint and printing ink industries utilize P.Bl.1 particularly where carbon blacks present processing problems or where a matt and velvety quality is required in a paint or print. P.Bl.1 is of interest in wood stains made from unsaturated polyester. In plastics, P.Bl.1 is used to advantage wherever carbon black cannot be used as a result of its inability to tolerate heat sealing.

4.5 Pigments With Hitherto Unknown Chemical Structure

The fact that many of these pigments have not even been assigned to any particular class makes it more convenient to list them by their shade in the order of increasing Colour Index Numbers.

Pigment Yellow 99

P.Y.99, which is derived from the anthraquinone structure, is produced in Japan. It affords very reddish shades of yellow, even redder but at the same time distinctly duller than those of P.Y.83. P.Y.99 is recommended especially for use in textile printing but is also used in plastics. Although HDPE systems are heat stable up to 300 °C, they noticeably lack tinctorial strength. 1/3 SD HDPE samples containing 1% TiO_2 are formulated at 0.53% pigment.

Pigment Yellow 187

P.Y.187, which was introduced in recent years, is a special-purpose pigment for polyamide. Thermostable up to 320 °C in this plastic material, the pigment furnishes greenish yellow shades. 1/1 SD samples score only as high as step 4 on the Blue Scale for lightfastness.

Incorporated in plasticized PVC, P.Y.187 demonstrates moderate fastness to bleeding. At a processing temperature of 130 °C, the pigment affords a greenish shade of yellow, while an orange shade is observed at 160 °C.

Pigment Red 204

The exact chemical constitution of P.R.204, a polycyclic compound, remains to be published. The pigment affords a dull, medium red full shade. In white reductions, the shade becomes much more bluish and comparatively cleaner. P.R.204 exhibits excellent fastness properties. Full shades (5%) in various paint systems equal step 7–8 on the Blue Scale for lightfastness, a value which is also reached by white reductions up to 1/25 SD. Moreover, P.R.204 also shows very good weatherfastness. The commercial grade demonstrates high hiding power and good rheology. Full shades and similar deep shades are of particular interest, as they are frequently combined with inorganic red pigments. The resulting blends are used to advantage in various types of industrial paints, including automotive finishes. P.R.204 provides good fastness to overcoating and shows a heat stability of 180°C for 30 minutes. Higher baking temperatures cause a color shift towards more bluish shades.

P.R.204 is also very lightfast in printing inks, although it is rarely encountered in these media. 1/1 to 1/25 SD letterpress proof prints, for instance, equal step 7 on the Blue Scale. The prints demonstrate very good fastness properties in application, despite the fact that they are not entirely fast to paraffin. Besides, they are not fast to sterilization.

Pigment Red 211

P.R.211 is offered only in the Japanese market. It is the calcium salt of a monoazo pigment whose exact chemical constitution has not yet been published. P.R.211 affords shades of scarlet which resemble those provided by Lake Red C types (P.R.53:1). The list of suitable media includes plastics and printing inks. Incorporated in these media, P.R.211 presents application and fastness properties which are very close to those of P.R.53:1. HDPE samples containing P.R.211, for instance, like those based on P.R.53:1, are heat stable up to approx. 260°C. Both pigments are fast to bleeding in plasticized PVC. In prints, however, the commercial grade of P.R.211 is tinctorially weaker than similarly colored Lake Red C types. Depending on the application medium, P.R.211 is weaker by 5 to 50%.

Pigment Black 20

The exact chemical structure of P.Bl.20, an anthraquinone pigment, is not known. P.Bl.20 is a specialty product used in camouflage paint. The pigment satisfies certain specifications regarding infrared reflection. It is employed in paints. 1/3 SD equal step 6 on the Blue Scale for lightfastness. Thermally stable up to 200°C, P.Bl.20 is not fast to overcoating, not even at low baking temperatures such as 120°C. Although alkali proof, P.Bl.20 is not entirely fast to acid. The pigment is insufficiently fast to common organic solvents such as esters and aromatic hydrocarbons. In this respect, P.Bl.20 equals step 2 on the 5-step scale.

References for Chapter 4

[1] B.K. Manukian and A. Mangini, Chimia 24 (1970) 328–339.

[2] A. Patil, G. Patkar and T. Vaidyanathan, Bombay Technol. 24 (1974) 26–33.

[3] DE-AS 1 770 960 (BASF) 1968.

[4] H. Sakai et al., J. Japan Soc. Colour Mat. 55 (1982) 685.

[5] DE-AS 2 706 872 (Teijin Ltd.) 1976.

[6] CH-OS 438 542 (Ciba) 1964.

[7] D.G. Farnum et al., Tetrahedron Letters 29 (1974) 2549.

[8] A. Iqubal et al., J. Coatings Technology 60 (1988) 37–45.

[9] US-PS 4 415 685 (Ciba-Geigy) 1983.

[10] P. Rys and H. Zollinger, Farbstoffchemie, 3. Ed., Verlag Chemie, Weinheim, 1982; H. Zollinger, Color Chemistry, VCH, Weinheim, 1991, p. 194.

[11] Pigment Handbook, ed. by P.A. Lewis, Vol. 1, p. 723 (1988), John Wiley, New York.

[12] B.L. Kaul, Soc. of Plastics Eng., Huron (Ohio), 213–226 (1989).

[13] B.L. Kaul, Congr. FATIPEC, Venedig 1986, Vol. 3, 73–93.

[14] H. Kanter, B. Ort, XVIII. Congr. FATIPEC, Venedig 1986, Congress book Vol. 3, 95–111.

5 Ecology, Toxicology, Legislation

5.1 General

As a starting point to discussing organic pigments, it should be noted that the specific physicochemical, toxicological, and ecological properties of these compounds present an uncommonly small risk potential.

The analysis of any substance or product which is to be marketed requires not just identification of its technical and application properties. Toxicological and ecological considerations have become a primary concern in today's industry. The pigment industry is no exception. Manufacturers, processers, and customers are always faced with the question of whether the production or use of a specific product might present an unacceptably high toxicological or ecological risk for man or environment. The problem of eliminating or at least reducing these risks to legally acceptable levels is a very urgent one. The question of acceptability is not only based on one's responsibility, but criteria must be developed for each individual case with respect to legal provisions.

It is of paramount importance to evaluate the physical, chemical, toxicological, and ecological properties of a pigment in order to be able to assess its influence on the environment and to estimate whether or not a given product presents a toxicological or ecological risk [1]. It is important also to consider the type of handling and use that a pigment is likely to undergo, the extent of human exposure that might be expected, and the amounts involved. Chemicals are typically assessed on the basis of a limited body of data, a task which is normally assigned to a specialist. More complex cases will even be judged by an interdisciplinary group of experts.

Prime considerations in pigment manufacture and use, as with any other chemical, include personnel safety, air emission and wastewater quality, and appropriate waste removal. Although basic information is supplied by the pigment manufacturer, it is the processer who is responsible for safe handling and use. He is also the one responsible for solving internal safety problems.

It has become increasingly common to provide information on safe pigment handling and use in concentrated form on safety data sheets. Apart from listing physical and chemical parameters and information as to safety and precautionary measures, fire and flammability data, first aid after contamination and accidents, waste disposal, these sheets also provide toxicological and ecological parameters.

Examining the "curriculum vitae" of a pigment comprises the following steps:
- production (starting materials, impurities, side products),
- working the pigment into its application medium,
- intended use of the finished article,
- destruction or removal of the article after use (waste management).

The single most important factor in evaluating the toxicological and ecological properties of an organic pigment is its extreme insolubility in water and in the application media.

Pigments are therefore processed largely as solid, crystalline, and therefore physiologically inert materials.

Since organic pigments are commonly combined with other materials, a pigmented system typically contains only a small percentage of actual pigment. It is therefore likely that other components, such as binders, solvents, and various agents may more severely affect the ecological and toxicological properties of the applied product.

5.2   Ecology

Although pigments, because they are practically insoluble in water and — if at all — only sparingly soluble in common solvents, are largely biologically inert, it is the responsibility of the pigment manufacturer and producer to comply with legal requirements.

Ecological considerations include air emissions and wastewater.

Air emissions: Organic pigments are basically subject to a general limit to particulate emission: the concentration of fine dust in the air may not exceed 6 mg/m^3 [2]. Although an organic pigment dust is not considered a health hazard, emission control includes filtering waste air after manufacture to remove the pigment dust. Suitable methods involve using dust filters or adsorption techniques. Generally, in order to avoid possible exposure to pigment dust, individuals involved in the manufacture of a pigment must wear additional protective wear (dust mask).

Wastewater: The danger of contaminating free-running water with organic pigments may be met with measures such as filtration, sedimentation, adsorption, or, if necessary, with biological treatment. By means of these techniques, wastewater pollution may be practically avoided. As a result of their negligible solubility, organic pigments are probably nontoxic to fish, at least there is no evidence to the contrary. Fish toxicity is conceivable only if the receiving water cannot offer a sufficiently high dilution factor and thus accumulates such quantities of pigment as to cause gill failure.

Although the bacteria used in biological treatment are not effective in degrading organic pigments, the compounds are likely to decompose slowly under anaerobic conditions.

Legal limitations placed upon the use of all newly launched nonbiodegradable chemicals in Japan require testing each product for accumulation in fish before it can be marketed. These regulations have stimulated corresponding studies about the bioaccumulation of pigments [3]. The distribution coefficient P_{ow} between n-octanol and water may be used to indicate the tendency of a chemical to accumulate in biological systems [4].

An empirical rule, which also applies to dyes, predicts a bioaccumulation factor of < 100 in fish if the P_{ow} is below 1000. It should be noted, however, that this rule does not apply to organic pigments, since they are only very sparingly soluble in water and in lipids (for instance in the lipid simulant solvent n-octanol) and consist of comparatively large molecules. Although the theoretical log P_{ow} is very high for pigments (up to 10), the compounds do not accumulate in fish.

5.3 Toxicology

Toxicological studies are concerned with a variety of aspects, primarily with the following factors:
- acute toxicity,
- irritation of skin and mucous membrane,
- toxicity after repeated application,
- sensitization,
- mutagenicity,
- chronic toxicity, especially
- carcinogenicity.

5.3.1 Acute Toxicity

Testing on animals may provide initial information on the effect of a possible short-term exposure on human health. Acute toxicity is defined as the toxic effect of a substance after a single oral, dermal, or inhalative application. For acute oral toxicity, for instance, LD_{50} is defined as the amount of substance expressed in mg per kg body weight which has a lethal effect on 50% of the test animals after a single oral application. Such tests are useful in that they assess the toxicity of a material relative to that of other known compounds.

More than 4000 colorants have been examined in this manner and the results have been summarized. The data were taken from studies published in safety data sheets by the member companies of ETAD* [5].

---

* Ecological and Toxicological Association of the Dyestuffs and Organic Pigments Manufacturers, Basel, Switzerland.

Another source [6] lists 108 organic pigments which have been examined concerning their acute oral lethal dose in rats. None of the studied pigments shows an
LD_{50} value of less than 5000 mg/kg body weight.

A third publication summarizes the results of testing 194 organic pigments for
toxicity. The most important chemical types were represented [7]. None of the
tested specimens was found to have a $LD_{50} < 5000$ mg/kg, with the exception of
four values which were located in the 2000 to 5000 mg/kg range. The EC Chemical
Law requires substances to be labeled as harmful if the LD_{50} for acute oral toxicity
is below 2000 mg/kg. For comparison, it might be mentioned that NaCl has an
LD_{50} of 3000 mg/kg.

Pigments are passed via the gastro-intestinal tract and not discharged via the
urethra. According to the results of these studies, organic pigments show practically no acute toxicity.

5.3.2   Irritation of Skin and Mucous Membrane

Publications summarizing the results of studies on the effect of pigments on the
skin and mucous membrane (conjunctiva of eyes) of rabbits describe similar observations. These sources refer to pigments as trade products which may contain
auxiliaries (Table 34).

Tab. 34: Irritation effects caused by organic pigments.

	Number of pigments	
	Skin	Mucous membrane (conjunctiva)
Nonirritating	186	168
Slightly irritating	5	20
Moderately irritating	1	1
Strongly irritating	0	3

To summarize the results of the two discussed chapters: it has been demonstrated that organic pigments exhibit very high LD_{50} values and only rarely cause
irritations of skin or mucous membranes.

5.3.3   Toxicity After Repeated Application

These so-called subacute or subchronic toxicity studies involve the repeated application of a test substance to animals, typically for a period of 30 or 90 days. The
time pattern is thus an intermediate one between acute and chronic toxicity. To test
a substance for subacute or subchronic toxicity, it is mainly applied by ingestion or

inhalation. Not one out of the large number of organic pigments which have thus been tested has demonstrated any irreversible toxic effect. No toxic response was observed in rats which were fed either Pigment Yellow 1 or Pigment Yellow 57:1 for 30 days [8].

5.3.4   Mutagenicity

A variety of methods are available to test a chemical for mutagenicity, i.e., its effect on the genetic material. Only very few of these methods, however, have gained more than minimum recognition. Heading the list is the Ames Test [9]. This is a bacterial test which allows fast performance and requires limited expense. Its correlation with the mutagenicity of mammals or even with a carcinogenic effect on mammals or humans has repeatedly been tested [5], but remains controversial.

Only two out of 24 Ames-tested organic pigments show a slightly positive result [6] (Table 35).

Tab. 35: Mutagenicity tests of organic pigments.

Pigment	C.I. Constitution No.	Test result
P.Y.1	11680	negative
P.Y.12	21090	negative
P.Y.74	11741	negative
P.O.5	12075	weakly positive
P.O.13	21110	negative
P.R.1	12070	weakly positive
P.R.4	12085	negative
P.R.22	12315	negative
P.R.23	12355	negative
P.R.48:1	15865:1	negative
P.R.48:2	15865:2	negative
P.R.49	15630	negative
P.R.49:1	15630:1	negative
P.R.49:2	15630:2	negative
P.R.53:1	15585:1	negative
P.R.57:1	15850:1	negative
P.R.63:1	15880:1	negative
P.Bl.15	74160	negative
P.Bl.15	74160:1	negative
P.Bl.15	74160:2	negative
P.Bl.15	74160:3	negative
P.Bl.15	74160:4	negative
P.Gr.7	74260	negative
P.Gr.36	74265	negative
P.V.19	73900	negative

5.3.5 Chronic Toxicity — Carcinogenicity

One of the prime concerns, apart from acute and subacute toxicity, is the question of whether a product causes chronic effects. In this context, carcinogenicity studies are of cardinal importance. A possible chronic hazard may be indicated by epidemological studies. Where such investigations are not available, experiments are performed on animals for the duration of their entire life span. The type of application depends on the exposition (perorally, dermally, per inhalation).

Dichlorobenzidine pigments have stimulated particular interest in this respect. Although reductive cleavage of the dichlorobenzidine moiety might conceivably afford 3,3'-dichlorobenzidine, this is not observed in animals. Despite the fact that all epidemological studies produced similarly negative results [10], the pigments in question were also tested in long-term feeding studies. Pigment Yellow 12, 16, and 83 were fed to rats and mice over a period of 2 years. The daily dosage for mice was up to 2 g pigment/kg body weight, while rats were fed up to 0.6 g/kg per day. No carcinogenic effect was observed [11]. The results were confirmed by two independent studies on Pigment Yellow 12 [12, 13].

There is a publication which, caused by an unsuitable experimental design, erroneously claims to have found 3,3'-dichlorobenzidine in the urine of rabbits—an observation which, if true, would have pointed to metabolism of P.Y.13 [14]. The results were challenged and contradicted later by comprehensive research on rats, mice [11], and rats, rabbits, and monkeys [15].

Recent studies investigating the bioavailability of P.Y.17 in rats, applied not only orally but also by inhalation, in no case showed the presence of any cleavage product of the pigment (e.g., 3,3-dichlorobenzidine or metabolites) in the urine or blood of the animals. Similar studies were carried act with P.Y.13 and 174 [16].

Table 36 lists all pigments which have thus far been tested for carcinogenicity or which are in the process of being examined [6].

Tab. 36: Chronic toxicity of organic pigments (long-term feeding studies).

C.I. Name	C.I. Constitution No.	Test animal	Result	References
P.Y.12	21090	mouse, rat	negative	[11, 12, 13]
P.Y.16	20040	mouse, rat	negative	[11]
P.Y.83	21108	mouse, rat	negative	[11]
P.O.5	12075	mouse, rat	equivocal	[17]
P.R.3	12120	mouse, rat	equivocal	[18]
P.R.4	12085	mouse, rat	in process	[17]
P.R.23	12355	mouse, rat	equivocal	[19]
P.R.49	15630	rat	negative	[20]
P.R.53:1	15585:1	mouse, rat	negative	[21]
P.Y.57:1	15850:1	rat	negative	
P.Bl.60	69800	rat	negative	

Organic pigment lakes, whose free acidic functions makes them at least partially water soluble, were tested in the form of their soluble sodium salts (Table 37) [22].

Tab. 37: Long-term feeding studies of water soluble Na salts of pigment lakes.

C.I. Name	C.I. Con-stitution No.	Tested as	Test animal	Result
P.Y.100	(Al),19140	Acid Yellow 23 (Na)	mouse, rat, dog	negative
P.Y.104	(Al),15985	Food Yellow 3 (Na)	mouse, rat	negative
P.R.172	(Al),45430	Food Red 4 (Na)	rat, dog	negative
P.Bl.24	(Ba),42090	Food Blue 9 (Na)	mouse, rat	negative
P.Bl.63	(Al),73015	Food Blue 1 (Na)	rat and others	negative

5.3.6   Impurities in Pigments

Impurities must be taken into consideration in all types of studies. Especially
- traces of aromatic amines,
- traces of heavy metals,
- polychlorobiphenyls,

may influence the outcome of toxicity tests in general.

Aromatic Amines

Only trace amounts of aromatic amines, which are used to manufacture azo pigments, are found in the finished products. Upper limits have been defined for certain areas of application. The content of aromatic amines in pigments which are to be used in materials for food packaging, for instance, may not exceed 500 ppm.

Heavy Metals

A comprehensive study, performed in the USA in 1973 [23] by DCMA*, has shown that the heavy metal content in organic pigments is markedly below legal standards [24]. For pigments to be used in consumer goods, including toys, legal limits have been implemented. For instance, 0.07 N hydrochloric acid may not extract more than 0.01% barium ions out of a pigmented system [20].

---

* Dry Color Manufacturers Association.

Polychlorobiphenyls (PCB)

Strict legal limits have been defined in some countries (for instance in the USA, and also throughout the EC) because of the persistency and the wide distribution of these compounds. These regulations have primarily been issued in order to protect the environment, less because of a direct hazard for humans.

PCB traces are mainly found in two groups of organic pigments, namely in
- azo pigments derived from chloroaniline or dichloro or tetrachlorodiaminodiphenyl, which may undergo various side reactions that produce traces of PCB,
- pigments which are manufactured in the presence of dichlorobenzene or trichlorobenzene as a solvent. In this case, PCB may be formed through radical reactions.

Polychlorinated Dioxins/Furans ("Dioxins")

Similar conditions as for the formation of PCB's might also lead to the formation of small trace amounts of dioxins. The German Hazardous Substances Ordinance prohibits to place products on the market for which very low limits of dioxin traces are exceeded.

5.4   Legislation

In recent years, the number of laws providing protection during the manufacture, use, and storage of chemicals has risen worldwide. The same is true for regulations concerning waste removal (air emission, wastewater, solid waste). Without exceeding the scope of this book, only a few examples can be mentioned in order to illustrate the legal situation, which may also affect organic pigments. Table 38 lists important regulations throughout Europe, the USA, Canada, Japan, and Australia.

Heading the list of regulations in Western industrialized countries are the Chemicals Acts in the EC countries. These regulations are derived from a basic Directive of the European Community (6th Amendment to Directive 67/548/EEC of the Control Directive of June 1967 on the approximation of laws, regulations, and administrative provisions relating to the classification, packaging, and labeling of dangerous substances, Sept 18, 1979). Besides, the list also includes the Toxic Substances Control Act in the USA (1976), the Canadian Environmental Protection Act (1988), the Chemical Substances Control Law in Japan (1973), and the Australian Chemicals Act (1985). These regulations, which also affect pigments, demand that all new products be registered. Various toxicological test results have

Table 38: Important environmental laws and ordinances.

Law	Country
● Guideline for the Classification, Packaging, and Labeling of Dangerous Substances with Several Amendments (67/548 EEC) (1967); particularly 91/325/EEC	EC
● General Preparations Directive (88/379/EEC)	EC
● Safety of Toys (87/C343/EC); (EN 71, part 3: chemical properties)	EC
● Directive for Materials and Articles Intended to Come into Contact with Foodstuff (89/109/EEC)	EC
● Directive for Plastic Material Intended to Come into Contact with Foodstuff (90/128/EEC)	EC
● Cosmetics Directive (76/768 EC, 82/368 EEC)	EC
● Austrian Chemicals Law (1989)	A
● Canadian Environmental Protection Act (1988)	CDN
● Poison Law (1969)	CH
● Act Concerning Environmentally Dangerous Substances (1986)	CH
● Chemicals Law (1980)	D
● Hazardous Substances Ordinance (1986)	D
● Amendment to the German Chemical Law (1990)	D
● 178th Communication of the BGA (1988)	D
● Law Concerning Chemical Products Control (1977)	F
● Safety of Toys (1984)	F
● Circulaire No. 176 (positive list of colorants for the packaging of foodstuffs (1959/84))	F
● Health and Safety at Work Act (1974)	GB
● Classification, Packaging, and Labeling Regulations (1984)	GB
● Control of Substances Hazardous to Health (1988)	GB
● Carcinogenic Substances Regulations (1976)	GB
● Toys (Safety) Regulation (1974)	GB
● Chemical Substances Control Law (1973)	J
● Poisonous and Deleterious Substances Control Law (1964)	J
● Industrial Safety and Health Law (1972)	J
● Fire Service Law (1990)	J
● Toxic Substances Control Act (TSCA) (1976)	USA
● Occupational Safety and Health Act (OSHA) (1970)	USA
● OSHA Hazard Communication Standard (1987)	USA
● Superfund Amendments and Reauthorization Act (SARA) (1986)	USA
● Colorants for Polymers, FDA (1982)	USA
● Right-to-Know Laws	USA
● Australian Chemicals Act (1985)	AUS
● National Chemicals Notification and Assessment Scheme (1989)	AUS

A = Austria	CH = Switzerland	F = France
AUS = Australia	D = Germany	GB = Great Britain
CDN = Canada	EC = European Community	J = Japan

to be presented, depending on the country. The discussed legal requirements are supplemented by many other more or less specific environmental acts, not only in the countries thus mentioned, but also in practically all other industrialized nations.

The primary concern of regulations which pertain to pigments in particular is to define upper limits for the impurities mentioned in Sec. 5.3.6. These limits, which are defined differently by individual countries, should not be exceeded. This is of paramount importance in pigment applications which involve particularly high potential human exposure, for instance in colors for the artists' market or for childrens' finger paints. The same is true for low exposure applications (pigments are incorporated in coatings or plastics and thus largely protected from human contact) which, however, affect a large number of persons. This is particularly important in areas such as toys or food packaging. In the FRG, the use of colorants for consumer goods is regulated by recommendation IX of the Federal Health Agency (Bundesgesundheitsamt), which stipulates that no colorant may enter the food at all, not even in traces [25].

Similar regulations are in operation in many other nations, including Belgium, France, Great Britain, the Netherlands, and Italy.

Toys are subject to the European Standard 71 concerning the "Safety of Toys". This standard, first introduced as a draft in 1971, has met with worldwide success and governs the pigment manufacturing and processing industry. It pertains primarily to paints and coatings, writing utensils, plastics, paper, and cardboard.

The insolubility and migration fastness of most organic pigments largely eliminates human health hazards. Despite this fact, the need for consumer protection and compliance with the legal regulations makes it necessary to regularly test not only the pigmented materials but also the pigments themselves. This is to ensure that no trace of pigment migrates from the printed or colored packaging into the food and that the stringent requirements, particularly regarding impurities (traces of heavy metal, aromatic amines, or PCB) are met. Several papers deal with the safe handling of colorants [26].

References for Chapter 5

[1] A.C.D. Cowley, Polymers Paint Colour J. (1985) August 7/21.
[2] Maximum Concentration at the Workplace and Biological Tolerance Values for Working Materials 1991 (DFG), VCH-Verlag Weinheim.
[3] R. Anliker and P. Moser, Chemosphere, Vol. 17, No. 8 (1988) 1631–1644.
[4] R. Anliker, E.A. Clarke and P. Moser, Chemosphere 10 (1981) 263–274.
[5] E.A. Clarke and R. Anliker, Organic Dyes and Pigments in: The Handbook of Environmental Chemistry, Vol. 3/Part A, edited by O. Hutzinger, Springer-Verlag, Berlin, 1980.
[6] NPIRI Raw Materials Data Handbook, Vol. 4 Pigments, Francis Mac Donald Sinclair Memorial Laboratory 7, Lehigh University Bethlehem, PA 18015, 1983.
[7] K.H. Leist, Toxicity of Pigments, Lecture given at the Nifab-Symposium in Stockholm, May 1980.

[8] K.H. Leist, Ecotoxicol. and Environm. Safety 6 (1982) 457–463.

[9] B.N. Ames, J. McCann and E. Yamasaki, Mut. Res. 31 (1975) 347–364.

[10] H.W. Gerarde and D.F. Gerarde, J. Occup. Med. 16 (1974), 322–324; T. Gadian, Chem. Ind. 19 (1975) 821–830; I. MacIntyre, J. Occup. Med. 17 (1975) 23–26.

[11] F. Leuschner, Toxicology Letters 2 (1978) 253–260.

[12] H. MacD. Smith, Am. Ink Maker 55 (6), (1977) 17–21.

[13] NCI/NIH Report: Bioassay of Diarylide Yellow for possible carcinogenicity, DHEW Publication No. (NIH) 77-830, 1977.

[14] T. Akiyama, Ikei Med. J. 17 (1970) 1–9.

[15] A. Mondino et al., La Medicina del Lavoro 69 (6) (1978) 693–697.

[16] Th. Hofmann, D. Schmidt, Arch. Toxicology, in press; P. Sagelsdorff et al., Intern. Cancer Congress, Hamburg 1990.

[17] Tox Tips Oct. 1978, June 1981. Toxicology Information Program, National Library of Medicine, Bethesda, MD 20014.

[18] NTP Technical Report TR 407, NIH Publication No. 92-3138, March 1992.

[19] NTP Technical Report TR 411, NIH Publication No. 91-3142, March 1991.

[20] K.J. Davis and O.G. Fitzhugh, Toxicol. Appl. Pharmacol. 5 (1963) 728–34.

[21] K.J. Davis and O.G. Fitzhugh, Toxicol. Appl. Pharmacol. 4 (1962) 200.

[22] R. Anliker and E.A. Clarke, Chemosphere Vol. 9 (1980) 595–609.

[23] Dry Color Manufacturers' Association. Trace metals in organic pigments, Am. Ink Maker 51 (10), (1973) 31–35.

[24] Trace Metals in Organic Pigments, National Association of Printing Ink Manufacturers, Am. Ink Maker 1973 (DCMA Annual Meeting White Sulphur Springs W.Va., June 18–20, 1973).

[25] BGA Empfehlung IX, Bundesgesundheitsblatt 31 (1988) 186. Mitt. Farbmittel zum Einfärben von Kunststoffen und anderen Polymeren für Bedarfsgegenstände.

[26] R. Anliker, D. Steinle, Verhütung von Risiken beim Gebrauch und bei der Handhabung von Farbmitteln, Textilveredlung 25 (1990) 2, 42–49; Safe handling of colorants, CIA (Great Britain) 1989; Colorants and the Environment, Guidance for user, ETAD, June 1992.

Review of Chemical Structures and Chemical Reactions

2.1 Starting Materials

1   **3,3′- or 3,3′-5,5′- subst. 4,4′-Diaminodiphenyls**

$$X : Cl, CH_3, OCH_3, OC_2H_5$$
$$Y : H, Cl$$

2   **Acetoacetic Acid Arylides**

$$CH_3COCH_2COOR$$

3   2-Naphthol / 2-Hydroxy-3-Naphthoic Acid / Naphthol AS

2.2   Synthesis of Azo Pigments

1   Diazotization

Preceding equilibria for the formation of XNO (X = Cl, Br, NO_2, HSO_4):

Diazotization (Ar: aromatic group):

Side reaction in the presence of an excess of $OH^{\ominus}$:

$$Ar-\overset{\oplus}{N}\equiv N + OH^{\ominus} \longrightarrow Ar-N \overset{\displaystyle \sim}{=} N-OH$$

$$Ar-N\overset{\displaystyle \sim}{=}N-OH + OH^{\ominus} \rightleftharpoons Ar-N\overset{\displaystyle \sim}{=}_N \rightleftharpoons ArN\overset{\displaystyle \sim}{=}_N-O^{\ominus}$$
$$\underset{O^{\ominus}}{|}$$

cis (syn)   trans (anti)

diazotate

Decomposition (thermal):

$$Ar-N\equiv N^{\oplus} \; Y^{\ominus} + H_2O \longrightarrow ArOH + N_2 + HY$$

2 Coupling (RH = Coupling Component)

$$ArN\equiv N^{\oplus} \; Y^{\ominus} + RH \longrightarrow Ar-N\overset{\displaystyle \sim}{=}_{N-R} + HY$$

Side reaction in the case of acid deficiency (esp. with weak amines):

$$Ar-N\equiv N^{\oplus} \; Y^{\ominus} + H_2NAr' \longrightarrow ArN\overset{\displaystyle \sim}{=}_{N-NHAr'}$$

diazoamino compound

2.3 Monoazo Yellow and Monoazo Orange Pigments

1 Non-Laked Monoazo Yellow and Monoazo Orange Pigments

R : CH_3, $COOCH_3$, $COOC_2H_5$
R' : H, CH_3

2   Monoazo Yellow and Monoazo Orange Pigment Lakes

R$_D$: substituent of the diazo component
R$_K$: substituent of the coupling component
m: 1 to 3; n: 1 to 3
M: Ca, Ba; X: Cl, NO$_3$

2.4 Disazo Pigments

1 Diarylide Yellow Pigments

2 Bisacetoacetarylide Pigments

3 Disazopyrazolone Pigments

R_D: substituent of the diazo component
R_K: substituent of the coupling component
X: H, Cl, CH_3, OCH_3   R = CH_3, COOCH_3, COOC_2H_5
Y: Cl, CH_3, OCH_3   R' = H, CH_3
m: 1 to 3
N: 1 or 2

2.5 / 2.6 β-Naphthol Pigments / Naphthol AS Pigments

2.5 β-Naphthol Pigments

2.6 Naphthol AS Pigments

R_D: substituent of the diazo component
R_K: substituent of the coupling component
m: 0 to 3
n: 0 to 3

2.7 Red Azo Pigment Lakes

1 β-Naphthol Pigment Lakes

Ar: phenyl or naphthyl
M: 2 Na, Ca, Ba, Sr, 2/3 Al

2   BONA Pigment Lakes

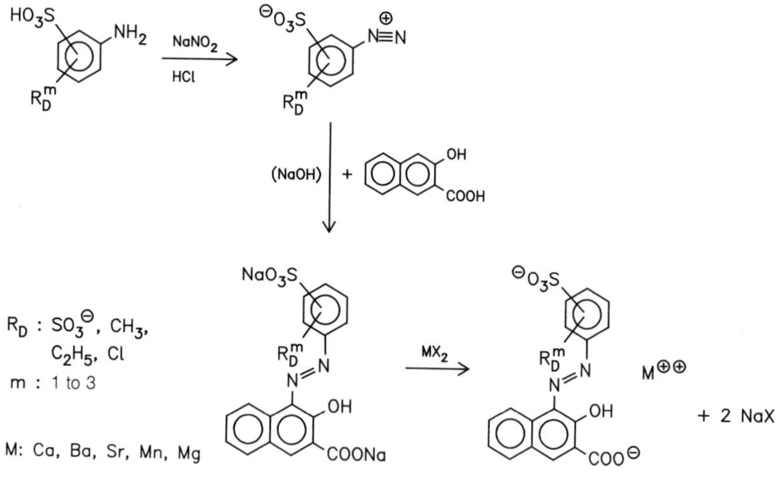

R_D : $SO_3^{\ominus}$, CH_3,
   C_2H_5, Cl

m : 1 to 3

M: Ca, Ba, Sr, Mn, Mg

3   Naphthol AS Pigment Lakes

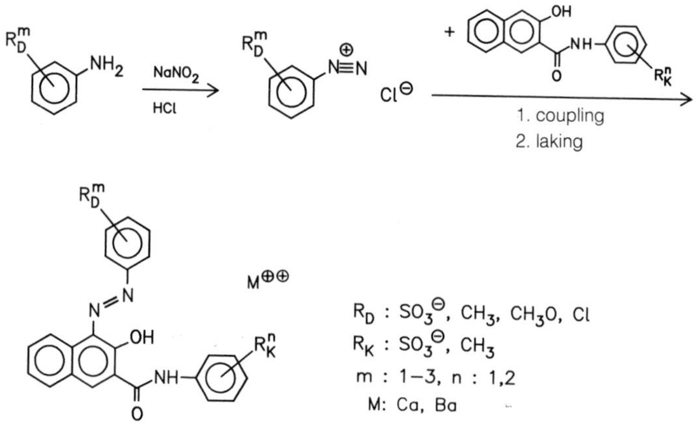

1. coupling
2. laking

R_D : $SO_3^{\ominus}$, CH_3, CH_3O, Cl
R_K : $SO_3^{\ominus}$, CH_3
m : 1–3, n : 1,2
M: Ca, Ba

4   Naphthalene Sulfonic Acid Pigment Lakes

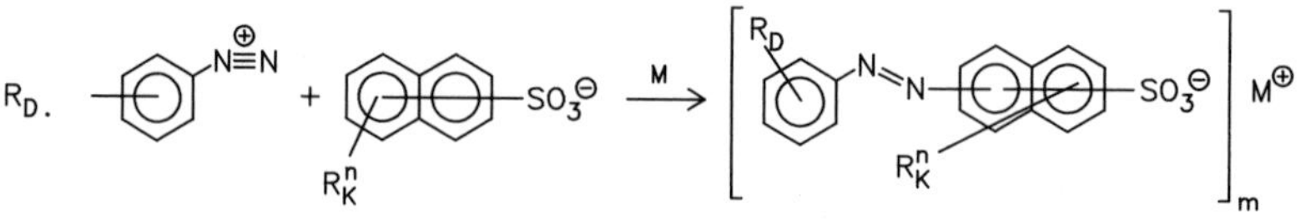

R_D: Cl, CH_3, CH_3O, $COO^{\ominus}$ or $SO_3^{\ominus}$

R_K^n : OH, HNCO—⟨◯⟩, $SO_3^{\ominus}$

n, m: 0 to 3
M: Ba, Al, Na

2.8 Benzimidazolone Pigments

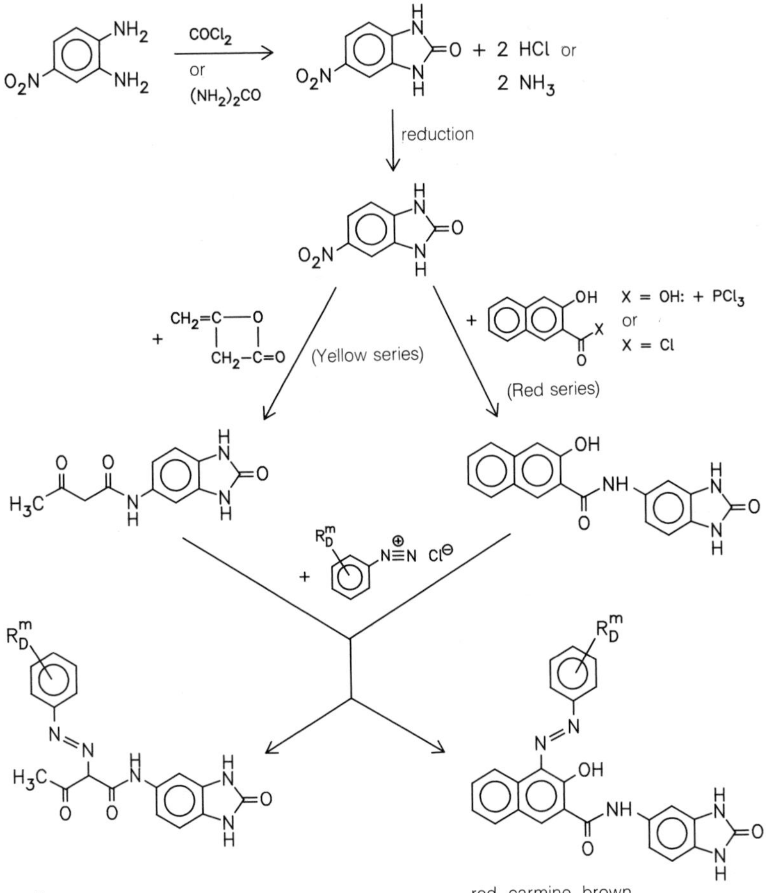

O_2N ... NH_2 / NH_2 $\xrightarrow{COCl_2 \text{ or } (NH_2)_2CO}$ ... $= O$ + 2 HCl or 2 NH$_3$

reduction

(Yellow series)

$CH_2=C-O$ / $CH_2-C=O$ +

+ ...OH X = OH: + PCl$_3$ or X = Cl

(Red series)

H_3C ... N ... $=O$

... OH ... NH ... $=O$

+ R_D^m ... $N\equiv N$ Cl$^\ominus$

R_D^m ... N=N ... H_3C ... $=O$

R_D^m ... N=N ... OH ... NH ... $=O$

yellow, orange

red, carmine, brown

R_D e.g., Cl, Br, F, CF$_3$, CH$_3$, NO$_2$, OCH$_3$, OC$_2$H$_5$, COOH, COOAlkyl, CONH$_2$,
CONHC$_6$H$_5$, SO$_2$NHAlkyl, SONHC$_6$H$_5$
m = 1 to 3; alkyl: C$_1$ to C$_4$

2.9 Disazo Condensation Pigments

1 Yellow Pigments

Type 1

Type 2

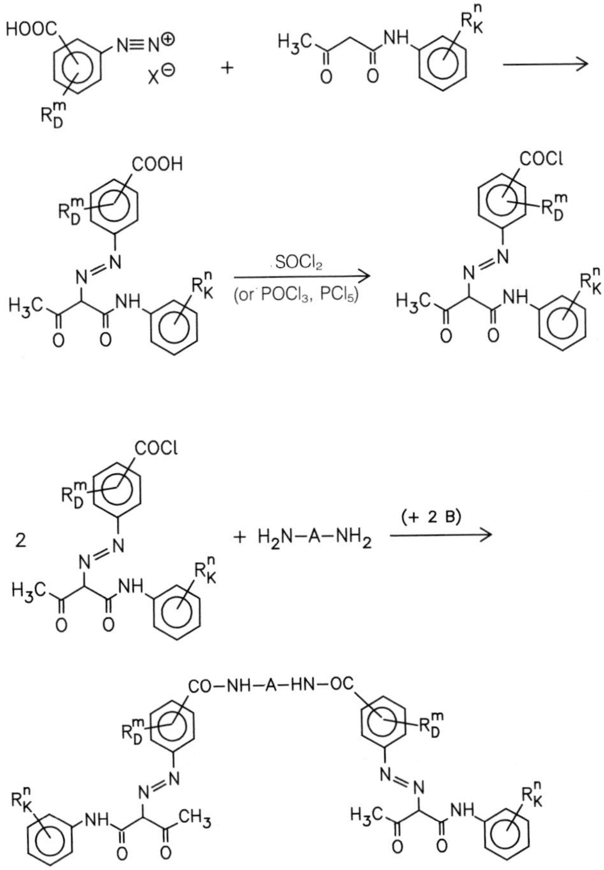

HOOC — ⬡ — N≡N⁺ X⁻ , R_D^m + H_3C — CO — CH₂ — CO — NH — ⬡ — R_K^n ⟶

COOH, R_D^m — ⬡ — N=N — C(COCH₃)(H_3CCO) ... NH — ⬡ — R_K^n $\xrightarrow[\text{(or POCl}_3\text{, PCl}_5)]{\text{SOCl}_2}$ COCl, R_D^m — ⬡ ... N=N ... NH — ⬡ — R_K^n

2   COCl, R_D^m — ⬡ — N=N — ... H_3CCO ... NH — ⬡ — R_K^n + H_2N—A—NH_2 $\xrightarrow{\text{(+ 2 B)}}$

CO—NH—A—HN—OC

R_K^n — ⬡ — NH — CO — CH(N=N—⬡—R_D^m) — CO — CH_3 H_3C — CO — CH(—N=N—⬡—R_D^m) — CO — NH — ⬡ — R_K^n

2 Red Pigments

Rᴅ: substituent at the moiety of the diazo component
Rᴋ: substituent at the moiety of the coupling component
m: 0 to 3; n : 0 to 3
A: phenylene or diphenylene group
Ar: (subst.), phenyl
B: base

2.10 Metal Complex Pigments

1 Azo Metal Complexes

Pigment Green 10

Complexes of azobarbituric acid

Ni-Complex

2 Azomethine Metal Complexes

from aromatic o-hydroxy aldehydes:

from diimino butyric acid anilides:

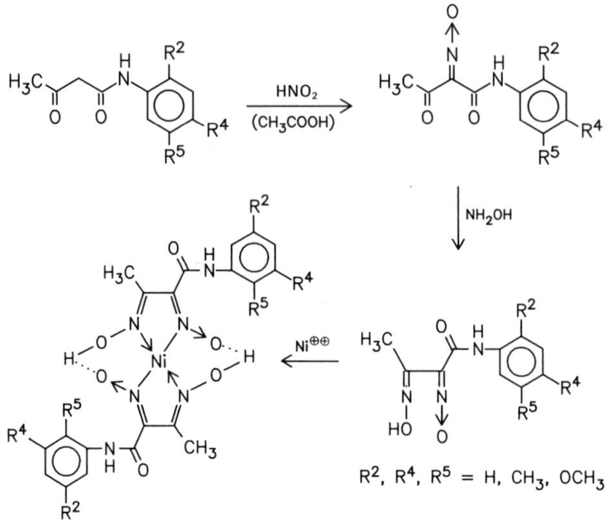
HNO₂
(CH₃COOH)
NH₂OH
Ni⊕⊕
R², R⁴, R⁵ = H, CH₃, OCH₃

from isoindolinones:

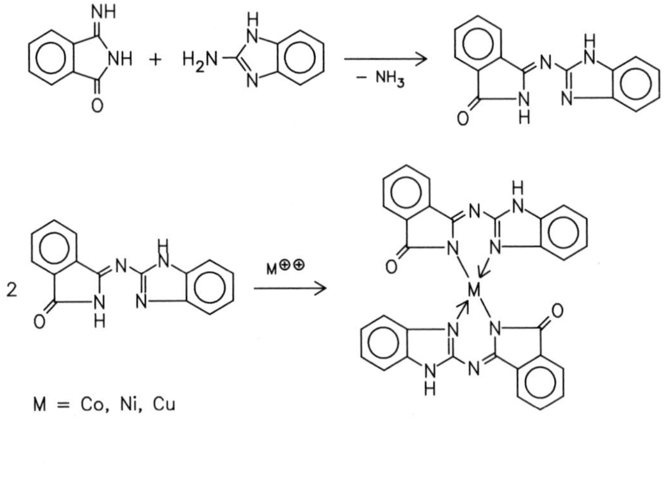
− NH₃
M⊕⊕
M = Co, Ni, Cu

2.11 Isoindolinone and Isoindoline Pigments

1   **Azomethine Type: Tetrachloroisoindolinone Pigments**

A: Cl, OCH₃

Dainippon Ink Process:

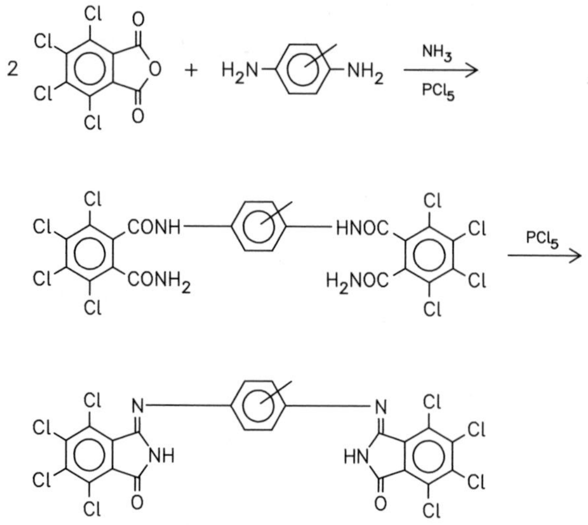

2 Methine Type: Isoindoline Pigments

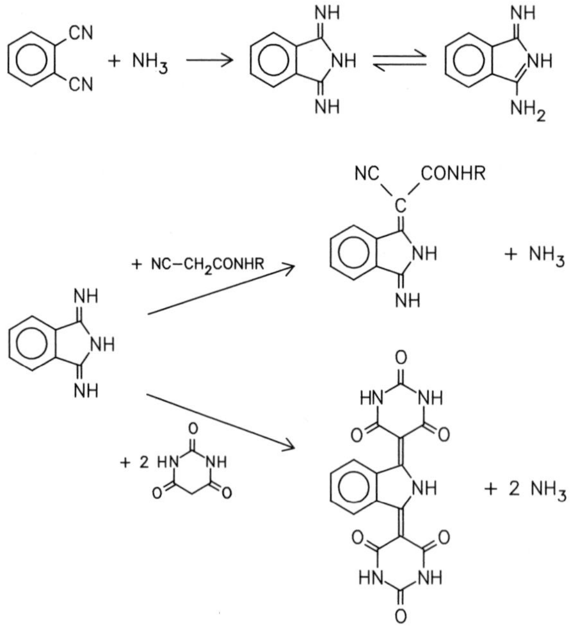

3.1 Phthalocyanine Pigments

1   **Phthalonitrile Process Ⓐ**

2   **Phthalic Anhydride / Urea Process Ⓑ**

polyisoindolenine

3.2 Quinacridone Pigments

1   Thermal Ring Closure

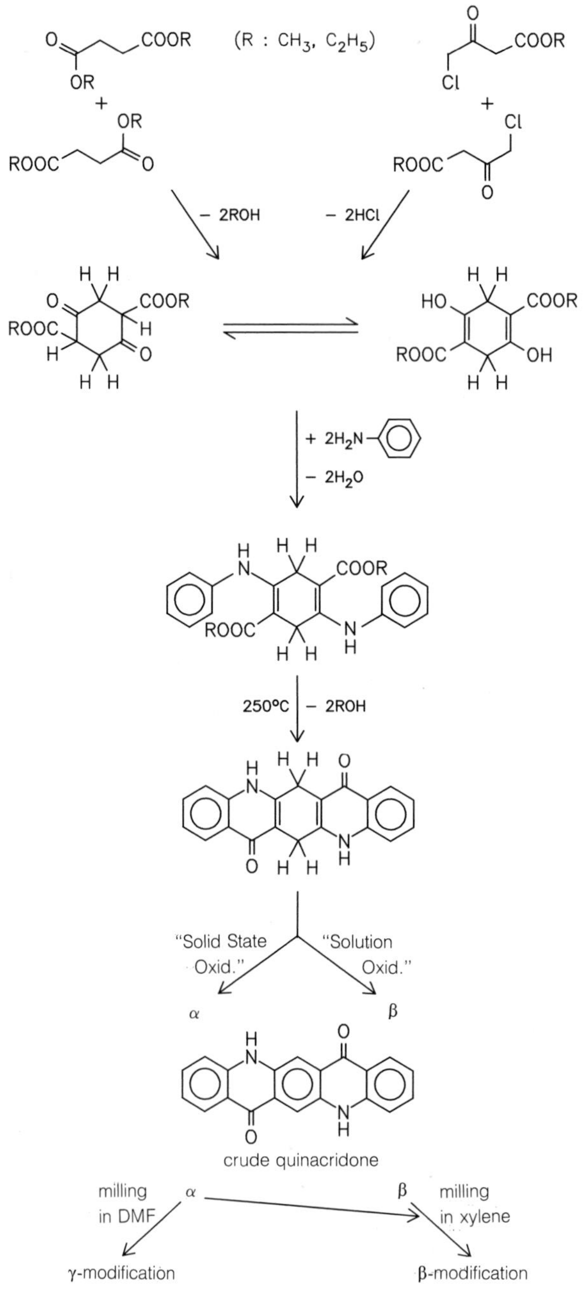

"Solid State Oxid." α

"Solution Oxid." β

crude quinacridone

milling in DMF α

β milling in xylene

γ-modification

β-modification

2   Acidic Ring Closure

R: CH$_3$, C$_2$H$_5$: Ar: (subst.) phenyl

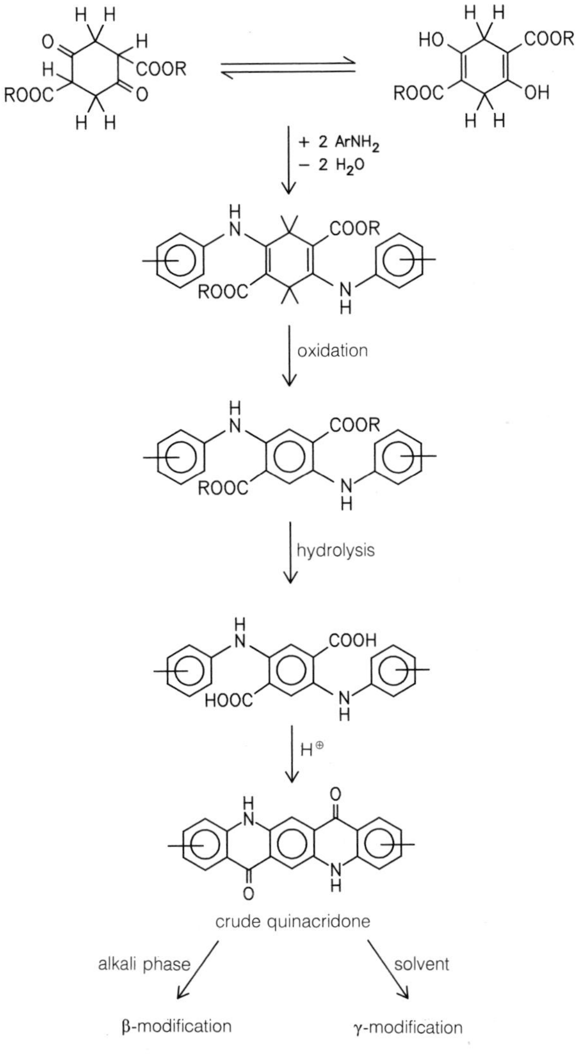

+ 2 ArNH$_2$
− 2 H$_2$O

oxidation

hydrolysis

H$^\oplus$

crude quinacridone

alkali phase / β-modification

solvent \ γ-modification

3 Dihalogenterephthalic Acid Process

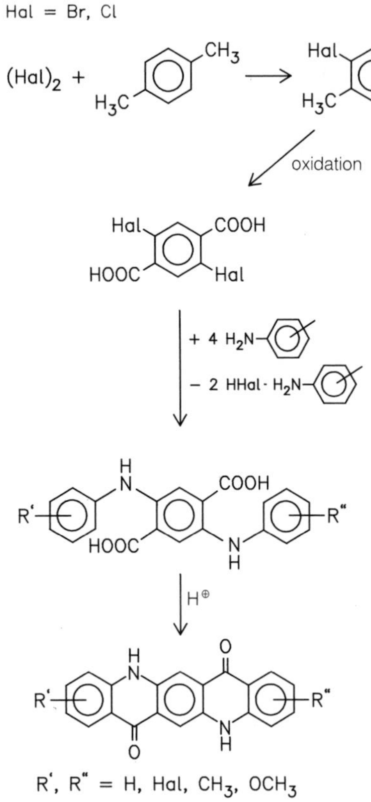
Hal = Br, Cl
$(Hal)_2$ +
CH_3
H_3C
Hal
CH_3
H_3C
Hal
oxidation
Hal
COOH
HOOC
Hal
+ 4 H_2N
− 2 HHal · H_2N
H
N
COOH
R'
R"
HOOC
N
H
$H^{\oplus}$
H
N
O
R'
R"
O
N
H
R', R" = H, Hal, CH_3, OCH_3

4   Hydroquinone Process

5 Substituted Quinacridone Pigments

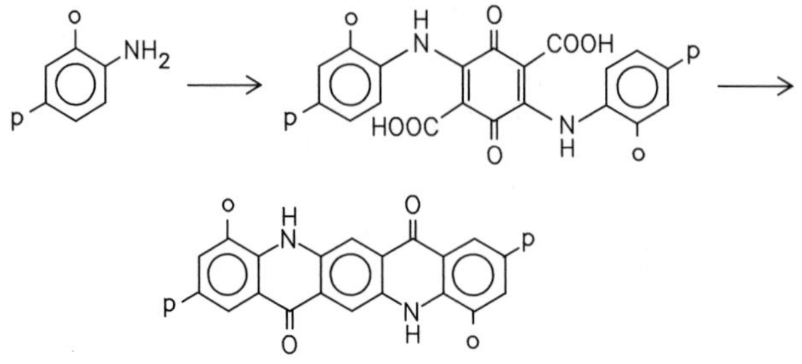

o: ortho (p = H): 4,11-disubstitution
p: para (o = H): 2,9-disubstitution

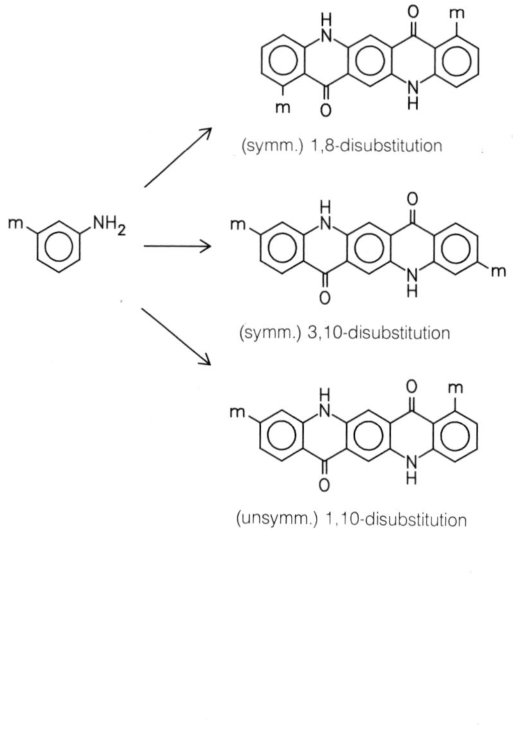

(symm.) 1,8-disubstitution

(symm.) 3,10-disubstitution

(unsymm.) 1,10-disubstitution

6   Quinacridone Quinone Synthesis

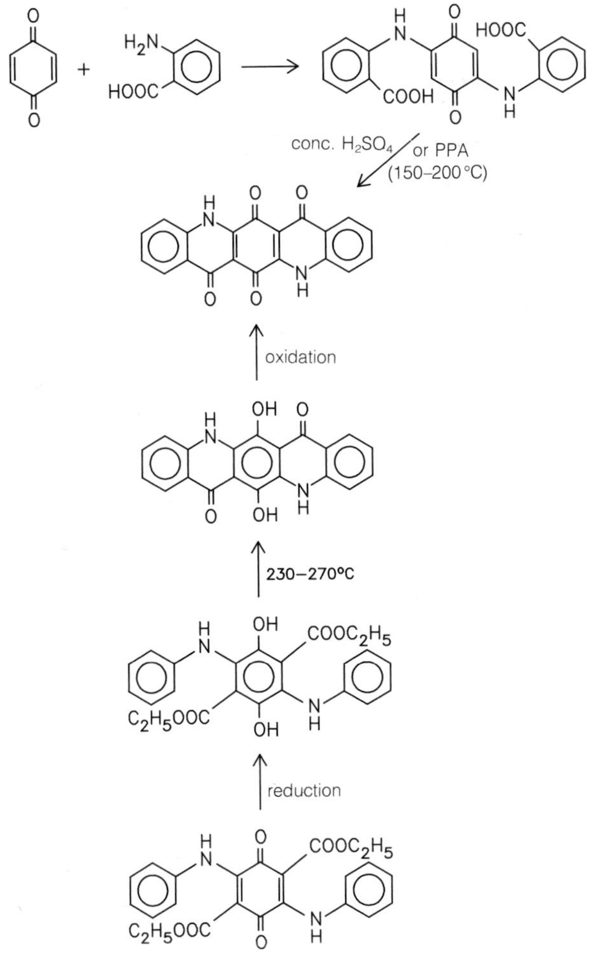

conc. H₂SO₄ or PPA (150–200 °C)

oxidation

230–270°C

reduction

PPA: polyphosphoric acid

3.4 Perylene and Perinone Pigments

1 Perylene Pigments

R: CH_3, subst. phenyl

Naphthaline Tetracarboxylic Acid Dianhydride

Pathway 1

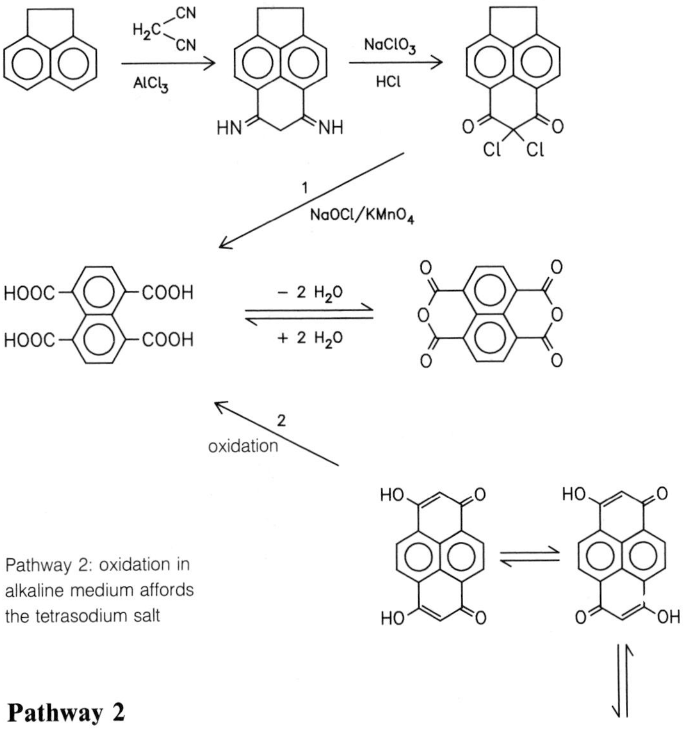

Pathway 2: oxidation in
alkaline medium affords
the tetrasodium salt

Pathway 2

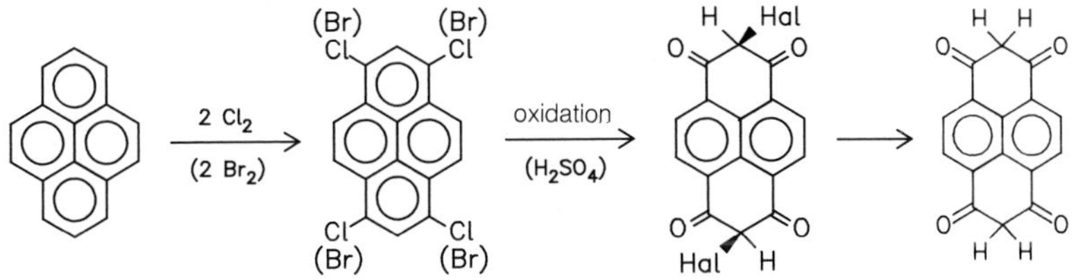

Perione Pigment

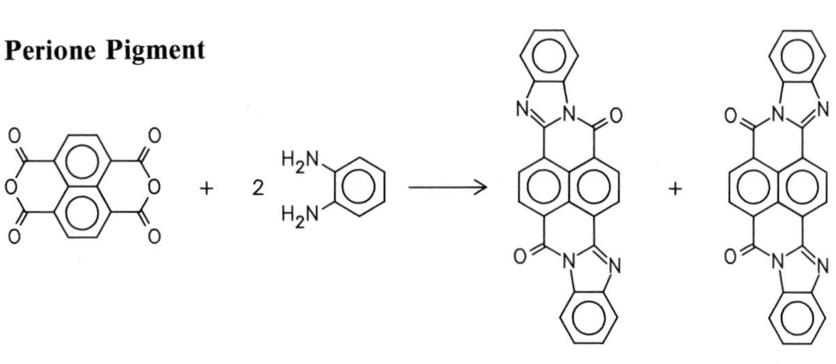

trans   cis

3.5 Thioindigo Pigments

3.6 Different Polycyclic Pigments

Heterocyclic Anthraquinone Pigments

1 Anthrapyrimidine

2 Indanthrone and Flavanthrone Pigments

Indanthrone

Flavanthrone

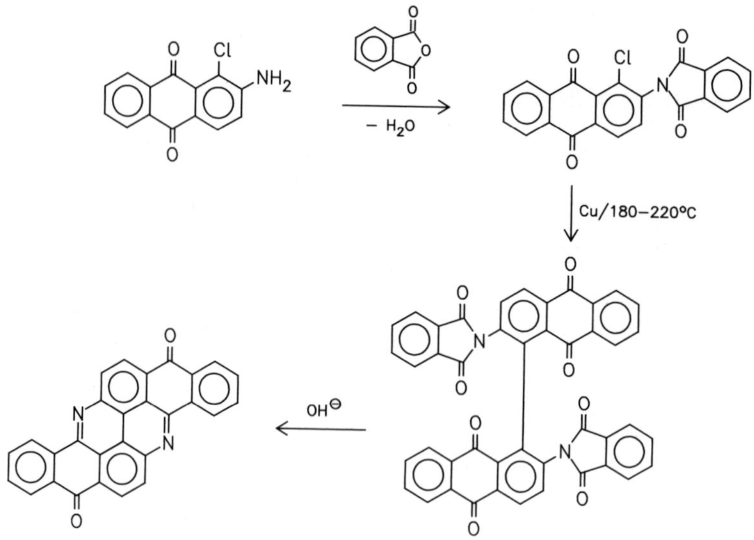

Polycarbocyclic Anthraquinone Pigments

1 Pyranthrone Pigments

X: H, Cl

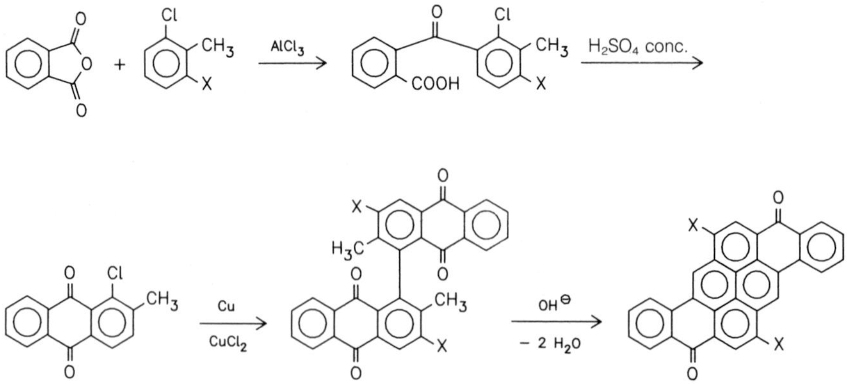

2 Anthanthrone Pigments

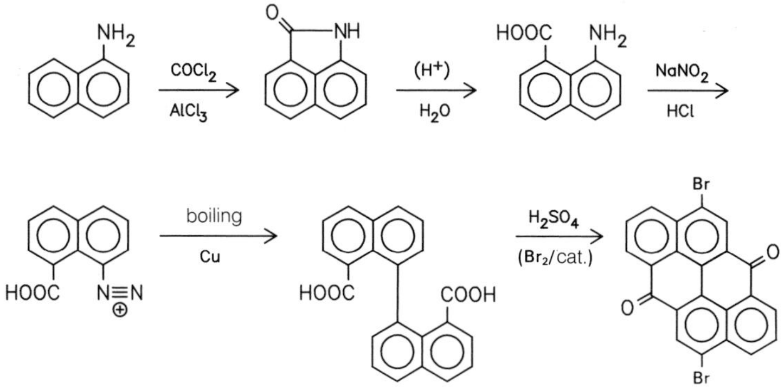

Mechanism:

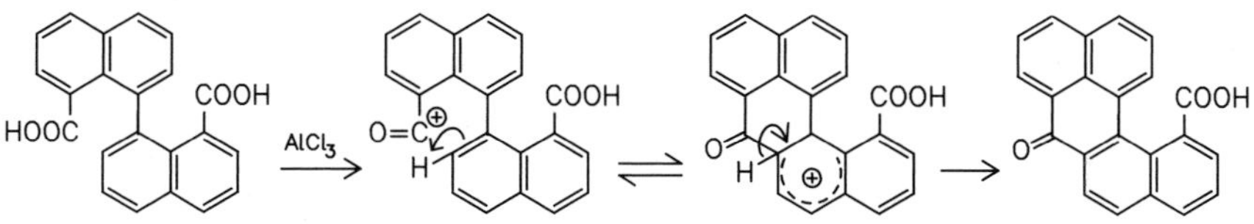

3 Isoviolanthrone Pigments

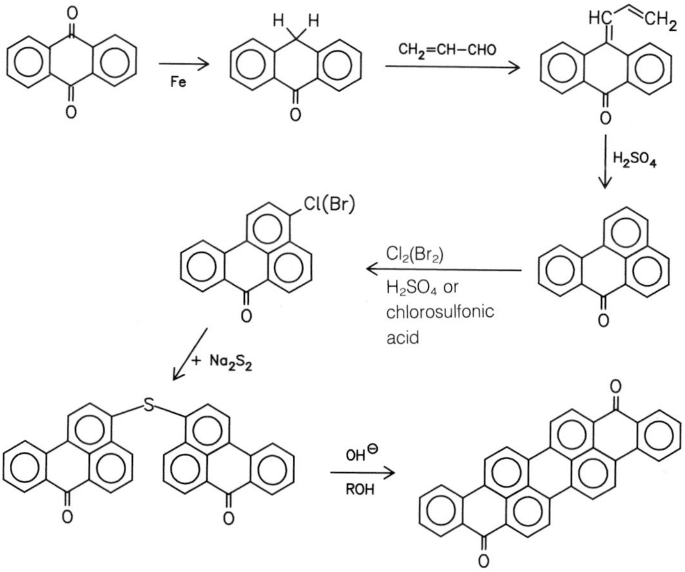

3.7 Dioxazine Pigments

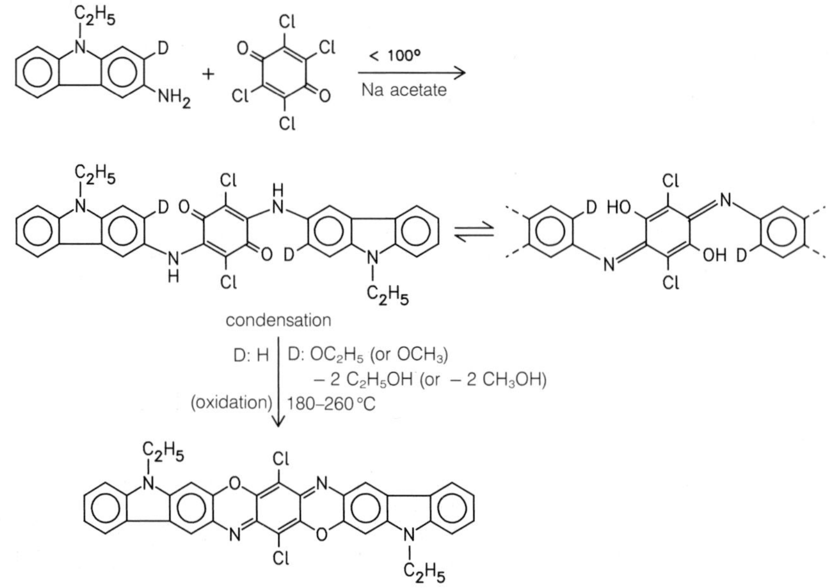
< 100°
Na acetate
condensation
D: H
(oxidation)
D: OC2H5 (or OCH3)
– 2 C2H5OH (or – 2 CH3OH)
180–260 °C

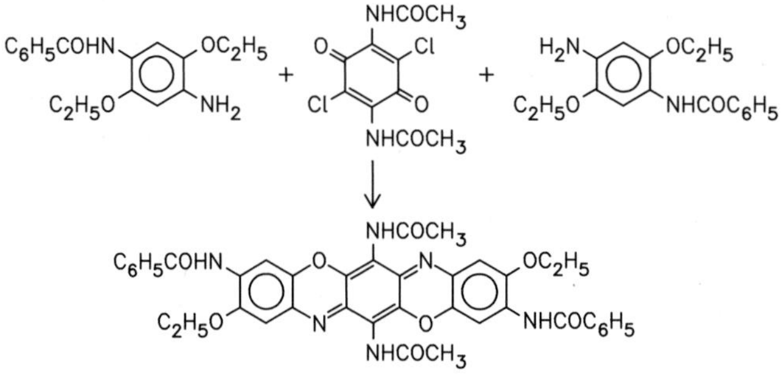

3.8 Triarylcarbonium Pigments

1 Inner Salts of Sulfonic Acids

Path A:

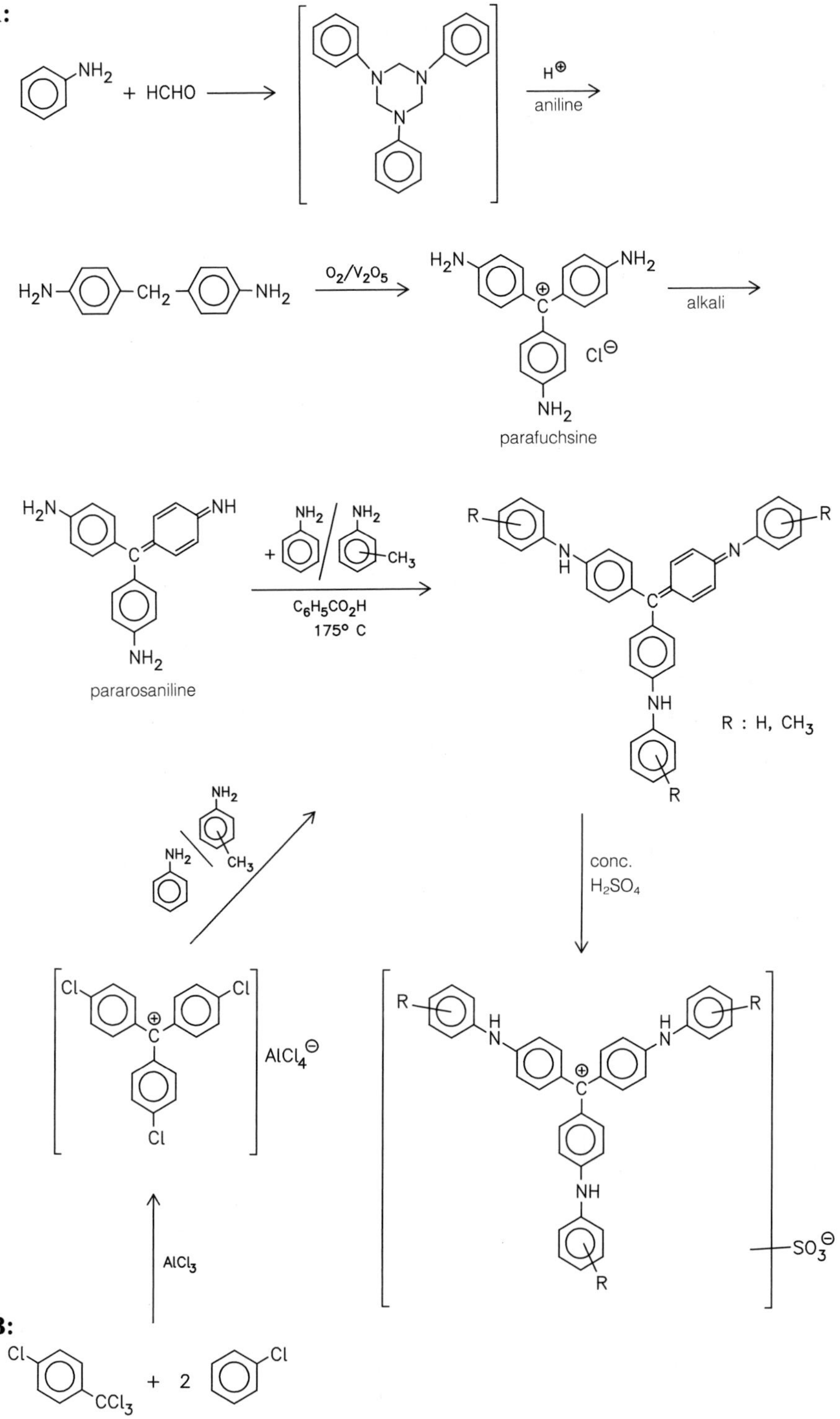

parafuchsine

pararosaniline

R : H, CH₃

conc.
H₂SO₄

Path B:

2   Dye Salts with Complex Anions

Dye component with the structure 123

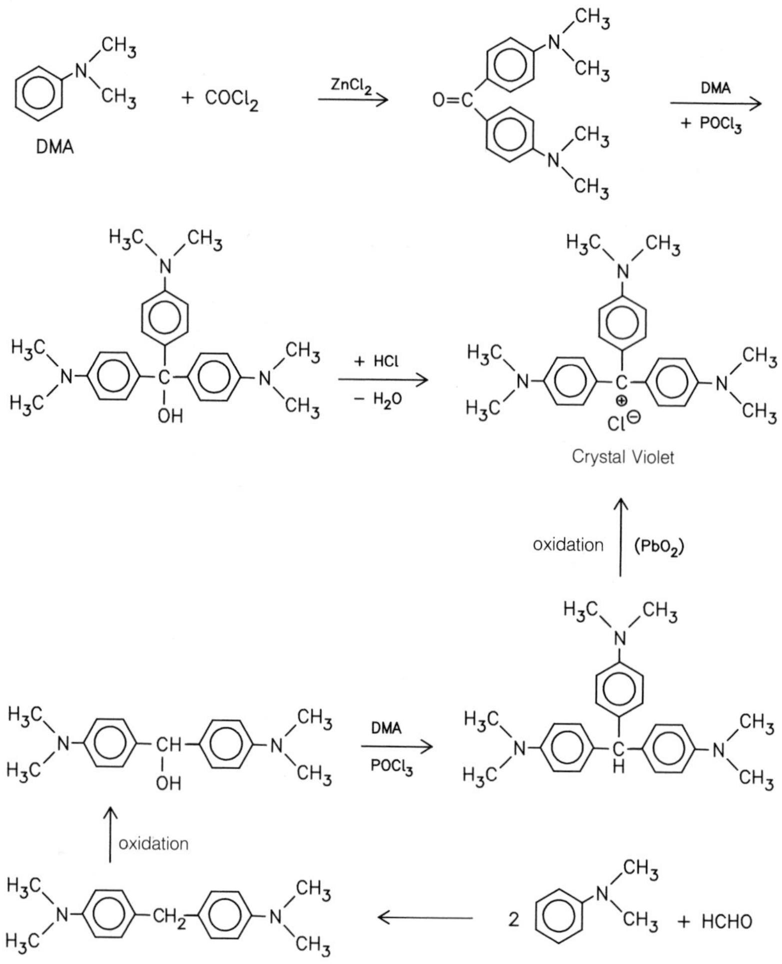

Crystal Violet

oxidation (PbO$_2$)

DMA POCl$_3$

oxidation

2 + HCHO

Substituting DMA in the above scheme by 1-N-ethylnapthylamine affords Victoria Blue types:

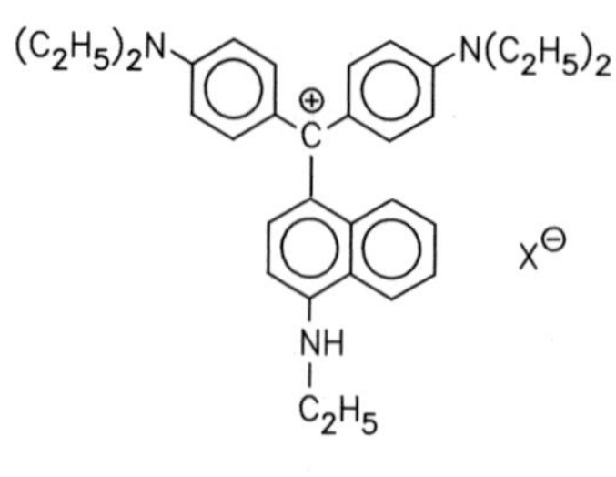

Dye component with the structure 124 (Xanthene dyes)

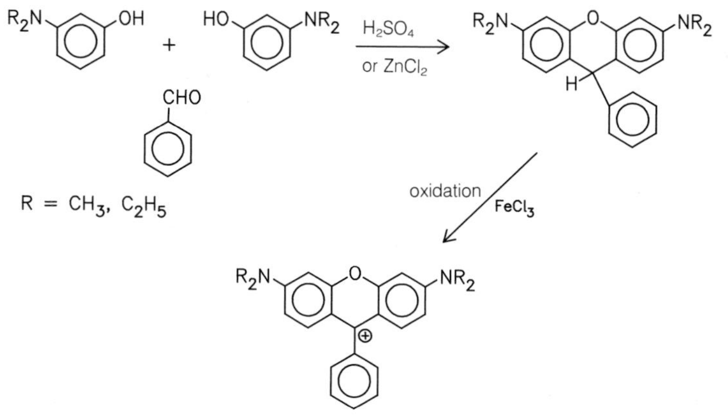

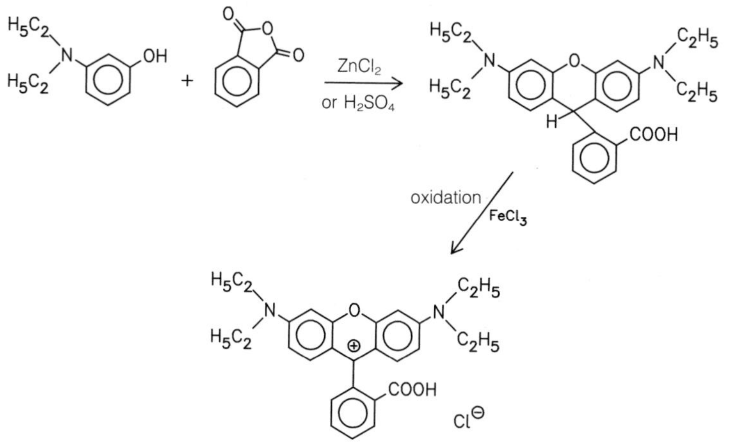

$R = CH_3, C_2H_5$

Rhodamine B

Dye component with the structure 125

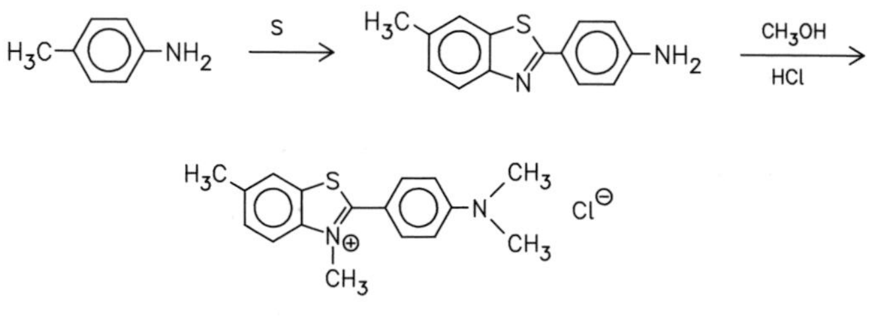

4.3 Diketopyrrolo Pyrrole (DPP) Pigments

$$R' = H, CH_3, CF_3, Cl,$$
$$Br, N(CH_3)_2$$

$$R'' = CH_3, C_2H_5$$

List of Commercially Available Pigments

Colour Index Name	C.I. Constitution No.	CAS No.	Pigment Class	Chap.	Tab.	Page
Yellow						
P.Y. 1	11680	2512-29-0	Monoazo Yellow	2.3	11	225
P.Y. 2	11730	6486-26-6	Monoazo Yellow	2.3	11	226
P.Y. 3	11710	6486-23-3	Monoazo Yellow	2.3	11	226
P.Y. 5	11660	4106-67-6	Monoazo Yellow	2.3	11	227
P.Y. 6	11670	4106-76-7	Monoazo Yellow	2.3	11	227
P.Y. 10	12710	6407-75-6	Monoazo Yellow	2.3	12	227
P.Y. 12	21090	6358-85-6	Diarylide Yellow	2.4	14	245
P.Y. 13	21100	5102-83-0	Diarylide Yellow	2.4	14	249
P.Y. 14	21095	5468-75-7	Diarylide Yellow	2.4	14	250
P.Y. 16	20040	5979-28-2	Bisacetoacetarylide	2.4	15	263
P.Y. 17	21105	4531-49-1	Diarylide Yellow	2.4	14	251
P.Y. 24	70600	475-71-8	Flavanthrone	3.6	–	504
P.Y. 49	11765	2904-04-3	Monoazo Yellow	2.3	11	227
P.Y. 55	21096	6358-37-8	Diarylide Yellow	2.4	14	252
P.Y. 60	21705	6407-74-5	Monoazo Yellow	2.3	12	228
P.Y. 61	13880	12286-65-6	Monoazo Yellow, Ca	2.3	13	234
P.Y. 62:1	13940:1	12286-66-7	Monoazo Yellow, Ca	2.3	13	234
P.Y. 63	21091	14569-54-1	Diarylide Yellow	2.4	14	253
P.Y. 65	11740	6528-34-3	Monoazo Yellow	2.3	11	228
P.Y. 73	11738	13515-40-7	Monoazo Yellow	2.3	11	228
P.Y. 74	11741	6358-31-2	Monoazo Yellow	2.3	11	227
P.Y. 75	11770	52320-66-8	Monoazo Yellow	2.3	11	230
P.Y. 81	21127	22094-93-5	Diarylide Yellow	2.4	14	253
P.Y. 83	21108	5567-15-7	Diarylide Yellow	2.4	14	254
P.Y. 87	21107:1	15110-84-6	Diarylide Yellow	2.4	14	256
P.Y. 90	–		Diarylide Yellow	2.4	14	256
P.Y. 93	20710	5580-57-4	Disazo Condensation	2.9	24	373
P.Y. 94	20038	5580-58-5	Disazo Condensation	2.9	24	376
P.Y. 95	20034	5280-80-8	Disazo Condensation	2.9	24	376
P.Y. 97	11767	12225-18-2	Monoazo Yellow	2.3	11	230
P.Y. 98	11727	12225-19-3	Monoazo Yellow	2.3	11	231
P.Y. 99	–	12225-20-6	Anthraquinone	4.4	–	563
P.Y. 100	19140:1	12225-21-7	Monoazo Pyrazolone, Al	2.3	13	235
P.Y. 101	48052	2387-03-3	Aldazine	4.3	–	555

Colour Index Name	C.I. Constitution No.	CAS No.	Pigment Class	Chap.	Tab.	Page
P.Y. 104	15985:1	15790-07-5	Naphth. Sulfonic Acid, Al	2.7	22	341
P.Y. 106	–	12225-23-9	Diarylide Yellow	2.4	14	256
P.Y. 108	68420	4216-01-7	Anthrapyrimidine	3.6	–	498
P.Y. 109	56284	12769-01-6	Isoindolinone	2.11	26	407
P.Y. 110	56280	5590-18-1	Isoindolinone	2.11	26	409
P.Y. 111	11745	69771-45-5	Monoazo Yellow	2.3	11	232
P.Y. 113	21126	14359-20-7	Diarylide Yellow	2.4	14	256
P.Y. 114	21092	71872-66-7	Diarylide Yellow	2.4	14	257
P.Y. 116	11790	30191-02-7	Monoazo Yellow	2.3	11	232
P.Y. 117	48043	21405-81-2	Metal Complex	2.10	25	394
P.Y. 120	11783	29920-31-8	Benzimidazolone	2.8	23	350
P.Y. 121	–	61968-85-2	Diarylide Yellow	2.4	14	257
P.Y. 123	65049	4028-94-8	Anthraquinone	3.6	–	491
P.Y. 124	21107	67828-22-2	Diarylide Yellow	2.4	14	257
P.Y. 126	21101	90268-23-8	Diarylide Yellow	2.4	14	257
P.Y. 127	21102	71872-67-8	Diarylide Yellow	2.4	14	258
P.Y. 128	20037	57971-97-8	Disazo Condensation	2.9	24	377
P.Y. 129	48042	68859-61-0	Metal Complex	2.10	25	395
P.Y. 130	–		Monoazo Yellow	2.3	11	233
P.Y. 133	–	132821-92-2	Monoazo Yellow, Sr	2.3	13	235
P.Y. 136	–		Diarylide Yellow	2.4	14	258
P.Y. 138	56300	56731-19-2	Quinophthalone	4.1	–	549
P.Y. 139	56298	36888-99-0	Isoindoline	2.11	26	410
P.Y. 147	60645	76168-75-7	Anthraquinone	3.6	–	493
P.Y. 148	59020	20572-37-6	–	4.4	–	556
P.Y. 150	12764	68511-62-6	Metal Complex	2.10	25	395
P.Y. 151	13980	61036-28-0	Benzimidazolone	2.8	23	352
P.Y. 152	21111	20139-66-6	Diarylide Yellow	2.4	14	259
P.Y. 153	48545	68859-51-8	Metal Complex	2.10	25	395
P.Y. 154	11781	63661-02-9	Benzimidazolone	2.8	23	354
P.Y. 155	–	68516-73-4	Bisacetoacetarylide	2.4	–	264
P.Y. 165	–		Monoazo Yellow	2.3	12	233
P.Y. 166	20035	76233-82-4	Disazo Condensation	2.9	24	378
P.Y. 167	11737	38489-24-6	Monoazo Yellow	2.3	12	233
P.Y. 168	13960	71832-85-4	Monoazo Yellow, Ca	2.3	13	235
P.Y. 169	13955	73385-03-2	Monoazo Yellow, Ca	2.3	13	235
P.Y. 170	21104	31775-16-3	Diarylide Yellow	2.4	14	259
P.Y. 171	21106	53815-04-6	Diarylide Yellow	2.4	14	259
P.Y. 172	21109	76233-80-2	Diarylide Yellow	2.4	14	259
P.Y. 173	–	51016-63-8	Isoindolinone	2.11	26	411
P.Y. 174	21098	78952-72-4	Diarylide Yellow	2.4	14	259
P.Y. 175	11784	35636-63-6	Benzimidazolone	2.8	23	355
P.Y. 176	21103	90268-24-9	Diarylide Yellow	2.4	14	260
P.Y. 177	48120	60109-88-8	Metal Complex	2.10	25	396
P.Y. 179	48125	63287-28-5	Metal Complex	2.10	25	396
P.Y. 180	21290	77804-81-0	Benzimidazolone	2.8	23	356

Colour Index Name	C.I. Constitution No.	CAS No.	Pigment Class	Chap.	Tab.	Page
P.Y. 181	11777	74441-05-7	Benzimidazolone	2.8	23	356
P.Y. 182	–	67906-31-4	Polycycl. Pigment	4.4	–	556
P.Y. 183	18792	65212-77-3	Monoazo Yellow, Ca	2.3	13	236
P.Y. 185	56290	76199-85-4	Isoindoline	2.11	26	412
P.Y. 187			Polycycl. Pigment	4.4	–	563
P.Y. 188	21094	23792-68-9	Diarylide Yellow	2.4	14	260
P.Y. 190			Monoazo Yellow, Ca	2.3	13	236
P.Y. 191	18795	129423-54-7	Monoazopyrazolone, Ca	2.3	13	236
P.Y. 192	–		Heterocycle	4.4	–	558
P.Y. 193	–		Anthraquinone	3.6	–	492
P.Y. 194	–	82199-12-0	Benzimidazolone	2.8	23	357

Orange

P.O. 1	11725	6371-96-6	Monoazo Yellow	2.3	11	223
P.O. 2	12060	6410-09-9	β-Naphthol	2.5	16	277
P.O. 5	12075	3468-63-1	β-Naphthol	2.5	16	277
P.O. 6	12730	6407-77-8	Monoazo Yellow	2.3	12	223
P.O. 13	21110	3520-72-7	Disazopyrazolone	2.4	15	267
P.O. 15	21130	6358-88-9	Diarylide Yellow	2.4	14	260
P.O. 16	21160	6505-28-8	Diarylide Yellow	2.4	14	260
P.O. 17	15510:1	15782-04-4	β-Naphthol, Ba	2.7	19	322
P.O. 17:1	15510:2	15876-51-4	β-Naphthol, Al	2.7	19	322
P.O. 19	15990	5858-88-8	Naphth. Sulfonic Acid, Ba	2.7	22	341
P.O. 22	12470	6358-48-1	Naphthol AS	2.6	18	299
P.O. 24	12305	6410-27-1	Naphthol AS	2.6	18	299
P.O. 31	20050	12286-58-7	Disazo Condensation	2.9	24	378
P.O. 34	21115	15793-73-4	Disazopyrazolone	2.4	15	268
P.O. 36	11780	12236-62-3	Benzimidazolone	2.8	23	357
P.O. 38	12367	12236-64-5	Naphthol AS	2.6	18	311
P.O. 40	59700	128-70-1	Pyranthrone	3.6	–	507
P.O. 43	71105	4424-06-0	Perinone	3.4	30	478
P.O. 44	–		Diarylide Orange	2.4	14	261
P.O. 46	15602	67801-01-8	β-Naphthol, Ba	2.7	19	322
P.O. 48	{ 73900	1047-16-1	Quinacridone/Quin. Quinone	3.2	28	464
	73920	1503-48-6				
P.O. 49	–	71819-75-5	Quinacridone/Quin. Quinone	3.2	28	464
P.O. 51	–	73309-48-5	Pyranthrone	3.6	–	508
P.O. 59	–		Metal Complex	2.10	–	396
P.O. 60	11782	68399-99-5	Benzimidazolone	2.8	23	359
P.O. 61	11265	76168-74-6	Isoindoline	2.11	–	412
P.O. 62	11775	75601-68-2	Benzimidazolone	2.8	23	359
P.O. 64	12760	72102-84-2	Azoheterocycle	4.4	–	558
P.O. 65	48053	20437-10-9	Metal Complex	2.10	25	397
P.O. 66	48210	68808-69-5	Isoindoline	2.11	–	413
P.O. 67	12915	74336-59-7	Pyrazoloquinazolone	4.4	–	559
P.O. 68	–	42844-93-9	Metal Complex	2.10	25	297
P.O. 69	56292	85959-60-0	Isoindoline	2.11	26	413

Colour Index Name	C.I. Constitution No.	CAS No.	Pigment Class	Chap.	Tab.	Page
Red						
P.R. 1	12070	6410-10-2	β-Naphthol	2.5	16	278
P.R. 2	12310	6041-94-7	Naphthol AS	2.6	18	289
P.R. 3	12120	2425-85-6	β-Naphthol	2.5	16	278
P.R. 4	12085	2814-77-9	β-Naphthol	2.5	16	279
P.R. 5	12490	6410-41-9	Naphthol AS	2.6	18	300
P.R. 6	12090	6410-13-5	β-Naphthol	2.5	16	280
P.R. 7	12420	6471-51-8	Naphthol AS	2.6	18	290
P.R. 8	12335	6410-30-6	Naphthol AS	2.6	18	290
P.R. 9	12460	6410-41-9	Naphthol AS	2.6	18	291
P.R. 10	12440	6410-35-1	Naphthol AS	2.6	18	291
P.R. 11	12430	6535-48-4	Naphthol AS	2.6	18	292
P.R. 12	12385	6410-32-8	Naphthol AS	2.6	18	292
P.R. 13	12395	6535-47-3	Naphthol AS	2.6	18	293
P.R. 14	12380	6471-50-7	Naphthol AS	2.6	18	293
P.R. 15	12465	6410-39-5	Naphthol AS	2.6	18	294
P.R. 16	12500	6407-71-2	Naphthol AS	2.6	18	294
P.R. 17	12390	6655-84-1	Naphthol AS	2.6	18	294
P.R. 18	12350	3564-22-5	Naphthol AS	2.6	18	294
P.R. 21	12300	6410-26-0	Naphthol AS	2.6	18	294
P.R. 22	12315	6448-95-9	Naphthol AS	2.6	18	295
P.R. 23	12355	6471-49-4	Naphthol AS	2.6	18	295
P.R. 31	12360	6448-96-0	Naphthol AS	2.6	18	301
P.R. 32	12320	6410-29-3	Naphthol AS	2.6	18	302
P.R. 37	21205	6883-91-6	Disazopyrazolone	2.4	15	268
P.R. 38	21210	6358-87-8	Disazopyrazolone	2.4	15	270
P.R. 41	21200	6505-29-9	Disazopyrazolone	2.4	15	270
P.R. 48:1	15865:1	7585-41-3	BONA, Ba	2.7	20	326
P.R. 48:2	15865:2	7023-61-2	BONA, Ca	2.7	20	327
P.R. 48:3	15865:3	15782-05-5	BONA, Sr	2.7	20	328
P.R. 48:4	15865:4	5280-66-0	BONA, Mn	2.7	20	328
P.R. 48:5	15865:5		BONA, Mg	2.7	20	329
P.R. 49	15630	1248-18-6	β-Naphthol, Na	2.7	19	318
P.R. 49:1	15630:1	1103-38-4	β-Naphthol, Ba	2.7	19	318
P.R. 49:2	15630:2	1103-39-5	β-Naphthol, Ca	2.7	19	318
P.R. 49:3	15630:3	6371-67-1	β-Naphthol, Sr	2.7	19	316
P.R. 50:1	15500:1	6372-81-2	β-Naphthol, Ba	2.7	19	318
P.R. 51:1	15580-1	5850-87-3	β-Naphthol, Ba	2.7	19	319
P.R. 52:1	15860:1	17852-99-2	BONA, Ca	2.7	20	330
P.R. 52:2	15860:2	12238-31-2	Bona, Mn	2.7	20	330
P.R. 53	15585	2092-56-0	β-Naphthol, Na	2.7	19	319
P.R. 53:1	15585:1	5160-02-1	β-Naphthol, Ba	2.7	19	319
P.R. 53:–	15585:–		β-Naphthol, Sr	2.7	19	321
P.R. 57:1	15850:1	5281-04-9	BONA, Ca	2.7	20	331
P.R. 58:2	15825:2	7538-59-2	BONA, Ca	2.7	20	333
P.R. 58:4	15825:4	52233-00-8	BONA, Mn	2.7	20	333

Colour Index Name	C.I. Constitution No.	CAS No.	Pigment Class	Chap.	Tab.	Page
P.R. 60:1	16105:1	15782-06-6	Naphth. Sulfonic Acid, Ba	2.7	22	341
P.R. 63:1	15880:1	6417-83-0	BONA, Ca	2.7	20	334
P.R. 63:2	15880:2	35355-77-2	BONA, Mn	2.7	20	334
P.R. 64	15800	16508-79-5	BONA, Ba	2.7	20	334
P.R. 64:1	15800:1	6371-76-2	BONA, Ca	2.7	20	335
P.R. 66	18000:1	68929-13-5	Naphth. Sulphonic Acid, Ba	2.7	22	342
P.R. 67	18025:1	68929-14-6	Naphth. Sulphonic Acid, Ba	2.7	22	342
P.R. 68	15525	5850-80-6	β-Naphthol, Ca	2.7	19	321
P.R. 81:1	45160:1	12224-98-5	Triarylcarbonium	3.8	33	544
P.R. 81:3	42160:3		Triarylcarbonium	3.8	33	544
P.R. 83:1	58000:1		Anthraquinone, Ca	3.6	–	496
P.R. 88	73312	14295-43-3	Thioindigo	3.5	31	484
P.R. 89	60745	6409-74-1	Anthraquinone	3.6	–	493
P.R. 90	45380:1	1326-05-2	Phloxine, Lead Salt	4.3	–	559
P.R. 95	15897	61968-79-4	Naphthol AS	2.6	18	295
P.R. 111	–	12224-99-6	Disazopyrazolone	2.4	15	271
P.R. 112	12370	6535-46-2	Naphthol AS	2.6	18	296
P.R. 114	12351	6358-47-0	Naphthol AS	2.6	18	297
P.R. 119	12469	72066-77-4	Naphthol AS	2.6	18	297
P.R. 122	73915	980-26-7	Quinacridone	3.2	28	460
P.R. 123	71145	24108-89-2	Perylene	3.4	29	471
P.R. 136	–		Naphthol AS	2.6	18	298
P.R. 144	20735	5280-78-4	Disazo Condensation	2.9	24	378
P.R. 146	12485	5280-68-2	Naphthol AS	2.6	18	302
P.R. 147	12433	68227-78-1	Naphthol AS	2.6	18	303
P.R. 148	12369	94276-08-1	Naphthol AS	2.6	18	299
P.R. 149	71137	4948-15-6	Perylene	3.4	29	471
P.R. 150	12290	56396-10-2	Naphthol AS	2.6	18	304
P.R. 151	15890	61013-97-6	Naphthol AS	2.7	21	336
P.R. 164	–	12216-95-4	Naphthol AS	2.6	18	304
P.R. 166	20035	12225-04-6	Disazo Condensation	2.9	24	379
P.R. 168	59300	4378-61-4	Anthanthrone	3.6	–	511
P.R. 169	45160:2	12224-98-5	Triarylcarbonium	3.8	33	544
P.R. 170	12475	2786-76-7	Naphthol AS	2.6	18	204
P.R. 171	12512	6985-95-1	Benzimidazolone	2.8	23	360
P.R. 172	45430:1	12227-78-0	Tetraiodine Fluorescein, Al	4.2	–	553
P.R. 175	12513	6985-92-8	Benzimidazolone	2.8	23	361
P.R. 176	12515	12225-06-8	Benzimidazolone	2.8	23	362
P.R. 177	65300	4051-63-2	Anthraquinone	3.6	–	493
P.R. 178	71155	3049-71-6	Perylene	3.4	29	473
P.R. 179	71130	5521-31-3	Perylene	3.4	29	473
P.R. 181	73360	2379-74-0	Thioindigo	3.5	31	486
P.R. 184	12487	99402-80-9	Naphthol AS	2.6	18	306
P.R. 185	12516	61951-98-2	Benzimidazolone	2.8	23	362
P.R. 187	12486	59487-23-9	Naphthol AS	2.6	18	307
P.R. 188	12467	61847-48-1	Naphthol AS	2.6	18	308

Colour Index Name	C.I. Constitution No.	CAS No.	Pigment Class	Chap.	Tab.	Page
P.R. 190	71140	6424-77-7	Perylene	3.4	29	474
P.R. 192	–	61968-81-8	Quinacridone	3.2	28	462
P.R. 194	71100	4216-02-8	Perinone	3.4	30	479
P.R. 200	15867	58067-05-3	BONA, Ca	2.7	20	335
P.R. 202	73907	68859-50-7	Quinacridone	3.2	28	462
P.R. 204	–		Polycycl. Pigment	4.5	–	563
P.R. 206	–	{ 1047-16-1 1503-48-6	Quinacridone	3.2	28	463
P.R. 207	73900	{ 1047-16-1 3089-16-5	Quinacridone	3.2	28	462
P.R. 208	12514	31778-10-6	Benzimidazolone	2.8	23	363
P.R. 209	73905	3089-17-6	Quinacridone	3.2	28	463
P.R. 210	12474	36968-27-1	Naphthol AS	2.6	18	309
P.R. 210	12477	61932-63-6	Naphthol AS	2.6	18	309
P.R. 211	–	107397-16-0	Monoazo, Ca	4.5	–	564
P.R. 212	–		Naphthol AS	2.6	18	309
P.R. 213	–		Naphthol AS	2.6	18	309
P.R. 214	–	40618-31-3	Disazo Condensation	2.9	24	380
P.R. 216	59710	1324-33-0	Pyranthrone	3.6	–	508
P.R. 220	20055	57971-99-0	Disazo Condensation	2.9	24	381
P.R. 221	20065	61815-09-6	Disazo Condensation	2.9	24	381
P.R. 222	–	71872-63-4	Naphthol AS	2.6	18	309
P.R. 223	–	82784-96-1	Naphthol AS	2.6	18	299
P.R. 224	71127	128-69-8	Perylene	3.4	29	474
P.R. 226	–	63589-04-8	Pyranthrone	3.6	–	509
P.R. 237	–		Naphthol AS	2.7	21	337
P.R. 238	–		Naphthol AS Lake	2.6	18	310
P.R. 239	–		Naphthol AS Lake	2.7	21	337
P.R. 240	–		Naphthol AS Lake	2.7	21	240
P.R. 242	20067	118440-67-8	Disazo Condensation	2.9	24	382
P.R. 243	15910	50326-33-5	Naphthol AS, Ba	2.7	21	338
P.R. 245	12317	68016-05-7	Naphthol AS	2.6	18	310
P.R. 247	15915	43035-18-3	Naphthol AS, Ca	2.7	21	338
P.R. 247:1	15915	43035-18-3	Naphthol AS, Ca	2.7	21	381
P.R. 248	–		Disazo Condensation	2.9	–	382
P.R. 251	12925	74336-60-0	Pyrazoloquinazolone	4.4	–	561
P.R. 252	–		Pyrazoloquinazolone	4.4	–	561
P.R. 253	12375		Naphthol AS	2.6	18	310
P.R. 254	56110	122390-98-1	DPP-Pigment	4.4	–	552
P.R. 255	–	120500-90-5	DPP-Pigment	4.4	–	553
P.R. 256	–		Naphthol AS	2.6	–	311
P.R. 257		70833-37-3	Metal Complex	2.10	25	397
P.R. 258	12318	57301-22-1	Naphthol AS	2.6	18	311
P.R. 260	56295	71552-60-8	Isoindoline	2.11	26	413
P.R. 261	12468	16195-23-6	Naphthol AS	2.6	18	311
P.R. 262	–		Disazo Condensation	2.9	–	383
Vat Red 74	–		Perione	3.4	30	480

Colour Index Name	C.I. Constitution No.	CAS No.	Pigment Class	Chap.	Tab.	Page
Violet						
P.V. 1	45170:2	1326-03-0	Triarylcarbonium	3.8	33	545
P.V. 2	45175:1	1326-04-1	Triarylcarbonium	3.8	33	545
P.V. 3	42535:2	1325-82-2	Triarylcarbonium	3.8	33	538
		67989-22-4				
P.V. 5:1	58055:1	1328-04-7	Anthraquinone	3.6	–	496
P.V. 13	–		Naphthol AS	2.6	18	300
P.V. 19	73900	1047-16-1	Quinacridone	3.2	28	457
P.V. 23	51319	6358-30-1	Dioxazine	3.7	–	518
P.V. 25	12321	6358-46-9	Naphthol AS	2.6	18	312
P.V. 27	42535:3	12237-62-6	Triarylcarbonium	3.8	33	540
P.V. 29	71129	12236-71-4	Perylene	3.4	29	475
P.V. 31	60010	1324-55-6	Isoviolanthrone	3.6	–	513
P.V. 32	12517	12225-08-0	Benzimidazolone	2.8	23	364
P.V. 37	51345	57971-98-9	Dioxazine	3.7	–	520
P.V. 39	42555:2	64070-98-0	Triarylcarbonium	3.8	33	541
P.V. 42	–	71819-79-9	Quinacridone	3.2	28	464
P.V. 44	–	87209-55-0	Naphthol AS	2.6	18	312
P.V. 50	12322	76233-81-3	Naphthol AS	2.6	18	313
Blue						
P.B. 1	42595:2	1325-87-7	Triarylcarbonium	3.8	33	541
P.B. 2	44045:2	1325-94-6	Triarylcarbonium	3.8	33	542
P.B. 9	42025:1	596-42-9	Triarylcarbonium	3.8	33	542
P.B. 10	44040:2	1325-93-5	Triarylcarbonium	3.8	33	542
P.B. 14	42600:1	1325-88-8	Triarylcarbonium	3.8	33	542
P.B. 15	74160	147-14-8	Cu Phthalo Blue, unstable	3.1	27	436
P.B. 15:1	74160	147-14-8	Cu Phthalo Blue, α-mod.	3.1	27	437
P.B. 15:2	74160	147-14-8	Cu Phthalo Blue, α-mod.	3.1	27	439
P.B. 15:3	74160	147-14-8	Cu Phthalo Blue, β-mod.	3.1	27	440
P.B. 15:4	74160	147-14-8	Cu Phthalo Blue, β-mod.	3.1	27	442
P.B. 15:6	74160	147-14-8	Cu Phthalo Blue, ϵ-mod.	3.1	27	442
P.B. 16	74100	574-93-6	Metal-free Phthalo Blue	3.1	27	443
P.B. 18	42770:1	1324-77-2	Triarylcarbonium	3.8	32	528
P.B. 19	42750	58569-23-6	Triarylcarbonium	3.8	32	528
P.B. 24:1	42090:1	654812-5	Triphenylmethane, Ba	4.2	–	554
P.B. 24:x	42090:2	15792-67-3	Triphenylmethane, Al	4.2	–	554
P.B. 25	21180	10127-03-4	Dianisidine/Naphthol AS	2.6	18	313
P.B. 56	42800	6417-46-5	Triarylcarbonium	3.8	32	528
P.B. 60	69800	81-77-6	Indanthrone	3.6	–	503
P.B. 61	42765:1	1324-76-1	Triarylcarbonium	3.8	32	528
P.B. 62	44084	57485-98-0	Triarylcarbonium	3.8	33	542
P.B. 63	73015:x	16521-38-3	Indigo Sulfonic Acid, Al	4.2	–	555
P.B. 64	69825	130-20-1	Indanthrone	3.6	–	500
P.B. 66	73000	482-89-3	Indigo, unsubst.	3.5	–	480

Colour Index Name	C.I. Con-stitution No.	CAS No.	Pigment Class	Chap.	Tab.	Page
Brown						
P.Br. 1	12480	6410-40-8	Naphthol AS	2.6	18	299
P.Br. 5	15800:2	16521-34-9	BONA, Cu	2.7	20	335
P.Br. 22	10407	12236-95-2	Nitro Pigment	4.3	–	561
P.Br. 23	20060	57972-00-6	Disazo Condensation	2.9	24	383
P.Br. 25	12510	6992-11-6	Benzimidazolone	2.8	23	365
P.Br. 38	–	126338-72-5	Isoindoline	2.11	–	414
P.Br. 41	–		Disazo Condensation	2.9	–	384
P.Br. 42	–		Disazo Condensation	2.9	–	385
Green						
P.G. 1	42040:1	1325-75-3	Triarylcarbonium	3.8	33	543
P.G. 2	42040:1 49005:1	1325-75-3 1326-11-0	Triarylcarbonium	3.8	33	546
P.G. 4	42000:2	61725-50-6	Triarylcarbonium	3.8	33	543
P.G. 7	74260	1328-53-6	Cu Phthalogreen	3.1	27	443
P.G. 8	10006	16143-80-9	Metal Complex	2.10	25	393
P.G. 10	12775	51931-46-5	Metal Complex	2.10	25	393
P.G. 36	74265	14302-13-7	Cu Phthalogreen	3.1	27	436
P.G. 45	–		Triarylcarbonium	3.8	33	543
Black						
P.Bl. 1	50440	13007-86-8	Aniline Black	4.3	–	562
P.Bl. 20	–	12216-93-2	Anthraquinone	4.4	–	564
P.Bl. 31	71132	67075-37-0	Perylene	3.4	–	475
P.Bl. 32	71133	83524-75-8	Perylene	3.4	–	475

Index